# Antibodies
# A LABORATORY MANUAL

## Ed Harlow
Cold Spring Harbor Laboratory

## David Lane
Imperial Cancer Research Fund Laboratories

**Cold Spring Harbor Laboratory**
**1988**

# Antibodies
## A LABORATORY MANUAL

All rights reserved
© 1988 by Cold Spring Harbor Laboratory
Printed in the United States of America
Book and cover design by Emily Harste

Cover: "Nature Abstracted," watercolor by Carl Molno

Library of Congress Cataloging-in-Publication Data

Antibodies : a laboratory manual / by Ed Harlow, David Lane.
    p.    cm.
    Bibliography: p.
    Includes index.
    ISBN 0-87969-314-2


    1. Immunoglobulins--Laboratory manuals.  2. Immunochemistry-
-Laboratory manuals.  I. Harlow, Ed.  II. Lane, David (David P.).
1952-
QR186.7.A53 1988
574.2'93'028--dc19

All Cold Spring Harbor Laboratory publications may be ordered directly from Cold Spring Harbor Laboratory, Box 100, Cold Spring Harbor, New York 11724. Phone: 1-800-843-4388. In New York (516) 367-8423.

# CONTENTS

A simple review of the immune response...definitions of standard terms.... Specific interactions between host proteins and foreign molecules control the strength and effectiveness of an immune response. Selective expansion or deletion of antigen-specific lymphocytes is the cellular basis of the response.

Structure of the antibody molecule...generation of a functional immunoglobulin heavy- or light-chain gene.... Specific mechanisms have evolved to allow the production of a vast repertoire of antigen recognition sites. This repertoire allows an organism to respond to an extensive array of foreign molecules.

Structure of antibody–antigen interactions...affinity...avidity.... Antibodies and antigens are held by a series of noncovalent bonds. The strength of the individual interactions and the overall stability of an antibody–antigen complex determines the ultimate success of every immunochemical test.

Molecular and cellular development of an antibody response...multiple steps of a primary or secondary antibody response.... The generation of a strong antibody response relies on cell-to-cell communication among B cells, helper T cells, and antigen presenting cells. Manipulating these interactions allows the tailoring of a response to a chosen antigen.

## 5 ■ IMMUNIZATIONS ▬▬▬▬▬▬▬▬▬▬▬ 53

Many molecules can be used as successful immunogens to raise useful antibodies . . . . In many cases, even poor immunogens can be altered to produce better responses.

**IMMUNOGENICITY 55**

## 6 ■ MONOCLONAL ANTIBODIES ▬▬▬▬▬▬▬ 139

Allelic exclusion ensures that a clonal population of cells arising from an individual B cell will secrete identical antibodies with a unique antigen recognition site. Techniques of cell fusion allow individual B cells to be converted into permanent antibody-secreting cell lines. These monoclonal antibodies can be used to test for the presence of a particular epitope.

## 7 ▪ GROWING HYBRIDOMAS                                         245

Hybridomas and myelomas can be grown under standard mammalian tissue culture conditions, and monoclonal antibodies can be collected as spent media or following the induction of ascites in animals.

## 11 ■ IMMUNOPRECIPITATION ▐▬▬▬▬▬▬▬▬▬▬▬▬▬▬ 421

Antibody–antigen complexes can be purified by collection on matrices that specifically bind antibodies. This is a versatile technique for determining many properties of soluble antigens.

## 12 ■ IMMUNOBLOTTING ▐▬▬▬▬▬▬▬▬▬▬▬▬▬▬▬▬▬ 471

Many antigens are easiest to study on immunoblots. Because the antigens are resolved prior to immunochemical detection, antibody binding is not limited to soluble molecules and can be used to detect and quantitate antigens from a wide variety of sources.

## 13 ■ IMMUNOAFFINITY PURIFICATION 511

When antibodies are covalently attached to a solid matrix, they can be used to purify large amounts of a particular antigen. Because of the specificity of the antibody–antigen interaction, these techniques provide excellent results, exceeding all other single-column methods in yield and purity.

## 14 ■ IMMUNOASSAY 553

A wide variety of immunoassays can be used to detect and quantitate antigens and antibodies, often well beyond the sensitivity of conventional methods. These assays are particularly useful when a large number of samples need to be analyzed or when extreme sensitivity is required.

## 15 ■ REAGENTS 613

# PREFACE

This manual was inspired by "the cloning manual," *Molecular Cloning: A Laboratory Manual* by Tom Maniatis, Ed Fritsch, and Joe Sambrook, a book that did so much to expand the community of scientists using molecular biology methods. The great joy of that manual was and is the confidence it gives to beginners and the way it tempts all workers to try new to approaches. At present, many of the methods of modern immunochemistry seem beyond the reach of nonimmunologists, and one of the goals of *Antibodies* is to make these methods accessible to a wider group. We hope that this manual is judged by how well it succeeds in breaking this barrier.

*Antibodies* has a long and complex genesis. The original ideas for an immunology manual were developed with Ron McKay, Steve Blose, and Jim Lin. Some portions of those early plans remain in this version, and we would like to thank Ron, Steve, and Jim for their contributions. Version 2, the eventual precursor to this book, was conceived in a naive burst of enthusiasm in the summer of 1986. In those early days, we talked of pooling our lab protocols and finishing quickly. As we worked and talked to our colleagues, two things became clear. First, there seemed to be a genuine need for a practical guide to immunochemical methods for the nonimmunologist (which was encouraging), and second, the gap between modern immunology and the backgrounds of most molecular biologists was larger than we had thought. This meant that more background information was needed if the book were to help a reasonable cross section of scientists (which was less encouraging).

The final version of *Antibodies* contains four introductory chapters that summarize the key features of the immune response, the structure of the antibody molecule, the activities of antibodies, and the mechanism of the antibody response. These summaries are designed to be an up-to-date consensus, and they will no doubt excite some ire among immunologists by their oversimplifications and omissions, but we hope that the nonspecialist will find them a helpful framework for the ideas and techniques presented in later chapters. The bulk of the book contains protocols for raising, purifying, and labeling monoclonal and polyclonal antibodies, as well as chapters describing ways of using antibodies to study antigens. Two biases are clear: we have concentrated on protein antigens, and we have excluded several of the classical methods of immunochemistry. In their place, we have concentrated on protocols for cell staining, immunoprecipitation, immunoblotting, immunoaffinity purification, and immunoassay, as these are the techniques most commonly required by the nonimmunologist.

The actual origin of a protocol is often hard to determine, and although we have given appropriate references wherever possible, there are many sins of omission. The methods have been derived mostly from those used in our labs or those of our close associates. Clearly, few protocols are definitive, and this is particularly true where so much depends on the individual qualities of a particular antibody. We have included both general notes on the procedures themselves plus more personalized comments. These are distinguished by different types of "boxes" in the text. The phrase "some workers find..." is used either when we could not reach an agreement among ourselves or when friends assured us the effect was significant but neither of us had any direct experience with it. We have attempted to give a clear explanation of the theoretical and practical basis of the techniques to provide an effective guide when problems arise.

For further general information, *Handbook of Experimental Immunology* edited by Weir, Herzenberg, Blackwell, and Herzenberg (1986) is the best source of technical advice, and *Monoclonal Antibodies: Principles and Practice* by Goding (1987) and *Hybridoma Technology in the Biosciences and Medicine* edited by Springer (1985) are excellent sources of information on monoclonal antibodies. More specific references are listed in the appropriate sections. For new protocols, the reader should scan the pages of the *Journal of Immunological Methods* and *Analytical Biochemistry*.

Many, many people helped us with these protocols, and our job has often been one of collation, checking, and interpretation. We would like to thank Joan Brugge and her lab (Michael DeMarco, Adele Filson, Lawrence Fox, Andy Golden, Joan Levy, Sally Lynch, Susan Nemeth, John Schmidt, and Susan Schuh) who used an early version of the manual and were immensely helpful, both in finding fault and in making powerful suggestions for improvement. Several members of our own labs, Carmelita Bautista, Karen Buchkovich, Margaret Falkowski, Julian Gannon, Richard Iggo, Margaret Raybuck, and Carmella Stephens, used the manual on a day-to-day basis and were very definite and precise in their criticisms!

The manual was read in its entirety by Lionel Crawford, Larry Banks, Mike Krangel, Margaret Raybuck, and David Chiswell and his colleagues at Amersham. Their many comments helped us resolve difficult decisions about content and presentation. Chapters 1–4 were read and critiqued by Winship Herr, Richard Iggo, and John Inglis. Steve Dilworth, Ann Harris, and Bob Knowles sorted out Chapters 6–9 and 15, and Jean Beggs and Birgitte Lane helped enormously to get Chapters 10–14 into shape. The appendices were read and amended by Carl Anderson and Mark Zoller. In addition to these main readers, many others kindly helped with specific sections. Advice on yeast came from John Kilmartin, Jean Beggs, Paul Nurse, and David Beach. Advice from Susan Alpert on cell staining was invaluable, as were Rebbeca Rowehl's and Jaqueline Bortzner's comments on Chapters 6 and 7. Ian Mohr forced us to get the immunoassay chapter into intelligible form, Seth Grant helped on Chapter 12, and Ella Wetzel read and reviewed an early version of Chapter 11. All these people helped to make this a better book. Usually, we took their advice, and we hope they will forgive us for those times that we didn't.

Special personal thanks to Birgitte Lane, Nicky Williamson, Gordon Peters, Brenda Marriot, Frank Fitzjohn, and Marilyn Goodwin who helped by getting us onto planes at the right time, by finding us places to sleep and write, and by fending off a wide range of tempting distractions. Somehow they managed still to be nice, even when we were at our most crabby. The excellent art work is by Mike Ockler, and looking at Carl Molno's suggestions for cover art was a wonderful break. Jim Pflugrath provided the computer graphics. Christy Kuret, Michele Ferguson, Inez Sialiano, and Susan Schaefer provided invaluable editorial assistance. The book was designed by Emily Harste (despite our interference), and Nancy Ford and Annette Kirk provided overall supervision and kind support throughout the work.

A final and special thanks to Judy Cuddihy, our editor, whose patient advice, encouragement, and fine judgement kept us going through the best and worst of times.

Hope it's fun,

**EH** and **DL**

# IMMUNE RESPONSE

The principal function of the immune system is to protect animals from infectious organisms and from their toxic products. This system has evolved a powerful range of mechanisms to locate foreign cells, viruses, or macromolecules; to neutralize these invaders; and to eliminate them from the body. This surveillance is performed by proteins and cells that circulate throughout the body. Many different mechanisms constitute this surveillance, and they can be divided into two broad categories—nonadaptive and adaptive immunity.

*Nonadaptive immunity* is mediated by cells that respond nonspecifically to foreign molecules and includes such systems as phagocytosis by macrophages, secretion of lysozyme by lacrimal cells, and cell lysis by natural killer cells. Nonadaptive immunity does not improve with repeated exposure to foreign molecules. In contrast, *adaptive immunity* is directed against specific molecules and is enhanced by reexposure. Adaptive immunity is mediated by cells called *lymphocytes*, which synthesize cell-surface receptors or secrete proteins that bind specifically to foreign molecules. These secreted proteins are known as *antibodies*. Any molecule that can bind to an antibody is known as an *antigen*. When a molecule is used to induce an adaptive response it is called an *immunogen*. The terms "antigen" and "immunogen" are used to describe different properties of a molecule. *Immunogenicity* is not an intrinsic property of any molecule, but is defined only by its ability to induce an adaptive response. *Antigenicity* also is not an intrinsic property of a molecule, but is defined by its ability to be bound by an antibody.

The term "immunoglobulin" is often used interchangeably with "antibody." Formally, an *antibody* is a molecule that binds to a known antigen, while *immunoglobulin* refers to this group of proteins irrespective of whether or not their binding target is known. For this manual, this distinction is trivial and the terms are used interchangeably.

## The immune system is composed of large numbers of specialized cells

The immune system contains more than $10^9$ lymphocytes distributed throughout the body, enabling them to respond promptly at any site. Lymphocytes arise continuously from progenitor stem cells in the bone marrow. They travel through the blood and lymphatic systems, pausing and accumulating in specialized structures known as lymphoid organs, which in mammals are the lymph nodes and spleen. During the course of an immune response, immunogens accumulate within one of the lymphoid organs, which then becomes the focus of the response.

Many types of lymphocytes with different functions have been identified. Most of the cellular functions of the immune system can be described by grouping lymphocytes into three basic types—B cells, cytotoxic T cells, and helper T cells. All three carry cell-surface receptors that can bind antigens. B cells secrete antibodies, and carry a modified form of the same antibody on their surface, where it acts as a receptor for antigens. Cytotoxic T cells lyse foreign or infected cells, and they bind to these target cells through their surface antigen receptor, known as the T-cell receptor. Helper T cells play a key regulatory role in controlling the response of B cells and cytotoxic T cells, and they also have T-cell receptors on their surface.

Adaptive immune responses generally are divided into two classes, humoral and cell mediated. The humoral response results in the generation of circulating antibodies that bind to foreign antigens. It is mediated by B lymphocytes in conjunction with helper T lymphocytes. The antibody response forms the basis for all of the techniques in this manual and is discussed in detail in Chapter 4. Cell-mediated responses are typified by the binding of cytotoxic T lymphocytes to foreign or infected cells and subsequent lysis of these cells. Cell-mediated immunity is performed primarily by cytotoxic T lymphocytes with some involvement of helper T cells. Detailed discussions of the control of these cell-mediated immune responses are found in the textbooks listed in the "Further Reading" section at the end of this chapter.

## The immune system can respond specifically to millions of different molecules

The immune system is challenged constantly by an enormous number of antigens. One of the key features of the immune system is that it can synthesize a vast repertoire of antibodies and cell-surface receptors, each with a different antigen binding site. The binding of the antibodies and T-cell receptors to foreign molecules provides the molecular basis for the specificity of the immune response (Fig. 1.1).

## Individual lymphocytes recognize individual antigens

The specificity of the immune response is controlled by a simple mechanism—one cell recognizes one antigen (Fig. 1.2)—because all of the antigen receptors on a single lymphocyte are identical. This is true for both T and B lymphocytes, even though the types of responses made by these cells are different.

All antigen receptors are glycoproteins found on the surface of mature lymphocytes. Somatic recombination, mutation, and other mechanisms generate more than $10^7$ different binding sites, and an-

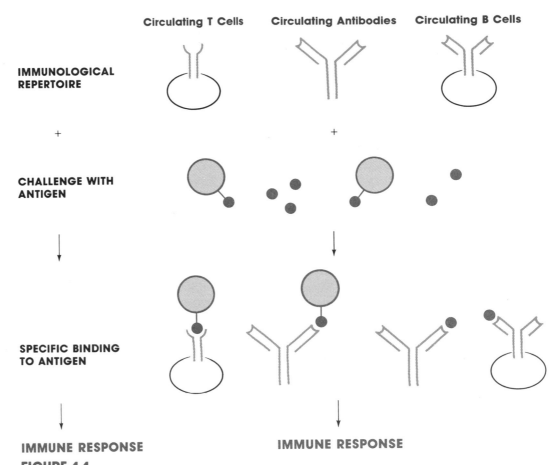

**FIGURE 1.1**
Specific binding of antigen triggers the immune response.

tigen specificity is maintained by processes that ensure that only one type of receptor is synthesized within any one cell. The production of antigen receptors occurs in the absence of antigen. Therefore, a diverse repertoire of antigen receptors is available before antigen is seen.

Although they share similar structural features, the surface antibodies on B cells and the T-cell receptors found on T cells are encoded by separate gene families; their expression is cell-type specific. The surface antibodies on B cells can bind to soluble antigens, while the

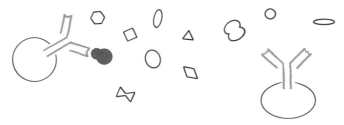

**FIGURE 1.2**
One cell recognizes one antigen.

T-cell receptors recognize antigens only when displayed on the surface of other cells.

## The immune system can distinguish foreign organisms and molecules from self

Since antibodies and cytotoxic T cells are very efficient at eliminating antigens, it is important that the immune system is able to distinguish foreign molecules from normal components of the body. Organisms typically do not mount immune responses against their own macromolecules. This lack of response to self is called *tolerance*. Tolerance is established and maintained by constantly eliminating lymphocytes that produce T-cell receptors or antibodies that bind to host molecules. If this selection process fails, the host produces an immune response against itself. These responses cause a range of disorders known collectively as *autoimmune diseases*.

## The immune system remembers each encounter with foreign antigens

When an animal first encounters a foreign antigen, the immune response is slow. As the immune response develops, the antigen is destroyed. If the antigen is a pathogen, this slow response may be insufficient to prevent disease. On second exposure to the same antigen, a stronger and more rapid response is elicited, often eliminating a pathogen before it can cause disease. This capacity to mount a strong secondary response is termed *immunological memory*. Immunological memory is specific; prior exposure to one antigen will not protect against an immunologically unrelated antigen. Immunological memory can last for as long as the animal lives.

## Many of the properties of the immune response are determined by clonal selection

Because receptors on any one lymphocyte can only bind to one antigen, it is easy to see how the immune system responds specifically to a particular antigen, activating one lymphocyte and not another. When B-cell surface antibodies bind antigen, the B lymphocyte is activated to secrete antibody and is stimulated to proliferate. T cells respond in a similar fashion. This burst of cell division increases the number of antigen-specific lymphocytes, and this clonal expansion is the first step in the development of an effective immune response. The primary site of this cellular division is in the lymphoid organs. As long as the antigen persists, the activation of lymphocytes continues, thus increasing the strength of the immune response (Fig. 1.3). After the antigen has been eliminated, some cells from the expanded pools of antigen-specific lymphocytes remain in circulation. These cells are primed to respond to any subsequent exposure to the same antigen, providing the cellular basis for immunological memory.

A second type of clonal selection accounts for tolerance. Here, lymphocytes whose cell-surface receptors bind to host components are eliminated from the immune system by a process known as clonal deletion. Self-reactive clones are specifically removed during their differentiation (Fig. 1.3). This process is particularly active while the immune system is developing, but since new lymphocytes are constantly being produced, this selection process must continue throughout life.

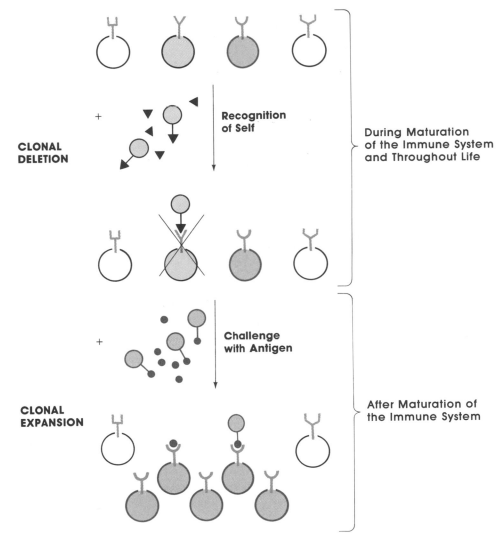

**FIGURE 1.3**
Clonal selection.

## The maturation of an effective immune response is regulated by extensive cell-to-cell communication

The immune system functions by extensive cell-to-cell communication, which is mediated by secretion of regulatory molecules and by physical contact between various lymphocytes and macrophages. These interactions coordinate and control the immune response. A large number of receptors and signaling molecules involved in the immune network have been identified, and they provide the best understood system of cell communication in a multicellular organism.

### Further Readings

Burnet, F.M. 1959. *The clonal selection theory of acquired immunity.* Cambridge University Press, London.

Cantor, H., and E.A. Boyse. 1975. Functional subclasses of T lymphocytes bearing different Ly antigens. *J. Exp. Med.* **141:** 1376–1389.

Dutton, R.W. and R.I. Mishell. 1967. Cellular events in the immune response of normal spleen cells to erythrocyte antigens. *Cold Spring Harbor Symp. Quant. Biol.* **32:** 407–414.

Gowans, J.L. and E.J. Knight. 1964. The route of recirculation of lymphocytes in the rat. *Proc. R. Soc. Lond. B. Biol. Sci.* **159:** 257–282.

Jerne, N.K. 1971. The somatic generation of immune recognition. *Eur. J. Immunol.* **1:** 1–9.

Moore, M.A.S. and J.J.T. Owen. 1967. Experimental studies on the development of the thymus. *J. Exp. Med.* **126:** 715–725.

Triplett, E.L. 1962. On the mechanism of immunologic self recognition. *J. Immunol.* **89:** 505–510.

Hood, L.E., I.L. Weissman, W.B. Wood, and J.H. Wilson. 1984. *Immunology,* 2nd edition. Benjamin/Cummings, Menlo Park, California.

Male, D., B. Champion, and A. Cooke. 1987. *Advanced immunology.* Gower, Lippencott Medical, Lippencott, Philadelphia.

Marchalonis, J.J., ed. 1977. Recirculating lymphocytes. *The lymphocyte: Structure and function.* M. Dekker, New York.

Nossal, G.J.V. 1983. Cellular mechanisms of immunologic tolerance. *Annu. Rev. Immunol.* **1:** 33–62.

Paul, W.E., ed. 1986. *Fundamental immunology.* Raven Press, New York.

Roitt, I., Brostoff, and D. Male. 1985. *Essential immunology.* Mosby, St. Louis, Missouri.

Sato, V.L. and M.L. Gefter, ed. 1982. *Cellular immunology: Selected readings.* Benjamin-Cummings.

# 2 ANTIBODY MOLECULES

Antibodies are host proteins produced in response to the presence of foreign molecules in the body. They are synthesized primarily by plasma cells, a terminally differentiated cell of the B-lymphocyte lineage, and circulate throughout the blood and lymph where they bind to foreign antigens. The antibody–antigen complexes are then removed from circulation primarily through phagocytosis by macrophages.

Antibodies are a large family of glycoproteins that share key structural and functional features. Functionally, they can be characterized by their ability to bind both to antigens and to specialized cells or proteins of the immune system. Structurally, antibodies are composed of one or more copies of a characteristic unit that can be visualized as forming a Y shape (Fig. 2.1). Each Y contains four polypeptides—two identical copies of a polypeptide known as the heavy chain and two identical copies of a polypeptide called the light chain. Antibodies are divided into five classes, IgG, IgM, IgA, IgE, and IgD, on the basis of the number of Y-like units and the type of heavy-chain polypeptide they contain.

Many of the important structural features of antibodies are easiest to discuss by considering IgG antibodies, which contain only one structural Y unit and are also the most abundant in serum. For simplicity, mouse antibodies are used as examples in the discussions below, but their key properties are similar to those from other species.

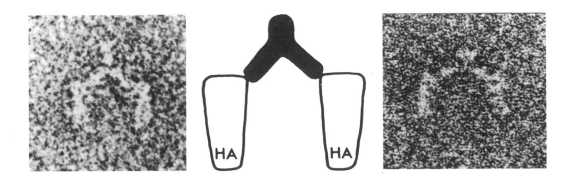

**FIGURE 2.1**
Two influenza hemagglutinin (HA) molecules bound by an IgG molecule. (Adapted from Wrigley et al. 1983.)

## Antibodies from the IgG class have two identical antigen binding sites

IgG molecules have three protein domains, and Figure 2.2 shows how these domains are organized. Two of the domains are identical and form the arms of the Y. Each arm contains a site that can bind to an antigen, making IgG molecules bivalent. The third domain forms the base of the Y, and this region is important in certain aspects of the immune response. The three domains may be separated from each other by cleavage with the protease papain. The two domains that carry the antigen binding sites are known as Fab fragments (named for the *fragment* having the *antigen binding* site), and the protein domain that is involved in immune regulation is termed the Fc fragment (for the *fragment* that *crystallizes*). The region between the Fab and Fc fragments is called the hinge. This segment allows lateral and rotational movement of the two antigen binding domains. In practice, this allows the binding domains freedom to interact with a large number of different antigen conformations (Fig. 2.3).

The two heavy-chain polypeptides in the Y structure are identical and are approximately 55,000 daltons. The two light chains are also identical and are about 25,000 daltons (Fig. 2.4). One light chain associates with the amino-terminal region of one heavy chain to form an antigen binding domain. The carboxy-terminal regions of the two

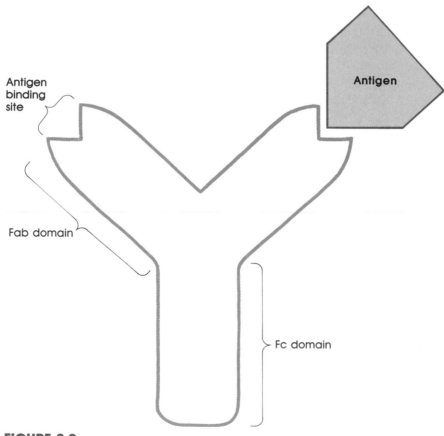

**FIGURE 2.2**
Antibody domains.

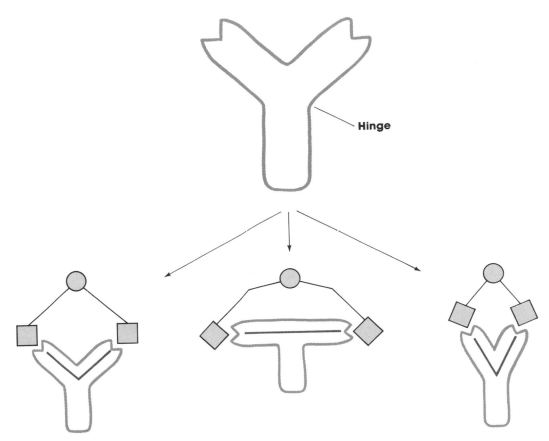

**FIGURE 2.3**
Hinge and flexibility.

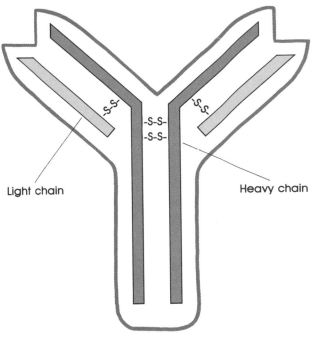

**FIGURE 2.4**
Heavy and light chains combine to form antibody.

heavy chains fold together to make the Fc domain. The four polypeptide chains are held together by disulfide bridges and noncovalent bonds.

## In addition to the IgG molecules, serum contains other classes of antibody molecules

The other antibody classes, IgM, IgA, IgE, and IgD, are distinguished by the type of heavy chain found in the molecule. Where IgG molecules have heavy chains known as $\gamma$-chains, IgMs have $\mu$-chains, IgAs have $\alpha$-chains, IgEs have $\epsilon$-chains, and IgDs have $\delta$-chains. The differences in the heavy-chain polypeptides allow these proteins to function in different types of immune responses and at particular stages of the maturation of the immune response. The protein sequences responsible for these differences are found primarily in the Fc fragment. Different classes of antibodies may also vary in the number of Y-like units that join to form the complete protein. IgM antibodies, for example, have five Y-shaped units. Because each Y unit has two antigen binding sites, IgMs have 10 identical antigen binding sites. The sites for association between the different Y units are also found in the Fc region. Table 2.1 summarizes many of the properties of these antibody classes.

While there are five different types of heavy chains, there are only two light chains, $\kappa$ and $\lambda$. One light chain always associates with one

**TABLE 2.1**
**Classes of Antibodies**

| Characteristics | IgG | IgM | IgA | IgE | IgD |
|---|---|---|---|---|---|
| **Heavy Chain** | $\gamma$ | $\mu$ | $\alpha$ | $\varepsilon$ | $\delta$ |
| **Light Chain** | $\kappa$ or $\lambda$ | $\kappa$ or $\lambda$ | $\kappa$ or $\lambda$ | $\kappa$ or $\lambda$ | $\kappa$ or $\lambda$ |
| **Molecular Formula** | $\gamma_2\kappa_2$ or $\gamma_2\lambda_2$ | $(\mu_2\kappa_2)_5$ or $(\mu_2\lambda_2)_5$ | $(\alpha_2\kappa_2)_n$ [a] or $(\alpha_2\lambda_2)_n$ | $\varepsilon_2\kappa_2$ or $\varepsilon_2\lambda_2$ | $\delta_2\kappa_2$ or $\delta_2\lambda_2$ |
| **Y Structure** | | | | | |
| **Valency** | 2 | 10 | 2, 4, or 6 | 2 | 2 |
| **Concentration in Serum** | 8–16 mg/ml | 0.5–2 mg/ml | 1–4 mg/ml | 10–400 ng/ml | 0–0.4 mg/ml |
| **Function** | Secondary response | Primary response | Protects mucous membranes | Protects against parasites (?) | ? |

[a] $n = 1, 2,$ or $3$.

heavy chain, so the total number of light chains will always equal the number of heavy chains. Since the basic structural Y unit has two heavy chains and two light chains, IgMs, which have five Y-units, have 10 light chains and 10 heavy chains. However, any one antibody molecule will have only one type of light chain and one type of heavy chain. There are no restrictions on which types of heavy or light chains can form antibodies, so antibodies of all classes (i.e., with different heavy chains) can contain either κ or λ light chains.

## Comparison of the primary amino acid sequences of light chains reveals a constant and a variable region

Light chains are approximately 220 amino acids long and can be divided into two regions, each about 110 amino acids in length (Fig. 2.5). When sequences from a number of light chains were first com-

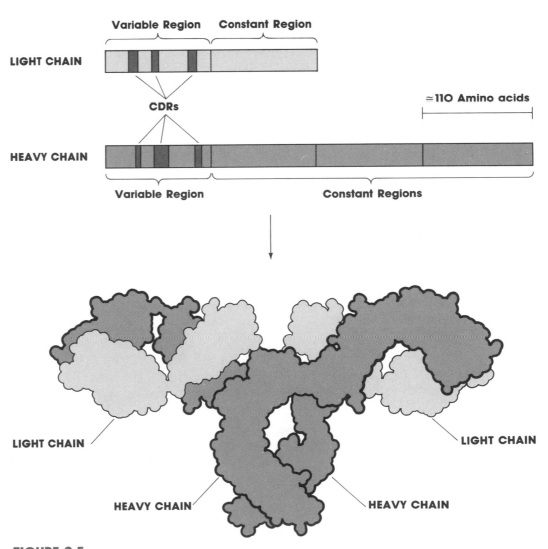

**FIGURE 2.5**
Light- and heavy-chain structure. (Adapted from Silverton et al. 1977.)

pared, it was found that the amino-terminal half was heterogeneous. This region is known as the variable (*V*) region. The carboxy-terminal half is known as the constant (*C*) region, and amino acid sequencing studies have shown that there are only two types of constant regions, one for *κ* light chains and one for *λ* light chains. In the mouse the gene for the *κ* light chain is located on chromosome 6, and the *λ* gene is found on chromosome 16.

## Comparison of the sequences of heavy chains also reveal variable and constant regions

The IgG heavy chains are approximately 440 amino acids long and are also divided into variable and constant regions (Fig. 2.5). However, IgG heavy chains contain one variable region and three constant regions. Like the light chain, each of these four regions is about 110 amino acids in length. Heavy chains from different classes may have additional constant regions. For example, $\mu$-chains from IgMs have one variable region but four constant regions. The sequences of the IgG heavy chains have also shown that there are four subclasses of $\gamma$-chains. In the mouse, these subclasses of IgG antibodies are known as $IgG_1$, $IgG_{2a}$, $IgG_{2b}$, and $IgG_3$. The coding region for the heavy-chain polypeptides is found on chromosome 12 in the mouse.

## The variable regions of the heavy and light chains form the antigen binding site

The variable regions of one heavy chain and one light chain combine to form one antigen binding site. The heterogeneity of the variable regions provides the structural basis for the large repertoire of binding sites used by an animal to mount an effective immune response. Perhaps not surprisingly, the sequence heterogeneity does not occur randomly throughout the variable region, but is concentrated in regions that form the sites of contact with the antigen. Most of the variability occurs in three short regions of each chain, giving three hypervariable regions for the light chain and three hypervariable regions for the heavy chain. These hypervariable regions form the majority of contact residues for the binding of the antibody to the antigen and are located on short loops extending into the region that interacts with the antigen. Because they are the actual binding site for the antigen, they are referred to as the complementarity determining regions or CDRs. More information on the structure of the antibody–antigen complex is found in Chapter 3.

The genetic basis for diversity in the variable region has been elucidated during recent years. These studies have shown how *κ*, *λ*, and heavy-chain genes are organized and how these genes are manipulated to generate different variable regions. The heterogeneity in the variable regions of both light and heavy chains is due to an intricate combination of recombination and mutation. The simplest case is found with the κ gene.

## DNA rearrangements are required for the formation of a functional κ gene

One of the remarkable features of the κ, λ, and heavy-chain genes is that the organization of these genes is not the same in all cells. For example, cells synthesizing functional κ chains have a different κ gene structure from cells in which the polypeptide is not expressed. In germ-line and non-B-cell DNA, the coding sequences for the variable and constant regions of the κ light chain are separated by several hundred kilobase pairs. The gene is silent in these cells. During B-cell differentiation, the DNA between these regions is removed, joining the variable and constant regions and producing a functional gene (Fig. 2.6). There are approximately 200 different κ variable regions and only one constant region. During the maturation of B cells, recombination brings one of the variable regions and the constant region together to produce a functional gene. The selection of the variable region is essentially random.

During studies of the fine structure of the recombination points between the variable and constant coding regions, it became clear that simple joining of these two regions did not account for all the variation that was detected. These studies led to the discovery of a third segment between the variable and constant regions called the *J* region for joining. Recombination brings the variable region just upstream of one of five *J* regions, producing a variable region–*J* region–constant region alignment. This alignment yields a functional κ gene. As discussed below, the recombination events do not always occur at exactly

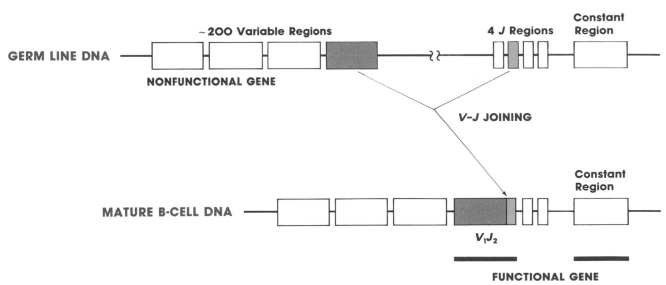

**FIGURE 2.6**
Kappa rearrangements in germ line and B lymphocytes.

the same nucleotides, and this allows for some variation in the amino acids found at the $V-J$ junction.

Although five different $J$ regions for $\kappa$-chains have been found in genomic DNA sequences, only four of these can be used to synthesize functional polypeptides. The other, $J_3$, has a mutated donor splice site and cannot produce a functional mRNA. Therefore, in the maturation of a functional $\kappa$ light-chain gene, recombination must occur between one of the approximately 200 variable regions and one of the four functional $J$ regions (Fig. 2.6), yielding approximately 800 different $\kappa$-chain polypeptides for homozygous mice.

## DNA rearrangements are also required to form a functional λ gene

Similar types of recombination events are required to produce a functional λ light-chain polypeptide (Fig. 2.7). In the germ-line DNA, there are two variable regions and four constant regions in the λ coding region. Each constant region is paired with a single $J$ region, and recombination occurs between the variable and $J$ regions. However, unlike the $\kappa$ gene, the two variable regions are not found upstream of the four constant regions, but are separated from each other. Each is paired with two $J$–constant regions. Analysis of the DNA sequences around the $J_4$ region has shown that it carries several mutations which block the successful production of mRNA. Also, no recombination events have ever been reported that join a downstream variable region to an upstream constant region. Therefore, this organization allows the production of five possible λ light chains.

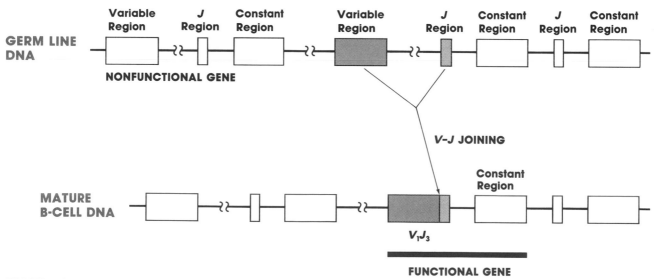

**FIGURE 2.7**
Lambda $V-J$ joining.

## Forming a functional heavy-chain gene requires two DNA rearrangements

In addition to the variable, *J*, and constant regions, heavy chains also contain a fourth coding region referred to as the *D*, or diversity, region (Fig. 2.8). The *D* region is located between the variable regions and the *J*–constant region. The production of a functional heavy chain requires two recombination events, one joining a *D* region to a *J*–constant region, and another joining a variable region to the *D* region. There are approximately 50–100 variable, 12 *D*, and 4 *J* chain regions that form the heavy-chain gene cluster on chromosome 12 in the mouse. However, the *D* regions produce functional proteins in all three reading frames, so, because the recombination events are not precise, the *D* regions can code for 36 different sequences. Most experimental work has suggested that the first recombination event joins a *D* region to a *J* region. Then the variable to *D* region recombination occurs. Having multiple regions helps to generate a large number of heavy chains, with at least 7200 different possibilities for homozygous mice.

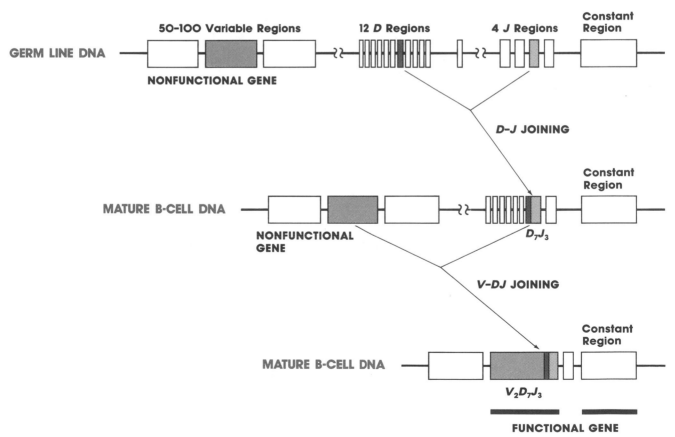

**FIGURE 2.8**
Heavy-chain *VDJ* joining.

## These rearrangements and additional special mechanisms create a vast number of antigen binding sites

As discussed above, recombination between the variable, *D*, and *J* regions produces a large number of different heavy- and light-chain polypeptide sequences. The variable regions themselves contain the coding sequences for CDR 1 and CDR 2 of both the heavy and light chains. As shown in Figure 2.9, the recombination junctions produce sequences that encode amino acids for the third CDR. As discussed in Chapter 3, the CDRs are the primary sites of antigen binding, and therefore, recombination provides both different assortments of the various CDRs and, because the recombination is not always precise, diversity in the third CDR.

For the mouse, there are at least 800 potential $\kappa$ chains, five $\lambda$ chains, and 7200 heavy chains. If all light chains have a chance to pair with all heavy chains in the production of the antigen binding site, there are potentially $5.8 \times 10^6$ binding domains ($[800 + 5] \times 7200$). Diversity at the recombination junctions raises the number of possible combinations. Sequencing of $\kappa$ chains has shown that a number of mutations can be introduced at the recombination sites. These types of mutations are also seen for heavy chains. Heavy chains show another type of mutation found at recombination sites in which a random number of nucleotides are introduced, further increasing the number of available coding variations. Recombination and random assortment of the light and heavy chains create the lower limit for the number of different antigen binding sites.

Another mechanism also increases antibody diversity. Greater than half of the heavy- or light-chain variable regions carry point mutations in the antigen binding regions. Although the mechanisms that generate these mutations are not known, they appear to provide a system that fine-tunes the binding site, thus creating a better fit for the antigen–antibody interaction. This process is known as affinity maturation and is discussed in more detail in Chapter 4.

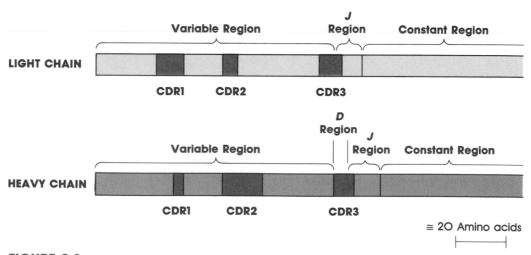

**FIGURE 2.9**
Heavy- and light-chain rearrangements both reassort and restructure CDRs.

Although calculating the number of different antigen binding sites that can be formed by recombination, junctional diversity, mutation, and association of different light and heavy chains quickly becomes an algebraic rather than an immunochemical problem, it does serve to highlight the ability of humoral immunity to respond to a large number of antigens.

## Allelic exclusion ensures that only one rearranged light-chain gene and one rearranged heavy-chain gene are expressed in any one B cell

The recombination events that produce different heavy and light chains do not always lead to a functional gene. During B-cell differentiation, light-chain rearrangements begin within the $\kappa$ region, and, if the first rearrangement of one of the diploid genes results in a nonfunctional allele, the other copy undergoes recombination. If both recombination events lead to nonfunctional genes, then recombination begins at the $\lambda$ locus. Similar mechanisms lead to the generation of a functional heavy-chain gene. Once recombination yields a functional antibody, an unknown mechanism prevents further recombination, fixing the antigen binding site for the life of the cell. This mechanism is called allelic exclusion and explains why B lymphocytes secrete antibodies with only one type of antigen binding site and why antibodies have only one type of light chain.

## Additional recombination events are used to generate the different classes and subclasses of antibodies

One fascinating aspect of the maturation of an immune response has been the discovery of a second set of recombination events that

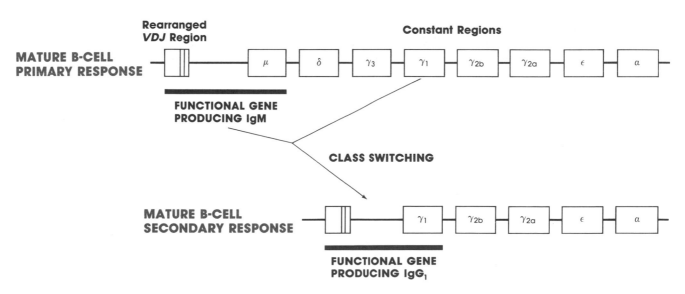

**FIGURE 2.10**
Class and subclass switching.

involve the heavy-chain constant regions. These rearrangements do not affect the variable region and do not alter the antigen binding domain. They do not contribute to the diversity of binding sites, but instead allow the replacement of one heavy-chain constant region with another. The rearrangement occurs downstream of the $J$ region. Aligned here in germ-line DNA is a tandem array of the heavy-chain constant region genes: $\mu$-chain (IgM), $\delta$-chain (IgD), $\gamma_3$-chain (IgG$_3$), $\gamma_1$-chain (IgG$_1$), $\gamma_{2b}$-chain (IgG$_{2b}$), $\gamma_{2a}$-chain (IgG$_{2a}$), $\epsilon$-chain (IgE), and $\alpha$-chain (IgA) (Fig. 2.10). In essence, these rearrangements move the same antigen binding site onto different antibody classes and subclasses. This process is known as class switching and is one of the characteristic events of the maturation of the antibody response (Chapter 4). Because the different classes and subclasses determine many of the functional properties of antibodies—for example, whether they will appear in serum or mucous secretions—this class switching creates an important mechanism for controlling where and how the antigen binding site will be used.

## Further Readings

### Structural Studies of Antibodies

Amzel, L.M., R.J. Poljak, F. Saul, J.M. Varga, and F.F. Richards. 1974. The three dimensional structure of a combining region-ligand complex of immunoglobulin NEW at 3.5-Å resolution. *Proc. Natl. Acad. Sci.* **71**: 1427–1430.

Capra, J.D. 1971. Hypervariable region of human immunoglobulin heavy-chains. *Nature* **230**: 62.

Edelman, G.M. 1959. Dissociation of gamma globulin. *J. Am. Chem. Soc.* **81**: 3155.

Edelman, G.M., B.A. Cunningham, W.E. Gall, P.D. Gottlieb, U. Rutishauser, and M.J. Waxdal. 1969. The covalent structure of an entire $\gamma$G1 immunoglobulin molecule. *Biochemistry* **63**: 78–85.

Hilschmann, N. and L.C. Craig. 1965. Amino acid sequence studies with Bence-Jones proteins. *Biochemistry* **53**: 1403–1409.

Kehoe, J.M. and J.D. Capra. 1971. Localization of two additional hypervariable regions in immunoglobulin heavy chains. *Proc. Natl. Acad. Sci.* **68**: 2019–2021.

Poljack, R.J., L.M. Amzel, H.P. Avey, B.L. Chen, R.P. Phizackerley, and F. Saul. 1973. Three-dimensional structure of the Fab' fragment of a human immunoglobulin at 2.8-Å resolution. *Proc. Natl. Acad. Sci.* **70**: 3305–3310.

Porter, et al. 1959. Hydrolysis of rabbit gamma globulin and antibodies with crystalline papain. *Bio. Chem. J.* **73**: 119–126.

Schiffer, M., R.L. Girling, K.R. Ely, and A.B. Edmundson. 1973. Structure of a $\lambda$-type Bence-Jones protein at 3.5-Å resolution. *Biochemistry* **12**: 4620–4631.

Segal, D.M., E.A. Padlan, G.H. Cohen, S. Rudikoff, M. Potter, and D.R. Davies. 1974. The three-dimensional structure of a phosphorylcholine-binding mouse immunoglobulin Fab and the nature of the antigen binding site. *Proc. Natl. Acad. Sci.* **71**: 4298–4302.

Titani, K., E. Whitley, Jr., L. Avogardo, and F.W. Putnam. 1965. Immunoglobulin structure: Partial amino acid sequence of a Bence-Jones protein. *Science* **149**: 1090–1092.

Wu, T.T. and E.A. Kabat. 1970. An analysis of variable regions of Bence-Jones proteins and myeloma light chains and their implications of antibody complementarity. *J. Exp. Med.* **132:** 211.

Amzel, L.M. and R.J. Poljak. 1979. Three-dimensional structure of immunoglobulins. *Annu. Rev. Biochem.* **48:** 961–997.

Davies, D.R. and H. Metzger. 1983. Structural basis of antibody function. *Annu. Rev. Biochem.* **1:** 87–117.

Gally, J.A. 1973. Structure of immunoglobulins. In *The antigens 1* (ed. M. Sela), pp. 162–298. Academic Press, New York.

Jeske, D.J. and J.D. Capra. 1984. Immunoglobulins: Structure and function. In *Fundamental immunology* (ed. W.E. Paul), pp. 131–165, Raven Press, New York.

### General Reviews on Antibody Genes and Diversity

Alt, F.W., T.K. Blackwell, and G.D. Yancopoulos. 1987. Development of the primary antibody repertoire. *Science* **238:** 1079–1087.

Hood, L., J.H. Campbell, and S.C.R. Elgin. 1975. The organization, expression and evolution of antibody genes and other multigene families. *Annu. Rev. Genet.* **9:** 305–353.

Tonegawa, S. 1983. Somatic generation of antibody diversity. *Nature* **302:** 575–581.

### Recombination in Immunoglobulin Genes

Dreyer, W.J. and J.C. Bennett. 1965. The molecular basis of antibody formation: A paradox. *Biochemistry* **54:** 864–869.

Hozumi, N. and S. Tonegawa. 1976. Evidence for somatic rearrangement of immunoglobulin genes coding for variable and constant regions. *Proc. Natl. Acad. Sci.* **73:** 3628–3632.

Tonegawa, S., N. Hozumi, G. Matthyssens, and R. Schuller. 1977. Somatic changes in the content and context of immunoglobulin genes. *Cold Spring Harbor Symp. Quant. Biol.* **41:** 877–889.

### VJ and VDJ Joining

Brack, C., M. Hirama, R. Lenhard-Schuller, and S. Tonegawa. 1978. A complete immunoglobulin gene is created by somatic recombination. *Cell* **15:** 1–14.

Davis, M.M., K. Calame, P.W. Early, D.L. Livant, R. Joho, I.L. Weissman, and L. Hood. 1980. An immunoglobulin heavy-chain gene is formed by at least two recombinational events. *Nature* **283:** 733–739.

Early, P., H. Huang, M. Davis, K. Calame, and L. Hood. 1980. An immunoglobulin heavy chain variable region gene is generated from three segments of DNA: $V_H$, D and $J_H$. *Cell* **19:** 981–992.

Max, E.E., J.G. Seidman, and P. Leder. 1979. Sequences of five potential recombination sites encoded close to an immunoglobulin $\kappa$ constant region gene. *Proc. Natl. Acad. Sci.* **76:** 3450–3454.

Sakano, H., K. Hüppi, G. Heinrich, and S. Tonegawa. 1979. Sequences at the somatic recombination sites of immunoglobulin light-chain genes. *Nature* **280:** 288–294.

Sakano, H., R. Maki, Y. Kurosawa, W. Roeder, and S. Tonegawa. 1980. Two types of somatic recombination are necessary for the generation of complete immunoglobulin heavy-chain genes. *Nature* **286:** 676–683.

Seidman, J.G. and P. Leder. 1978. The arrangement and rearrangement of antibody genes. *Nature* **276:** 790–795.

Seidman, J.G., E.E. Max, and P. Leder. 1979. A $\kappa$-immunoglobulin gene is formed by site-specific recombination without further somatic mutation. *Nature* **280:** 370–375.

### Multiple V, D, and J Regions as a Source of Diversity

Max, E.E., J.G. Seidman, and P. Leder. 1979. Sequences of five potential recombination sites encoded close to an immunoglobulin κ constant region gene. *Proc. Natl. Acad. Sci.* **76:** 3450–3454.

Seidman, J.G., A. Leder, M.H. Edgell, F. Polsky, S.M. Tilghman, D.C. Tiemeier, and P. Leder. 1978. Multiple related immunoglobulin variable-region genes identified by cloning and sequence analysis. *Proc. Natl. Acad. Sci.* **75:** 3881–3885.

Weigert, M., L. Gatmaitan, E. Loh, J. Schilling, and L. Hood. 1978. Rearrangement of genetic information may produce immunoglobulin diversity. *Nature* **276:** 785–790.

### Junctional Diversity

Early, P., H. Huang, M. Davis, K. Calame, and L. Hood. 1980. An immunoglobulin heavy chain variable region gene is generated from three segments of DNA: $V_H$, D and $J_H$. *Cell* **19:** 981–992.

Sakano, H., K. Hüppi, G. Heinrich, and S. Tonegawa. 1979. Sequences at the somatic recombination sites of immunoglobulin light-chain genes. *Nature* **280:** 288–294.

Weigert, M., R. Perry, D. Kelley, T. Hunkapiller, J. Schilling, and L. Hood. 1980. The joining of V and J gene segments creates antibody diversity. *Nature* **283:** 497–499.

### Somatic Mutation

Bernard, O., N. Hozumi, and S. Tonegawa. 1978. Sequences of mouse immunoglobulin light chain genes before and after somatic changes. *Cell* **15:** 1133–1144.

Bothwell, A.L.M., M. Paskind, M. Reth, T. Imanishi-Kari, K. Rajewsky, and D. Baltimore. 1981. Heavy chain variable region contribution to the $NB^b$ family of antibodies: Somatic mutation evident in a γ2a variable region. *Cell* **24:** 625–637.

Crews, S., J. Griffin, H. Huang, K. Calame, and L. Hood. 1981. A single $V_H$ gene segment encodes the immune response to phosphorylcholine: Somatic mutation is correlated with the class of the antibody. *Cell* **25:** 59–66.

Gearhart, P.J., N.D. Johnson, R. Douglas, and L. Hood. 1981. IgG antibodies to phosphorylcholine exhibit more diversity than their IgM counterparts. *Nature* **291:** 29–34.

Gershenfeld, H.K., A. Tsukamoto, I.L. Weissman, and R. Joho. 1981. Somatic diversification is required to generate the Vκ genes of MOPC 511 and MOPC 167 myeloma proteins. *Proc. Natl. Acad. Sci.* **78:** 7674–7678.

Kim, S., M. Davis, E. Sinn, P. Patten, and L. Hood. 1981. Antibody diversity: Somatic hypermutation of rearranged $V_H$ genes. *Cell* **27:** 573–581.

Pech, M., J. Höchtl, H. Schnell, and H.G. Zachau. 1981. Differences between germ-line and rearranged immunoglobulin $V_κ$ coding sequences suggest a localized mutation mechanism. *Nature* **291:** 668–670.

Selsing, E. and U. Storb. 1981. Somatic mutation of immunoglobulin light-chain variable-region genes. *Cell* **25:** 47–58.

Siekevitz, M., S.Y. Huang, and M.L. Gefter. 1983. The genetic basis of antibody production: A single heavy chain variable region gene encodes all molecules bearing the dominant anti-arsonate idiotype in the strain A mouse. *Eur. J. Immunol.* **13:** 123–132.

Valbuena, O., K.B. Marcu, M. Weigert, and R.P. Perry. 1978. Multiplicity of germline genes specifying a group of related mouse κ chains with implications for the generation of immunoglobulin diversity. *Nature* **276**: 780–784.

Weigert, M.G., I.M. Cesari, S.J. Yonkovich, and M. Cohn. 1970. Variability in the lambda light chain sequences of mouse antibody. *Nature* **228**: 1045–1047.

### Mechanism of VJ/VDJ Joining

Alt, F.W. and D. Baltimore. 1982. Joining of immunoglobulin heavy chain gene segments: Implications from a chromosome with evidence of three D-J$_H$ fusions. *Proc. Natl. Acad. Sci.* **79**: 4118–4122.

Early, P., H. Huang, M. Davis, K. Calame, and L. Hood. 1980. An immunoglobulin heavy chain variable region gene is generated from three segments of DNA: V$_H$, D and J$_H$. *Cell* **19**: 981–992.

Kurosawa, Y., H. von Boehmer, W. Haas, H. Sakano, A. Trauneker, and S. Tonegawa. 1981. Identification of D segments of immunoglobulin heavy-chain genes and their rearrangement in T lymphocytes. *Nature* **290**: 565–570.

Max, E.E., J.G. Seidman, and P. Leder. 1979. Sequences of five potential recombination sites encoded close to an immunoglobulin κ constant region gene. *Proc. Natl. Acad. Sci.* **76**: 3450–3454.

Sakano, H., K. Hüppi, G. Heinrich, and S. Tonegawa. 1979. Sequences at the somatic recombination sites of immunoglobulin light-chain genes. *Nature* **280**: 288–294.

Sakano, H., Y. Kurosawa, M. Weigert, and S. Tonegawa. 1981. Identification and nucleotide sequence of a diversity DNA segment (D) of immunoglobulin heavy-chain genes. *Nature* **290**: 562–565.

Sakano, H., R. Maki, Y. Kurosawa, W. Roeder, and S. Tonegawa. 1980. Two types of somatic recombination are necessary for the generation of complete immunoglobulin heavy-chain genes. *Nature* **286**: 676–683.

### Sequence of VJ/VDJ Joining and Allelic Exclusion

Alt, F.W., V. Enea, A.L.M. Bothwell, and D. Baltimore. 1980. Activity of multiple light chain genes in murine myeloma cells producing a single, functional light chain. *Cell* **21**: 1–12.

Alt, F., N. Rosenberg, S. Lewis, E. Thomas, and D. Baltimore. 1981. Organization and reorganization of immunoglobulin genes in A-MuLV-transformed cells: Rearrangement of heavy but not light chain genes. *Cell* **27**: 381–390.

Alt, F.W., G.D. Yancopoulos, T.K. Blackwell, C. Wood, E. Thomas, M. Boss, R. Coffman, N. Rosenberg, S. Tonegawa, and D. Baltimore. 1984. Ordered rearrangement of immunoglobulin heavy chain variable region segments. *EMBO J.* **3**: 1209–1219.

Cebra, J.J., J.E. Colberg, and S. Dray. 1966. Rabbit lymphoid cells differentiated with respect to α-, γ- and μ-heavy polypeptide chains and to allotypic markers AA1 and AA2. *J. Exp. Med.* **123**: 547–558.

Coleclough, C., R.P. Perry, K. Karjalainen, and M. Weigert. 1981. Aberrant rearrangements contribute significantly to the allelic exclusion of immunoglobulin gene expression. *Nature* **290**: 372–378.

Hieter, P.A., S.J. Korsmeyer, T.A. Waldmann, and P. Leder. 1981. Human immunoglobulin κ light-chain genes are deleted or rearranged in λ-producing B cells. *Nature* **290**: 368–372.

Maki, R., J. Kearney, C. Paige, and S. Tonegawa. 1980. Immunoglobulin gene rearrangement in immature B cells. *Science* **209**: 1366–1369.

Pernis, B., G. Chiappino, A.S. Kelus, and P.G.H. Gell. 1965. Cellular localization of immunoglobulins with different allotypic specificities in rabbit lymphoid tissues. *J. Exp. Med.* **122:** 853–875.

Perry, R.P., D.E. Kelley, C. Coleclough, and J.F. Kearney. 1981. Organization and expression of immunoglobulin genes in fetal liver hybridomas. *Proc. Natl. Acad. Sci.* **78:** 247–251.

Ritchie, K.A., R.L. Brinster, and U. Storb. 1984. Allelic exclusion and control of endogenous immunoglobulin gene rearrangement in κ transgenic mice. *Nature* **312:** 517–520.

Rusconi, S. and G. Köhler. 1985. Transmission and expression of a specific pair of rearranged immunoglobulin μ and κ genes in a transgenic mouse line. *Nature* **314:** 330–334.

Weaver, D., F. Costantini, T. Imanishi-Kari, and D. Baltimore. 1985. A transgenic immunoglobulin Mu gene prevents rearrangement of endogenous genes. *Cell* **42:** 117–127.

Early, P. and L. Hood. 1981. Allelic exclusion and nonproductive immunoglobulin gene rearrangements. *Cell* **24:** 1–3.

### Class Switch

Coleclough, C., D. Cooper, and R.P. Perry. 1980. Rearrangement of immunoglobulin heavy chain genes during B-lymphocyte development as revealed by studies of mouse plasmacytoma cells. *Proc. Natl. Acad. Sci.* **77:** 1422–1426.

Cory, S. and J.M. Adams. 1980. Deletions are associated with somatic rearrangements of immunoglobulin heavy chain genes. *Cell* **19:** 37–51.

Davis, M.M., K. Calame, P.W. Early, D.L. Livant, R. Joho, I.L. Weissman, and L. Hood. 1980. An immunoglobulin heavy-chain gene is formed by at least two recombinational events. *Nature* **283:** 733–739.

Honjo, T. and T. Kataoka. 1978. Organization of immunoglobulin heavy chain genes and allelic deletion model. *Proc. Natl. Acad. Sci.* **75:** 2140–2144.

Hurwitz, J.L., C. Coleclough, and J.J. Cebra. 1980. C$_H$ gene rearrangements in IgM-bearing B cells and in the normal splenic DNA component of Hybridomas making different isotypes of antibody. *Cell* **22:** 349–359.

Kataoka, T., T. Kawakami, N. Takahashi, and T. Honjo. 1980. Rearrangement of immunoglobulin γ1-chain gene and mechanism for heavy-chain class switch. *Proc. Natl. Acad. Sci.* **77:** 919–923.

Lawton, A.R., P.W. Kincade, and M.D. Cooper. 1975. Sequential expression of germ line genes in development of immunoglobulin class diversity. *Fed. Proc.* **34:** 33–39.

Marcu, K.B., R.B. Lang, L.W. Stanton, and L.J. Harris. 1982. A model for the molecular requirements of immunoglobulin heavy chain class switching. *Nature* **298:** 87–89.

Nossal, G.J.V., N. Warner, and H. Lewis. 1971. Incidence of cell simultaneously secreting IgM and IgG antibody to sheep erythrocytes. *Cell. Immunol.* **2:** 41–53.

Rabbitts, T.H., A. Forster, W. Dunnick, and D.L. Bentley. 1980. The role of gene deletion in the immunoglobulin heavy chain switch. *Nature* **283:** 351–356.

Sakano, H., R. Maki, Y. Kurosawa, W. Roeder, and S. Tonegawa. 1980. Two types of somatic recombination are necessary for the generation of complete immunoglobulin heavy-chain genes. *Nature* **286:** 676–683.

# 3 ANTIBODY–ANTIGEN INTERACTIONS

The interaction of an antibody with an antigen forms the basis of all immunochemical techniques. This chapter discusses the properties of the antibody–antigen interaction and is divided into three sections. The first summarizes the structure of the antibody–antigen bonds, the second covers the strength of these interactions, a characteristic known as affinity, and the third presents the factors that contribute to the overall stability of immune complexes, a property called avidity.

## STRUCTURE OF THE ANTIBODY–ANTIGEN COMPLEX

The structure of the antibody–antigen complex has been studied by measuring the affinity of binding between an antibody and a series of related antigens, by using affinity labeling reagents, by site-directed mutagenesis of the antibody combining site, by molecular modeling, and, most compellingly, by X-ray diffraction studies of antibody–antigen cocrystals. Together, these techniques have delineated the region of the antibody molecule that is involved in antigen binding, the region of the antigen molecule that interacts with the antibody, and the molecular basis for antibody specificity.

## The antigen binding site of an antibody is formed by the variable regions of the heavy and light chains

Affinity labeling and X-ray crystallography of immune complexes have established that the antigen binding site is formed by the heavy- and light-chain variable regions (see Fig. 2.5). The two variable regions are closely associated and are bound to each other by noncovalent interactions. The remainder of the heavy and light chains forms other domains that are not involved in antigen binding (see Chapter 2). The amino acids forming the antigen binding site are derived from both the heavy and light chains and correspond to the amino acids of the hypervariable regions determined from protein sequencing. The hypervariable regions are known as the complementarity determining regions (CDRs). There are six CDRs, three on each chain, and they form discrete loops anchored and oriented by the framework residues of the variable domains (Fig. 3.1).

## The region of an antigen that binds to an antibody is called an epitope

The region of an antigen that interacts with an antibody is defined as an epitope. An epitope is not an intrinsic property of any particular structure, as it is defined only by reference to the binding site of an antibody. The size of an epitope is governed by the size of the combining site. From X-ray studies of the structures of cocrystals between small antigens bound to antibodies, the size of the combining site was thought to be relatively small. The site was visualized as a cleft or pocket into which the epitope docked. Relatively few of the amino acid side chains of the CDR were in close contact with the antigen. Recently, the structures of three protein antigen–antibody complexes have been solved. These structures have changed our view of the potential size and shape of epitopes. The two antigens used in these studies were lysozyme and influenza neuraminidase. The area of these antigens in close apposition to the antibody is large, occupying approximately 500–750 $Å^2$ and involving contacts with all six CDRs. Although these studies have shown that epitopes can be much larger than originally thought, it is still clear that high-affinity antibodies can be raised to small epitopes.

Because antibodies can recognize relatively small regions of antigens, occasionally they can find similar epitopes on other molecules. This forms the molecular basis for cross-reaction. Although similar epitopes can occur on related molecules, the presence of similar epitopes does not necessarily imply a functional relationship.

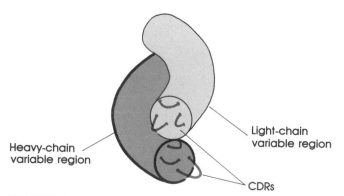

**FIGURE 3.1**
The six CDRs form the binding sites for antigen-antibody association.

## Epitopes on protein antigens are local surface structures that can be formed by contiguous or noncontiguous amino acid sequences

In one of the lysozyme–antibody cocrystals, the amino acids of lysozyme that form the epitope come from two distant stretches of the primary sequence (residues 18–27 and residues 116–129). Though separated from each other in the primary sequence, these stretches of amino acids are adjacent on the protein surface. At the interface between the antigen and the antibody, a total of 16 amino acids of the antigen make close contacts with 17 amino acids of the antibody, the latter involving all six CDRs. The whole interface is tightly packed and excludes solvent. Strikingly, 748 $Å^2$ or 11% of the surface of lysozyme is covered by the antibody. Similar conclusions come from the study of the second lysozyme–antibody cocrystals and the neuraminidase–antibody cocrystals. Here, either three (lysozyme) or four (neuraminidase) stretches of distant primary sequence form portions of the epitope structure.

Although no cocrystals of epitopes derived from contiguous stretches of amino acids have been solved, work with antibodies raised against synthetic peptides or other small antigens has shown that these interactions exist.

## Some immune complexes show no alterations in the structure of the antibody or antigen, whereas others show large conformational changes

Antibody–antigen interactions can occur either with large structural changes in the antibody or the antigen or with no detectable changes. From the structure of the first three antibody–protein antigen cocrystals, it is clear that both flexible and rigid structures can form good epitopes. In the crystal structure of one of the lysozyme–antibody complexes, no distortion of either the antigen or antibody could be detected, even at the 2.8 Å level. In sharp contrast, the crystal structure of the neuraminidase–antibody complex revealed substantial structural alterations of both the antigen and antibody. Because the crystallization process can induce structural alterations itself, it is difficult to prove that these changes are due to antibody binding. However, many other studies have shown that antibodies can induce structural changes in antigens. Good examples of this are the removal of heme from myoglobin and the activation of enzymes by antibody binding. More structural analyses are needed to determine how frequently such alterations accompany complex formation.

## The antibody–antigen complex is held together by multiple, noncovalent bonds

The binding of the antibody to the antigen is entirely dependent on noncovalent interactions, and the antibody–antigen complex is in equilibrium with the free components. The immune complex is stabilized by the combination of weak interactions that depend on the precise alignment of the antigen and antibody. These noncovalent interactions include hydrogen bonds, van der Waals forces, coulombic interactions, and hydrophobic bonds. These bonds can occur between side chains or the polypeptide backbones.

Small changes in antigen structure can affect profoundly the strength of the antibody–antigen interaction. The loss of a single hydrogen bond at the interface can reduce the strength of interaction 1000-fold. The overall interaction is a balance of many attractive and repulsive interactions at the interface. This can be demonstrated in vitro by site-directed mutagenesis. Changing the amino acid residues that form the binding site can alter the strength of an antibody–antigen interaction. This is performed elegantly in vivo by the selection of cells secreting higher-affinity antibodies. By an unknown process, the CDR residues from differentiating clones of B cells undergo extensive mutation, yielding antibodies that differ widely in the microstructure of their antigen binding sites. Cells that express antibodies with higher affinity are stimulated preferentially to divide. This process continues during the exposure and reexposure to antigen and results in a stronger and more specific antibody response (see Chapter 4).

## Antibodies can bind to a wide range of chemical structures and can discriminate among related compounds

The microenvironment of the combining site can accommodate highly charged as well as hydrophobic molecules. Epitopes composed of carbohydrates, lipids, nucleic acids, amino acids, and a wide range of synthetic organic chemicals have all been identified. The repertoire of possible binding sites is enormous, and antibodies that are specific to novel compounds have been derived readily.

The specificity of antibodies has been demonstrated by a large number of experiments showing that small changes in the epitope structure can prevent antigen recognition. For example, antibodies have been isolated that will differentiate between conformations of protein antigens, detect single amino acid substitutions, or act as weak enzymes by stabilizing transition forms.

AFFINITY

Affinity is a measure of the strength of the binding of an epitope to an antibody.

# The binding of antibodies to antigens is reversible, and the strength of the interaction can be described in terms of an equilibrium reaction

Antibody binding to antigen is noncovalent and reversible. The binding of antibody to antigen follows the basic thermodynamic principles of any reversible bimolecular interaction. Thus, if [Ab] = molar concentration of the unoccupied antibody binding sites, [Ag] = molar concentration of the unoccupied antigen binding sites, and [Ab–Ag] is the molar concentration of the antibody–antigen complex, the affinity constant $K_A$ = [Ab–Ag]/[Ab]·[Ag]. In practical terms, affinity describes the amount of antibody–antigen complex that will be found at equilibrium.

The time taken to reach equilibrium is dependent on the rate of diffusion and does not vary from one antibody to another. However, high-affinity antibodies will bind larger amounts of antigen in a shorter period of time than low-affinity antibodies. In practice, this means that high-affinity interactions are substantially complete well before low-affinity interactions. High-affinity antibodies perform better in all immunochemical techniques. This is due not only to their higher capacity but also to the stability of the complex. For example, the half-time for dissociation of an antibody binding to a small protein antigen with high affinity is 30 min or more, while for a low-affinity antibody this time may be a few minutes or less.

# The affinity of the antibody–antigen interaction varies over a wide range

The range of measured values of affinity constants for antibody–antigen binding is enormous and extends from below $10^5$ mol$^{-1}$ to above $10^{12}$ mol$^{-1}$. For comparison, the affinity of trypsin for its substrate is approximately $1.25 \times 10^4$ mol$^{-1}$, the affinity of $\lambda$ repressor converting from monomer to dimer is $5 \times 10^7$ mol$^{-1}$, and the affinity of $\lambda$ repressor for DNA is $10^{10}$ mol$^{-1}$. Like all equilibrium reactions, the affinity constant for antibody–antigen interactions is affected by temperature, pH, and solvent. Changes in these may increase or decrease the number of antibody–antigen complexes found at equilibrium. These alterations will change the affinity constant, either driving the reaction toward complete binding or releasing bound antigen.

The affinity of monoclonal antibodies can be determined exactly, but the affinity of polyclonal antibodies cannot. Because monoclonal antibodies are homogeneous, the exact measurement of their affinity is possible using a range of techniques. Polyclonal sera contain complex mixtures of antibodies of different affinities, therefore the affinity of such sera cannot be exactly determined.

## Although the antibody–antigen affinity in a particular environment does not vary, the extent of complex formation can be manipulated

The easiest way to control the extent of complex formation is to vary the concentration of the antibody or antigen. Provided neither component is saturated, adding more antibody to a constant volume will increase the amount of antigen that is bound. Similarly, adding more antigen will increase the bound antibody. With suitably high affinities, the addition of excess antibody can be used to bind essentially all of the available antigen, but with low-affinity antibodies a significant fraction of the antigen will remain free. Increasing the reaction volume, thus lowering the concentration of both antibody and antigen, will decrease the amount of complex. The amount of complex decreases approximately with the square of the volume, when neither component is saturated. For low-affinity antibodies, reaction volume will greatly influence the amount of complex formed.

Table 3.1 lists typical affinity values for antibodies and compares these values with the required affinities for several of the common immunochemical techniques.

**TABLE 3.1**
**Factors Affecting the Strength of Antibody Binding**

| | |
|---|---|
| **Cell Staining** | 1. Affinity, $10^6$ mol$^{-1}$ (weak signal) to $10^8$ mol$^{-1}$ (strong signal) |
| | 2. Possible bivalent binding |
| | 3. Possible multivalent binding to secondary reagent |
| | 4. Possible local concentration effects |
| **Immunoprecipitation** | 1. Affinity, $10^7$ mol$^{-1}$ (weak signal) to $10^9$ mol$^{-1}$ (strong signal) |
| | 2. Possible polyclonal binding |
| | 3. Possible multivalent binding to secondary reagents |
| **Immunoblotting** | 1. Affinity, $10^6$ mol$^{-1}$ (weak signal) to $10^8$ mol$^{-1}$ (strong signal) |
| | 2. Possible bivalent binding |
| | 3. Possible multivalent binding to secondary reagents |
| | 4. Possible local concentration effects |

## AVIDITY

Avidity is a measure of the overall stability of the complex between antibodies and antigens. The overall strength of an antibody–antigen interaction is governed by three major factors, the intrinsic affinity of the antibody for the epitope, the valency of the antibody and antigen, and the geometric arrangement of the interacting components. Since the avidity describes the complete reaction, this value ultimately determines the success of all immunochemical techniques.

## When the antibody and antigen can form multivalent complexes, the strength of the interaction is greatly increased

Most immunochemical procedures involve multivalent interactions. Antibodies are multivalent; IgGs and most IgAs are bivalent, and IgMs are decavalent. Antigens can be multivalent either because they contain multiple copies of the same epitope, as in the case of homopolymers, or because they contain multiple epitopes recognized by different antibodies. Multivalent interactions can greatly stabilize immune complexes, rendering the reactions practically irreversible.

Multimeric interactions allow low-affinity antibodies to bind tightly. Because some techniques are more likely to allow multivalent interactions, low-affinity antibodies may work well in one technique, but not in another. When antibodies show this variation among different techniques, it is often due to differences in multivalent binding. Similarly, a cross-reaction that is undetectable in one technique may dominate the results of another assay.

It is not always easy to predict the effects of multivalency, as the reactions involve geometric arrangements that impose steric constraints. However, the effects of multivalency can be illustrated by considering several simple cases as described below: (1) where a monoclonal antibody interacts with a homopolymeric antigen (Fig. 3.2); (2) where polyclonal antibodies bind to an antigen that has multiple epitopes creating large multimeric complexes (Fig. 3.3); (3) where polyclonal antibodies bind to an antigen that has multiple epitopes creating a good target for multimeric binding by secondary reagents (Fig. 3.4); and (4) where an antibody binds to an antigen immobilized on a solid support (Fig. 3.5). There are other variations on these patterns, but they share the basic features illustrated by these examples.

## Homopolymeric antigens present identical, repeating epitopes that encourage bivalent binding

When a monoclonal antibody binds to a multimeric antigen, the initial reaction is identical to the bimolecular interactions discussed above. The antibody finds the antigen by diffusion. However, the second step of the reaction links the unoccupied combining site of the antibody with an identical epitope on the same antigen molecule. This reaction is an intramolecular conversion and does not depend on diffusion. It is restrained only by conformation (Fig. 3.2). Once this molecular complex is assembled, the rate of dissociation of the individual antibody–epitope interactions is similar to the normal bimolecular complex. However, since the antigen will still be held by the other interaction,

**HIGH AVIDITY**

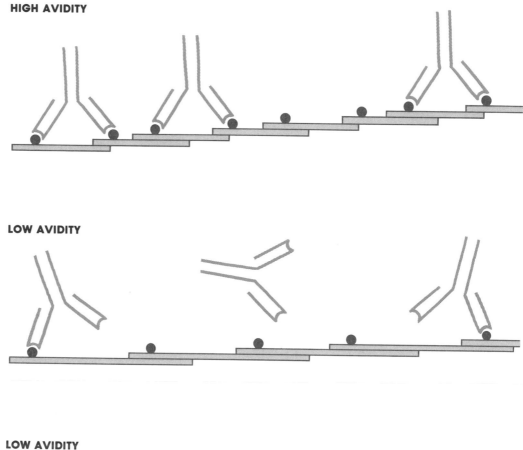

**LOW AVIDITY**

**LOW AVIDITY**

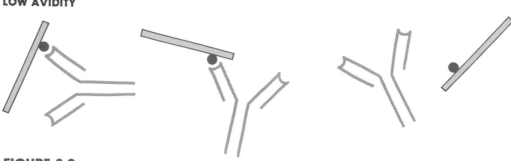

**FIGURE 3.2**
Bivalent binding of antibodies to polymeric antigens increases the avidity.

CYCLIC COMPLEX

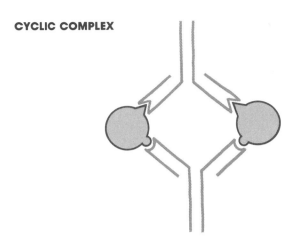

LATTICE COMPLEX

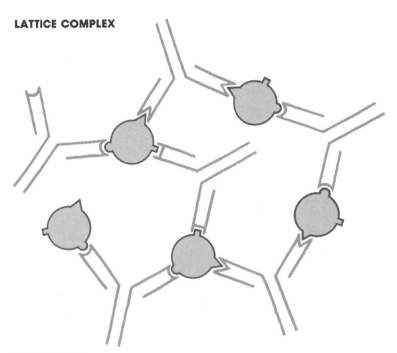

**FIGURE 3.3**
Polyclonal antibodies binding to multivalent antigens.

the observed rate of dissociation will be much slower, thus forming a more stable antibody–antigen complex.

This type of multimeric interaction can occur in all immunochemical techniques.

## Antibodies that bind to multiple sites on an antigen can form large, stable, multimeric complexes

When an antigen with more than one epitope is mixed with polyclonal antibodies, the complexes that form can be stabilized by intermolecular bridges (Fig. 3.3). During complex formation, one antibody binds more than one antigen molecule. If these antigen molecules are linked by the binding of a second antibody to other epitopes, a cyclic or lattice structure can be formed. The rate of dissociation of any one antibody–epitope binding is the same as for a simple interaction, but because the antigen is still held by other interactions, the overall rate of dissociation is very slow. These interactions form stable and often very large complexes. The formation of this type of complex is dependent on the relative molar ratios of the antigens and antibodies. Either used in excess will restrict the extent of cross-linking.

This type of multimeric interaction can occur in immunoprecipitation.

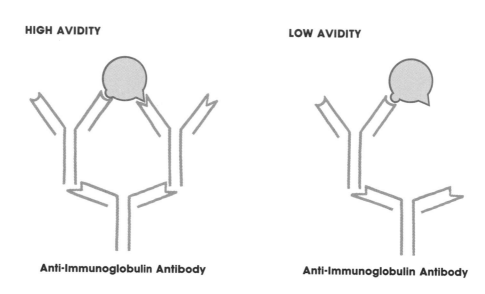

**HIGH AVIDITY**  **LOW AVIDITY**

Anti-Immunoglobulin Antibody  Anti-Immunoglobulin Antibody

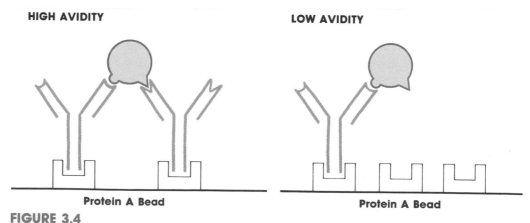

**HIGH AVIDITY**  **LOW AVIDITY**

Protein A Bead  Protein A Bead

**FIGURE 3.4**
Polyclonal antibodies binding to multivalent antigens provide good targets for secondary reagents.

## Antibodies that bind to multiple sites on an antigen provide an excellent target for secondary reagents

When an antigen is coated by many antibodies, the multimeric complex provides multiple binding sites in a flexible geometric arrangement that allows stable multimeric interactions with secondary reagents such as anti-immunoglobulin antibodies or protein A beads (Fig. 3.4).

This type of multimeric interaction can occur in most immunochemical techniques using secondary reagents.

## Antigens immobilized on solid supports at high concentrations promote high-avidity, bivalent binding

When an antibody binds to an antigen on a solid phase, the interaction is biphasic, and two factors, in addition to the intrinsic affinity, control the strength of the interaction. These are the high local concentration of the antigen and the possibility of bivalent binding. The initial

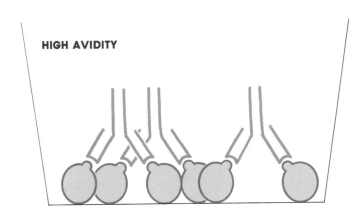

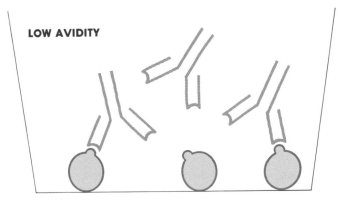

**FIGURE 3.5**
Bivalent binding to antigens immobilized on a solid phase increases the avidity.

binding of the antibody to the immobilized antigen is limited by diffusion, but after the first antibody–epitope interaction occurs, the formation of the second bond may be an intramolecular conversion if sterically possible (Fig. 3.5). In addition, the high local concentration of antigen increases the chance that any dissociated antibodies will rebind to neighboring antigens. In essence, diffusion occurs, but the high concentration of antigen acts as a trap to hold the antibody to the solid phase. These factors combine to yield a high avidity.

This type of multimeric interaction can occur in cell staining, immunoblotting, and many types of immunoassays.

## Further Readings

### Antibody/Antigen Interactions

Berzofsky, J.A. and I.J. Berkower. 1984. Antigen-antibody interaction. In *Fundamental immunology* (ed. W.E. Paul), pp. 595–644. Raven Press, New York.

### Antigenic Structure

Crumpton, M.J. and J.M. Wilkinson. 1965. The immunological activity of some of the chymotryptic peptides of sperm-whale myoglobin. *Biochem. J.* **94**: 545–556.

Atassi, M.Z. 1975. Antigenic structure of myoglobin: The complete immunochemical anatomy of a protein and conclusions relating to antigenic structures of proteins. *Immunochemistry* **12**: 423–438.

Benjamin. D.C., J.A. Berzofsky, I.J. East, F.R.N. Gurd, C. Hannum, S.J. Leach, E. Margoliash, J.G. Michael, A. Miller, E.M. Prager, M. Reichlin, E.E. Sercarz, S.J. Smith-Gill, P.E. Todd, and A.C. Wilson. 1984. The antigenic structure of proteins: A reappraisal. *Annu. Rev. Immunol.* **2**: 67–101.

Landsteiner, K. 1936. *The specificity of serological reactions.* C.C. Thomas, Springfield, Illinois.

### Structural Studies of Antibodies with Haptens

Marquart, M., J. Deisenhofer, R. Huber, and W. Palm. 1980. Crystallographic refinement and atomic models of the intact immunoglobulin molecule Kol and its antigen-binding fragment at 3·Å and 1·Å resolution. *J. Mol. Biol.* **141**: 369–391.

Satow, Y., G.H. Cohen, E.A. Padlan, and D.R. Davies. 1986. Phosphocholine binding immunoglobulin Fab McPC603: An X-ray diffraction study of 2.7Å. *J. Mol. Biol.* **190**: 593–604.

Saul, F.A., L.M. Amzel, and R.J. Poljak. 1978. Preliminary refinement and structural analysis of the Fab fragment from human immunoglobulin new at 2.0 Å resolution. *J. Biol. Chem.* **253**: 585–597.

### Structural Studies of Antibodies with Antigens

Amit, A.G., R.A. Mariuzza, S.E.V. Phillips, and R.J. Poljak. 1986. Three-dimensional structure of an antigen-antibody complex at 2.8 Å resolution. *Science* **233**: 747–753.

Colman, P.M., W.G. Laver, J.N. Varghese, A.T. Baker, P.A. Tulloch, G.M. Air, and R.G. Webster. 1987. Three-dimensional structure of a complex of antibody with influenza virus neuraminidase. *Nature* **326:** 358–363.

Sheriff, S., E.W. Silverton, E.A. Padlan, G.H. Cohen, S.J. Smith-Gill, B.C. Finzel, and D.R. Davies. 1987. Three-dimensional structure of an antibody–antigen complex. *Proc. Natl. Acad. Sci.* **84:** 8075–8079.

Colman, P.M., G.M. Air, R.G. Webster, J.N. Varghese, A.T. Baker, M.R. Lentz, P.A. Tulloch, and W.G. Laver. 1987. How antibodies recognize virus proteins. *Immunol. Today* **8:** 323–326.

Rees, A.R. 1987. The antibody combining site: Retrospect and prospect. *Immunol. Today* **8:** 44–45

# 4 ANTIBODY RESPONSE

An antibody response is the culmination of a series of interactions between macrophages, T lymphocytes, and B lymphocytes, all reacting to the presence of a foreign antigen. The end product of this response is the production of large numbers of antibody molecules that bind specifically to the antigen and speed its removal from the animal. The first exposure to an antigen induces a relatively weak reaction known as the primary response. A series of specialized events during the primary response prepare the animal for any subsequent exposure to the same antigen. When the antigen is reintroduced, the secondary response is typically rapid and intense. For both the primary and secondary responses, specialized cells known as plasma cells are the main sites of antibody production. These cells are terminally differentiated B cells. Another cell from the B-cell lineage, the memory B cell, is instrumental in promoting the rapid and vigorous secondary response.

The control of the antibody response is maintained by regulating the differentiation of B cells into plasma cells and memory cells. This differentiation is mediated by the secretion of soluble polypeptides and by a series of cell-to-cell contacts between macrophages, T lymphocytes, and the B-cell precursors.

## When an antigen is introduced into a naive animal, the first step in the generation of the primary response is phagocytosis of the antigen

As the first step in mounting an antibody response, foreign antigens are engulfed nonspecifically by macrophages or other cells of the reticuloendothelial system. These cells include Langerhans cells in the skin, dendritic cells in the spleen and lymph nodes, and monocytes in the blood. As all of these cells process antigens in the same manner, they are collectively referred to as antigen-presenting cells (APCs). For most antigens, this nonspecific uptake is essential to initiate an effective response (Fig. 4.1).

Many of the common techniques designed to optimize immunizations work by increasing the efficiency of phagocytosis. They include choosing an appropriate injection site, altering the form of the antigen, and using adjuvant. Not all sites of injection are equally good at targeting antigens for phagocytosis. Optimal sites have high numbers of APCs and low rates of antigen degradation. Commonly used routes of injection are subcutaneous, intradermal, intramuscular, intraperitoneal, and intravenous. The efficiency of phagocytosis also depends on the

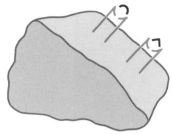

**Nonspecific Phagocytosis**                    **Antigen Processing**                    **Antigen Presentation**

**FIGURE 4.1**
Antigen processing by APCs.

size and form of the antigen. Particulate antigens are easily phagocytosed, whereas small soluble molecules are not. Therefore, converting small soluble molecules into large particulate aggregates usually is an effective way to increase their immunogenicity. The third commonly used method to increase phagocytosis is to mix the antigen with adjuvants. Adjuvants help in two ways. First, they protect the antigen from rapid dispersal by trapping it in a local deposit. Second, they contain substances that stimulate the secretion of host factors that recruit macrophages to the site of antigen deposition and increase the local rate of phagocytosis. Protocols for using these techniques are found in Chapter 5 along with a more detailed discussion of their mechanism of action.

## Antigen-presenting cells degrade the antigen and display fragments of it on the cell surface, bound to an MHC class II protein

After an antigen is engulfed by an APC, the phagocytic vesicle fuses with a lysosome, and the antigen is partially degraded (Fig. 4.1). By a poorly understood mechanism, fragments of the antigen appear on the cell surface of the APC. Here, they are found in a complex with a cell-surface glycoprotein known as the MHC (major histocompatibility complex) class II protein. Class II molecules are heterodimers, containing polypeptides of approximately 25,000 and 33,000 daltons. Although the crystal structure of the class II protein has not yet been solved, an approximation of its structure can be determined by comparison with the related class I protein. These extrapolations suggest that class II proteins have a cleft near the top external face of the protein, and this cleft binds the antigen fragments. This structure can be thought of as cradling the antigen fragment and presenting it to the extracellular environment.

Sequencing studies have confirmed that both polypeptides in the class II protein are highly polymorphic and, depending on the species, more than one copy of the gene for each polypeptide may be found. In total, any one animal will produce somewhere between 6 and 40 different class II molecules. Many of the polymorphic differences in the amino acid sequences of class II polypeptides are in the area of the cleft.

Although the cleft in the class II protein is able to bind a wide range of structures, some antigen fragments will not bind. If none of the fragments bind to the class II proteins, the antigen may be masked from the immune system. Animals with class II proteins that do not bind to fragments of an antigen are known as nonresponders for this antigen. This type of nonresponsiveness is a genetic trait that maps to the class II genes. This block can be overcome in one of two ways: first, by immunizing animals with different alleles at the class II loci or, second, by chemically modifying the antigen to create sites that will generate suitable fragments for binding to class II proteins. Protocols for these techniques are found in Chapter 5 along with a more detailed discussion of their mode of action.

## Helper T cells associate with antigen-presenting cells by binding to the antigen fragment–class II protein complex

In the next step of the antibody response, helper T cells bind to APCs (Fig. 4.2). This interaction is mediated by binding of the antigen fragment–class II protein complexes on the APC with T-cell receptors on the surface of the helper T cell. The T-cell receptor recognizes the antigen fragment and the class II protein, and it is specific for both components. This binding ensures that helper T cells do not respond to free antigen or to unoccupied class II proteins.

The T-cell receptor is a heterodimer containing polypeptides known

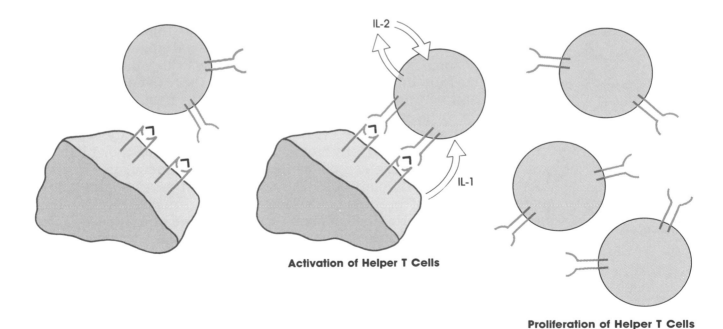

IL-2

IL-1

**Activation of Helper T Cells**

**Proliferation of Helper T Cells**

**FIGURE 4.2**
Interactions between APCs and helper T cells.

as $\alpha$- and $\beta$-chains.[1] This receptor has many similarities to an antibody molecule. Both $\alpha$- and $\beta$-chains contain variable and constant regions, with the variable regions of the two chains folding together to create the binding site. There are over $10^7$ different T-cell receptors, and this diversity is generated by special genetic mechanisms. Many of the mechanisms used to generate antibody diversity in B cells are also used in T cells for the generation of T-cell receptor diversity (see Chapter 2). Like antibodies, a large number of variable regions are located upstream of the constant region in germ-line DNA. During differentiation, recombination selects one variable region and, by removing intervening sequences, places it adjacent to a constant region to yield a functional gene. The $\alpha$-chain resembles the antibody light chain in that the recombination events link a variable region to a J region. The $\beta$-chain resembles the antibody heavy chain in that two recombination events, a variable region to D region rearrangement and a D region to J region rearrangement, are needed to generate a functional gene. Since any $\alpha$-chain can associate with any $\beta$-chain, a large number of potential binding sites are possible. However, unlike antibodies, no evidence for somatic mutation in the generation of diversity has been reported for the T-cell receptor. The generation of T-cell receptors occurs without stimulation by antigens, and therefore, the repertoire of T-cell receptors is present before immunization.

Each helper T cell expresses only one $\alpha$- and one $\beta$-chain, thus ensuring that one T cell can only respond to a particular antigen fragment–class II complex. Since all of the T-cell receptors on an individual T cell are identical, only a tiny fraction of the total helper T cells in an animal responds to a given antigen fragment–class II protein complex.

Because T-cell receptors are generated randomly during T-cell differentiation, cells bearing receptors that react with self-components are produced. By a poorly understood process, these self-reactive clones are removed. The result of these clonal deletions is self-tolerance. Clonal deletion occurs in the thymus during the development of the immune system. The absence of these T-cell clones is a major reason why it is difficult to raise antibodies against self-components. Immunizations with antigens that are closely related to or mimic self-components often fail to induce a good antibody response. The more an antigen differs from self-components, the more likely a good antibody response can be achieved. Two approaches can be used to overcome T-cell tolerance. The easiest solution is to immunize animals from a different species. Even switching mouse strains helps for some antigens. Because of the useful properties of antibodies from some species and because some antigens are highly conserved between species, this approach is not always possible. The second approach is to modify the antigen structurally, creating new sites for T cell recognition. Protocols for antigen modifications are discussed in Chapter 5.

---

[1]A second T-cell receptor has been described recently. The chains that combine to form this receptor are known as $\gamma$ and $\delta$. More information on $\gamma$ - $\delta$ receptors can be found in the "Further Readings" at the end of this chapter.

## Binding of a helper T cell to an antigen-presenting cell leads to proliferation of helper T cells

Binding of a helper T cell to an APC starts a cascade of events leading to T-cell proliferation and differentiation (Fig. 4.2). The first stage of this cascade is the activation of helper T cells. Activation requires two signals. First, helper T cells must bind to the correct antigen fragment–class II protein complex. This binding ensures that only helper T cells able to recognize the correct antigen are stimulated. The second signal is mediated by the binding of a soluble polypeptide factor known as interleukin 1 (IL-1). IL-1 is produced by the APC as well as by many other cell types. The production of IL-1 is nonspecific, but within the helper T-cell population, only T cells that are bound to an APC will respond.

The next step in the proliferation and differentiation of helper T cells involves a second well-characterized polypeptide growth factor, interleukin 2 (IL-2). IL-2 is a 15,000-dalton glycoprotein that is secreted by T cells in response to the combined stimulation of antigen binding and IL-1. This combined stimulus also induces the helper T cells to synthesize IL-2 receptors. In the presence of IL-2 these cells will now divide exponentially. This proliferation requires continuous stimulation by the growth factor, and, in the absence of IL-2, the cells stop dividing. The system is controlled by the level of expression of the IL-2 receptor on the surface of proliferating T cells. In the absence of antigen-specific stimulation by the APC, the number of IL-2 receptors declines rapidly. Within 14 days of the original response, no T cells expressing the receptor remain, and proliferation of the helper T cell ceases, even if IL-2 is still present. Resting T cells cannot respond to IL-2, because they lack IL-2 receptors.

IL-2 is a more specific growth factor than IL-1. T cells are the major cell type that express the receptor. Because helper T cells also produce the IL-2, at least in vitro, these cells can proliferate by autocrine stimulation. As mentioned above, the proliferative response in vivo is controlled by the continued presence of the IL-2 receptor. On the other hand, IL-1 stimulates a broad range of responses in many cell types. Neither IL-1 nor IL-2, however, is antigen specific, and any T cell in the animal bearing the relevant receptor will proliferate in the presence of the growth factor. The specificity of the response is maintained by the requirement for antigen recognition to induce and maintain receptor expression. This system of growth factor regulation allows rapid amplification of the helper T-cell response, yet retains a high degree of regulation and specificity.

## B cells also process antigens, but do so in an antigen-specific manner

While antigen-specific helper T cells are proliferating, similar events lead to the division of antigen-specific B cells. Antigens are processed by B cells in much the same way as by APCs (Fig. 4.3). The antigen is

**TABLE 4.1**
**Interleukins**

| Interleukin | Molecular weight | Source | Target | Receptor | Action |
|---|---|---|---|---|---|
| Interleukin 1 (IL-1α, IL-1β) | 17,000 17,000 | Macrophages, many other sources | T helper cells, many other targets | p79 | T-cell activation |
| Interleukin 2 (IL-2) | 15,000 | Activated T cells | Activated T cells | p58/p75 dimer | T-cell proliferation |
| Interleukin 4 (IL-4) | 20,000 | Activated T cells | Activated B cells | p60 | B-cell proliferation |
| Interleukin 5 (IL-5) | 50,000 | Activated T cells | Activated B cells | ? | B-cell differentiation |

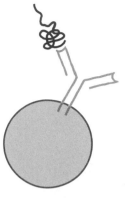

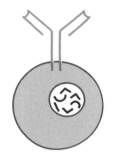
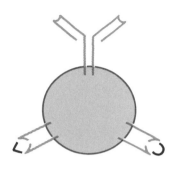

**Specific Antigen Binding**     **Antigen Processing**     **Antigen Presentation**

**FIGURE 4.3**
Antigen processing by B cells.

degraded and fragments appear on the cell surface bound to class II proteins. Unlike APCs, the uptake of antigen by B cells is specific. Virgin B cells have a modified antibody on their cell surface that is free to bind to antigens. The antibody–antigen complexes are internalized, and the antigen processed. The complex of class II protein and antigen fragment is now a target for binding by helper T cells whose T-cell receptors are specific for the complex. The complex of antigen fragment and class II protein on B cells is identical to the analogous complex on APCs, so the same helper T cells that were stimulated to proliferate after binding to the APC can now bind to the B cell.

## Binding of helper T cells to B cells is required for a strong antibody response

The requirement for contact between B cells and helper T cells is a major factor in the regulation of antibody production. This contact provides a key stimulus that leads to B-cell division and the production of antibody (Fig. 4.4). Three key steps in the T cell–B cell interaction require the formation of specific complexes between host proteins and the antigen. First, the antigen must bind to the surface antibody on the B cell. The surface antibody has the same antigen binding site as the secreted antibody. Only cells that bind the antigen are stimulated, thus ensuring the selection of the correct antibody-secreting clones. Second, the processed antigen fragments must bind to the class II proteins. This complex forms the basis of recognition by T cells. Third, the antigen fragment–class II protein complex must bind to the T-cell receptor. This binding ensures that B-cell stimulation by helper T cells is specific.

The properties of these physical interactions explain why some compounds make good antigens and others do not. For a compound to induce an antibody response, it must have an epitope that can bind to the surface antibody on the virgin B cell, and on degradation must generate a fragment that can bind simultaneously to both the class II protein and the T-cell receptor. These sites on the antigen do not need

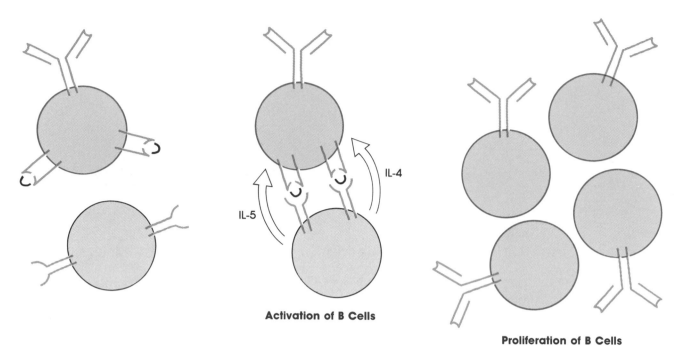

**Activation of B Cells**

**Proliferation of B Cells**

**FIGURE 4.4**
Interactions between helper T cells and B cells.

to be related. One class II protein–T-cell receptor binding site per antigen molecule is sufficient to stimulate all of the available antigen-specific B cells. Because the binding occurs with degraded fragments of the antigen, the class II binding site may not be exposed on the undegraded antigen. However, the B cells will bind directly to the injected antigen and therefore only epitopes that are displayed on the intact antigen can stimulate antibody production. All of these interactions must occur to generate a strong response.

The failure to form these complexes can be caused by three mechanisms. First, some molecules will lack one or more of the appropriate binding sites. The best example of this is in regard to size. Small molecules frequently provide good epitopes for binding to B cells but are unable to induce a response because they are too small to contain the determinants necessary for simultaneous binding with class II proteins and T-cell receptors. Such molecules are known as haptens. To elicit a good antibody response, haptens must be coupled to other molecules that can be processed to provide suitable sites for T-cell receptor binding. Such molecules are called carriers. Second, compounds may fail to induce an antibody response if the genetic background of the animal does not provide class II proteins that can bind to the antigen fragments. This can often be circumvented experimentally by immunizing a different strain or species of animal. The third way in which B cell–T cell collaboration can fail is through tolerance. If either the B or T cell expressing the correct receptor has been eliminated during maturation of the immune system, no response will be seen. Removal of self-reactive B cells appears to be much less efficient than removal of self-reactive T cells. This may be because higher concentrations of antigens are required to eliminate B cells or because somatic mutation of the antibody variable region genes acts to generate self-reactive B cells at a point in their differentiation where they are not removed by the clonal deletion mechanism. If an animal lacks antigen-specific B-cell clones, there are no tricks to overcome this lack of responsiveness. However, if tolerance is due to the lack of the appropriate helper T cell, then modifying the compound to create a binding site for the T-cell receptor will break tolerance. Techniques for antigen modification are discussed in Chapter 5.

## Binding of helper T cells to B cells leads to proliferation and differentiation of the B cells

The binding of helper T cells to B cells causes the B cells to synthesize a cell-surface receptor for a B-cell growth factor called interleukin-4 (IL-4). IL-4 is synthesized by activated helper T cells. Once B cells have received all the appropriate signals, they start a process of exponential division (Fig. 4.5). After binding IL-4, B cells start to synthesize the receptor for a B-cell differentiation factor known as interleukin-5 (IL-5). IL-5 is also produced by activated helper T cells. Binding of IL-5 in conjunction with other signals that are not well understood induces differentiation of the dividing B cells into plasma cells and memory cells.

Plasma cells are highly specialized to secrete large amounts of antibody. Typically 40% or more of the total protein synthesis in

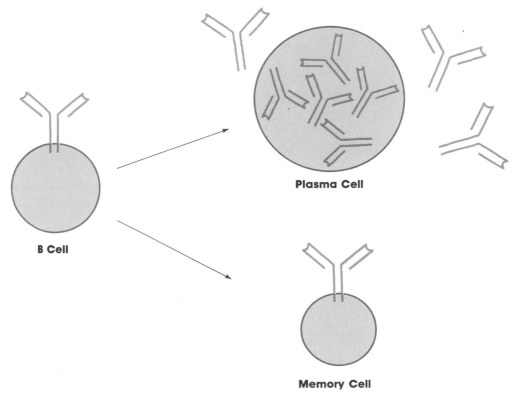

Plasma Cell

B Cell

Memory Cell

**FIGURE 4.5**
B-cell differentiation.

plasma cells is devoted to the production of antibodies. They are terminally differentiated and live for only approximately 3–4 days, primarily in the lymphoid organs.

B memory cells are long-lived cells that do not secrete antibody but retain cell-surface antibody as their specific antigen receptor. These cells remain in circulation and are primed to respond to subsequent exposure to the antigen.

## Differentiation of B cells leads to the production of higher-affinity antibodies and to changes in the class of heavy chains synthesized

Two processes accompany the differentiation of B cells. The first is somatic mutation of the variable regions of the functional antibody light and heavy chains (see Chapter 2). The second is the process of class shifting. Somatic mutations produce a large pool of B cells with antibodies having different affinities for the antigen. Some mutations yield antibodies with higher affinities than the original B cell clone. These cells are selected preferentially to proliferate because their cell-surface antibody will bind more efficiently to the available antigen. The higher-affinity cells follow the same differentiation pathway as other B cells, some becoming plasma cells and some becoming memory cells.

The second event that accompanies division and differentiation of B cells is class shifting. The heavy chains produced during the early stages of B-cell stimulation are from the $\mu$-class. By recombination of the chromosomal DNA encoding the heavy-chain polypeptide, a functional *VDJ* unit is moved from one constant region to another (see Chapter 2). This allows the same variable region to be expressed with different constant regions and leads to the production of antibodies that are better able to achieve certain purposes. For example, only some antibodies can cross the placenta, be secreted into mucous membranes, or fix complement. During the maturation of a typical response, the IgM antibodies that are characteristic of the early response are replaced by IgG antibodies. Because of the affinity maturation discussed above, IgG molecules will normally have higher affinities for the antigen than IgMs.

## At the end of the primary response, the antigen is cleared, leaving primed memory cells

The antibodies produced during a primary response are mostly IgMs, and the level of antibody detected in the serum peaks around 7–10 days after immunization. The secreted antibody will bind and enhance the rate of clearance of any remaining antigen. If no further antigen is introduced, the response declines as the plasma cells die, but the helper T cells and memory B cells remain in circulation.

## Subsequent injections of antigen induce a more potent response

When an animal is given a second injection with the same antigen, a much faster, more potent, and more persistent response occurs. Because memory cells are long-lived, the secondary response can take place months or years after the primary response. The cellular and molecular events of the secondary response are similar to the primary response. However, the antibodies produced are now predominantly of the IgG class, often contain somatic mutations of their light- and heavy-chain variable regions, and bind to the antigen with a higher affinity than those produced during the primary response. These differences in the nature of the secondary response are readily explained by the presence of higher numbers of helper T cells and memory B cells in the immune system of the "primed" animal.

### T-Cell-Independent Antigens

In addition to the T helper cell-dependent antibody response described on pp. 39–45, some antigens can induce a humoral response without the involvement of T cells. These compounds are referred to as T-cell-independent antigens and induce a characteristic type of response. Typically they are polymers and present multiple copies of a particular epitope to the virgin B cell. The binding of the polymeric antigen to multiple antibody molecules on the B cells is sufficient to stimulate the B cells to proliferate and secrete antibodies. However, no differentiation of the B cells occurs, and the response has all the characteristics of a primary response. Even after repeated immunizations with these antigens, no secondary response is seen. Therefore, T-cell-independent antigens do not appear to yield immunological memory and do not produce high-affinity antibodies.

## Further Readings

### Major Histocompatibility Complex

Klein, J. 1975. *Biology of mouse histocompatibility complex.* Springer-Verlag, Berlin.

Klein, J. 1979. The major histocompatibility complex of the mouse. *Science* **203:** 516–521.

Klein, J., F. Figueroa, and C.S. David. 1983. *H-2* haplotypes, genes and antigens: Second listing. II. The *H-2* complex. *Immunogenetics* **17:** 553–596.

Ploegh, H.L., H.T. Orr, and J.L. Strominger. 1981. Major histocompatibility antigens: The human HLA-A, -B, -C) and murine (H-sK, H-2D) class I molecules. *Cell* **24:** 287–299.

Snell, G.D., J. Dauset, and S. Nathenson. 1976. *Histocompatibility.* Academic Press, New York.

Steinmetz, M. and L. Hood. 1983. Genes of the major histocompatibility complex in mouse and man. *Science* **222:** 727–733.


### Structural Studies of MHC Proteins

Bjorkman, P.J., M.A. Saper, B. Samraoui, W.S. Bennett, J.L. Strominger, and D.C. Wiley. 1987. Structure of the human class I histocompatibility antigen, HLA-A2. *Nature* **329:** 506–512.

Bjorkman, P.J., M.A. Saper, B. Samraoui, W.S. Bennett, J.L. Strominger, and D.C. Wiley. 1987. The foreign antigen binding site and T cell recognition regions of class I histocompatibility antigens. *Nature* **329:** 512–518.

Brown, J.H., T. Jardetzky, M.A. Saper, B. Samraoui, P.J. Bjorkman, and D.C. Wiley. 1988. A hypothetical model of the foreign antigen binding site of class II histocompatibility molecules. *Nature* **332:** 845–850.


### Requirements for Cell-to-cell Interactions for Antibody Production

Claman, H.N., E.A. Chaperon, and R.F. Triplet. 1966. Thymus-marrow cell combinations-synergism in antibody production. *Proc. Soc. Exp. Biol. Med.* **122:** 1167.

Davies, A.J.S., E. Leuchars, V. Wallis, R. Marchant, and E.V. Elliott. 1967. *Transplantation* **5:** 222–231.

Hersh, E.M. and J.E. Harris. 1968. Macrophage-lymphocyte interaction in the antigen-induced blastogenic response of human peripheral blood leukocytes. *J. Immunol.* **100:** 1184.

Katz, D.H., T. Hamaoka, and B. Benacerraf. 1973. Cell interactions between histoincompatible T and B lymphocytes. II. Failure of physiologic cooperative interactions between T and B lymphocytes from allogeneic donor strains in humoral response to hapten-protein conjugates. *J. Exp. Med.* **137:** 1405–1418.

Katz, D.H., M. Graves, M.E. Dorf, H. Dimuzio, and B. Benacerraf. 1975. Cell interactions between histoincompatible T and B lymphocytes. 7. Cooperative responses between lymphocytes are controlled by genes in 1 region of H-2 complex. *J. Exp. Med.* **141:** 263.

Katz, D.H., T. Hamaoka, M.E. Dorf, and B. Benacerraf. 1973. Cell interactions between histoincompatible T and B lymphocytes. The H-2 gene complex determines successful physiologic lymphocyte interactions. *Proc. Natl. Acad. Sci.* **70:** 2624–2628.

Kindred, B. and D.C. Shreffler. 1972. Dependence of co-operation between T and B cells in vivo. *Immunology* **109:** 938–943.

Mitchison, N.A. 1971. The carrier effect in the secondary response to hapten-protein conjugates. II. Cellular cooperation. *Eur. J. Immunol.* **1**: 18–27.

Mosier, D.E. 1967. A requirement for two cell types for antibody formation in vitro. *Science* **158**: 1573–1575.

Paul, W.E., E.M. Shevach, D.W. Thomas, S.F. Pickeral, and A.S Rosenthal. 1977. Genetic restriction in T-lymphocyte activation by antigen-pulsed peritoneal exudate cells. *Cold Spring Harbor Symp. Quant. Biol.* **41**: 571–578.

Rajewsky, K., V. Schirrmacher, S. Nase, and N.K. Jerne. 1969. The requirement of more than one antigenic determinant for immunogenicity. *J. Exp. Med.* **129**: 1131–1143.

Rosenthal, A.S. and E.M. Shevach. 1973. Function of macrophages in antigen recognition by guinea pig T lymphocytes. *J. Exp. Med.* **138**: 1194–1212.

Schwartz, R.H., C.S. David, D.H. Sachs, and W.E. Paul. 1976. T lymphocyte-enriched murine peritoneal exudate cells. III. Inhibition of antigen-induced T lymphocyte proliferation with anti-Ia antisera. *J. Immunol.* **117**: 531–540.

Shevach, E.M., W.E. Paul, and I. Green. 1972. Histocompatibility-linked immune response gene function in guinea pigs. *J. Exp. Med.* **136**: 1207–1219.

Shevach, E.M. and A.S. Rosenthal. 1973. Function of macrophages in antigen recognition by guinea pig T lymphocytes. *J. Exp. Med.* **138**: 1213–1229.

Zinkernagel, R.M. and P.C. Doherty. 1974. Immunological surveillance against altered self components by sensitised T lymphocytes of lymphocytic choriomeningitis. *Nature* **251**: 547–548.

Benacerraf, B. and H.O. McDevitt. 1972. Histocompatibility-linked immune response genes. *Science* **175**: 273–279.

Schwartz, R.H. 1986. Immune response (*Ir*) genes of the murine major histocompatibility complex. *Adv. Immunol.* **38**: 31–201.

Sprent, J. 1978. Role of H-2 gene products in the function of T helper cells from normal and chimeric mice *in vivo*. *Immunol. Rev.* **42**: 108–137.

### Antigen Presentation by APC

Babbitt, B.P., P.M. Allen, G. Matsueda, E. Haber, and E.R. Unanue. 1985. Binding of immunogenic peptides to Ia histocompatibility molecules. *Nature* **317**: 359–360.

Benacerraf, B. 1978. A hypothesis to relate the specificity of T lymphocytes and the activity of I region-specific Ir genes in macrophages and B lymphocytes. *J. Immunol.* **120**: 1809–1812.

Buus, S., S. Colon, C. Smith, J.H. Freed, C. Miles, and H.M. Grey. 1986. Interaction between a "processed" ovalbumin peptide and Ia molecules. *Proc. Natl. Acad. Sci.* **83**: 3968–3971.

Buus, S., A. Sette, S.M. Colon, C. Miles, and H.M. Grey. 1987. The relation between major histocompatibility complex (MHC) restriction and the capacity of Ia to bind immunogenic peptides. *Science* **235**: 1353–1358.

Guillet, J.-G., M.-Z. Lai, T.J. Briner, S. Buus, A. Sette, H.M. Grey, J.A. Smith, and M.L. Gefter. 1987. Immunological self, nonself discrimination. *Science* **235**: 865–870.

Shimonkevitz, R., J. Kappler, P. Marrack, and H. Grey. 1983. Antigen recognition by H-2 restricted cells. I. Cell-free antigen processing. *J. Exp. Med.* **158**: 303–316.

Watts, T.H., A.A. Brian, J.W. Kappler, P. Marrack, and H.M. McConnell. 1984. Antigen presentation by supported planar membranes containing affinity-purified I-A$^d$. *Proc. Natl. Acad. Sci.* **81**: 7564–7568.

Werdelin, O. 1982. Chemically related antigens compete for presentation by accessory cells to T cells. *J. Immunol.* **129**: 1883–1891.

Ziegler, K. and E.R. Unanue. 1981. Identification of a macrophage antigen-processing event required for I-region-restricted antigen presentation to T lymphocytes. *J. Immunol.* **127:** 1869–1875.

Ziegler, H.K. and E.R. Unanue. 1982. Decrease in macrophage antigen catabolism caused by ammonia and chloroquine is associated with inhibition of antigen presentation to T cells. *Proc. Natl. Acad. Sci.* **79:** 175–178.

Allen, P.M. 1987. Antigen processing at the molecular level. *Immunol. Today* **8:** 270–273.

Rosenthal, A.S. 1978. Determinant selection and macrophage function in genetic control of the immune response. *Immunol. Rev.* **40:** 135–152.

Schwartz, R.H. 1985. T-lymphocyte recognition of antigen in association with gene products of the major histocompatibility complex. *Annu. Rev. Immunol.* **3:** 237–261.

Unanue, E.R. 1984. Antigen-presenting function of the macrophage. *Annu. Rev. Immunol.* **2:** 395–428.

Unanue, E.R. and P.M. Allen. 1987. The basis for the immunoregulatory role of macrophages and other accessory cells. *Science* **236:** 551–557.

### Antigen Presentation by B Cells

Chestnut, R.W. and H.M. Grey. 1981. Studies on the capacity of B cells to serve as antigen-presenting cells. *J. Immunol.* **126:** 1075–1079.

Grey, H.M., S.M. Colon, and R.W. Chestnut. 1982. Requirements for the processing of antigen by antigen-presenting B cells. II. Biochemical comparison of the fate of antigen in B cell tumors and macrophages. *J. Immunol.* **129:** 2389–2395.

Kakiuchi, T., R.W. Chestnut, and H.M. Grey. 1983. B cells as antigen-presenting cells: The requirement for B cell activation. *J. Immunol.* **131:** 109–114.

Lanzavecchia, A. 1985. Antigen-specific interaction between T and B cells. *Nature* **314:** 533–535.

Rock, K.L. and B. Benacerraf. 1984. Selective modification of a private I-A allo-stimulating determinant(s) upon association of antigen with an antigen-presenting cell. *J. Exp. Med.* **159:** 1238–1252.

### T Cell Receptor and Antigen Recognition by T cells

Allison, J.P., B.W. McIntrye, and D. Bloch. 1982. Tumor-specific antigen of murine T-lymphoma defined with monoclonal antibody. *J. Immunol.* **129:** 2293–2300.

Bluestone, J.A., D. Pardoll, S.O. Sharrow, and B.J. Fowlkes. 1987. Characterization of murine thymocytes with CD3-associated T-cell receptor structures. *Nature* **326:** 82–84.

Brenner, M.B., J. McLean, D.P. Dialynas, J.L. Strominger, J.A. Smith, F.L. Owen, J.G. Seidman, S. Ip, F. Rosen, and M.S. Krangel. 1986. Identification of a putative second T-cell receptor. *Nature* **322:** 145–149.

Chien, Y.-H., M. Iwashima, K.B. Kaplan, J.F. Elliott, and M.M. Davis. 1987. A new T-cell receptor gene located within the alpha locus and expressed early in T-cell differentiation. *Nature* **327:** 677–682.

Clark, R.B., J. Chiba, S.E. Zweig, and E.M. Shevach. 1982. T-cell colonies recognize antigen in associatoin with specific epitopes on Ia molecules. *Nature* **295:** 412–414.

Goding, J.W. and A.W. Harris. 1981. Subunit structure of cell surface proteins: Disulfide bonding in antigen receptors, Ly-2/3 antigens, and transferrin receptors of murine T and B lymphocytes. *Proc. Natl. Acad. Sci.* **78:** 4530–4534.

Haskins, K., R. Kubo, J. White, M. Pigeon, J. Kappler, and P. Marrack. 1983. The major histocompatibility complex-restricted antigen receptor on T cells. *J. Exp. Med.* **157:** 1149–1169.

Hedrick, S.M., L.A. Matis, T.T. Hecht, L.W. Samelson, D.L. Longo, E. Heber-Katz, and R.H. Schwartz. 1982. The fine specificity of antigen and Ia determinant recognition by T cell hybridoma clones specific for pigeon cytochrome c. *Cell* **30:** 141–152.

Hedrick, S.M., D.I. Cohen, E.A. Nielsen, and M.M. Davis. 1984. Isolation of cDNA clones encoding T cell-specific membrane-associated proteins. *Nature* **308:** 149–153.

Kappler, J.W., B. Skidmore, J. White, and P. Marrack. 1981. Antigen-inducible, *H-2*-restricted interleukin-2-producing T cell hybridomas. *J. Exp. Med.* **153:** 1198–1214.

Meuer, S.C., K.A. Fitzgerald, R.E. Hussey, J.C. Hodgdon, S.F. Schlossman, and E.L. Reinherz. 1983. Clonotypic structures involved in antigen-specific human T cell function: Relationship to the T3 molecular complex. *J. Exp. Med.* **157:** 705–719.

Pardoll, D.M., B.J. Fowlkes, J.A. Bluestone, A. Kruisbeek, W.L. Maloy, J.F. Coligan, and R.H. Schwartz. 1987. Differential expression of two distinct T-cell receptors during thymocyte development. *Nature* **326:** 79–81.

Reinherz, E.L., S.C. Meuer, K.A. Fitzgerald, R.E. Hussey, J.C. Hodgdon, O. Acuto, and S.F. Schlossman. 1983. Comparison of T3-associated 49- and 43-kilodalton cell surface molecules on individual human T-cell clones: Evidence for peptide variability in T-cell receptor structures. *Proc. Natl. Acad. Sci.* **80:** 4104–4108.

Saito, H., D.M. Kranz, Y. Takagaki, A.C. Hayday, H.N. Eisen, and S. Tonegawa. 1984. Complete primary structure of a heterodimeric T-cell receptor deduced from cDNA sequences. *Nature* **309:** 757–762.

Samelson, L.E., R.N. Germain, and R.H. Schwartz. 1983. Monoclonal antibodies against the antigen receptor on a cloned T-cell hybrid. *Proc. Natl. Acad. Sci.* **80:** 6972–6976.

Yagüe, J., J. White, C. Coleclough, J. Kappler, E. Palmer, and P. Marrack. 1985. The T cell receptor: The α and β chains define idiotype, and antigen and MHC specificity. *Cell* **42:** 81–87.

Yanagi, Y., Y. Yoshikai, K. Leggett, S.P. Clark, I. Aleksander, and T.W. Mak. 1984. A human T cell-specific cDNA clone encodes a protein having extensive homology to immunoglobulin chains. *Nature* **308:** 145–149.

Allison, J.P. and L.L. Lanier. 1987. The T-cell antigen receptor gamma gene: Rearrangement and cell lineages. *Immunol. Today* **8:** 293–296.

Kronenberg, M., G. Sui, L.E. Hood, and N. Shastri. 1986. The molecular genetics of the T-cell antigen receptor and T-cell antigen recognition. *Annu. Rev. Immunol.* **4:** 529–591.

Marrack, P. and J. Kappler. 1987. The T-cell receptor. *Science* **238:** 1073–1078.

Weiss, A., J. Imboden, K. Hardy, B. Manger, C. Terhorst, and J. Stobo. 1986. The role of the T3/antigen receptor complex in T-cell activation. *Annu. Rev. Immunol.* **4:** 593–619.

### Interleukins

Auron, P.E., A.C. Webb, L.J. Rosenwasser, S.F. Mucci, A. Rich, S.M. Wolff, and C.A. Dinarello. 1984. Nucleotide sequence of human monocyte interleukin 1 precursor cDNA. *Proc. Natl. Acad. Sci.* **81:** 7907–7911.

Dutton, R.W., R. Falkoff, J.A. Hirst, M. Hoffmann, J.W. Kappler, J.R. Kettman, J.F. Lesley, and D. Vann. 1971. Is there evidence for a non-antigen specific disfusable chemical mediator from the thymus-derived cell in the initiation of the immune response? *Prog. Immunol.* **1:** 355–368.

Gery, I., R.K. Gerson, and B.H. Waksman. 1972. Potentiation of the T-lymphocyte response to mitogens. *J. Exp. Med.* **136:** 128–142.

Howard, M., J. Farrar, M. Hilfiker, B. Johnson, K. Takatsu, T. Hamaoka, and W.E. Paul. 1982. Identification of a T cell-derived B cell growth factor distinct from interleukin 2. *J. Exp. Mol.* **155:** 914–923.

Kinashi, T., N. Harada, E. Severinson, T. Tanabe, P. Sideras, M. Konishi, C. Azuma, A. Tominaga, S. Bergstedt-Lindqvist, M. Takahashi, F. Matsuda, Y. Yaoita, K. Takatsu, and T. Honjo. 1986. Cloning of complementary DNA encoding T-cell replacing factor and identity with B-cell growth factor II. *Nature* **324:** 70–73.

Lomedico, P.T., U. Gubler, C.P. Hellmann, M. Dukovich, J.G. Giri, Y.-C.E. Pan, K. Collier, R. Semionow, A.O. Chua, and S.B. Mizel. 1984. Cloning and expression of murine interleukin-1 cDNA in *Escherichia coli*. *Nature* **312:** 458–462.

March, C.J., B. Mosley, A. Larsen, D.P. Cerretti, G. Braedt, V. Price, S. Gillis, C.S. Henney, S.R. Kronheim, K. Grabstein, P.J. Conlon, T.P. Hoff, and D. Cosman. 1985. Cloning, sequence and expression of two distinct human interleukin-1 complementary DNAs. *Nature* **315:** 641–647.

Noma, T., P. Sideras, T. Naito, S. Bergstedt-Lindquist, C. Azuma, E. Severinson, T. Tanabe, T. Kinashi, F. Matsuda, Y. Yaoita, and T. Honjo. 1986. Cloning of cDNA encoding the murine IgG1 induction factor by a novel strategy using SP6 promoter. *Nature* **319:** 640–646.

Schimpl, A. and E. Wecker. 1972. Replacement of T-cell function by a T-cell product. *Nat. New Biol.* **237:** 15–17.

Swain, S.I., M. Howard, J. Kappler, P. Marrack, J. Watson, R. Booth, G.D. Wetzel, and R.W. Dutton. 1983. Evidence for two distinct classes of murine B cell growth factors with activities in different functional assays. *J. Exp. Med.* **158:** 822–835.


Aarden, L.A.. 1979. Revised nomenclature for antigen-nonspecific T cell proliferation and helper factors. *J. Immunol.* **123:** 2928–2929.

Dower, S.K. and D.L. Urdal. 1987. The interleukin-1 receptor. *Immunol. Today* **8:** 46–51.

Kishimoto, T. 1985. Factors affecting B-cell growth and differentiation. *Annu. Rev. Immunol.* **3:** 133–157.

Melchers, F. and J. Andersson. 1986. Factors controlling the B-cell cycle. *Annu. Rev. Immunol.* **4:** 13–36.

Oppenheim, J.J., E.K. Kovacs, K. Matsushima, and S.K. Durum. 1986. There is more than one interleukin 1. *Immunol. Today* **7:** 45–56.

Smith, K.A. 1987. The two-chain structure of high-affinity IL-2 receptors. *Immunol. Today* **8:** 11–13.

# 5 IMMUNIZATIONS

The production of antibodies relies on the in vivo humoral response of animals. This chapter describes the factors that influence the strength of an antibody response and presents protocols for the production of antibodies in laboratory animals. Chapter 4 provided a discussion of our current knowledge of the molecular and cellular events that lead to an antibody response. This chapter is divided into five sections: (1) an overview of immunogenicity, (2) techniques for preparing the immunogen, (3) techniques for injecting the immunogen into animals, (4) methods for collecting and sampling antibodies, and (5) methods for modifying antigens to make them more immunogenic.

## ■ IMMUNOGENICITY

Immunogenicity, the ability of a molecule to induce an immune response, is determined both by the intrinsic chemical structure of the injected molecule and by whether or not the host animal can recognize the compound. These two factors are intimately related and represent two facets of the same question—Can the immune system respond to a given molecule? Simple injections of many foreign molecules, viruses, or cells elicit a strong antibody response in laboratory animals. However, some substances fail to induce a strong response. In these cases, the immune system can often be manipulated to increase the response by modifying either the antigen or the host. Until recently these modifications have been discovered empirically, but our increased understanding of the cellular and molecular basis of the immune response now allows a more rational approach to these manipulations. This knowledge of the controlling events of the immune system is just beginning to have an impact on the practical design of immunizations, and it is likely that many new improvements will follow quickly. The potential repertoire of antigen-combining sites is vast, and therefore, with the correct choice of immunogen and manipulation, it is theoretically possible to generate antibodies with almost any specificity.

Antibodies are synthesized primarily by specialized lymphocytes known as plasma cells. These cells are one of the final products of B-lymphocyte differentiation. The production of a strong antibody response is controlled by inducing and regulating the differentiation of B cells into plasma cells. During this differentiation, B cells go from virgin B cells, which have a modified antibody as a cell-surface antigen receptor and do not secrete antibodies, to activated B cells, which both secrete antibodies and have cell-surface antibodies, to plasma cells, which are highly specialized antibody factories with no surface antigen receptors. The steps in this differentiation process are controlled by the presence of the antigen and by cell-to-cell communication between B cells and helper T cells. These events are discussed in detail in Chapter 4. The discussion below is presented in conjunction with that introduction.

## What makes a molecule immunogenic?

Proteins, peptides, carbohydrates, nucleic acids, lipids, and many other naturally occurring or synthetic compounds can act as successful immunogens. In general, for a compound to elicit a primary antibody response and a strong secondary response, it must contain an epitope that can bind to the cell-surface antibody of a virgin B cell and it must promote cell-to-cell communication between B cells and helper T cells (Table 5.1).

Binding of the antigen to the surface antibody molecule of a virgin B cell is an absolute requirement for an antibody response. This binding determines the specificity of the resulting antibodies, as the antigen binding site on the surface antibody molecule will be identical to the binding site on the secreted antibodies. An immunocompetent animal can produce antibodies with well over $10^7$ different antigen-combining sites, and most compounds can be bound tightly by at least one of these surface antibodies.

The second property of a molecule that allows it to be a good immunogen is that it must promote cell-to-cell communication between helper T cells and B cells. This is achieved by providing a physical link between these two cells. When an immunogen enters the B cell, it is partially degraded, and these fragments migrate to the surface, where they bind to receptors known as class II proteins. This complex of antigen fragment and class II protein is then bound by a receptor on the surface of helper T cells known as the T-cell receptor. The physical link between these two cells is formed by the antigen fragment and is an essential step in the differentiation of a B cell into a plasma cell. The site for class II–T-cell receptor binding does not need to be related structurally to the epitope for surface antibody binding, but the two sites must occur on the same molecule or physical complex. The link between these sites can be covalent or noncovalent, so long as the sites remain linked until the immunogen enters the B cell. Only one good class II–T-cell receptor binding site is required per immunizing molecule.

Similar types of cell-to-cell contact are needed in other stages of an antibody response, particularly in activating helper T cells. These interactions occur between helper T cells and a group of cells known as antigen-presenting cells. The molecular requirements for these cell-to-cell interactions are identical to those needed for B cells and helper T cells, and therefore, immunogens that work well for one type of interaction work well for the other.

The requirement for both an epitope and a class II–T-cell receptor binding site imposes a minimum size limit on an immunogen, and molecules of less than about 3000–5000 daltons are generally not good immunogens. Compounds smaller than this can often bind to surface antibodies on the B cell but may not have suitable sites for the simultaneous binding of a class II protein and a T-cell receptor. Physical coupling of small molecules to larger immunogenic molecules overcomes this problem by providing the missing class II–T-cell receptor binding sites and allows the induction of a good antibody

**TABLE 5.1**
**Chemical Nature of Good Immunogens**

| Compound property | Defect | Classification of defect | Result | Remedy |
|---|---|---|---|---|
| **Lacks an Epitope** | No antigen binding to B cells | Normally B-cell tolerance | No response | None |
| **Lacks Class II Protein Binding Site** | No antigen fragment presentation | Nonresponsiveness | Either no response or primary response only | Conjugate with good class II binding site, switch animals |
| **Lacks T-Cell Receptor Binding Site** | No T helper cell involvement | T-cell tolerance | Either no response or primary response only | Conjugate with good T-cell receptor binding site, switch animals |
| **Nondegradable** | No antigen fragment presentation | Nonresponsiveness | Either no response or primary response only | No remedy |
| **Small Size** | Often no T helper cell activation | Nonresponsiveness | No response | Conjugate with carrier |
| **Nonparticulate Form** | Poor phagocytosis | — | Weaker response | Self-polymerize, couple to beads or RBC |

response against the small molecules. Small molecules showing this behavior are defined as haptens. The large molecules that render them immunogenic are defined as carriers.

In summary, a good immunogen has three chemical features: (1) it must have an epitope that can be recognized by the cell-surface antibody found on B cells, (2) it must have at least one site that can be recognized simultaneously by a class II protein and by a T-cell receptor, and (3) usually, it must be degradable. These three properties are the only intrinsic chemical features needed for a molecule to elicit a strong antibody response. However, to force an animal to respond to an antigen, other factors need to be considered. These include the dose and form of the immunogen, the use of adjuvants, and potential modifications. Practical advice on these aspects is presented below.

## What stops an animal from responding?

For an animal to respond strongly to an antigen, it must synthesize cell-surface antibodies, T-cell receptors, and class II proteins that can bind the antigen or processed antigen fragments. Because there is a theoretical limit for the total number of different antigen binding sites, molecules must exist that have structures or compositions that cannot be recognized by the repertoire of available antigen binding sites. However, many types of molecules that were thought to be nonimmunogenic have been converted into good immunogens in recent years, so the actual members of this group remain a point of contention.

One well-documented reason why an animal may fail to respond to a given molecule is that the appropriate B and/or T cells have been eliminated during the development of self-tolerance. An animal avoids responding against itself by the elimination of B cells or T cells that produce receptors that can bind to host molecules. Therefore, if an injected molecule is identical or very similar to a host molecule, it is ignored by the immune system. Tolerance can be induced by the removal of a particular antigen-specific clone of either B cells or T cells, although T-cell tolerance is more common. B-cell tolerance cannot be overcome; animals that lack the appropriate B cells for a given molecule cannot be induced to respond. T-cell tolerance can be circumvented by modification of the antigen.

Another reason an animal may not respond to a particular compound is the failure of class II proteins to bind to the antigen fragments. This failure is relatively rare. However, as there are seldom more than 40 different class II proteins synthesized by any one animal, some animals may not respond to some antigens because of the lack of the appropriate class II molecule. Failure to induce an immune response because of the absence of the correct class II protein is more prevalent in inbred animals, such as strains of laboratory mice, than in outbred animals such as rabbits. Most antigens will have at least one fragment that can bind to a class II protein, and one fragment per immunizing molecule is all that is needed to allow all the epitopes on the immunogen to be recognized. The lack of an antibody response due to this reason can be overcome by modifying the antigen to add new sites for class II protein binding.

## ■ SOURCES OF ANTIGEN

The object of preparing the antigen for immunization is to present it to the animal in a form that will induce the strongest and most appropriate response. This section presents protocols for preparing proteins, peptides, and haptens for immunization.

In considering the preparation of an antigen, the major decision to be made is how pure does the antigen need to be before starting an immunization schedule. This decision should be based on the intended use of the resulting antibodies. If highly specific antibodies that will only recognize the appropriate antigen are essential, then the antigen must be purified to homogeneity, or the antigen preparation should be used to prepare monoclonal antibodies (see Chapter 6). If a mixed reaction is acceptable, then further purification may not be warranted. If further purification is needed, standard techniques, including column chromatography, differential extraction, and subcellular fractionation, are the most useful. When using protein antigens, if the polypeptide of interest can be seen as a unique band on a SDS-polyacrylamide gel, then the gel can be used as a final purification step (p. 61).

When mixed populations of antigens are used for immunizations, an antibody response to several components of the preparation is expected. If all the compounds in a preparation are equally immunogenic, then the resulting antibodies will mimic this distribution. However, in the more likely case that the immunogenicity of the compounds varies, antibodies to some of the compounds will dominate the response. This can be only determined empirically. There are numerous examples of immunizations succeeding particularly well or particularly poorly because of immunodominant compounds. The most useful advice is to prepare as much of the antigen as possible, try a simple immunization, and use the results from the test bleed to determine the relative dominance of your antigen.

## ■ Pure Antigens

Having pure antigens provides the best case for the production of antibodies. However, not all pure antigens will elicit a good response, and there are no tests to determine whether an antigen will be immunogenic, other than the obvious empirical test of injecting the preparation and testing the sera. As long as the antigen is larger than the minimal size range, the preparation should be inoculated into an appropriate animal. Even if only small quantities of pure protein are available, the best advice is to proceed with the injections and determine the immunogenicity. Immunization methods and advice on choosing the correct animal are found below (see p. 92).

Depending on the source of the pure antigen, individual preparations may have unusual buffers, high or low pH, or denaturants. For routes of immunization other than intravenous, the presence of small amounts of detergents and nonphysiological pH or salt concentration is not a serious problem; however, if the antigen is available in reasonable supply, changing the buffer to a neutral physiological buffer such as PBS by dialysis or gel filtration is seldom a disadvantage.

### Inject or Sequence?

When studying a new protein antigen, particularly a rare species, many workers reach a stage where, after a long period of work, they have purified a small amount of their protein and must decide whether to inject the protein into animals or attempt to determine its amino acid sequence. Which is best? Similar amounts (1–10 $\mu$g of a 50,000-dalton protein) will be needed to determine whether the protein is immunogenic or whether the amino terminus is blocked. If more protein than this is available, both methods can be tried. If neither approach is more appropriate because of the experimental goals, then a rational decision might be based on the competence of the groups that will be doing the antibody production or the sequencing. If good support in both areas is available, try sequencing. Any amino acid sequence can be used for both the synthesis of oligonucleotides and/or peptides. The extra advantages of two potential routes to proceed outweighs the production of a relatively small and limited amount of antibodies.

## ■ Purifying Antigens from Polyacrylamide Gels

If the antigen is not available in pure form, one simple method for purifying a protein antigen further is to separate the samples on SDS-polyacrylamide gels. Antigens purified this way often induce good antibody responses. Since the antigen is denatured by this route of preparation, the resulting antibodies are usually particularly good for techniques that need or benefit from denaturation-specific antibodies. However, some of the resulting antibodies will not bind to the native antigen.

Run your protein sample on any of the standard gels (p. 636). Locate the antigen using any of the methods below and process for injection.

### Locating the Antigen after Electrophoresis

After electrophoresis, the protein band of interest must be located in the gel, and the gel slice excised for injection. A variety of identification methods can be used, all of which are designed to avoid excessive fixation of the protein in the gel matrix. The choice of method depends partly on the abundance of the polypeptide. Three methods are commonly used: (1) staining of side strips cut from the edge of the gel, (2) light staining of the gel itself, and (3) locating the band by radioactive labeling of the antigen.

Staining of strips of the gel cut from the side is valuable because it avoids fixation of the gel. When isolating abundant proteins that are well separated from other bands, staining of side strips is a useful method. If the protein is not abundant or a careful separation from a nearby contaminating band must be made, one of the other methods should be used. If the protein is reasonably abundant, then a light staining of the proteins in the gel with Coomassie blue will permit localization without fixing. Alternatively, the bands in the gel can be visualized by immersing the gel in sodium acetate or copper chloride. If the protein is radiolabeled with $^{125}$I, $^{32}$P, or $^{35}$S, then use an autoradiogram as a template to excise the band of interest.

## SIDE-STRIP METHOD

1. After running the gel, remove one of the plates. Place a straight edge or ruler on the gel so that it runs vertically up the gel about 1 cm from one edge.

2. With a small cork borer, remove a series of plugs about 4–5 mm in diameter down this line. Remove the plugs at random, placing about 10 holes in a line. Repeat the same process down the other side of the gel, making sure that the two series of holes do not line up.

3. Cut down the two vertical lines dissecting the holes. Remove the two side strips and stain them using one of the techniques described on p. 649 to 654.

4. Use the guide holes to line up the stained strip with the rest of the gel.

5. Excise the band of interest using a scalpel. Sensitivity is approximately 0.1–0.5 µg/band.

6. Stain the remainder of the gel to check the accuracy of excision (p. 649).

The proteins can be injected directly into the polyacylamide gel if the immunizations will be done in larger laboratory animals such as rabbits. See p. 67 for the protocols for preparing the gel slices for injection. When injecting mice or other small animals, the proteins should be electroeluted prior to use (see p. 70).

## COOMASSIE BRILLIANT BLUE STAINING

1. After electrophoresis, the gel should be washed with three or four changes of deionized water.

2. Add several gel volumes of 0.05% Coomassie brilliant blue R-250 prepared in water.

3. Incubate for 10 min at room temperature with shaking.

4. Wash the gel with numerous changes of water over the next few hours.

5. When the appropriate band is visible, excise with a scalpel. Sensitivity is approximately 1–2 $\mu$g/band.

The proteins can be injected in the polyacrylamide gel if the immunizations will be done in larger laboratory animals such as rabbits. See p. 67 for the protocols for preparing the gel slices for injection. When injecting mice or other small animals, the proteins should be electroeluted prior to use. See p. 70.

## SODIUM ACETATE STAIN*

1. Wash the gel briefly after electrophoresis in several changes of deionized water.

2. Transfer gel to 4 M sodium acetate (10–12 gel volumes).

3. Shake gently for 40–80 min.

4. View bands against a black background and excise the relevant area of the gel using a scalpel. Sensitivity is approximately 1–2 $\mu$g/band.

   The gel fragment can now be processed for injection. If the sample is to be prepared by electroelution it must be reequilibrated into the elution buffer to remove excess acetate.

*Higgins and Dahmus (1979).

## COPPER CHLORIDE STAIN*

1. Wash the gel briefly after electrophoresis in several changes of deionized water.

2. Transfer gel to 0.3 M $CuCl_2$.

3. Shake gently for 5 min.

4. View bands against a black background and excise the relevant area of the gel using a scalpel. Sensitivity is approximately 0.1 $\mu$g/band for 0.5-mm gels and ~1 $\mu$g/band for 1-mm gels.

The gel fragment can now be processed for injection. If the sample is to be prepared by electroelution, it must be destained by incubation in 0.25 M EDTA, 0.25 M Tris (pH 9.0) with three changes of 10 min each, and then reequilibrated in elution buffer.

*Lee et al. (1987).

## AUTORADIOGRAPHY

Satisfactory autoradiographs are readily obtained from wet, unfixed polyacrylamide gels containing $^{125}$I, $^{32}$P, or high-specific-activity $^{35}$S-labeled proteins. The entire preparation of the proteins can be labeled or the sample can be spiked with a small amount of labeled protein.

1. After electrophoresis, the top plate is removed, and the gel on the bottom plate is wrapped tightly in plastic wrap.

2. Place small pieces of tape or labels in each corner of the gel. Mark each piece with a small symbol using radioactive ink. Allow the ink to dry completely. A good source of radioactive ink is to add 100 $\mu$l of India ink to an empty bottle of methionine. The radioactive methionine that remains will be sufficient for most labeling.

3. Place a piece of X-ray film over the gel and expose the film. Close contact is essential for a good autoradiograph, but it is also important not to distort the gel. If the gel is kept on the plate with the spacers or if spacers are placed along two edges of the gel, the spacers can be used to keep tight contact with the X-ray film without distorting the gel. The X-ray film is placed over the wrapped gel and then another glass plate is carefully laid over the top; the glass sandwich is clamped with clips over the gel spacers. Diffusion in polyacrylamide gels is slow, so that exposure times up to several days are acceptable.

4. Develop the film and use the exposed film as a guide to excise the band of interest. The remainder of the gel should be stained, destained, dried, and autoradiographed to check the accuracy of the excision.

   The exposed film can be used as the template to remove the correct area of the gel or the film can be used to make a direct copy on a piece of clear plastic, and the plastic used as a guide to cut out the corresponding regions.

   The proteins can be injected in the polyacrylamide gel if the immunizations will be done in larger laboratory animals such as rabbits. See p. 67 for the protocols for preparing the gel slices for injection. When injecting mice or other small animals, the proteins should be electroeluted prior to use. See p. 70.

### NOTE

i. If a small amount of purified or nearly purified antigen is available, it can often be iodinated and used to spike the sample before running on the gel. Because of the high specific activities that are possible using iodination, this variation often allows the easy detection and localization of antigens.

## *Processing of the Gel Fragments for Immunization*

For injections, many different approaches can be used to process a gel fragment. First, the gel fragment can be injected with the protein intact. The gel should be fragmented prior to injection. Two methods are commonly used: The gel is fragmented by passing it repeatedly through a syringe, or the whole gel slice is dried and ground into a powder. Injections using the whole gel fragment should only be used with larger animals such as rabbits. For mice or other small animals, other methods should be used. These include the electroelution of the protein and the electrophoretic transfer of the protein to a suitable membrane.

## FRAGMENTING A WET GEL SLICE

1. Rinse the gel slice in deionized water for a few minutes with several changes of water. Place the gel slice on a piece of parafilm or plastic wrap. Use the edge of a paper towel to remove the standing water by capillary action.

2. Remove the plungers from the barrels of two 5-ml syringes. Place the gel fragment into one of the barrels.

3. Replace the plunger and place the syringe outlet in the barrel of the second syringe. Using firm and rapid pressure on the plunger, push the gel into the second syringe.

4. Repeat the process at least five times, passing the gel fragments back and forth between the two syringes.

5. Place 21-gauge needles onto the outlet of the syringes, and repeat the process. A small amount of buffer may be necessary to keep the small fragments passing back and forth between the syringes.

The samples are now ready for injection (p. 92).

## LYOPHILIZATION OF A GEL SLICE

1. Rinse the gel slice in deionized water for a few minutes with several changes of water, and then place the gel slice on a piece of parafilm or plastic wrap. Use the edge of a paper towel to remove the standing water by capillary action.

2. Transfer the gel slice to a suitable container and lyophilize for 24 to 48 hr.

3. Remove the dried gel slices to a mortar and pestle and grind them into a fine powder. Transfer the powder to a test tube.

4. Add a volume of water equal to one-half of the original gel volume. Incubate at room temperature for 1 hr.

The samples are now ready for injection (p. 92).

## *ELECTROELUTION OF PROTEIN ANTIGENS FROM POLYACRYLAMIDE GEL SLICES\**

1. Transfer the gel slice containing your protein to a clean dialysis tube containing 1.0 ml of 0.2 M Tris/acetate (pH 7.4), 1.0% SDS, 100 mM dithiothreitol per 0.1 gram of wet polyacrylamide gel.

2. Place the dialysis tubing crosswise in a horizontal electrophoresis chamber. The running buffer is 50 mM Tris/acetate (pH 7.4), 0.1% SDS, 0.5 mM sodium thioglycolate. Run at 100 volts (about 100 mA).

3. After 3 hr, open one end of the dialysis tubing and remove the gel band. Stain the band with Coomassie blue to ensure that the electrophoresis is complete. Reclose the tubing and dialyze with several changes against 0.2 M sodium bicarbonate, 0.02% SDS.

4. Remove the protein solution and lyophilyze. Resuspend protein as best as possible in 10 ml of water (heat if necessary) and lyophilyze a second time.

5. Resuspend in 0.1 volume of the original gel volume in water. Check the purity by electrophoresis and Coomassie blue staining (p. 649).

Samples are now ready for injection (p. 92).

### NOTES

i. If the gel slice contains sodium acetate (see p. 64) or CuCl$_2$ (see p. 65), it must be equilibrated into the elution buffer (three changes, 10 min each) before starting the electroelution process.

ii. Several commercial companies now supply suitable apparatuses for electroelution. The types that elute into a small chamber covered by dialysis tubing work very well.

*Leppard et al. (1983).

## ELECTROPHORETIC TRANSFER TO NITROCELLULOSE MEMBRANES*

Proteins can be transferred to nitrocellulose membranes, and then the nitrocellulose can be processed for injection. This method is simple, retains the high resolution of the gel, and allows very precise excision of the relevant polypeptide. In the limited number of cases tried, the antibody response varies with the particular antigen. In some cases, the technique works very well, and in other cases, little or no response is seen.

1. Separate proteins by electrophoresis.

2. Transfer proteins electrophoretically to nitrocellulose membranes using one of the methods described in Chapter 12.

3. Stain the nitrocellulose paper with India ink or Ponceau S (see pp. 494 or 495).

4. Locate the band of interest and excise using a scalpel.

   For proteins at the desired level, the bands will be easily visible using the Ponceau S or India ink staining. Proteins that are present at lower levels are normally not sufficient to induce an antibody response.

5. Cut the piece of nitrocelluose into very fine pieces, add 200 $\mu$l of PBS, and sonicate for 10 bursts at 10 sec. Cool between each sonication step. Alternatively, dissociate the nitrocellulose using one of the methods described on p. 162.

   The fine powder suspension can now be injected with or without adjuvant (p. 96).

*Knudsen (1985); Diano et al. (1987).

## ▓ Haptens

Many small chemicals can be used to raise antibodies, if they are coupled to larger protein molecules. The small compounds are known as haptens, while the proteins to which they are coupled to are called carriers. The haptens themselves serve as epitopes for binding to the antibodies on the B-cell surface, and the carriers provide the class II–T-cell receptor binding sites. In general, haptens should be coupled to soluble carriers such as bovine serum albumin (BSA) or keyhole limpet hemacyanin (KLH). The coupling mechanism will vary with each hapten, but many of the bifunctional coupling reagents listed in Table 5.5 (p. 130) will be helpful. Also, the techniques on the coupling of synthetic peptides to carriers on p. 78 may be applied. In general, approximately 1 mole of hapten per 50 amino acids of carrier is a reasonable coupling ratio.

## ▓ Synthetic Peptides

The use of synthetic peptides as immunogens has been an important technique in the elucidation of the properties of an antibody response (e.g., Goebel 1938; Anderer 1963; Anderer and Schlumberger 1965; Sela 1966, 1969; Arnon et al. 1971). Recently, as more DNA sequences

### Choosing between Bacterial Expression and Peptides for Immunogen Production

When a cloned DNA sequence in available, antibodies can be prepared either using peptides or bacterially expressed proteins. There are strong proponents for both approaches, both groups citing their experiences favoring one method over the other. Both have advantages and disadvantages, and for a particular antigen one may be better suited than another. However, if both approaches are available to a researcher, both should be used.

For anti-peptide antibodies, a good response to the desired peptide usually can be generated with careful selection of the sequence and coupling method. Because of the way in which the peptide is displayed to the immune system, most peptides elicit a good response. Therefore, the anti-peptide approach has major advantages if the antigen is known to be highly conserved. Likewise, if antibodies need to be raised against a particular region, anti-peptide antibodies have many advantages. The major disadvantage with anti-peptide antibodies is that they may not recognize the native antigen. The percentage of antibodies raised against peptides that will bind to the native protein will vary from antigen to antigen. Values reported in the and their corresponding protein sequences have become known, synthetic peptides have been used to prepare antibodies specific for previously uncharacterized proteins (Sutcliffe et al. 1980; Walter et al. 1980; and reviewed in Lerner 1982, 1984; Walter 1986; Doolittle 1976; and in Ciba Foundation 1986). Peptides are normally synthesized using the solid-phase techniques pioneered by Merrifield (1963). The synthetic peptides are purified and coupled to carrier proteins, and these conjugates are then used to immunize animals. In these cases, the peptides serve as haptens with the carrier proteins, providing good sites for class II–T-cell receptor binding. Peptide–carrier conjugates seldom fail to elicit a response because of tolerance. Consequently, the peptides can usually be seen as epitopes, and high-titered antisera commonly are prepared. Characteristically, these antibodies will bind well to denatured proteins, but may or may not recognize the native protein.

The two most important advantages of anti-peptide antibodies are that they can be prepared immediately after determining the amino acid sequence of a protein (either from protein sequencing or from DNA sequencing) and that particular regions of a protein can be targeted specifically for antibody production. Rapid conversion from DNA sequence information to antibodies has enormous potential for application in molecular biology. Similarly, the production of site-specific antibodies has immediate implications for functional and clinical studies.

literature range from 0/4 to 3/4 of anti-peptide antibodies will bind to the native antigen. Synthetic peptide antigens are also more expensive to produce than bacterial fusion protein antigens.

Bacterially expressed antigens present a different set of problems. Some will be difficult to express in *E. coli*, presumably because of their toxic side effects. In these cases, inducible systems such as the T7 systems of Studier (see Rosenberg 1987; Studier and Moffatt 1986) are recommended. Even when high levels of the antigen of interest can be produced, there may be some instances where the protein will not be immunogenic or where the antibodies will not recognize the native protein. However, because of the larger size of the bacterially expressed protein, there is a better chance that the antibodies will bind to the native protein.

A reasonable compromise for antibody production would be: (1) If the budget is limited and/or antibodies for the native protein are essential, use fusion proteins or full-length expression in *E. coli*. (2) If the budget is large enough, try both bacterially produced immunogens and peptides. (3) If the protein is highly conserved, use peptides. (4) If site-directed antibodies are needed, use peptides or prepare large banks of monoclonal antibodies against the bacterially produced immunogens.

The major problem that is encountered when preparing anti-peptide antibodies is whether they will recognize the native protein. Assays that need or benefit from anti-native antibodies, such as immuno-precipitation, many cell staining techniques, or immunoaffinity purification, will succeed only when the peptide sequence is displayed on the surface of the native molecule in a conformation similar to the peptide–carrier conjugate. Therefore, the successful production of anti-peptide antibodies is often determined by the researcher's ability to predict the location of certain peptide sequences in the three-dimensional structure of the protein.

Because of their size, peptides may not be immunogenic on their own. To elicit an antibody response directly, they must contain all of the features of any immunogen, notably they must have an epitope for B-cell binding and a site for class II–T-cell receptor binding. Some peptides, even surprisingly small ones, contain both these sites (or more properly, one sequence that can serve both functions), and these peptides can be used without carriers (e.g., see Beachy et al. 1981; Lerner et al. 1981; Dreesman 1982; Jackson 1982; Atassi and Webster 1983; Young et al. 1983). Unfortunately, there are no methods, short of immunization, to test this, and therefore, most peptides are coupled to carrier proteins before injection. An exciting recent development is the use of synthetic class II–T-cell receptor sites synthesized directly with the desired epitope (Francis et al. 1987; see also, Good et al. 1987; Borras-Cuesta et al. 1987; Leclerc et al. 1987). Although there are not enough cases to determine how widely applicable this approach will be, the concept is provocative. With this strategy, the peptide of interest is synthesized as either an amino- or carboxy-terminal extension of a known class II–T-cell receptor site. The synthetic peptide, now containing both sites, is injected without coupling and used to induce an antibody response. The first experiments using this approach look very promising, and this may become an important alternative to coupling with carrier proteins.

Peptides usually are synthesized with an automated machine using solid-phase techniques. The methods for synthesis and purification of the peptide are beyond the scope of this book. However, to judge the success of the coupling reaction and to determine the number of moles of peptide bound to the carrier, a small proportion of the peptide needs to be labeled. This can be done by including a small amount of [14]C-labeled amino acid in the synthesis or by iodinating a sample of the peptides on a tyrosine or histidine residue (see p. 324) after the synthesis. A small sample of these iodinated peptides can then be added to the coupling reactions to ascertain the success of the coupling.

During immunization, antibodies to the carrier proteins or the coupling agent will also be produced, and these are normally removed by affinity-purifying the anti-peptide antibodies on a column prepared with conjugates of the peptide and a second carrier molecule. Techniques for affinity purification of the antibodies are described in general on p. 313.

*Designing the Peptide*

Probably the most frequently asked question concerning synthetic peptides is what sequence should be used for the immunogen (reviewed in Doolittle 1986). Although there is no one correct answer, enough anti-peptide antibodies have been raised to make suggestions for peptide choices. However, preparing anti-peptide antibodies is still an empirical exercise. What works well for one immunogen may fail completely for another.

*Choosing the Appropriate Peptide Sequence*

With careful synthesis, coupling, and immunizations, most sequences can be used to induce antibodies specific for the peptide itself. When considering which sequence to use, most people actually want to know how likely will it be that the anti-peptide antibodies will recognize the native protein. Early work suggested that peptides containing hydrophilic amino acids (Hopp and Woods 1981, 1983; Kyte and Doolittle 1982) and proline residues were more likely to be exposed on the surface of the native protein than other sequences, and many peptides have been prepared using these criteria. In assessing the value of these criteria, hydrophilicity is required but is not sufficient to predict the surface location of a particular sequence. Many strongly hydrophilic amino acid sequences are buried in water pockets or form inter- or intramolecular bonds and are thus excluded from interactions with anti-native antibodies. Therefore, hydrophilicity can be thought of as required but not sufficient for choosing peptide sequences (see p. 661 for hydrophilicity values). Hydrophilic peptides are also more likely to be soluble for coupling reactions.

The presence of proline residues in synthetic peptides originally was suggested because $\beta$-turns often form portions of known epitopes. However, the presence of proline residues in peptides does not have much predictive value when antisera are tested for binding to the surface of native proteins. Although many excellent anti-peptide antisera have been prepared against sequences with proline residues, there is not sufficient evidence to target prolines when designing peptides.

More recently, several workers have noted that carboxy-terminal sequences often are exposed and can be targeted for anti-peptide sequences. Although using carboxy-terminal sequences does not guarantee that the resulting antibodies will recognize the native protein, a surprisingly high percentage will. Similarly, many amino-terminal regions are exposed, and these also may make good targets.

Another potentially useful parameter for selecting peptide sequences is the "mobility" of the amino acid residues. Originally, it was noted that the regions of a protein that become epitopes often have a higher temperature than other regions, as determined by NMR and X-ray structure (Moore and Williams 1980; Robinson et al. 1983; Tainer et al. 1984; Westhof et al. 1984). Higher temperature in crystallography and NMR distinguishes regions that are more mobile from regions that are more static. These observations have led to the suggestion that stretches of amino acids that are more flexible are more likely to be epitopes. In the preparation of anti-peptide antibodies, when a peptide is coupled to a carrier molecule, it has a different local environment than in the original protein. When choosing a sequence for antibody production, a region of the protein that is more flexible will be more likely over time to form a structure that is similar to the peptide–carrier conjugate. Although the measure of mobility may become a useful criterion for selecting good peptide sequences, it has not been tested in enough detail to determine whether it will have any predictive value.

At present, a reasonable order of suggestions for choosing peptide sequences would be:

1. If possible, use more than one peptide.
2. Use the carboxyl-terminal sequence if it is hydrophilic and if a suitable coupling group is available or can be added.
3. Use the amino-terminal sequence if it is hydrophilic and if a suitable coupling group is available or can be added.
4. Use internal hydrophilic regions; perhaps using longer peptides.

### Size of the Peptide

The smallest synthetic peptides that will consistently elicit antibodies that bind to the original protein are 6 residues in length. Responses to smaller peptides are typically weak or will not recognize the protein of interest, either in a native or denatured state. Since epitopes consisting of smaller regions have been reported, the lower limit presumably reflects the difficulty of recognizing the smaller peptides coupled to carriers. With peptides of 6 amino acids or slightly larger, the responses vary. Some will generate good antibodies and some will not. Generally, peptides of approximately 10 residues should be used as a lower limit for coupling.

In the literature two strategies are suggested for peptide length. One school suggests using long peptides (up to 40 amino acids long) to increase the number of possible epitopes, while other authors argue that smaller peptides are adequate and their use ensures that the site-specific character of anti-peptide antibodies is retained. Both strategies have been used successfully. Two important preliminary questions to consider are: (1) Does the anti-peptide serum need to recognize the native protein? If so, use longer peptides or prepare anti-peptide antisera against multiple peptides. (2) How good is your peptide synthesis facility? Peptides over 20 residues in length are increasingly difficult to synthesize, yielding products with inappropriate side reactions. Longer peptides also are more likely to contain residues that make the coupling to carrier molecules more difficult. The correct decison between peptides with 10–15 residues and longer peptides will depend on the experimental design and will normally be a compromise between these factors. The safest choice, but also the most expensive, will be to prepare multiple small peptides of 10–15 amino acids in length from various regions of the sequence.

## Coupling Strategy

When choosing the sequence for a synthetic peptide, one factor that often is overlooked is the method of coupling. Most coupling methods rely on the presence of free amino, sulfhydryl, phenolic, or carboxylic acid groups. Free amino groups used for coupling will be found on lysine side chains or on the amino-terminal residue. Sulfhydryl groups are found on cysteine side chains, phenolic groups on tyrosines, and carboxylic acid groups on aspartic acids, glutamic acids, and the carboxy-terminal residue. Coupling methods should be used that link the peptide to the carrier via either the carboxy- or amino-terminal residue. When preparing antibodies against the carboxy-terminal region of the protein, the coupling should be done through the amino terminus of the peptide. Similarly, the coupling for amino terminal fragments should be done through the carboxy-terminal region of the peptide. For internal fragments, the major consideration is that the peptide be coupled by an end and not through a central residue.

The easiest strategy to manipulate the type of coupling is to add an extra amino acid on either the amino or carboxyl terminus to allow simple, one-site coupling to the carrier. Any coupling method that potentially can bind to an internal residue should be avoided. Similarly, coupling methods should be chosen that will bind to only one amino acid, if possible. If multiple coupling sites are possible, they should be localized to either the amino or carboxyl terminus, and the coupling should be adjusted to link only through one site per peptide on average. It is important to remember that it is often easier to use different peptides than design elaborate coupling schemes.

## Choosing the Appropriate Carrier

Many different carrier proteins can be used for coupling with synthetic peptides. The two most commonly used are keyhole limpet hemacyanin (KLH) and bovine serum albumin (BSA). Both work well in most cases, but each has disadvantages. Because of its large size, KLH is more likely to precipitate during cross-linking, and this can make handling KLH difficult in some cases. On the other hand, BSA is very soluble, but often is a good immunogen in its own right. For most purposes, either carrier will be adequate. Use whichever is more convenient.

Three other carriers that are used occasionally are ovalbumin, mouse serum albumin, or rabbit serum albumin. Ovalbumin can be used as a good carrier for most purposes. It is also a good choice for a second carrier when checking that antibodies are specific for the peptide itself and not the carrier. MSA or RSA may be used when the antibody response to the carrier molecule must be kept to a minimum.

BSA has 59 lysine (30–35 are available for coupling), 19 tyrosine, 35 cysteine, 39 aspartic acid, and 59 glutamic acid residues. Ovalbumin has 20 lysine, 10 tyrosine, 6 cysteine, 14 aspartic acid, and 33 glutamic acid residues.

## COUPLING PEPTIDES TO CARRIER PROTEINS USING GLUTARALDEHYDE*

Glutaraldehyde is a bifunctional coupling reagent that links two compounds primarily through their amino groups. Glutaraldehyde cross-linking is one of the most stable linkages used; however, the glutaraldehyde bridge will often form a portion of an epitope recognized by the immunized animal. Positive sera should always be screened in assays using a second glutaraldehyde-coupled peptide. In some cases antisera may be judged mistakenly as positive for the peptide when they react primarily against the glutaraldehyde.

Glutaraldehyde cross-linking is particularly useful when the synthetic peptide contains only a single free amino group at its amino terminus. When peptides with more than one free amino group are used, large multimeric complexes can be formed. In these cases other cross-linking reagents should be considered before using glutaraldehyde. If glutaraldehyde must be used, the ratios between the synthetic peptide and the carrier protein need to adjusted carefully to minimize overcoupling. Also, lowering the pH well below the pK of the amino group (p. 660) will slow the rate of coupling, as the $NH_2$ group is the target and not $NH_3$. Other amino acids such as cysteine are also linked with glutaraldehyde, although at a considerably lower frequency.

Two approaches can be used to couple the peptide to the carrier. In the single-step coupling method, the peptide and carrier are mixed at the appropriate ratios and a limiting amount of glutaraldehyde is added. In this case overcoupling is limited by the amount of glutaraldehyde. This method is recommended when more than one free amino group is found on the peptide.

If the peptide is suitably large and contains only one free amino group, a two-step method offers many advantages. The peptide is treated with an excess of glutaraldehyde and then separated from the free glutaraldehyde. By having the coupling agent present only on the peptide, the number of moles of peptide bound to the carrier is easy to control. Also, the carrier molecules cannot be cross-linked to one another. If this approach is used, the glutaraldehyde should be added to the peptide in a large molar excess. This will lower the number of peptide-to-peptide coupling events.

As a general starting point, for each 50 amino acids in the carrier protein, 1–2 moles of peptide are added to the coupling reaction.

*Reichlin (1980).

**For single-step coupling:**

1. Prepare a solution of 5 mg/ml of the synthetic peptide in PBS (1 mg of a 10-mer is approximately 1 $\mu$mole).

2. Add the carrier protein to the peptide at the correct molar ratio (approximately 1 mole of peptide per 50 amino acids of carrier). For a normal-sized reaction, a total volume of peptide and carrier should be approximately 2 ml. Add a stir bar to the solution and place on a magnetic stirrer in the fume hood.

3. Prepare a solution of 0.2% glutaraldehyde in PBS. Slowly, add an equal volume of glutaraldehyde to the peptide/protein solution with constant agitation. One milliliter of a 0.2% solution is approximately 20 $\mu$moles.

> **Caution** Glutaraldehyde is toxic. Work in a fume hood and use proper methods for disposal.

4. Incubate at room temperature for 1 hr.

5. Add glycine from a 1 M stock in PBS (check pH and readjust to 7.2, if necessary) to a final concentration of 200 mM. Incubate with stirring for 1 hr.

6. Separate the peptide–protein conjugate from the peptide by dialysis against PBS. Four changes overnight will normally be sufficient.

   Longer peptides may not dialyze efficiently, and in this case, gel filtration can be used to separate the conjugates from the free peptides.

**For two-step coupling:**

1. Prepare a solution of 10 mg/ml of peptide in PBS (10 mg of a 10-mer is 10 $\mu$moles). Add a stir bar to the solution and place on a magnetic stirrer in the fume hood.

2. Prepare a 10% solution of glutaraldehyde in PBS.

> **Caution**   Glutaraldehyde is toxic. Work in a fume hood and use proper methods for disposal.

3. Add an equal volume of 10% glutaraldehyde to the peptide solution.

4. Incubate at room temperature for 30 min with stirring.

5. Add glycerol to a final concentration of 5% and apply to a G10 gel filtration column equilibrated with PBS. Glutaraldehyde-coupled peptide should elute at or near the excluded peak. The location of the peptide can be followed by radioactivity or by absorbance at 206 nm.

6. Add a stir bar to the eluted peptide, and place on a magnetic stirrer. Add carrier to the desired final molar ratio (normally 1 mole of peptide per 50 amino acids of carrier is appropriate). Consider the yield from the column to be 100%.

7. Incubate for 1 hr at room temperature with stirring.

8. Add glycine from a 1 M stock in PBS (check pH and adjust to 7.2 if necessary) to a final concentration of 200 mM. Incubate with stirring for 1 hr.

9. Separate the peptide–protein conjugate from the peptide by dialysis against PBS. Four changes overnight will normally be sufficient.

   Longer peptides may not dialyze efficiently, and in this case, gel filtration can be used to separate the conjugates from the free peptides.

## NOTES

i. The molar ratio of peptide to protein can be determined by measuring the amount of radioactivity bound to the final protein product.

ii. The free coupling groups on the glutaraldehyde can be blocked by replacing the glycine with sodium borohydride. Use 10 mg/ml for 15 min. Dialyze as per normal.

## COUPLING PEPTIDES TO CARRIER PROTEINS WITH
## m-MALEIMIDOBENZOYL-N-HYDROXYSUCCINIMIDE ESTER*

*m*-Maleimidobenzoyl-*N*-hydroxysuccinimide ester (MBS) is a heterobifunctional reagent that can be used to link peptides to carrier proteins through cysteines and free amino groups. Many different strategies have been used successfully for MBS coupling. Most employ a two-step procedure, either activating the peptide and then purifying it from MBS or activating the carrier and purifying it.

**For activation of the peptide:**

1. Peptides with one amino group and no cysteines can be activated by direct treatment with MBS (start at step 3). Peptides with one cysteine that will be the coupling group should be treated with citraconic anhydride prior to MBS treatment.

2. (**Optional**)  To block free amino groups in the peptide, treat with citraconic anhydride (Atassi and Habeeb 1972). Dissolve the peptide in water at a concentration of 10 mg/ml. Adjust the pH to 8.5. Add a stir bar and place on a magnetic stirrer at room temperature. Prepare a solution of 10 mg/ml of citraconic anhydride in water. Slowly add an equal volume of the citraconic anhydride to the peptide. Monitor the pH often, and add NaOH to keep the pH between 8 and 9. After the addition of citraconic anhydride, incubate at room temperature for 1 hr. Add 0.1 volume of 1 M sodium phosphate (pH 7.2) to the peptide solution.

3. Dissolve MBS in dimethyl formamide (DMF) to a concentration of 25 mg/ml. Water-soluble variants of MBS are also available (see p. 130).

4. For peptides that have not been blocked with citraconic anhydride, dissolve at 5 mg/ml in 100 mM sodium phosphate (pH 7.2).

5. Add a stir bar and place the peptide solution on a magnetic stirrer. Dropwise and with constant agitation, add MBS to a final concentration of 5 mg/ml. Incubate at room temperature for 30 min.

6. Two approaches can be used at this point. If the peptide is long enough (greater than approximately 12–15 residues), it can be separated by gel filtration from the free MBS on a G10 column, preequilibrated with 100 mM sodium phosphate (pH 7.2). Alternatively, add a sulfhydryl-containing compound such as β-mercaptoethanol to a final concentration of 35 mM (the final molarity of the MBS solution is 17 mM) and incubate for 1 hr at room temperature. This will kill any remaining maleimide groups on the MBS, but not

*Kitagawa and Aikawa (1976); Liu et al. (1979); Green et al. (1982).

affect the succinimide ester. Another variation for removing excess maleimide is to add agarose beads containing a free thiol group. Add enough beads to bind all available MBS (1 mg of MBS is approximately 3.4 $\mu$moles). Incubate at room temperature for 1 hr with agitation. Remove by filtration.

7. Dissolve carrier protein in PBS. Add enough carrier protein to make a final ratio of approximately 1 mole of peptide for each 50 amino acids of carrier.

8. Incubate for 3 hr at room temperature with stirring.

9. Dialyze versus PBS to remove the uncoupled peptide. Four changes overnight will normally be sufficient.

Longer peptides may not dialyze efficiently and, in this case, gel filtration can be used to separate the conjugates from the free peptides.

**For activation of the carrier:**

1. Dissolve the carrier protein to a concentration of 10 mg/ml in PBS.

2. Prepare 25 mg/ml of MBS in dimethylformamide.

3. Add 0.1 volume of MBS dropwise with stirring to avoid a high local concentration, and stir the complete reaction at room temperature for 30 min.

4. Separate the activated carrier from free MBS by gel filtration on Sephadex G25 in PBS.

5. Dissolve the peptide in PBS, and add to a final concentration of 1 mole of peptide for each 50 amino acids of carrier.

6. Adjust the final pH to 7–7.5, and stir the reaction for 3 hr at room temperature.

7. Dialyze versus PBS to remove the uncoupled peptide. Four changes overnight normally will be sufficient.

Longer peptides may not dialyze efficiently, and in this case, gel filtration can be used to separate the conjugates from the free peptides.

## NOTE

i. The molar ratio of peptide to protein can be determined by measuring the amount of radioactivity bound to the final protein product.

## COUPLING PEPTIDES TO CARRIER PROTEINS
## USING CARBODIIMIDE*

Carbodiimides attack carboxylic groups to change them into reactive sites for free amino groups. In general, peptides are treated with a carbodiimide first and then carrier is added. To stop the cross-linking of one peptide to another, they are treated with citraconic anhydride before coupling to block their amino groups. After the free amino groups are blocked with citraconic anhydride, peptides are treated with 1-ethyl-3-(3-dimethylaminopropyl) carbodiimide hydrochloride (EDC), which then binds to free carboxylic acids. The reactive peptides are then mixed with the carrier proteins to produce the final conjugates.

Carbodiimides react with the side chains of aspartic and glutamic acid as well as the carboxyl terminal group.

1. (**Optional**)  If desired, block free amino groups in the peptide by treating with citraconic anhydride (Atassi and Habeeb 1972). Dissolve the peptide in water at a concentration of 10 mg/ml. Adjust the pH to 8.5. Add a stir bar and place on a magnetic stirrer at room temperature. Prepare a solution of 10 mg/ml of citraconic anhydride in water. Slowly, add an equal volume of the citraconic anhydride to the peptide. Monitor the pH often, and add NaOH to keep the pH between 8 and 9. After the addition of citraconic anhydride, incubate at room temperature for 1 hr.

   Separate the blocked peptide from the free anhydride. One of the final steps in the purification of the peptide can be repeated or try a G10 gel filtration equilibrated in $H_2O$. Avoid amino or carboxylic acid groups in the purification.

   One disadvantage of this method is that the acylated product created during blocking will have a free carboxylic acid group that can be attacked by the carbodiimide. During deblocking this structure is lost, and any peptide bound to the carrier through these sites will be lost.

2. Dilute the peptide to 1 mg/ml in water.

3. Add EDC to a final concentration of 10 mg/ml. Adjust the pH to 5, unless citraconic anhydride has been used, then adjust to pH 8.

4. Incubate for 5 min at room temperature. Check the pH during activation and add NaOH to maintain pH, if necessary.

---

*Hoare and Koshland (1967); Bauminger and Wilchek (1980); Goodfriend et al. (1964)

5. Add an equal volume of carrier protein to yield a final molar ratio of approximately 1 mole of peptide to each 50 amino acids of protein. One milligram of a 10-mer peptide is approximately 1 $\mu$moles.

6. Incubate at room temperature for 4 hr.

7. Stop the reaction by adding sodium acetate (pH 4.2) to a final concentration of 100 mM. Incubate at room temperature for 1 hr.

8. Separate the peptide–protein conjugate from the peptide by dialysis. For conjugates with blocked amino groups, first dialyze with several changes over 3 hr versus sodium acetate (pH 4.2) and then against PBS. For unblocked peptides, dialyze directly against PBS. Four changes overnight will normally be sufficient.

   Longer peptides may not dialyze efficiently, and in this case gel filtration can be used to separate the conjugates from the free peptides.

## NOTE

i. The molar ratio of peptide to protein can be determined by measuring the amount of radioactivity bound to the final protein product.

## COUPLING PEPTIDES TO CARRIER PROTEINS
## USING BIS-DIAZOTIZED BENZIDINE*

Bis-diazobenzidine (BDB) is used to cross-link peptides to carriers primarily through tyrosine side chains. However, the BDB cross-linkers are not as specific as other cross-linkers and multiple interactions are possible, particularly with lysine, cysteine, and histidine side chains. The reaction is normally done using a simple one-step procedure that relies on careful adjustment of the ratio of the peptide, carrier, and BDB. While the molar ratio of peptide and carrier is normally easy to adjust, the amount of BDB is often more difficult to judge. If overcoupling occurs, the amount of BDB added should be titrated to determine the most effective level.

---

**Caution**   Benzidine may be carcinogenic. Wear gloves and handle with care.

---

1. (**Optional**)   If desired, block free amino groups in the peptide by treating with citraconic anhydride (Atassi and Habeeb 1972). Dissolve the peptide in water at a concentration of 10 mg/ml. Adjust the pH to 8.5. Add a stir bar and place on a magnetic stirrer at room temperature. Prepare a solution of 10 mg/ml of citraconic anhydride in water. Slowly add an equal volume of the citraconic anhydride to the peptide. Monitor the pH often, and add NaOH to keep the pH between 8 and 9. After the addition of citraconic anhydride, incubate at room temperature for 1 hr.

   Separate the blocked peptide from the free anhydride. One of the final steps in the purification of the peptide can be repeated or try a G10 gel filtration column equilibrated in $H_2O$.

2. Prepare fresh BDB. Dissolve benzidine-HCl in 0.2 M HCl at 5 mg/ml. Add 35 mg of $NaNO_2$ per 1 ml of benzidine. Stir at 4°C for 1 hr.

3. Prepare a solution of peptide and carrier at 1 mole of peptide to each 50 amino acids of carrier in 100 mM NaCl, 100 mM borate (pH 9.0). The final concentration of carrier should be approximately 5 mg/ml. Place on ice.

---

*Gordon et al. (1958); Bassiri et al. (1979).

4. For each 1 ml of peptide–protein solution, add 200 $\mu$l of BDB solution.

5. Incubate with stirring at 4°C for 1 hr. Monitor the pH and adjust to pH 9.0 with occasional addition of 0.5 M NaOH.

6. (**Optional**)  Add glycyl-L-tyrosine hydrate to 100 mM. Incubate at 4°C for 30 min with stirring. The tyrosine group in this chemical will terminate the reaction immediate and stop any further coupling.

7. Dialyze versus PBS to remove the uncoupled peptide. Four changes overnight will normally be sufficient.

   Longer peptides may not dialyze efficiently, and, in this case, gel filtration can be used to separate the conjugates from the free peptides.

## NOTE

i. The molar ratio of peptide to protein can be determined by measuring the amount of radioactivity bound to the final protein product.

## ■ Preparing Antigens from Bacterial Overexpression Vectors

When antibodies are needed against a cloned gene, one excellent source of antigen is overexpression in bacteria (reviewed in Carroll and Laughon 1987; Marston 1987). A wide variety of coding regions can be expressed in bacteria either on their own or as fusion proteins. There are a number of vectors available for both types of cloning, some of which are listed in Appendix IV (p. 694). For fusion proteins, vectors producing $\beta$-galactosidase or tryptophan (trpE) fusion proteins are the most commonly used, but many others are available as well. For expression of a full-length protein in bacteria, there is an extensive collection of vectors available, driven by a range of bacterial or viral promoters. Some of these will yield inducible expression, some fully constitutive expression. Methods for cloning these fragments can be found in one of the molecular cloning manuals, such as Sambrook et al. (1989). For proteins over 10,000 daltons, it is recommended that full-length clones be prepared whenever possible. If no preference is suggested from previous work, the T7 plasmids from Studier and colleagues (Studier and Moffatt 1986; Rosenberg et al. 1987) are recommended. They are excellent for high-level expression and have one of the few promoter systems that are truly off when not induced.

After inserting a coding region into a vector, two steps need to be taken before large-scale isolation. The first requirement is to determine the molecular weight of the newly expressed protein, and the second is to determine whether the expressed protein is deposited in an inclusion body. Determining the molecular weight is done by comparing the patterns of proteins on one-dimensional SDS-polyacrylamide gels from bacteria either expressing or not expressing the protein of interest. Bacteria with or without the plasmids are compared along with samples prepared with or without inducer, if the plasmid has an inducible promoter. Proteins that cannot be seen as individual bands on gels normally will not produce high enough yields to make purification possible. So, if the bands cannot be detected by this technique, change the vector used for expression and/or check the construction. This strategy can also be used to determine the time points for maximal induction of protein to allow the collection of larger amounts of antigen. Second, to determine whether the protein is found in an inclusion body, lyse the cells with lysozyme as below. Spin at 12,000$g$. Check pellet and supernatant for the protein by SDS-polyacrylamide electrophoresis and staining (p. 635).

## ISOLATION OF PROTEIN ANTIGENS FROM BACTERIAL OVEREXPRESSION SYSTEMS—SOLUBLE PROTEINS

Soluble proteins normally will be purified on an SDS-polyacrylamide gel before injection.

1. Centrifuge the bacteria at 7000g for 10 min. Remove and discard the supernatant. Resuspend the cell pellet in 100 mM NaCl, 1 mM EDTA, 50 mM Tris, (pH 8.0) to approximately 5% (vol/vol). Centrifuge again and discard the supernatant.

2. Resuspend the cell pellet in sample buffer (for Tris/glycine SDS-polyacrylamide gel electrophoresis, the 1× sample buffer is 2% SDS, 100 mM dithiothreitol, 60 mM Tris, pH 6.8, 0.001% bromphenol blue) to a final concentration of approximately 10% (vol/vol). Mix vigorously.

3. Most samples should be boiled for 5 min. Some antigens will come out of solution under these conditions; if no signal is seen, check different temperatures.

4. Shear the chromosomal DNA by sonicating the sample. Sonicators with either an immersible tip or cuphorn are both suitable; several bursts of approximately 15–30 sec at full force are sufficient to shear the DNA.

5. Spin the sample at 10,000g for 10 min. Recover the supernatant and discard any pellet.

   After the final spin, samples may be frozen at −20°C. They are stable at this step for months.

   The samples are now ready for electrophoresis. (p. 636).

## ISOLATION OF PROTEIN ANTIGENS FROM BACTERIAL OVEREXPRESSION SYSTEMS — INCLUSION BODIES*

1. Centrifuge the bacteria at 7000g for 5 min.

2. Resuspend the cell pellet in 100 mM NaCl, 1 mM EDTA, 50 mM Tris (pH 8.0) to a final concentration of 10% (vol/vol). Add lysozyme to 1 mg/ml.

3. Incubate at room temperature for 20 min.

4. Spin at 5000g for 10 min. Remove and discard the supernatant. Transfer the pellet to ice. Pellets can be stored at this stage by quick-freezing the centrifuge tube in a dry ice–ethanol bath and storing at −70°C.

5. Resuspend the spheroplasts in ice-cold 100 mM NaCl, 1 mM EDTA, 0.1% sodium deoxycholate, 50 mM Tris (pH 8.0).

6. Incubate on ice with occasional mixing for 10 min.

7. Add $MgCl_2$ to a final concentration of 8 mM and DNase I to a final concentration of 10 $\mu$g/ml. Incubate at 4°C with occasional mixing until the viscosity disappears. Add fresh DNase I, if necessary.

8. Remove the inclusion bodies from suspension by centrifugation. For most, 10,000g for 10 min will be satisfactory. Slower speeds may be suitable for some samples, and the slower speeds may be useful in lowering the background.

*Adapted from Kleid et al. (1981) and Marston (1987).

9. Wash the pellet once by resuspending in 1% NP-40, 100 mM NaCl, 1 mM EDTA, 50 mM Tris (pH 8.0), and centrifuging. Many pellets can be washed in 1 M urea, 50 mM Tris (pH 8.0), without dissolving.

10. Wash the pellet with 100 mM NaCl, 1 mM EDTA, 50 mM Tris (pH 8.0).

The antigen is now ready for injection or further purification by SDS-polyacrylamide electrophoresis.

## NOTE

i. If the final product is to be injected directly, the large particles should be broken into as small pieces as possible. Sonication is useful. Complete solubilization is not important, as particulate antigens make excellent immunogens. However, if solubilization is desired, dissolve the particles in a strong denaturant such as 8 M urea, and then dialyze versus PBS prior to injection.

## IMMUNIZING ANIMALS

There are only a limited number of places at which a researcher can intervene to manipulate and tailor the response to a particular antigen. The types of intervention can be divided into two broad categories: either modifying the antigen or altering the injection conditions. Methods for modifying the antigen itself are discussed below on pp. 124 and 128. In general, these methods should only be tried after an initial immunization regime has failed. The second class of interventions includes the choice of animal, the dose and form of the antigen, the use of adjuvants, the route and number of injections, and the period left between injections. Advice and methods for each of these are given below.

### Using Animals

Immunizing and bleeding animals for the production of polyclonal sera or to act as donors for hybridoma fusions (see p. 146) is a technically straightforward and humane procedure. To obtain good antibody responses, healthy and well-cared-for animals are essential. To inject and bleed animals safely and painlessly requires skill, patience, and practical training. In most countries these procedures and the care and maintenence of laboratory animals are governed by specific legal requirements. These regulations vary from country to country but are designed to ensure the welfare of animals and to ensure that the operator is skilled and the manipulation justified. The use of animals is also clearly governed by moral considerations and is the subject of intense public debate. It is the duty of the investigator to acquire a thorough knowledge of the relevant regulations and to abide by them. The investigator must take responsibility for the animals he or she uses or directs to be used. In the sections of this book dealing with animals, simple procedures are described to aid the investigator. However, they are only guides, and the correct handling of animals can only be learned by specific training. All investigators should receive this training or ensure that the procedures are carried out by those that have. The local legality of any procedure described here must also be verified.

## ■ Choice of Animal

The choice of animal for immunizations is determined by four points, how much serum is needed, from what species is the antigen isolated, are monoclonal antibodies needed, and how much antigen is available.

### Which Species and Which Strain?

A wide range of vertebrate species can be used for the production of antisera. The five most commonly used laboratory animals are rabbits, mice, rats, hamsters, and guinea pigs (Table 5.2). Typically, a single sample bleed from these animals will yield 25 ml of serum from a rabbit, 100–200 $\mu$l from a mouse, and 1–2 ml from a rat, hamster, or guinea pig. Only when large volumes of sera are required are larger animals needed. Pigs, horses, sheep and donkeys are all used commercially and can be used if large volumes of serum are needed.

For practical reasons, rabbits represent a good choice for the routine production of polyclonal sera. They are easy to keep and handle, can be safely and repeatedly bled, and the antibodies they produce are well characterized and easily purified. With careful management, at least 500 ml of serum can be obtained from one rabbit through the course of an immunization regime. Most laboratory rabbits are outbred and relatively little is known of the genetics of the immune response in this

**TABLE 5.2**
**Choice of Animal**

| Animal | Maximal amount of sera (ml) | Monoclonal antibodies possible | Inbred | Comments |
|---|---|---|---|---|
| **Rabbits** | 500 | No | No | Often best choice for polyclonal antibody production even when antigen limiting |
| **Mice** | 2 | Yes | Yes | Often best choice for monoclonal antibody production, excellent genetics of immune response |
| **Rats** | 20 | Yes | Yes | Good choice for monoclonal antibody production |
| **Hamsters** | 20 | No | Mostly not | Good choice for polyclonal antibody production when antigen is limiting |
| **Guinea Pigs** | 30 | No | No | Hard to bleed |
| **Chickens** | 50 | No | No | Good for highly conserved mammalian antigens |

species, but because they are outbred, they will have a wider range of class II proteins or other immune response proteins than inbred animals.

For most purposes, smaller animals such as rats, mice, hamsters, and guinea pigs are not normally the animal of choice for polyclonal antibody production, because only small volumes of serum can be obtained. This problem can be reduced by inducing the formation of ascites in mice; these procedures allow up to 10 ml of ascites fluid to be obtained from one animal (see p. 274). Antibody titers in ascites fluids are almost as high as serum titers. Small rodents can be used for polyclonal antibody production when only small amounts of antibody are needed or when only tiny amounts of antigen are available, because small rodents generally respond better to low doses than larger animals. For monoclonal antibody production, both mice and rats can be used (see Chapter 6). Most laboratories use inbred strains of mice and rats. Here, the genetics of the immune response have been intensively studied. Genes that effect both the quality and quantity of the antibody response occur at many distinct loci, but those associated with the major histocompatibility complex class II proteins are often the most important. The function of these proteins is discussed in Chapter 4. If a weak response to an antigen is seen, changing the strain of animal that is immunized may help overcome the low production. The table on p. 686 lists the class II alleles of many of the standard strains of mice. A reasonable approach used to deal with the class II genetic background is to immunize one of the standard strains kept in your animal facility, and if no or only a weak response is seen, then consider alternative strains. In particular, $F_1$ hybrid mice and/or laboratory "outbred strains" such as Swiss mice should be tried.

All immune systems display the property of self-tolerance. This actively acquired state protects the animal from autoimmune damage and can have a dominant role in determining the immunogenicity of a given antigen. Thus, at its simplest, mouse serum albumin is not immunogenic in mice but is in rabbits. The generation of tolerance and its consequences are discussed in Chapter 4. In practice, immunizations should be done in animals that are as far in evolutionary distance from the source of the antigen as possible. For weakly immunogenic, highly conserved proteins isolated from mammalian sources, chickens or other fowl provide a valuable alternative, since their self-tolerance profile to such molecules is considerably different from that of mammals. If mice need to be used for immunizations with well-conserved antigens, NZB mice often are a good choice, as they tend to make autoantibodies more easily than other strains.

Animals from various species reach immune maturation at different times post birth. Mice are commonly considered to be mature at 6 weeks, rats at 6–8 weeks, and rabbits 12 weeks post birth.

## How Many Animals?

Even in genetically identical animals, a single preparation of antigen will elicit different antibodies. When outbred animals such as rabbits are used, these differences are heightened. If the amount of antigen is not limiting, several animals should be used for any immunization scheme, and individual animals should always be screened separately. For rabbits, two animals should be used as a minimum, with three to four being preferable. For mice or other rodents, use three to six.

## Anesthesia

Anesthesia is used to sedate animals during injections or during operations where unexpected or rapid movements would endanger the animal. For laboratory animals the most common anesthetics are ether, carbon dioxide, sodium pentobarbitone, fentanyl/fluanisone, and fentanyl/droperidol.

Diethyl ether is administered by inhalation either by placing the animals (often mice) in a closed container permeated with ether vapor or by using mask (for larger animals). Animals sedated by ether can be handled for injections or simple operations. The depth of anesthesia is difficult to control and to assess during long treatments, and therefore ether is only used for short-term anesthesia. Ether should always be used in a vented explosion-proof area. Carbon dioxide is also administered by inhalation, and is useful for short periods of 1–2 minutes. $CO_2$ should be mixed 1:1 with oxygen delivered either in a closed container or with a mask (can use syringe barrel to continue anesthesia). $CO_2$ treatment is useful for injections, bleedings, or ascites tapping.

Sodium pentobarbitone (Nembutal, Sagatal) is delivered by injection. Mice and rats are normally injected ip, while rabbits are injected iv. Sodium pentobarbitone can be used for operations or more complicated injections. Fentanyl/fluanisone and fentanyl/droperidol combinations are used for heavy sedation needed for major surgery. For the mouse and rat, injections are done either ip or im, and for the rabbit, im. These injections are often preceded by an ip injection of diazepam to potentiate the anesthetics.

Suggested doses for these drugs vary over a surprisingly wide range. Consult your local authorities for appropriate dose.

To judge the depth of anesthesia, pull the animal's hind leg out straight, and pinch the foot hard using your thumb and forefinger. Animals should not be operated on until the withdrawal reflex has abated.

## ■ Adjuvants

Nonspecific stimulators of the immune response are known as adjuvants. The judicious use of adjuvants is essential to induce a strong antibody response to soluble antigens, though they may not always be required for particulate or whole-cell antigens. The action of adjuvants is not fully understood, but most adjuvants incorporate two components. One is a substance designed to form a deposit protecting the antigen from rapid catabolism. The two traditional methods of forming a deposit are to use mineral oils or aluminium hydroxide precipitates (Glenny et al. 1926). With mineral oils, such as those used in Freund's adjuvant, the immunogen is prepared in a water-in-oil emulsion. For aluminum hydroxide, the immunogen is either adsorbed to preformed precipitants or is trapped during precipitation. Alternatives to the delivery systems include liposomes or synthetic surfactants (Hunter et al. 1981). Liposomes are only effective when the immunogen is incorporated into the outer lipid layer; entrapped molecules are not seen by the immune system. The most effective of the synthetic surfactants are the pluronic polyols.

The second component needed for an effective adjuvant is a substance that will stimulate the immune response nonspecifically. These substances act by raising the level of a large set of soluble peptide growth factors known as lymphokines. Lymphokines stimulate the activity of antigen-processing cells directly and cause a local inflammatory reaction at the site of injection. Early work relied entirely on heat-killed bacteria (Dienes 1936) or lipopolysaccharide (LPS) (Johnson et al. 1956). LPS is reasonably toxic, and, through analysis of its structural components, most of its properties as an adjuvant have been shown to lie in a portion known as lipid A. Lipid A is available in a number of synthetic and natural forms that are much less toxic than LPS, but still retains most of the better adjuvant properties of the parental LPS molecule. Lipid A compounds are often delivered using liposomes. The two bacteria that are commonly used in adjuvants are *Bordetalla pertussis* and *Mycobacterium tuberculosis*. When used as whole bacteria, they must be heat-killed prior to use. The immunomodulatory mediators of *B. pertussis* include a LPS component and the pertussis toxin. The pertussis toxin has been purified and is available commercially. *M. tuberculosis* is most commonly found in complete Freund's adjuvant. The most active component of *M. tuberculosis* has been localized to muramyl dipeptide (MDP) (Ellouz et al. 1974). MDP is available in a number of forms.

The overall effect of adjuvants is dramatic and their importance cannot be overemphasized. The depot action means that much smaller doses of antigen can be used and that antibody responses are more persistent. The nonspecific activation of the immune response often can spell the difference between success and failure in obtaining an immune response. Adjuvants should always be used for the first injections unless there is some very specific reason to avoid this.

Recently an intense drive to find potent adjuvants with more acceptable side effects has led to the production of new synthetic adjuvants.

Two that are available and offer many advantages are SAF-1 (available for research purposes from Vaccine Adjuvant Project, Syntex Research, 3401 Hillview Avenue, Palo Alto, California 94304) and RAS (Ribi Adjuvant System, Ribi Immunochem Research, Inc., P.O. Box 1409, Hamilton, Montana 59840). While these formulations look very promising, they have not yet found wide enough use for us to comment on their efficacy in all situations.

When beginning an immunization, choosing the correct adjuvant can be difficult. As a general suggestion, Freund's adjuvant should be used when small amounts of the immunogen are available. If large amounts are available or if the compound is known to be highly immunogenic, then other adjuvants can be used. Freund's must never be given intravenously.

## FREUND'S ADJUVANT*

The most commonly used adjuvant for research work is Freund's adjuvant. Freund's adjuvant is a water-in-oil emulsion prepared with nonmetabolizable oils. If the mixture contains killed *M. tuberculosis*, it is referred to as complete Freund's adjuvant (CFA). Without the bacteria it is known as incomplete Freund's adjuvant (IFA). Freund's is one of the best adjuvants for stimulating strong and prolonged responses. The principal disadvantage of Freund's adjuvant is that it can invoke very aggressive and persistent granulomas. It is not used for injections of humans for this reason, and the possible side effects should be monitored carefully during antibody production in animals. To avoid the majority of the side effects, the primary injection should be given in CFA, while all boosts should be done in IFA.

1. If using complete Freund's adjuvant, resuspend the *M. tuberculosis* by vortexing or shaking.

2. Protein antigens, preferably in saline, are mixed with an equal volume of the adjuvant oil and an emulsion is formed. To generate this emulsion, vigorous and prolonged mixing is needed. Three methods can be used, depending on the volume of adjuvant to be made.

   For small volumes, add the adjuvant to the tube. While vortexing, add an equal volume of the antigen solution. Vortex vigorously until a thick emulsion develops.

   For intermediate volumes, use two syringes connected through a luer fitting. A three-way valve is appropriate. Using glass syringes, take equal volumes of the aqueous and the adjuvant into two different syringes, but do not fill over one-half of the syringe capacity. Remove all air and connect the syringes through the luer fitting. Depress the plunger from the aqueous solution first, driving the antigen into the oil of the adjuvant. Alternately push the plungers, mixing the adjuvant and the immunogen solution into an emulsion. Continue until the syringe plungers are difficult to push.

   For large volumes, use a tissue homogenizer. Add the adjuvant to the homogenizer first. Run the homogenizer for a short period to coat the inside with the adjuvant. Add the aqueous solution and run until a thick emulsion develops.

3. The emulsion should be very thick and not disperse when a drop of it is placed on the surface of a saline solution.

4. Transfer to a syringe (or remove one of the syringes from the luer fitting). Remove all the air. Add an appropriately sized needle.

   The samples are now ready for injection (p. 92).

*Freund and McDermott 1942; Freund 1956.

## ALUMINUM HYDROXIDE ADJUVANT*

A common alternative to the use of Freund's adjuvant is to adsorb the immunogen onto an aluminum salt (Glenny 1926). Aluminum hydroxide is the most commonly used, but all types avoid many of the more harmful side effects of Freund's adjuvants. The immunogen is either allowed to adsorb to the preformed aluminum salt or can be trapped in the salt during precipitation. When injected, the precipitate provides the depot effect. If desired, nonspecific stimulation can be provided by adding killed bacteria to the preparation; most often, heat-killed *B. pertussis* is used. However, the addition of *B. pertussis* reintroduces some of the potential side effects seen with complete Freund's adjuvant.

Aluminum hydroxide adjuvants, when used without the *B. pertussis* stimulants, can be injected into all sites. When *B. pertussis* or other agents are used, all sites are satisfactory except intravenous.

1. Prepare 10% potassium alum [aluminum potassium sulfate, $AlK(SO_4)_2 \cdot 12H_2O$] in distilled water. For long-term storage, filter-sterilize and dispense in convenient volumes.

2. In a 50-ml conical tube, add 10 ml of 10% potassium alum. While vortexing, add 22.8 ml of 0.25 N NaOH, dropwise.

3. Incubate at room temperature for 10 min.

4. Centrifuge at 1000g for 10 min. Remove and discard the supernatant. Add 50 ml of distilled water to the pellet and resuspend the $Al(OH)_3$.

5. Centrifuge at 1000g for 10 min.

6. One milligram of $Al(OH)_3$ will bind approximately 50–200 $\mu$g of protein antigen. Dilutions can be done in 0.9% saline. If the antigen is abundant, set up a titration of different amounts of the $Al(OH)_3$ versus a constant amount of antigen.

7. Incubate at room temperature for 20 min. Spin at 10,000g for 10 min. Test the supernatant for the presence of the antigen to be certain it has bound.

8. (**Optional**) Nonspecific stimulation of the immune response can be induced in this system by mixing in *B. pertussis*, heat-killed vaccine organism, around $4 \times 10^9$ organisms per 100 $\mu$l of material to be injected.

The samples are now ready for injection (p. 92).

*Chase (1967).

## ■ Dose of the Antigen

Two different criteria are important to consider in determining the proper dose: first, the optimum dose to achieve the strongest response, and, second, the minimum dose likely to induce the production of a useful polyclonal antiserum or donor animals for hybridoma fusions. Much of the injected material will be catabolized and cleared before reaching an appropriate target cell. The efficiency of this process will vary with host factors, the route of injection, the use of adjuvants, and the intrinsic nature of the antigen. Thus, the effective dose delivered to the immune system may bear little relationship to the introduced dose, and so descriptions of dose requirements are inevitably empirical. For rabbits, if a pure, soluble protein antigen is being used and is abundant, then a dose of 0.5–1 mg in adjuvant at each immunization is a sensible general recommendation; for mice this figure can be reduced 10-fold to 50–100 $\mu$g. This 10-fold difference between rabbits and mice is a good general rule for determining the amount of antigen needed to produce a strong response, when comparing these two animals.

The minimum amount of antigen capable of inducing a response will depend on the nature of the antigen and on the host, but in some mouse systems it may be possible to obtain a good response to fractions of a microgram of antigen. (See Chapter 6 for further discussions of dose and routes of injection of mice.) For rabbits, the minimum dose will be in the range of 10 $\mu$g per injection, but 100 $\mu$g per injection will be used more commonly. Tables 5.3 and 5.4 list suggested amounts and routes for various immunizations. If no information is available concerning the immunogenicity of an antigen, the best advice is to do the injection and see. If antibody responses are not required following the primary injection, effective initial doses can be kept quite low. Secondary injections and later boosts can be given with amounts similar to the priming injection. These recommendations are somewhat contradictory to the standard traditional suggestions for doses. Traditional suggestions propose that primary injections should use two to three times more antigen than boosts. However, most of these recommendations are based on good immunogens and in cases where the immunogen is available in large supplies. For rare antigens, one or two injections with a low dose to prime the animals followed by a larger secondary boost have often been shown to be effective for antibody production.

## ■ Form of the Antigen

The form of an antigen is also of crucial importance. Particulate antigens are usually much better immunogens than soluble molecules. Whole cells, bacteria, and viruses are usually very immunogenic. Soluble monomeric fractions of many proteins induce poor responses. Yet, when the same protein molecules are aggregated, they induce a good antibody response. There are also many examples of weakly antigenic proteins that have been made highly antigenic by coupling to large matrices such as agarose beads. One of the reasons for these effects is the rapid phagocytosis of particulate antigens compared to soluble material. Most soluble antigens can be made more immunogenic by coupling them chemically to beads or cells or by converting them to larger compounds by self-polymerization (chemical cross-linkers, partial denaturation) or binding to carrier proteins (p. 128).

The valency of an antigen can also have a strong effect on its immunogenicity. This is particularly true for large complex carbohydrates which are much more immunogenic than simple compounds. Apparently, the repeating epitopes on these immunogens make better targets for binding to B cells. However, polymeric antigens with nondegradable backbones can paralyze the B cell, blocking its surface receptors and not being processed.

**TABLE 5.3**
**Suggested Doses of Immunogens for Rabbits**

| Form of Antigen | Examples | Possible routes | Dose ($\mu$g) |
|---|---|---|---|
| **Soluble Proteins** | Enzymes<br>Carrier proteins<br>   conjugated w/peptides<br>Immune complexes | sc[a]<br>im[b]<br>id[c]<br>iv[d] | 50–1000 |
| **Particulate Proteins** | Viruses (killed)<br>Yeast (killed)<br>Bacteria (killed)<br>Structural proteins | sc<br>im<br>id | 50–1000 |
| **Insoluble Proteins** | Bacterially produced<br>   from inclusion bodies<br>Immunopurified proteins<br>   bound to beads | sc<br>im<br>id | 50–1000 |
| **Carbohydrates** | Polysaccharides<br>Glycoproteins | sc<br>im<br>id<br>iv | 50–1000 |
| **Nucleic Acids** | Carrier proteins<br>   conjugated w/N.A. | sc<br>im<br>id<br>iv | 50–1000 |

[a]Subcutaneous.
[b]Intramuscular.
[c]Intradermal.
[d]Intravenous.

**TABLE 5.4**
**Routes of Injection for Rabbits**

| Route | Abbr. | Maximum volume | Adjuvant | Immunogen requirement | Comments |
|---|---|---|---|---|---|
| Subcutaneous | sc | 800 $\mu$l per site, 10 sites/rabbit | +/− | Soluble or insoluble | Easy injections |
| Intramuscular | im | 0.5 ml | +/− | Soluble or insoluble | Slow release |
| Intradermal | id | 100 $\mu$l per site, 40 sites/rabbit | +/− | Soluble or insoluble | Injections more difficult, slow release |
| Intravenous | iv | 1 ml | No Freund's | Soluble, ionic detergent < 0.2% Nonionic detergent < 0.5% Salt < 0.3 M Urea < 1 M | Not effective for primary immunizations |
| Lymph Node | — | Special uses | No Freund's | Soluble/or insoluble | Good applications for experienced workers |
| Intraperitoneal | ip | | | | Normally not recommended for rabbits |

## ■ Routes of Injection

The route of injection is guided by three practical decisions: (1) what volume must be delivered, (2) what buffers and other components will be injected with the immunogen, and (3) how quickly should the immunogen be released into the lymphatics or circulation (see Allen 1967 for review of lymphatic tissues). For rabbits, large-volume injections normally are given at multiple subcutaneous sites. For mice, large volumes are only possible with intraperitoneal injections. If adjuvants or particulate matter are included in the injection, the immunogen should not be delivered intravenously. If slow release of the inoculum is desired, the injections should be done either intramuscularly or intradermally. For immediate release, use intravenous injections. There are numerous reviews and summaries on routes and techniques of injections. Two useful general sources are Herbert and Fristensen (1986) and Poole (1987).

---

### Injections

For most injections disposable syringes are best. Plastic syringes are appropriate for all injections except when using Freund's adjuvant, in which case disposable glass syringes are recommended. Syringes can be purchased with a number of different tips. The most useful are the luer tip and luer lock. For most work the luer tip is fine. Use the luer lock for samples that are viscous. After removing the syringe from its container, push the plunger in completely to loosen the seal that has formed between the plunger and the barrel.

For removing samples from a closed vial through a septum, fill the syringe with air to approximately the total volume to be removed from the vial. Rub the top of the septum with an alcohol-soaked pad. Insert the needle through the septum and inject the air into the vial. Lift the vial vertically above the syringe. Adjust the point of the needle until it is within the liquid in the vial. Withdraw the plunger and depress several times to wet the inside surfaces of the syringe and to dislodge any air. Withdraw the plunger to the appropriate amount. Remove the needle from the vial. Check to be certain there is no air in the needle by depressing the plunger until a drop is seen at the tip. The sample are now ready for injections as described on p. 92.

For removing samples from an open vial, depress the plunger completely. Place the needle in the sample and withdraw slightly more than the total volume to be injected. Remove the syringe from the sample and place in a vertical position with the needle topmost. Withdraw the plunger until an air space can be seen in the barrel. Dislodge any air bubbles trapped on the side of the syringe by tapping the barrel sharply with a finger or with any object. Depress the plunger gently until a drop is seen at the tip of the needle. The samples are now ready for injection as described on p. 92.

## Subcutaneous Injections

Subcutaneous injections (sc) are used widely for the immunization of laboratory animals, especially rabbits. Injections can include both particulate immunogens and/or adjuvants. Injected material will drain quickly into the local lymphatic system and will become concentrated in the lymph nodes closest to the injected sites. Larger animals such as rabbits normally are given multiple sc injections on the back, while mice are injected on the back of the neck or on one or both sides in the groin. Each injection should be relatively small, often 50–100 $\mu$l for mice, up to 400 $\mu$l for rats, and up to 800 $\mu$l for rabbits. Large injection volumes containing Freund's adjuvant should not be injected into one site, as they can cause severe granulomas. The inoculum is far better delivered by multiple (up to 10 sites for a rabbit) doses in separate sites. Anesthesia is normally recommended for larger animals, but is not normally required for mice. It is used only to allow easier handling during injections.

Techniques for sc injections of mice are described on pp. 164–166.

## SUBCUTANEOUS INJECTIONS FOR RABBITS

1. The rabbit is anesthesized as described on p. 95. A normal sc injection for a rabbit is approximately 400 $\mu$l.

2. Starting on one side near the back of the neck, pinch the skin between the thumb and forefinger. Pull the skin away from the body and insert the needle (25-gauge) into the space that has been created (Fig. 5.1).

3. Move the end of the needle a short distance from side-to-side to ensure it is not inserted into muscle or the body wall.

4. Depress the plunger to inject the desired amount. Pause a few seconds. As the needle is withdrawn, the hole should be gently rubbed with the forefinger and the thumb to stop any of the inoculum from escaping.

5. Move to the next site and repeat. Up to 10 sites per animal are common.

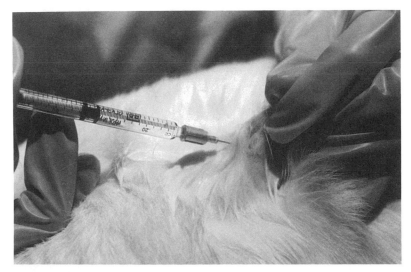

**FIGURE 5.1**
Subcutaneous injection of rabbits.

## *Intramuscular Injections*

Intramuscular injection (im) is one of the routes of inoculation used to produce a slow release of an antigen. The inoculum is deposited directly into the muscle tissue, and the antigen is seen as it drains into the nearby interstitial spaces. The inoculum will then drain into the local lymph nodes. Because of the difficulty of the injecting into the muscle tissue of smaller rodents, im injections are not recommended for mice. Rabbits can be injected with volumes up to 0.5 ml, and both adjuvant and particulate inoculi can be used.

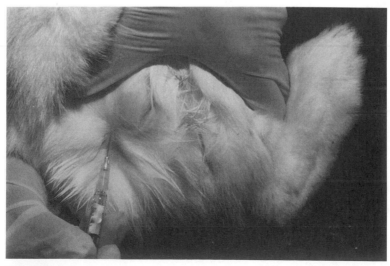

**FIGURE 5.2**
Intramuscular injection of rabbits.

## *INTRAMUSCULAR INJECTIONS FOR RABBITS*

1. It may be easier to anesthetize the rabbit before injection, but if a second worker is available to immobilize the rabbit's leg, then this may not be necessary. A normal im injection is 200–400 $\mu$l.

2. Remove the rabbit from its cage and place it on a wire or solid surface that is rough enough for the rabbit to feel comfortable. Laboratory benches should be avoided as the rabbit will constrict its muscles on any smooth surface. Wrapping the rabbit in a towel is often sufficent, and this may serve as a restricting device.

3. Injections are normally done in the thigh muscle of a rear leg. Grasp the leg from the front and make the injection into the rear of the thigh muscle near the hip (Fig. 5.2). Insert the needle (25-gauge) into the muscle and gently try to withdraw the plunger. If there is resistance, the injection can proceed. If blood appears, the needle has found one of the small veins or arteries of the leg. Withdraw the syringe and move to a nearby site and try again.

4. The inoculum should be added slowly in a steady motion. Withdraw the needle and gently massage the site of injection.

### Intradermal Injections

Intradermal injection (id) is used quite commonly for the immunization of larger animals. Like im injections, id injections are used to give a very slow rate of release of the immunogen. The inoculum is injected between the layers of the skin and is slowly absorbed into the body. Intradermal injections can be used with particulate and adjuvant containing inoculi.

Only small injection volumes can be used, thus id injection is only suitable if the antigen is available in a concentrated form. It is normal practice to deliver the injection over multiple sites. The procedure requires more skill than other injections. It is technically difficult to administer with small rodents and so is seldom used for mice.

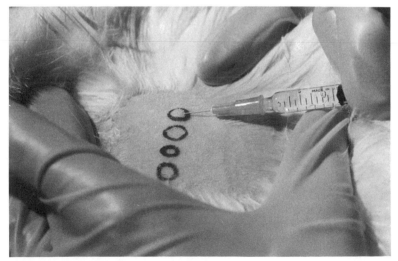

**FIGURE 5.3**
Intradermal injection of rabbits.

## *INTRADERMAL INJECTIONS FOR RABBITS**

1. Rabbits should be anesthetized prior to injection (p. 95).

2. Injections can either be done high on the flank (between the ribs and hip) or across the back. Shave an appropriately sized area to expose the skin (Fig. 5.3).

3. Hold the skin between the forefinger and thumb.

4. Check to be certain the needle is secured tightly to the syringe. Insert the needle (25-gauge) with the bevel side up slightly under the skin. Continue pushing the needle between the layers of skin and guide it forward for at least 0.5–0.7 cm. Inject no more than 100 $\mu$l. The inoculum should form a tight blister under skin (Fig. 5.3).

5. Withdraw the needle, holding the needle track with the thumb and forefinger. Compress the skin gently as the needle comes out.

6. Repeat the procedure at the next site. Injecting at up to 40 sites is common.

### NOTES

i. If complete Freund's adjuvant is included, small sores will appear at the sites of injection. These will clear within a few weeks.

ii. Under no circumstances should complete Freund's adjuvant be used for more than the first injection when using intradermal injections.

*Vaitukaitis (1981); Vaitukaitis et al. (1971).

## Intravenous Injections

For most laboratory animals the intravenous injection (iv) is a practical method of delivering a secondary or later boost. It is seldom used as for a primary injection. Antigen delivered directly in the blood stream will be processed rapidly by the reticuloendothelial system, primarily in the liver, lungs, and spleen. Direct iv injection carries serious risk to the host animal, as it can readily induce pulmonary embolisms or lethal anaphylactic shock in sensitized animals. Any toxic material present in the inoculum, such as bacterial endotoxins, is particularly dangerous when delivered iv. It is not safe or appropriate to use Freund's adjuvant with iv injections. Physiological buffers and salt conditions should be used if at all possible. Detergent concentration should never be allowed to exceed 0.1% for ionic detergents or 0.2% for nonionic.

Methods for iv injections of mice are given on p. 168.

## Anaphylactic Shock

Anaphylatic shock is one of the types of antibody-mediated hypersensitivity responses, i.e., allergies. When the injection of an immunogen induces a strong IgE response, the animal is often sensitized to the antigen and will develop an allergic response upon subsequent exposure. When the injections are given sc, id, or im, although an allergic response may develop, the animal is normally in no danger. However, when the injection is given iv, anaphylactic shock (systemic anaphylaxis) may ensue. Immediate hypersensitive responses are characterized by the release of histamine and vasoconstriction, followed by dilation of peripheral blood vessels. In most cases, if left untreated, the animal will die from suffocation. When using iv injections, a supply of epinephrine (adrenalin) must be handy to guard against death from anaphylactic shock. Epinephrine is given iv. Consult your local authorities for appropriate dose.

## *INTRAVENOUS INJECTIONS FOR RABBITS*

1. Remove the rabbit from its cage and place in a restraining device.

2. Intravenous injections are normally done in the marginal ear vein. A normal iv injection is 500–1000 $\mu$l. Pull the ear out straight. The vein is on the far inside edge of the ear. By pressing lightly on the base of the ear, the return of the blood to the body can be restricted and the vein will stand out strongly. Unless the hair is very long, it seldom needs to be clipped or shaved. Do not use xylene to help dilate the vein.

3. Insert the needle (25-gauge) into the vein (Fig. 5.4). When the needle appears to be in the vein, gently pull back on the plunger. If blood appears in the syringe, proceed with the injection. If no blood is seen, move to another location on the vein and try again.

4. Slowly and evenly inject the antigen solution in the vein. Pause for a few seconds and remove the needle. Place a piece of cotton or sterile gauze over the injection site as the needle is removed. Hold the site tightly for a few seconds to stop any bleeding.

### NOTE

i. Injection by this route into immunized animals may cause an anaphylactic reaction. This can be prevented by a prior injection of an antihistamine. Contact your local animal committee for guidance.

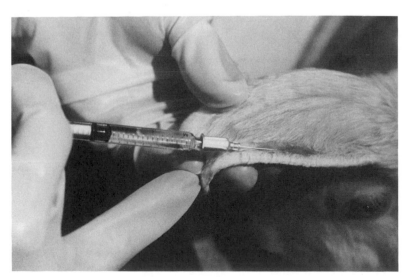

**FIGURE 5.4**
Intravenous injection of rabbits.

## Intraperitoneal Injections

Direct injection into the peritonal cavity (ip) is particularly easy and appropriate in small rodents. Antigens injected into the peritoneum will drain into the thoracic lymphatics and then to the vena cava. Particulate antigens or antigens in adjuvant can both be given safely by this route. Repeated injection of complete Freund's adjuvant by this route will induce granulomas and fibrous adhesions within the peritoneal cavity that can make subsequent disruption of the spleen difficult for hybridoma production. It may also cause the production of ascites fluid and even plasmacytomas in some animals. Ascites fluid induction can be an extremely useful maneuver to obtain large quantities of polyclonal mouse antisera. It can be enhanced by a carefully timed injection of plasmacytoma cells (see p 123).

Intraperitoneal injections are not often used for inoculating rabbits. Because of their larger size, other routes such as subcutaneous or intramuscular are normally used.

Techniques for ip injections with mice are described on pp. 158–162.

## Injections into Lymphoid Organs

Injections directly into a lymphoid organ offer some strong theoretical advantages over other injection routes (see Sigel et al. 1983, and Nilsson et al. 1987 for examples). The total dose of the immunogen is concentrated in one region that is specialized in dealing with it. The major difficulties are that the injection is technically more difficult than other injections, often requiring an operation, and many of the advantages of adjuvants are lost because they must be avoided due to their harmful side effects in the delicate lymphoid organ. As better adjuvants are becoming available, some of these second problems may be less worrisome. However, the technical difficulties in delivering the immunogens are still severe, particularly for novices. Two compromises are to inject into the lower region of the leg or to mark draining lymph nodes with dyes to make them easier to visualize during surgery.

No methods are given here for these techniques. They should be done only under the guidance of a trained animal worker or veterinarian. We have intentionally not given detailed protocols in order to force workers to consult the proper authorities for these techniques.

### Injecting Rabbits in the Popliteal Lymph Node

The lower leg of a rabbit or rodent has some advantages for the injection of an immunogen. In general, injections into the lower region of the foot drain into the popliteal gland near the back of the knee or into the inguinal gland in the groin. The injection can be targeted directly to the lymph node by marking the node with an appropriate dye. The injection of a dye into the region between the toes will stain the draining lymphatics and can be used as a guide in locating the nodes. One day after the injection, an operation on the rabbit is performed to locate the appropriate lymph node. Injections are done with one of the newer synthetic adjuvants or in PBS alone.

### Injecting into the Footpads of Rabbits or Rodents

Footpad injections are common sites of injections in some countries, but are banned as inhumane in others. Consult your local authorities for advice. Unless there are particular reasons for performing injections into this site, we recommend using other routes. The injections should never be done in the major weight-bearing regions of the pad. Also, boosts given in the footpad should never contain complete Freund's adjuvant. Always use anesthetics for footpad injections.

# ■ Boosts

An increase in B cells bearing surface antibodies specific for the inoculated antigens is first detected 5–6 days after the primary injection of antigen. Antibody is usually detected in the serum from around 7 days after the injection and persists at a low level for a few days, typically reaching peak titer around day 10. Primary responses often are very weak, particularly for readily catabolized, soluble antigens. In many cases efficient priming can take place in the absence of detectable antibody production, so for practical purposes, it will not be of much value to monitor this response.

A delay is required before reintroducing the antigen into a primed animal. A minimum of 2 or 3 weeks is recommended, but there is no penalty in leaving much greater intervals. Mice and rabbits will remain effectively primed for at least a year after receiving the first injection.

The response to the second injection of the same antigen is dramatically different. The number of B cells bearing antigen-specific cell-surface antibodies increases exponentially after the secondary injection, reaching a peak between days 3 and 4. Antibodies in the serum are also detectable at this time but peak levels are usually achieved around days 10–14. Typically, high levels of antibody persist for about 2–4 weeks after the second injection.

The response to the third and subsequent injections broadly mirrors that of the secondary injection. Higher titers of antibody are reached, but more importantly the nature and quality of the antibodies present in the serum changes. These changes are known as the maturation of the immune response and have considerable practical importance because they yield high-affinity antibody preparations. The intervals between secondary, tertiary, and subsequent injections may also be varied, but usually need to be extended to allow the circulating level of antibody to drop enough to prevent rapid clearance of the newly injected antigen.

For rabbits, injections are normally given every 4–6 weeks. One common error in immunizing rabbits is to boost every 2 weeks. Although this will continually reexpose the animal to the immunogen, the effective size of the boost will be reduced by the circulating antibody. For mice, secondary and later boosts should be spaced at least at 3-week intervals.

## Hyperimmunization

Polyclonal antibodies are often raised by hyperimmunizing animals. Hyperimmunization refers to injection schedules in which animals repeatedly are boosted with the same antigen. The resulting antibodies have many useful properties, including:

**Class shift**   Sera from primary injections contain a substantial proportion of IgMs, whereas sera from hyperimmunized animals contain mostly IgG antibodies. Hyperimmune sera often have higher levels of IgG, a substantial part of which is specific antibody. Levels of 1 mg/ml of antigen-specific IgG are possible. This is equivalent to 10% of the total IgG content of serum.

**Affinity maturation**   The average affinity of antibodies for an antigen increases with repeated injections. This change continues through multiple rounds of immunization. Antibody affinity affects the sensitivity of many immunochemical procedures.

The basis for affinity maturation is fairly well understood in terms of clonal selection and somatic mutation. In the presence of limiting amounts of antigen, those B cells with the highest-affinity antigen receptors will compete most successfully for antigen. Thus, clones of B cells secreting high-affinity antibodies will be selected for proliferation. This process is extended by somatic mutation of variable region genes, and under continuous selective pressure, higher-affinity antibodies will result. Repeated injection of low doses of antigen is often best to select for sera of high affinity, though these sera may not contain concentrations of antibody as high as those elicited by larger doses.

**Clonal dominance**   The selection of high-affinity clones can also affect the quality of the response in other ways. A common finding when immunizing with large proteins is that the response becomes less diverse with repeated injections. A few clones secreting antibodies to a limited number of epitopes will tend to dominate the later stages of hyperimmunization protocols. Because the process that selects these clones involves a number of random factors, variation is often seen from animal to animal. One way to avoid this problem is to immunize multiple animals and to monitor test bleeds regularly.

## ■ SAMPLING SERUM

After an injection, samples of serum are taken to check the production of specific antibodies. Comparing the titers of antibodies isolated after successive injections allows the antibody response to be monitored. Often the first test bleeds are taken before the immunizations begin to prepare a suitable control antibody for future tests. Although test bleeds following primary injections may be collected, they normally do not produce useful antibodies, as these antibodies typically have lower affinities than later samples.

## ■ Test Bleeds

Samples of the serum should be taken 7–14 days after an injection. This timing will correspond with the peak of antibody titers for most injection routes. Normally, small samples are collected until the desired antibodies are detected and the levels have reached acceptable levels. Test bleeds normally are done from the ear vein for rabbits and from the tail vein for mice or rats. These sites are used because they are easily accessible and do not have high numbers of nerve endings. Taking serum samples should be relatively painless for the animal, considerably less painful than giving blood is for humans. For rabbits, 5–10 ml can be collected conveniently, and this will provide more than enough serum for most tests. For mice, 200–400 $\mu$l is the usual size for most test bleeds.

Test bleeds normally are assayed against the immunogen itself by one of the techniques described in Chapters 10–14. When assessing the quality of the sera, be sure to test the antibodies in assays that resemble as closely as possible the techniques for which the antibodies are being raised. The sera should be compared to other test bleeds in titrations to determine the strength of the antibody response. Comparing titrations of various test bleeds will give an accurate measure of the course of the immunization and will provide a good measure of the correct time to begin collecting large volumes of serum.

Once a good titer has developed against the antigen of interest, regular boosts and bleeds are performed to collect the maximum amount of serum. For rabbits, boosts should be spaced every 6 weeks and serum samples of 20–40 ml should be collected approximately 10 days after each boost.

## TEST BLEED ON RABBITS—MARGINAL EAR VEIN

1. Place rabbit in restraining device, or wrap in a towel. The marginal vein is found on the inner edge of the ear and should be visible. If the vein is not visible, view the ear against the light from a lamp or other source. Gently shave a patch around the vein about two-thirds of the distance from the head to the tip of the ear.

2. If the marginal vein is not visible or easily found, heat the ear under a lamp.

3. The easiest method for collecting blood from the ear is to cut the vein with a clean razor or sterile scalpel. With either, the cut is made at a 45° angle to the vein and not directly transverse. The incision is made just through the top of the vein. Do not sever. Hold the scalpel or razor close to the blade to ensure that the cut does not go too deep.

   If the rabbit is scared or if the cut startles the animal, the normal response will be to constrict the ear artery, thus stopping the flow of blood to the ear. If this occurs, keep the rabbit warm and comfortable. As soon as the rabbit relaxes, the blood flow will start again and the serum can be collected.

4. Collect the blood by allowing it to drip into a clean test tube. If the blood clots on the cut before the desired amount is collected, wipe the ear with warm clean water and continue to collect. The maximum volume to be collected from a rabbit at one time is 50 ml. Do not attempt to take more, as this will make the rabbit anemic.

5. The blood flow can be stopped by gentle pressure to the cut with a sterile piece of gauze or bandage. Apply pressure for 10 to 20 seconds and then check to be sure the flow has stopped.

### NOTES

 i. Warmth is essential for this method. Also, the rabbit must not be placed on a smooth surface, as this tends to incite panic.

ii. Xylene on ears is cruel. Don't do it.

For test bleed on mouse—tail bleed, see next chapter (p. 172).

## TEST BLEED ON RATS—TAIL VEIN

1. Place the rat in small cage or container. Heat with an infrared lamp. The heat will increase the flow of blood to the tail, but is not meant to hurt the animal. If it's too hot for your hand, it's too hot for the rat.

2. Place the rat in a restraining device.

3. Cut a small section off tip of tail with surgical scissors.

4. Hold the tail over a centrifuge tube and gently stroke tail. Collect up to 2 ml of blood.

5. The bleeding will stop spontaneously.

## SERUM PREPARATION

1. After collection, blood should be allowed to clot for 30–60 min at 37°C.

2. The clot should then be separated from the sides of the collection vessel (ringing) using a Pasteur pipette, straightened paper clip, or similar instrument. Place the clot at 4°C overnight to allow it to contract.

3. The serum should then be removed from the clot and any remaining insoluble material removed by centrifugation at 10,000g for 10 min at 4°C.

   After preparation, serum can be stored for many years at −20°C or below. See p. 276 for more information regarding antibody storage.

## ■ Exsanguination

Exsanguination is the recommended method for the collection of large volumes of serum from an animal that will be sacrificed. Consult your local authorities for proper methods and instructions on the techniques.

## ▦ Inducing Ascites Fluid in Mice

Most strains of mice can be induced to produce ascites fluid by repeated injection of complete Freund's adjuvant. In BALB/c mice this process can be improved by injecting myeloma cells into the immune animal. Up to 10 ml or more of ascitic fluid can be produced by each mouse. As these fluids contain roughly the same level of specific antibody as the animal's serum, they provide a valuable source of polyclonal antibody and may provide an alternative to using larger animals for antibody production. The induction of the ascites must be timed to coincide with the presence of high levels of specific antibody production.

## *INDUCTION OF ASCITES USING FREUND'S ADJUVANT**

1. Prepare antigen in complete Freund's adjuvant using a ratio of 1 volume of antigen to 9 volumes of adjuvant.

2. Inject 0.2 ml of the emulsion on days 0, 14, 21, 28, and 35. All injections should be done ip.

3. Abdominal swelling usually will occur after the third or fourth injection. If no ascites develops, repeat the injection weekly.

4. Tap the ascites every 3–7 days as described on p. 274.

The ascites can be stored as described on p. 276.

### NOTE

i. Strain differences and effects of the animal's age and rate of response to the antigen need to be considered in timing the injection. Two injections 28 days apart followed by weekly injections, once a good serum response to the antigen has been detected, should be tried for new or weak antigens. See Tung et al. for specific recommendations.

* Tung et al. (1976).

## INDUCTION OF ASCITES IN BALB/c MICE USING MYELOMA CELLS*

1. Inject 0.5 ml of pristane ip. Pristane is 2,6,10,14-tetramethyl-decanoic acid.

2. After 10–30 days, inject the antigen in complete Freund's adjuvant (p. 98).

3. Repeat the injection on days 7 and 21 after the first injection.

4. Test bleed 4 days after the final injection. If the titer is satisfactory, proceed to step 5. If not, repeat injection of antigen.

5. Inject $10^6$ Sp2 or NS1 cells ip.

6. Harvest ascites as it develops. See p. 274.

* Lacy and Voss (1986).

## ◼ MAKING WEAK ANTIGENS STRONG ▬▬▬▬▬▬▬

If no response is detected after three or four injections, an alternate animal species should be tried and the dose of antigen increased, if possible. If neither of these manipulations works, then one of the procedures below should be tried.

Many compounds on their own do not have all of the properties needed to induce a strong antibody response. By altering the molecules with any of a variety of agents, they often can gain these properties. The common methods for doing this include the addition of small modifying groups such as dinitrophenol or arsanate to the molecules, denaturing the antigen by treating with heat and/or SDS, coupling the antigens to proteins that are known to be good immunogens, coupling the antigen with small synthetic peptides that are good sites for T-cell receptor–class II protein binding, coupling the antigens to large particles such as sheep red blood cells or agarose beads, and purifying the antigens with other antibodies and injecting the immune complexes.

## ◼ Modifying Antigens

Small changes in the structure of an antigen can often greatly alter the immunogenicity of a compound. This has been used extensively as a general procedure to increase the chances of raising an antisera, particularly against well-conserved antigens. These modification techniques either alter regions of the immunogen to provide better sites for T-cell binding or expose new epitopes for B-cell binding. The techniques are rapid and easy, but all antibody preparations should be tested for activity against the native antigen.

## MODIFYING ANTIGENS BY DINITROPHENOL COUPLING*

Adding dinitrophenol (DNP) groups to antigens is a common method of modifying antigens to make them more immunogenic. This method has been used successfully to modify well-conserved proteins such as actin to raise a strong antibody response.

1. Dialyze the antigen extensively against 0.5 M sodium carbonate (pH 9.5). Other methods to change buffer systems such as gel filtration chromatography are fine. The final concentration of the antigen should be approximately 2 mg/ml.

2. Just prior to use, prepare a solution of 4 mg/ml of sodium dinitrobenzene sulfonic acid (DNBS) in 0.5 M sodium carbonate (pH 9.5).

3. Mix the antigen and DNBS solution 1:1. Stir overnight at 4°C in the dark.

4. Transfer the reaction to dialysis tubing and dialyze extensively against PBS.

The modified antigen is now ready for injection (p. 92).

* Amkraut et al. (1966).

## MODIFIYING ANTIGENS BY ARSYNYL COUPLING*

In this technique the protein to be used as an immunogen is modified by coupling with arsanilic acid. In the first step of the reaction, diazotized arsanilic acid is prepared, and in the second step this reagent is coupled to the protein. The diazotized arsanyl group reacts primarily with tyrosine side chains, but also binds less frequently to histidine, free amino groups, and sulfhydryl groups. Arsanilation appears to be a particularly good modification to create new sites for T cell recognition and thus to break T-cell tolerance.

In an alternative procedure, arsanilic acid is coupled to tyrosine and the *n*-hydroxysuccinimide ester synthesized. This compound is available commercially and will bind to free amino groups on the protein.

1. Dialyze the protein antigen against several changes of 100 mM sodium borate (pH 9.0). Adjust the concentration to approximately 5 mg/ml.

2. Dissolve 1 mmole of *p*-arsanilic acid (0.21 gram) in 30 ml of 80 mM HCl containing 8 mM sodium bromide (0.25 gram/30 ml).

3. Place the acidic arsanilic acid solution in an ice/salt bath. While mixing slowly add 10 ml of 0.7% sodium nitrite (prepared fresh in water and prechilled before addition).

4. Incubate on ice for 30 min. Add 60 ml of cold water.

5. Slowly add 0.5 ml of the cold diazotized arsanilic acid per milliliter of the protein solution. Check the pH and readjust to pH 9.0, if necessary.

6. Incubate at 4°C for 4 hr. Periodically check the pH and readjust to 9.0 with 0.2 N NaOH if necessary.

7. Dialyze against PBS or saline with several changes overnight.

Samples are now ready for injection.

## NOTE

i. The *p*-azobenzenearsonate-L-tyrosine-*t*-BOC-*n*-succinicmide ester is available from Biosearch (San Rafael, California).

* Howard and Wild (1957); Amkraut et al. (1966); Tabachnick and Sobotka (1959, 1960).

## MODIFYING PROTEIN ANTIGENS BY DENATURATION

Many molecules can be made more immunogenic by denaturation. This treatment will change the structure of many compounds, particularly proteins, and expose new epitopes. Also, heating will often cause protein antigens to aggregate, and, as aggregated antigens are often more immunogenic, this may increase the antibody response. Either heating alone or heating with SDS are the usual treatments.

Injecting denatured antigens will be more likely to produce an antibody response against epitopes that are not found on the native antigen. If antibodies against totally denatured proteins are desired, the SDS–heat treatment is normally best. These treatments will generate antibodies that should be particularly good for immunoblots, for screening bacterial expression libraries, and for immunoprecipitation of proteins synthesized during in vitro translations.

1. Adjust the concentration of the antigen to 0.5–2 mg/ml in any convenient buffer. If no preference is indicated, PBS is fine. For SDS denaturation, add to 0.5%.

2. Heat to 80°C for 10 min.

3. Allow the antigen solution to cool to room temperature. If large aggregates occur, they can be broken by pipetting. Do not spin. For SDS treatment, dilute 1 in 5 with PBS.

   The samples are now ready for injection (p. 92). Samples with large aggregates should not be injected iv.

### NOTE

i. If detergents should be avoided, use 8 M urea for 30 min at room temperature. Dilute 1 in 8 with PBS and inject as normal.

## ▪ Coupling Antigens

One of the requirements for the production of a strong antibody response is the cell-to-cell contact between B cells and helper T cells and between helper T cells and antigen-presenting cells. This contact is mediated by a fragment of the antigen that has binding sites for both class II protein binding and T-cell receptor binding. Any antigen that does not have a region that can bind to both these proteins will not elicit a strong response. However, these sites can be added to the protein by coupling them to molecules that do have the proper site. Three methods for these additions are given below: either coupling protein antigens to other proteins that have good class II–T-cell receptor sites, coupling peptides that are known to be good class II–T-cell receptor sites directly to the antigen, or coupling the antigen to red blood cells.

A second method that can be used to increase the immunogenicity of some antigens is to make it easier to phagocytose by the antigen-presenting cells. This can be done by coupling the antigen to large particles such as red blood cells or agarose beads.

## COUPLING ANTIGENS TO PROTEIN CARRIERS

Some proteins will not be immunogenic because of T-cell tolerance or because of the lack of the appropriate class II proteins. This lack of response is most readily overcome by choosing a different strain or species to immunize. For very highly conserved proteins or in situations where immunization of a given species or strain is desired, the protein can often be made immunogenic by coupling it to a carrier protein. Useful carriers include keyhole limpet hemacyanin (KLH), bovine serum albumin (BSA), ovalbumin, or PPD (purified protein derivative of tuberculin).

Coupling can be achieved with any bifunctional reagent. Table 5.5 lists several of the most commonly use homo- and heterobifunctional reagents. In general, the coupling should be done to try and keep the antigen of interest in as native a condition as possible.

When coupling two proteins, a general strategy that is useful is, first, to treat one of the proteins (often the carrier) with a saturating amount of the coupling agent. High concentrations are used to stop the cross-linking of the first component to itself. Next, the cross-linker is removed by dialysis or gel filtration. Then an equal molar amount of the second component is added and allowed to bind. Last, if necessary, the conjugates can be purified from the starting components on the basis of size or binding to an ion-exchange column. In the cases where one of the proteins is known not to display accessible groups for the cross-linking, heterobifunctional reagents can be used to an advantage. Also, adjusting the pH can control the number of $NH_2$ groups that are available for cross-linking. At pHs well below the pK of the $NH_2$ groups, the amino groups will react only slowly. It may also be helpful to block some of the reactive groups with removable blocking groups such as citraconic anhydride (used for free amines, p. 82).

Because each coupling must be approached as an individual problem, no specific techniques are given.

## COUPLING ANTIGENS WITH T-CELL RECEPTOR– CLASS II PROTEIN BINDING SITES

Class II–T-cell receptor binding sites can be added to an immunogen by coupling a synthetic peptide to the molecule under study. This approach has only been used in the study of helper T-cell induction and is not in wide use for raising antibodies. However, a number of small antigens of defined sequence have been made immunogenic by coupling a known class II–T-cell receptor site to them, and the technique promises to be extremely valuable in boosting the immunogenicity of troublesome molecules.

**TABLE 5.5**
**Bifunctional Cross-linkers**

| | |
|---|---|
| Bis(sulfosuccinimidyl) Suberate | Homobifunctional |
| Bis(diazobenzidine) | Homobifunctional |
| Dimethyl Adipimidate | Homobifunctional |
| Dimethyl Pimelimidate | Homobifunctional |
| Dimethyl Suberimidate | Homobifunctional |
| Disuccinimidyl Suberate | Homobifunctional |
| Glutaraldehyde | Homobifunctional |
| m-Maleimidobenzoyl-N-Hydroxysuccinimide | Heterobifunctional |
| Sulfo-m-Maleimidobenzoyl-N-Hydroxysuccinimide | Heterobifunctional |
| Sulfosuccinidyl 4-(N-Maleimidomethyl) Cycloheane-1-carboxylate | Heterobifunctional |
| Sulfosuccinimidyl 4-(p-Maleimido-Phenyl) Butyrate | Heterobifunctional |

Most of the information about synthetic class II–T-cell receptor sites has been gathered in studies of inbred mice. These studies have shown that the success of a particular site is primarily dependent on the genetics of the available class II proteins. Because the class II genes differ from strain to strain and from species to species, the choice of peptide that can serve as good class II–receptor sites will vary. In practice, this means that a good choice of synthetic class II–T-cell receptor site can be made only for laboratory animals whose genetic background is known. Unfortunately, there are not enough good examples to cover all commonly used laboratory animals. Table 5.6 lists some known sites for mice. For other strains or species, coupling the antigen to a larger-molecular-weight carrier protein probably offers more advantages, at least at present.

| Links $NH_2$ to $NH_2$ | $H_2O$ soluble | pH between 7 and 9 |
|---|---|---|
| Links Tyr to Tyr | $H_2O$ soluble | Prepare fresh (p. 86) |
| Links $NH_2$ to $NH_2$ | $H_2O$ soluble | Keep pH above 8.3, use immediately |
| Links $NH_2$ to $NH_2$ | $H_2O$ soluble | Keep pH above 8.3 use immediately |
| Links $NH_2$ to $NH_2$ | $H_2O$ soluble | Keep pH above 8.3 use immediately |
| Links $NH_2$ to $NH_2$ | | pH between 7 and 9 |
| Links $NH_2$ to $NH_2$ | $H_2O$ soluble | pH between 7 and 8 |
| Links $NH_2$ and SH | Soluble in DMF | For SH/maleimide use pH 7 for $NH_2$/hydroxysuccinimide use pH 7–9 |
| Links $NH_2$ and SH | $H_2O$ soluble | For SH/maleimide use pH 7 for $NH_2$/hydroxysuccinimide use pH 7–9 |
| Links $NH_2$ and SH | $H_2O$ soluble | For SH/maleimide use pH 7 for $NH_2$/hydroxysuccinimide use pH 7–9 |
| Links $NH_2$ and SH | $H_2O$ soluble | For SH/maleimide use pH 7 for $NH_2$/hydroxysuccinimide use pH 7–9 |

**TABLE 5.6**
**Possible Synthetic Class II–T-Cell Receptor Sites**

| Protein | Residues | Sequence | Reference |
|---|---|---|---|
| Sperm Whale Myoglobin | 68–78 | VLTALGAILKK | Livingstone and Fathman (1987) |
| | 102–118 | KYLEFISEAIIHVLHSR | Cease et al. (1986) |
| | 132–146 | NKALELFRKDIAAKY | Berkower et al. (1986) |
| Chicken Lysozyme | 46–61 | NTDGSTDYGILQINSR | Allen et al. (1985) |
| | 74–86 | NLCNIPCSALLSS | Shastri et al. (1985) |
| | 81–96 | SALLSSDITASVNCAK | Shastri et al. (1985) |
| | 108–119 | WVAWRNRCKGRD | Katz et al. (1982) |
| Chicken Ovalbumin | 323–339 | ISQQVHAAHAEINEAGR | Shimonkevitz et al. (1984) |
| Pig Cytochrome *c* | 94–104 | LIAYLKQATAK | Schwartz et al. (1985) |
| Influenza HA from A/PR8/48 | 109–119 | SSFERFEIFPK | Hackett et al. (1983) |
| | 129–140 | NGVTAACSHEGK | Hurwitz et al. (1984) |
| | 302–313 | CPKYVRSAKLRM | Hurwitz et al. (1984) |
| Hepatitis B S Antigen | 38–52 | SLNFLGGTTVCLGQN | Milich et al. (1985) |
| | 95–109 | LVLLDYQGMLPVCPL | Milich et al. (1985) |
| | 140–154 | TKPSDGNCTCIPITS | Milich et al. (1985) |
| Lambda Repressor | 12–26 | QLEDARRLKAIYEKK | Guillet et al. (1986) |

Adapted from Spouge et al. (1987).

## *COUPLING ANTIGENS TO RED BLOOD CELLS**

Coupling antigens to red blood cells (erythrocytes) solves many of the requirements for strengthening an immune response. They are large and particulate, making them good targets for phagocytosis. Their size also slows the dispersal. If the source of the cells is different from the animal to be injected, they can provide good targets for class II–T-cell receptor binding. They are also easy to handle for coupling and storage.

Most workers prefer sheep red blood cells (RBCs), which are available commercially in citrate buffer. Because of the complexity of proteins found on the surface of the RBCs, almost all chemical groups are available for coupling. Choosing the coupling method should depend on the antigen. Commonly used coupling methods include tannic acid, chromic chloride, and glutaraldehyde. Any of the bifunctional reagents listed in Table 5.5 are fine as well.

1. Wash the RBCs three times by centrifugation at 800g for 5 min and resuspend in PBS.

2. After the last wash, resuspend the RBCs to a final concentration of 5% (vol/vol).

3. **Either**: Add an equal volume of freshly prepared 0.005% tannic acid. Mix well and incubate at 37°C for 15 min with occasional mixing. Centrifuge at 800g for 5 min. Remove the supernatant and add the antigen in PBS. Incubate for 15 min at 37°C with occasional mixing.

   **Or**: Add an equal volume of 1% glutaraldehyde in PBS. Mix well and add the antigen dissolved in PBS. Incubate at room temperature for 1 hr with shaking.

---

**Caution**   Glutaraldehyde is toxic. Work in a fume hood and use proper methods for disposal.

---

4. Centrifuge the cells at 800g for 5 min. Wash twice with PBS.

   The samples are now ready for injection (p. 92).

* Adapted from Adler and Adler (1980).

## COUPLING ANTIGENS TO BEADS

Soluble antigens that are poorly phagocytosed can be converted into excellent targets for antigen-presenting cells by coupling them to agarose beads, activated carbon, or bentonite. Agarose beads can be activated by treating them with chemical reagents that allow the simple coupling of protein antigens. Common methods for activation include treatment with carbodiimidazole, cyanogen bromide, glutaraldehyde, hydroxysuccinimide, or tosyl chloride. Protocols for activation are described on p. 528, or the activated beads can be purchased commercially. Activated beads are coupled to protein antigens through the antigen's free amino groups. Alternatively, the antigen can be adsorbed to carbon or bentonite. Titrate the amount of antigen that can bind to a fixed volume of beads by adding increasing amount of beads to a constant volume and amount of beads. Incubate to allow binding. Spin to remove the beads and test the supernatant for the presence of antigen.

## ▪ Immune Complexes As Antigens

Antigens purified by immunoprecipitation often show enhanced immunogenicity. If a source of antibody is available for the purification of an antigen, the compound can be purified readily and efficiently and used to immunize further animals. Purified immune complexes can be injected directly or can be injected coupled to beads. The presence of antibodies and/or the beads will stimulate phagocytosis of the antigen and usually will not harm the response. The antibodies should be as close as possible to the species used for the immunizations.

This method can provide a useful step in antigen enrichment for a wide variety of applications. It is particularly valuable when antibodies have been raised to one form of an antigen and now a different sort of reagent is needed. For example, the original antibodies may have been raised against a denatured antigen and now anti-native antibodies are needed. If there are epitopes displayed on the native antigen, then the antibodies can be used to collect a fully native antigen for further immunizations. This method can also be useful when using a monoclonal antibody as an intermediate to the preparation of a good polyclonal antisera. Injecting antibody-coated antigens has also been used to mask a particularly immunodominant epitope on an antigen, and thereby develop a response against other epitopes.

## PREPARING IMMUNE COMPLEXES FOR INJECTION

The amount of antigen needed to elicit a strong response using immune complexes will vary from one compound to another. Doses of as low as 50 ng of antigen have been used successfully when delivered this way.

1. First titrate the amount of antibody needed to remove the antigen from solution. Add increasing amounts of antibodies to identical samples of antigen. Antibody can be added as either crude serum, monoclonal antibody ascites fluid, monoclonal antibody tissue culture supernatant, or purified antibody. In general, the purer the antibody preparation, the less chance of raising spurious anti-immunoglobulin antibodies.

2. After a suitable incubation time, process the samples as for an immunoprecipitation reaction (p. 429). Collect the immune complexes on protein A beads (p. 468).

3. Next titrate the amount of protein A beads needed to clear the antibody–antigen complexes from circulation. Prepare a large batch of antibody–antigen complexes using the ratios determined in step 1. Dispense in equal amounts and add increasing amount of protein A beads to each tube.

4. Incubate for 2 hr with rocking at 4°C. Process as usual for an immunoprecipitation.

5. After determining the appropriate amount of antibody and protein A beads, mix the antigen-containing solution with the appropriate amount of antibody. Incubate on ice for 2 hr.

6. Add the protein A beads. Incubate for 2 hr with rocking at 4°C.

7. Wash thoroughly. The beads can be collected by centrifugation or allowed to settle by gravity. Wash the beads four times by resuspension with the buffer in which the antigen was stored.

8. Run a small sample of the immune complexes on an SDS-polyacrylamide gel to check for purity. If the antibody–antigen immune complexes will be used directly to immunize animals, the beads and immune complexes can be injected as described on p. 103.

9. To elute the antibody–antigen complexes prior to injection, transfer the beads to a column.

10. Wash the beads with 10 column volumes of 100 mM Tris (pH 8.0).

11. Wash the beads with 10 column volumes of 10 mM Tris (pH 8.0).

12. Prepare a solution of 100 mM acetic acid (a 1 in 180 dilution of glacial acetic acid). Check the pH and adjust to approximately 2.5–3.0 by either adding more acetic acid or by diluting with water. Elute the column with the diluted acetic acid.

13. Identify the fractions that contain the immune complexes by using a spot test (p. 671 or 672). Neutralize the sample by adding NaOH or dialysis against PBS. If the samples are dilute and need to be concentrated, either neutralize or dialyze with a volatile buffer such as ammonium bicarbonate, and then lyophilyze.

14. Run a small sample of the immune complexes on an SDS-polyacrylamide gel to check for purity.

The immune complexes are now ready for injection (p. 92).

## NOTES

i. If the antibodies used to purify the antigen do not bind well to protein A (p. 616), the buffers should be changed to the high salt variations described on p. 311 or the antibodies should be purified and covalently attached to the beads (p. 528).

ii. If the background is unacceptably high, any of the variations discussed in Chapter 11 can be used.

# 6 MONOCLONAL ANTIBODIES

Serum contains many different types of antibodies that are specific for many different antigens. Even in hyperimmune animals, seldom are more than one-tenth of the circulating antibodies specific for one antigen. The use of these mixed populations of antibodies creates a variety of different problems in immunochemical techniques. Therefore, the preparation of homogeneous antibodies with a defined specificity was a long-standing goal of immunochemical research. This goal was achieved with the development of the technology for hybridoma production.

The first isolation of a homogeneous population of antibodies came from studies of B-cell tumors. Clonal populations of these cells can be propagated as tumors in animals or grown in tissue culture. Because all of the antibodies secreted by a B-cell clone are identical, these tumor cells provide a source of homogeneous antibodies. Unfortunately, B-cell tumors secreting antibodies of a predefined specificity cannot be isolated conveniently.

In the animal, antibodies are synthesized primarily by plasma cells, a type of terminally differentiated B lymphocyte. Because plasma cells cannot be grown in tissue culture, they cannot be used as an in vitro source of antibodies. Köhler and Milstein (1975) developed a technique that allows the growth of clonal populations of cells secreting antibodies with a defined specificity. In this technique an antibody-secreting cell, isolated from an immunized animal, is fused with a myeloma cell, a type of B-cell tumor. These hybrid cells or *hybridomas* can be maintained in vitro and will continue to secrete antibodies with a defined specificity. Antibodies that are produced by hybridomas are known as *monoclonal antibodies*.

## Monoclonal antibodies are powerful immunochemical tools

The usefulness of monoclonal antibodies stems from three characteristics—their specificity of binding, their homogeneity, and their ability to be produced in unlimited quantities. The production of monoclonal antibodies allows the isolation of reagents with a unique, chosen specificity. Because all of the antibodies produced by descendants of one hybridoma cell are identical, monoclonal antibodies are powerful reagents for testing for the presence of a desired epitope. Hybridoma cell lines also provide an unlimited supply of antibodies. Even the most farsighted researchers have found that large supplies of valuable antisera eventually run out. Hybridomas overcome these difficulties. In addition, one unique advantage of hybridoma production is that impure antigens can be used to produce specific antibodies. Because hybridomas are single-cell cloned prior to use, monospecific antibodies can be produced after immunizations with complex mixtures of antigens.

Hybridomas secreting monoclonal antibodies specific for a wide range of epitopes have been prepared. Any substance that can elicit a humoral response can be used to prepare monoclonal antibodies. Their specificities range from proteins to carbohydrates to nucleic acids. However, monoclonal antibodies are often more time-consuming and costly to prepare than polyclonal antibodies, and they are not necessarily the best choice for certain immunochemical techniques. In theory, either as single antibody preparations or as pools, monoclonal antibodies can be used for all of the tasks that require or benefit from the use of polyclonal antibodies. In practice, however, producing exactly the right set of monoclonal antibodies is often a difficult and laborious job. Researchers should be certain that they need these types of reagents before they begin constructing hybridoma cell lines. Table 6.1 summarizes some of the uses of antibodies and some general suggestions for choosing the best reagents.

## Hybridomas are immortal somatic cell hybrids that secrete antibodies

In the early 1970s, a number of research groups worked on different methods to extend the life span of antibody-secreting cells in vitro. For murine cells, the practical aspects of this goal were solved by applying techniques used in somatic cell genetics. By fusing two cells, each having properties necessary for a successful hybrid cell line, Köhler and Milstein (1975) showed that antibody-secreting cell lines could be established routinely and maintained in vitro. The two cells that are commonly used as partners in these fusions are antibody-secreting cells isolated from immunized animals and myeloma cells. The myeloma cells provide the correct genes for continued cell division in tissue culture, and the antibody-secreting cells provide the functional immunoglobulin genes.

Early work solved the three technical problems for achieving a successful fusion: (1) finding appropriate fusion partners, (2) defining conditions for efficient fusion, and (3) choosing an appropriate system to select for hybrid cells against the background of unfused cells.

**TABLE 6.1**
**Immunochemical Techniques, Polyclonal versus Monoclonal Antibodies**

| Technique | Polyclonal antibodies | Monoclonal antibodies | Pooled monoclonal antibodies |
|---|---|---|---|
| **Cell Staining** | Usually good | Antibody dependent | Excellent |
| **Immunoprecipitation** | Usually good | Antibody dependent | Excellent |
| **Immunoblots** | Usually good | Antibody dependent | Excellent |
| **Immunoaffinity Purification** | Poor | Antibody dependent | Poor |
| **Immunoassays** | | | |
|   **Labeled Antibody** | Difficult | Good | Excellent |
|   **Labeled Antigen** | Usually good | Antibody dependent | Excellent |

## Myelomas from BALB/c mice are good cells for fusion

Myelomas can be induced in a few strains of mice by injecting mineral oil into the peritoneum. Many of the first examples of these myelomas were isolated from BALB/c mice by Potter (1972), and these cells are referred to by the abbreviation MOPC (for mineral oil plasmacytoma). Derivatives of BALB/c myelomas have become the most commonly used partners for fusions. Table 6.2 lists some of the myeloma cell lines used for hybridoma construction. Myelomas have all the cellular machinery necessary for the secretion of antibodies, and many secrete these proteins. To avoid the production of hybridomas that secrete more than one type of antibody, myelomas that are used for fusions have been selected for the lack of production of functional antibodies. Figure 6.1 shows the derivation of many of the commonly used myeloma cell lines.

The other cell for the fusion is isolated from immunized animals. These cells must carry the rearranged immunoglobulin genes that specify the desired antibody. Because of the difficulties in purifying cells that can serve as appropriate partners, fusions are normally performed with a mixed population of cells isolated from a lymphoid organ of the immunized animal. Although a number of studies have helped to characterize the nature of this B-cell-derived partner, the exact state of differentiation of this cell is still unclear.

Hybridomas can be prepared by fusing myelomas and antibody-secreting cells isolated from different species, but the number of viable hybridomas increases dramatically when closely related species are used. Therefore, fusions are normally done with cells from the same species. All commonly used mouse strains can serve as successful fusion partners with BALB/c myelomas; however, immunizations are normally done in BALB/c mice, as this allows the resulting hybridomas to be grown as tumors in this mouse strain.

## Polyethylene glycol is the most commonly used agent to fuse mammalian cells

In theory, the fusion between the myeloma cell and the antibody-secreting cell can be effected by any fusogen. In practice, hybridoma fusions became routine after the introduction of the use of polyethylene glycol (PEG). The use of PEG as a fusing agent for mammalian cells was first demonstrated by Pontecorvo (1975), and was quickly adopted by somatic cell geneticists. PEG is the method of choice for hybridoma production, allowing the rapid and manageable fusion of mammalian cells.

PEG fuses the plasma membranes of adjacent myeloma and/or antibody-secreting cells, forming a single cell with two or more nuclei. This heterokaryon retains these nuclei until the nuclear membranes dissolve prior to mitosis. During mitosis and further rounds of division, the individual chromosomes are segregated into daughter cells.

**TABLE 6.2**
**Myeloma Cell Lines Used as Fusion Parents**

| Cell line | Reference | Derived from | Chains expressed | Secreting | Comments |
|---|---|---|---|---|---|
| **Mouse Lines** | | | | | |
| P3-X63Ag8 | Köhler and Milstein (1975) | P3K | $\gamma 1, \kappa$ | $IgG_1$ | Not recommended |
| X63Ag8.653 | Kearney et al. (1979) | P3-X63Ag8 | None | No | Recommended |
| Sp2/0-Ag14 | Köhler and Milstein (1976); Shulman et al. (1978) | P3-X63Ag8 × BALB/c | None | No | Recommended |
| FO | de St. Groth and Scheidegger (1980) | Sp2/0-Ag14 | None | No | Recommended |
| NSI/1-Ag4-1 | Köhler et al. (1976) | P3-X63Ag8 | Kappa | No | Recommended |
| NSO/1 | Galfre and Milstein (1981) | NSI/1-Ag4-1 | None | No | Recommended |
| FOX-NY | Taggart and Samloff (1984) | NSI/1-Ag4-1 | Kappa (?) | No | |
| **Rat Lines** | | | | | |
| Y3-Ag1.2.3 | Galfre et al. (1979) | Y3 | Kappa | No | Not recommended |
| YB2/0 | Kilmartin et al. (1982) | YB2/3HL | None | No | Recommended |
| IR983F | Bazin (1982) | LOU/c rats | None | No | Recommended |

Because of the abnormal number of chromosomes, segregation does not always deliver identical sets of chromosomes to daughter cells, and chromosomes may be lost. If one of the chromosomes that carries a functional, rearranged immunoglobulin heavy- or light-chain gene is lost, production of the antibody will stop. In a culture of hybridoma cells, this will be seen phenotypically as a decrease in antibody titer and will result in unstable lines. If the chromosome that is lost contains a gene used in drug selection (see below), then the growth of the hybridoma will be unstable, and cells will continue to die during selection. In practice, the selection for the stable segregation of the drug selection marker is so strong that within a short time the hybridoma is either lost completely or a variant is isolated that stably retains the selectable marker.

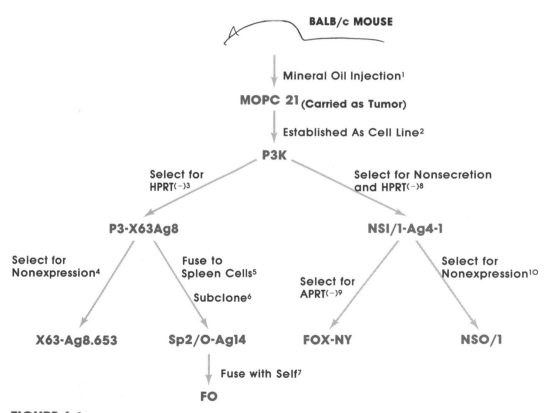

**FIGURE 6.1**

Myeloma family tree. [1]Potter (1972); [2]Horibata and Harris (1970); [3]Kohler and Milstein (1975); [4]Kearney et al. (1979); [5]Kohler and Milstein (1975); [6]Shulman et al. (1978); [7]de St. Groth and Scheidegger (1980); [8]Kohler et al. (1976); [9]Taggart and Samloff (1982); [10]Galfre and Milstein (1981).

## Unfused myeloma cells are eliminated by drug selection

Even in the most efficient hybridoma fusions, only about 1% of the starting cells are fused, and only about 1 in $10^5$ form viable hybrids. This leaves a large number of unfused cells still in the culture. The cells from the immunized animal do not continue to grow in tissue culture, and so do not confuse further work. However, the myeloma cells are well adapted to tissue culture and must be killed. Most hybridoma constructions achieve this by drug selection. Commonly, the myeloma partner has a mutation in one of the enzymes of the salvage pathway of purine nucleotide biosynthesis (first reported by Littlefield 1964). For example, selection with 8-azaguanine often yields a cell line harboring a mutated hypoxanthine-guanine phosphoribosyl transferase gene (HPRT). The addition of any compound that blocks the de novo nucleotide synthesis pathway will force cells to use the salvage pathway. Cells containing a nonfunctional HPRT protein will die in these conditions. Hybrids between myelomas with a nonfunctional HPRT and cells with a functional HPRT will be able to grow. Selections are commonly done with aminopterin, methotrexate, or azaserine (Fig. 6.2).

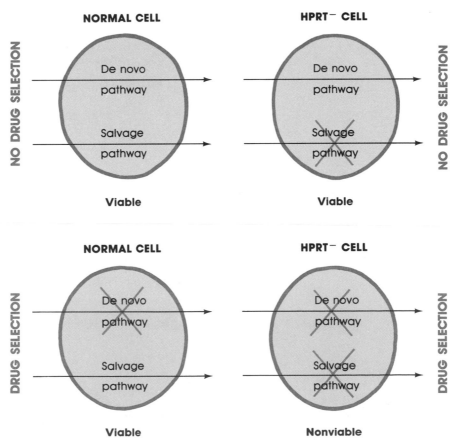

**FIGURE 6.2A**
Pathways of nucleotide synthesis.

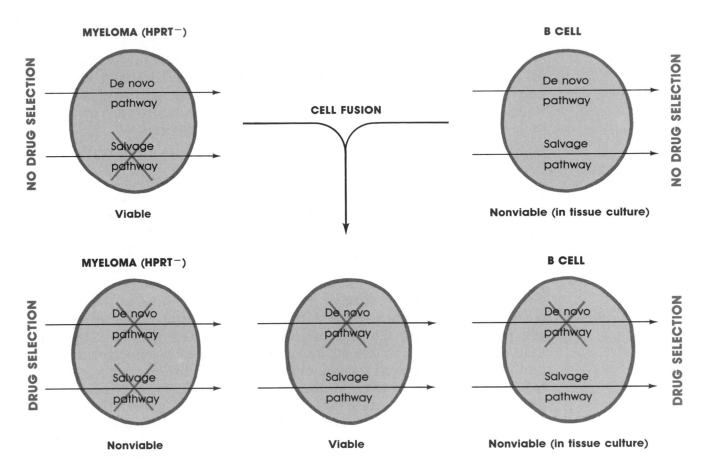

**FIGURE 6.2B**
Drug selection for viable hybridomas.

# ▦ PRODUCTION OF MONOCLONAL ANTIBODIES ▦

Although hybridomas can be prepared from animals other than mice, all of the techniques below use mice as examples. Similar techniques can be used for fusions of rat myelomas and rat antibody-secreting cells. More specialized fusions using interspecies crosses or human cells are discussed briefly on p. 240 and 241, respectively.

## ▦ Stages of Hybridoma Production

Figure 6.3 outlines the steps in the production of monoclonal antibodies. Animals are injected with an antigen preparation, and once a good humoral response has appeared in the immunized animal, an appropriate screening procedure is developed. The sera from test bleeds are used to develop and validate the screening procedure. After an appropriate screen has been established, the actual production of the hybridomas can begin. Several days prior to the fusion, animals are boosted with a sample of the antigen. For the fusion, antibody-secreting cells are prepared from the immunized animal, mixed with the myeloma cells, and fused. After the fusion, cells are diluted in selective medium and plated in multiwell tissue culture dishes. Hybridomas are ready to test beginning about 1 week after the fusion. Cells from positive wells are grown and then single-cell cloned. Hybridoma production seldom takes less than 2 months from start to finish, and it can take well over a year. It is convenient to divide the production of monoclonal antibodies into three stages: (1) immunizing mice, (2) developing the screening procedure, and (3) producing hybridomas. Any one of these stages may proceed very quickly, but all have inherent problems that should be considered prior to the start of the project, and these areas are discussed separately below.

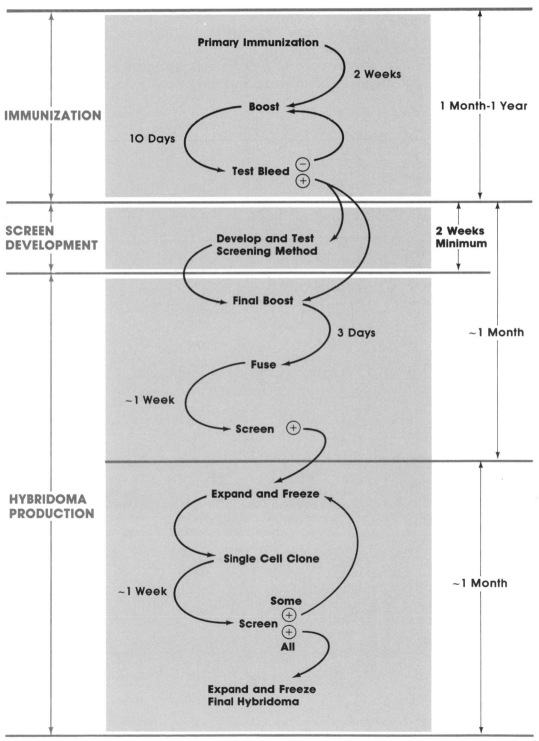

**FIGURE 6.3**
Stages of hybridoma production.

## ■ IMMUNIZING MICE

Figure 6.4 shows a typical antibody response to multiple injections with a good immunogen. Also included in this figure is a description of the characteristics of a typical monoclonal antibody that might be isolated following one of the immunizations. Although this simplified view can only serve as a rough guide, it does give an indication of the potential time frame for the production of antibodies with particular properties.

Chapters 4 and 5 discuss in detail both the theoretical and practical aspects of immunizing laboratory animals.

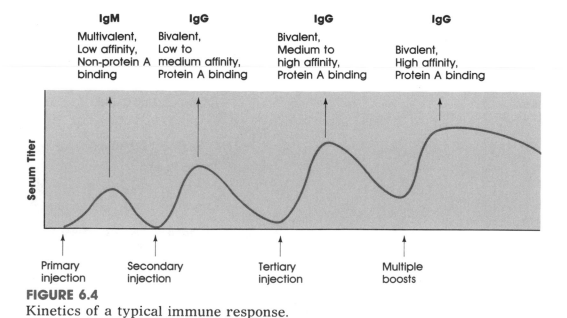

**FIGURE 6.4**
Kinetics of a typical immune response.

## ■ Dose and Form of the Antigen

The amount of antigen necessary to induce a good immune response will depend on the individual antigen and host animal (see Chapter 5). Suggested doses for mice are summarized in Table 6.3.

### Soluble Proteins

Soluble protein antigens can yield strong responses and good monoclonal antibodies with doses of as low as 1 $\mu$g/injection. More commonly, injections are adjusted to deliver 10–20 $\mu$g. If the antigen is available in large quantities, 50 $\mu$g should be used. Except for special cases it is seldom worthwhile to use more than 200 $\mu$g of a protein antigen per injection. Even if the antigen is not pure, the total dose should not normally exceed 500 $\mu$g. When highly conserved proteins are being used to raise antibodies, it is often necessary to modify these antigens prior to injection. This can be done by covalently adding small immunogenic haptens to proteins. Modifying proteins by binding them to large immunogenic proteins such as the hemocyanins has also been shown to be an effective way of breaking T cell tolerance. These methods are discussed in detail in Chapter 5.

## An Example of a Typical Immunization Schedule

1. For each mouse, mix 250 $\mu$l of antigen solution with 250 $\mu$l of complete Freund's adjuvant. Inject six BALB/c female mice ip.

2. After 14 days, repeat the injections but use incomplete Freund's adjuvant.

3. Collect tail bleeds from immunized mice on day 24. Do 1 in 5 dilutions in PBS and test all samples by comparison with similar dilutions of normal mouse serum in a dot blot.

4. On day 35, inject all animals ip with incomplete Freund's.

5. Day 45, do tail bleeds and test by dot blot. All serum samples checked by immunoprecipitation against in vivo radiolabeled antigen preparation.

6. Day 56, inject best responder, 100 $\mu$l iv and 100 $\mu$l ip. All others get ip injection with incomplete Freund's.

7. Day 59, fuse splenocytes from best responder.

8. Day 66, hard work starts, but champagne for first positives.

**TABLE 6.3**
**Suggested Doses of Immunogens for Mice**

| Form of antigen | Examples | Primary injections and boosts | | | Final boosts | | |
|---|---|---|---|---|---|---|---|
| | | Possible routes | Dose | Adjuvant | Possible routes | Dose | Adjuvant |
| Soluble Proteins | Enzymes<br>Carrier proteins conjugated with peptides<br>Immune complexes | ip[a]<br>sc[b] | 5–50 μg | + | iv[c] | 5–50 μg | — |
| Particulate Proteins | Viruses (killed)<br>Yeast (killed)<br>Bacteria (killed)<br>Structural proteins | ip<br>sc | 5–50 μg | + | iv | 5–50 μg | — |
| Insoluble Proteins | Bacterially produced from inclusion bodies<br>Immunopurified proteins bound to beads | ip<br>sc | 5–50 μg | + | ip | 5–50 μg | — |
| Live Cells | Mammalian cells | ip | $10^5$–$10^7$ cells | — | iv | $10^6$ cells | — |
| Live Tumorigenic Cells | Oncogenic mammalian cells | ip<br>sc | $10^4$–$10^6$ cells | — | iv | $10^6$ cells | — |
| Carbohydrates | Polysaccharides<br>Glycoproteins | ip<br>sc | 10–50 μg | +/– | iv | 10–50 μg | — |
| Nucleic Acids | Carrier proteins conjugated with N.A. | ip | 10–50 μg | + | iv | 10–50 μg | — |

[a] Intraperitoneal.
[b] Subcutaneous.
[c] Intravenous.

## Particulate Proteins

In general, particulate antigens make excellent immunogens, because they are readily phagocytosed (see Chapter 4). Soluble proteins may be converted to particulate antigens by self-polymerization or by binding them to solid substrates such as agarose beads (p. 528). Large insoluble antigens should not be injected intravenously (iv) due the possible development of embolisms.

## Proteins Produced by Overexpression

Recent advances in recombinant DNA technology have made the production of many protein antigens simple. Overexpression of fusion proteins or full-length polypeptide chains using both prokaryotic and eukaryotic vectors has become routine. These proteins are often excellent antigens and can be produced in large quantities. They normally present few problems for the production of monoclonal antibodies. These proteins can be purified and injected as soluble or insoluble antigens (p. 88).

## Synthetic Peptides

A second source of immunogens, based on the availability of a coding sequence, is the in vitro synthesis of peptides. Synthetic peptides, when coupled to carrier proteins such as bovine serum albumin or keyhole limpet hemocyanin, normally elicit a strong humoral response. Constructing these carrier complexes and the production of anti-peptide sera are described in Chapter 5 (p. 73). Using peptide–carrier protein complexes for the production of monoclonal antibodies normally is done only for specific reasons. Because these peptides are relatively short, many of the advantages of monoclonal antibody specificity are lost. Monoclonal antibodies do provide two advantages over polyclonal anti-peptide sera. The first is that the source of the antibodies will be unlimited, and the second is that monoclonal antibodies may be more useful in immunoaffinity purifications. Like all immunizations using peptide antigens, the major difficulty will be in preparing antibodies that will bind to the native protein.

## Live Cells

A number of studies have used live cells as immunogens for generating antibodies to surface antigens. Except in unusual circumstances, injections of cells should not include live bacteria or yeast. Although mice are normally capable of killing and clearing bacteria and yeast infections, the possibility of infecting an entire mouse colony is too great to risk these types of injections.

Although large numbers of hybridomas have been prepared to surface antigens of mammalian cells, these antibodies may be of low affinity and care should be taken to ensure that the immune response includes antibodies that will be useful in later studies. When raising antibodies to live tumorigenic cells, it is easy to pass the cells as tumors and thereby eliminate any activity against tissue culture reagents, including proteins in bovine serum.

### Nucleic Acids

Nucleic acids normally are not good antigens, and antibodies to them usually are raised against small haptens bound to carrier proteins. Because nucleic acids are weak antigens, it is particularly important to test sera for antibodies that will work in all assays for which the monoclonal antibodies are being raised.

### Carbohydrates

Simple carbohydrates usually are weak immunogens. These compounds should be coupled to carrier proteins. Large complex carbohydrates ( > 50,000) will induce a moderate response, but often without a secondary response. High doses readily induce tolerance, so the injected amount should be controlled carefully. These immunogens are best injected as a portion of a larger particle, such as a bacterial cell wall

**TABLE 6.4**
**Routes of Injection**

| Route | Abbreviation | Uses | | Maximum volume |
| | | Primary injections and boosts | Final boost | |
|---|---|---|---|---|
| **Intraperitoneal** | ip | Good | Fair | 0.5 ml |
| **Subcutaneous** | sc | Good | Poor | 0.2 ml |
| **Intravenous** | iv | Poor | Good | 0.2 ml |
| **Intramuscular** | im | Not recommended for mice | | |
| **Intradermal** | id | Not recommended for mice | | |
| **Lymph node** | | Special uses | | 0.1 ml |

or equivalent. Coupling these larger carbohydrates to carrier proteins can be beneficial. For glycoproteins, the polypeptide backbone can function as an effective carrier.

## ■ Route of Inoculation

Table 6.4 gives a summary of the potential routes of introducing an antigen into mice. Most injections for hybridoma production are done in female mice, because they are somewhat easier to handle than male mice.

> Prior to beginning an immunization, contact your local safety and animal committees for advice on animal care and handling, local regulations, and proper procedures for immunization.

| Adjuvant | Immunogen requirement | Comments | Route |
|---|---|---|---|
| +/− | Soluble/or insoluble | If used for final boost, wait 5 days before fusion | **Intraperitoneal** |
| +/− | Soluble/or insoluble | Local response, Serum levels slower to increase | **Subcutaneous** |
| No Freund's | Soluble, Ionic detergent <0.2% Nonionic detergent <0.5% Salt <0.3 M Urea <1 M | Poor for immunizing | **Intravenous** |
| | | | **Intramuscular** |
| | | | **Intradermal** |
| No Freund's | Soluble/or insoluble | Good applications for experienced workers | **Lymph node** |

## COMMENTS ■ Immunizations

There are several points that are important in designing an immunization regime that will produce the appropriate monoclonal antibodies.

• Choose the appropriate animal or strain for the desired antibody. Important points to consider are (1) tolerance, (2) amount of antigen available, and (3) specific properties (including ease of purification) of the resultant antibodies. (See Chapters 4 and 5 for details.)

  If no preference in the choice of animal is dictated, then start the immunizations in female BALB/c mice (6 weeks old). In general, mice are cheaper to maintain, are easier to handle, and will respond to lower antigen levels than other laboratory animals. In addition, BALB/c × BALB/c hybridomas can be grown as ascites in BALB/c mice (see p. 274). This can be valuable both in the production of large quantities of monoclonal antibodies and in the eradication of contaminating microorganisms from cultures of hybridoma cells grown in vitro.

- Individual animals, even from the same genetic background, will often respond to identical antigen preparations in completely different ways. Therefore, immunizing more than one animal is a major advantage. In addition, laboratory animals occasionally die, so starting immunizations with several animals may save valuable time.

- Hyperimmunization (multiple immunizations with the same antigen) will yield antibodies with higher affinity for the antigen, especially when the immunizations are widely spaced over a period of weeks to months (see Chapter 5). However, multiple immunizations will not continue to increase the number of epitopes that are recognized.

- Except in unusual circumstances, **do not** start the fusion until the serum from the test bleed contains antibodies with the desired specificity. This may mean extensive testing of the serum in a number of assays, but do not expect to recover antibody activities from the fusion that are not found in the test serum.

- If the animal responds weakly or not at all, consult Chapter 5 for suggestions.

## INTRAPERITONEAL INJECTIONS—WITH ADJUVANT

Intraperitoneal injections (ip) are the most commonly used method for introducing antigens into mice. Because of the large volume of the peritoneal cavity, the volume of the immunogen can be larger for ip injections than for other sites. Also because the injections do not deliver the antigen directly into the blood system, particulate antigens can be used.

> **Caution**  Freund's adjuvants are potentially harmful. Be careful during preparation and injection.

1. The antigen should be dispersed in approximately 250 $\mu$l of buffer. Draw the solution into a 1.0-ml syringe. Draw an equal volume of Freund's adjuvant into a second syringe. For a primary immunization the adjuvant should be complete Freund's (see p. 98). Complete Freund's should be vortexed before use to resuspend the killed *Mycobacterium tuberculosis* bacteria. All other injections should be in incomplete Freund's.

2. Aqueous antigen solutions and oil-based adjuvants are immiscible, but when mixed will form an emulsion. This is most easily done by pushing the mixture between the two 1-ml syringes via a two- or three-way valve as shown in Figure 6.5. Remove all air from both

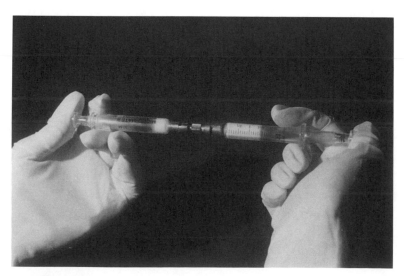

**FIGURE 6.5**
Preparing Freund's adjuvant for injection.

syringes before connecting to the valve. Make sure the connections are tightened securely. Depress the plunger on the syringe containing the aqueous antigen solution first. Push the mixture between the two syringes until it becomes difficult to continue (approximately 10–20 times). Then push the emulsion into one syringe. Remove the valve and the other syringe, and place a 23- or 25- gauge needle on the syringe. Displace any air that may be held in the barrel or in the needle by carefully depressing the plunger until the emulsion reaches the needle.

3. Hold the mouse as shown in Figure 6.6. Inject the antigen–adjuvant emulsion into the peritoneal cavity.

### NOTES

i. If the volume of the antigen solution is small (100 $\mu$l or less), the emulsion between the Freund's and aqueous solutions may be prepared by vortexing or sonicating in a 1.5-ml conical tube.

ii. Plastic syringes may be difficult to use when preparing the emulsion; therefore, most workers use glass syringes (glass disposable syringes with luer locks are best).

iii. Freund's adjuvants can be replaced by other adjuvants (see p. 96).

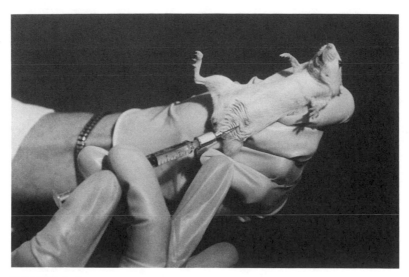

**FIGURE 6.6**
Intraperitoneal injection of a mouse.

## INTRAPERITONEAL INJECTIONS—WITHOUT ADJUVANTS

Intraperitoneal injections can also be used to deliver live cells into the peritoneal cavity. These immunizations normally are used to introduce live mammalian cells into mice and prepare anti–cell-surface antibodies. In general, adjuvants should not be used.

1. Cells should be washed extensively in PBS or other isotonic solutions prior to injection to remove as many extraneous proteins as possible. For example, many of the components of bovine serum are highly immunogenic, and if they are injected with the cells, can be antigenically dominant. Even washed cells will have a large number of extraneous proteins bound to the plasma membrane. If this remains a problem, it may be necessary to transfer the cells to low serum, serum-free medium, or mouse serum prior to the injection.

2. The cells should be taken up in 500 $\mu$l of PBS and injected using a 25-gauge needle. Hold the mouse as shown in Figure 6.6.

3. Inject the cells into the peritoneum.

### NOTES

i. Normal doses of mammalian cells will be between $10^5$ and $10^7$ cells/ injection.

ii. Because of potential infections of the mouse colony, injections of live viruses, bacteria, or fungi normally are not recommended. These antigens are commonly killed or inactivated prior to injection.

## *INTRAPERITONEAL INJECTION—ANTIGEN BOUND TO BEADS*

One excellent method to increase the chances of an antigen being phagocytosed is to bind it to beads. Protein antigens can be bound by free amino groups to any of a number of agarose or polyacrylamide beads. The methods for coupling are discussed on p. 528. After coupling the beads are injected ip with adjuvant as described on p. 158. These preparations should never be injected iv because the chance of forming embolisms is too great.

## INTRAPERITONEAL INJECTION—NITROCELLULOSE

Protein antigens can be bound to nitrocellulose and injected or implanted into the peritoneum. Two approaches can be used. If the entire piece of nitrocellulose will be implanted, the subcutaneous route is suggested (see p. 166). For intrapertioneal injections, the nitrocellulose should either be fragmented or dissolved.

> **Caution**  Freund's adjuvants are potentially harmful. Be careful during preparation and injection.

1. Incubate a solution of the antigen (no more than 1.0 mg/ml in PBS) with a sheet of nitrocellulose (0.1 ml/cm$^2$) at room temperature for 1 hr in a humid atmosphere. For some more abundant antigens or partially purified antigens, the proteins can be transferred directly from an SDS-polyacrylamide gel using standard blotting techniques (see p. 479).

2. Wash the sheet three times with PBS.

3. **Either**: Drain the paper well and freeze at −70°C for 10 min (liquid nitrogen also works well). Transfer to a clean, cold mortar and pestle and quickly grind the paper into small pieces. Remove the plunger from a 1.0-ml syringe, and transfer the pieces into the barrel. Use 250 μl of PBS to help in the transfer.

   **Or**: Allow the nitrocellulose to dry completely. Remove the plunger from a 1.0-ml syringe. Push the nitrocellulose (use pure nitrocellulose without acetate) to the bottom of barrel. Reinsert the plunger and depress completely. Draw up 125 μl of dimethylsulfoxide into the syringe. Alllow to sit at room temperature for 30 min. Draw up 125 μl of PBS.

   **Or**: Allow the nitrocellulose to dry completely. Remove the plunger from a 1.0-ml syringe. Push the nitrocellulose (use pure nitrocellulose without acetate) to the bottom of the barrel. Push the luer lock of the syringe into a three-way valve with the valve opening to the syringe closed. Add 250 μl of acetone to the syringe. Tap the syringe several times to mix the acetone with the nitrocellulose. Leave the syringe open and allow to dry (several hours). Remove the three-way valve and replace the plunger. Push the plunger to the end and draw 250 μl of PBS into the syringe.

4. Draw 250 μl of Freund's adjuvant into a second syringe. For a primary immunization the adjuvant should be complete Freund's. (Complete Freund's should be vortexed before use to resuspend the killed *Mycobacterium tuberculosis* bacteria.) All other injections should be in incomplete Freund's.

5. The nitrocellulose mixture and the oil-based adjuvant are immiscible, but when mixed will form an emulsion. This is most easily done by pushing the mixture between the two 1-ml syringes via a two- or three-way valve as shown in Figure 6.5. Remove all air from both syringes before connecting to the valve. Make sure the connections are tightened securely. Depress the plunger on the syringe containing the PBS solution first. Continue to push the mixture between the two syringes until it becomes difficult (approximately 10–20 times). Push the emulsion into one syringe. Remove the valve and the other syringe, and place a 23- to 25-gauge needle on the syringe. Displace any air that may be held in the barrel or in the needle by carefully depressing the plunger until the emulsion reaches the needle.

6. Hold the mouse as shown in Figure 6.6. Inject the antigen–adjuvant emulsion into the peritoneal cavity.

## NOTES

i. Plastic syringes may be difficult to use when preparing the emulsion; therefore, most workers use glass syringes (glass disposable syringes with luer locks are best).

ii. Freund's adjuvants can be replaced by other adjuvants (see p. 96).

## *SUBCUTANEOUS INJECTIONS—WITH ADJUVANTS*

Subcutaneous injections (sc) are used to deliver soluble or insoluble antigens into a local environment that is a good site of lymphoid draining. Maximum volumes for sc injections are about one-fifth the maximum used for ip injections (100 $\mu$l compared to 500 $\mu$l). Subcutaneous injections normally are done at more than one site to help ensure that the antigen is detected.

> **Caution** Freund's adjuvants are potentially harmful. Be careful during preparation and injection.

1.  The antigen should be dispersed in approximately 50–100 $\mu$l per site of injection. Take up into a 1.0-ml syringe. Take up an equal volume of Freund's adjuvant into a second syringe barrel. For a primary immunization the adjuvant should be complete Freund's. (Complete Freund's should be vortexed before use to resuspend the killed *Mycobacterium tuberculosis* bacteria.) All other injections should be in incomplete Freund's.

2.  The adjuvant, which is oil based, and the aqueous antigen solutions are immiscible, but with mixing will form an emulsion. This is most easily done by pushing the mixture between the two 1-ml syringes via a two- or three-way valve as shown in Figure 6.5. Remove all air from both syringes before connecting to the valve. Make sure the connections are securely tightened. Depress the plunger on the syringe containing the aqueous antigen solution first. Continue to push the mixture between the two syringes until it becomes difficult to push (approximately 10–20 times). Push the emulsion into one syringe. Remove the valve and the other syringe, and place a 23- or 25-gauge needle on the syringe. Displace any air that may be held in the barrel or in the needle by carefully depressing the plunger until the emulsion reaches the needle.

3.  Hold the mouse as shown in Figure 6.7. Inject approximately 200 $\mu$l total under the skin.

### NOTES

i.  If the volume of the antigen solution is small (100 $\mu$l or less), the emulsion between the Freund's and aqueous solutions may be prepared by vortexing or sonicating in a 1.5-ml conical tube.

ii. Plastic syringes may be difficult to use when preparing the emulsion; therefore, most workers use glass syringes (glass disposable syringes are best).

iii. Freund's adjuvants can be replaced by other adjuvants (see p. 96).

## SUBCUTANEOUS INJECTIONS—WITHOUT ADJUVANTS

Like the peritoneal injections, subcutaneous (sc) injections that do not use adjuvants normally are used for delivering live cells to the mouse. This route is often used for tumorigenic cells.

1. Cells should be washed carefully prior to injection to remove proteins from the growth medium. For example, many of the components of bovine serum are highly immunogenic, and, if they are injected with the cells, they can be antigenically dominant. Even washed cells will have a large number of extraneous proteins bound to the plasma membrane. If this remains a problem, it may be necessary to transfer the cells to low serum, serum-free medium, or mouse serum prior to the injection.

2. The cells should be resuspended in approximately 100 $\mu$l of PBS per site of injection. Use a 25-gauge needle.

3. Hold the mouse as shown in Figure 6.7. Inject under the skin.

### NOTES

i. Normal doses of mammalian cells will be between $10^5$ and $10^7$ cells/injection.

ii. Because of potential infections of the mouse colony, injections of live viruses, bacteria, or fungi normally are not recommended. These antigens are commonly killed or inactivated prior to injection.

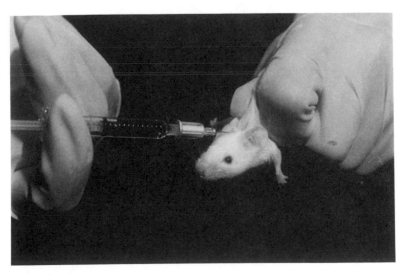

**FIGURE 6.7**
Subcutaneous injection of a mouse.

## SUBCUTANEOUS IMPLANTS—NITROCELLULOSE

Protein antigens immobilized on nitrocellulose often make exceptionally good immunogens. This is probably due to their slow release from the paper, thus behaving somewhat like an adjuvant (p. 96). Not all antigens show increased immunogenicity using this methodology, but some do. The antigen is bound to paper and is implanted on the back of the mouse's neck, a location that makes it difficult for the mouse to disturb the surgical clip.

1. Incubate a solution of the antigen (no more than 1.0 mg/ml in PBS) with a sheet of nitrocellulose (0.1 ml/cm$^2$) at room temperature for 1 hr in a humid atmosphere.

2. Wash the sheet three times with PBS.

3. Anesthetize the mouse by injecting a suitable drug (see Chapter 5, p. 95). For mice, 0.05 ml of Nembutal ip (sodium pentobarbitone, 40–85 mg/kg) is appropriate. The mouse will be ready for the operation in about 1–2 min.

4. Swab the skin on the mouse's back, just below the base of the neck, with alcohol. Raise the skin with a forceps and make a 1.5-cm incision with sterile scissors. Pull the skin from the inner body wall and insert a 0.5-cm$^2$ piece of the nitrocellulose. Close the incision with a skin clamp and return the mouse to its cage.

5. Take weekly serum samples beginning about 14 days after the implant. About 10 days after the serum titer drops, the final boost can be given. The final boost will still be an iv injection.

## INTRAVENOUS INJECTIONS

Intravenous injections (iv) can be used for two purposes. When immunizing mice, their main use will be to deliver the final boost just before a hybridoma fusion (p. 210). However, iv injections also are useful to ensure that the antigen is seen by the immune system. A rapid and strong response can be expected, as the antigen will be collected quickly in the spleen, liver, and lungs. The antigen will be processed quickly and no continued release of the antigen into the immune system can be expected. Consequently, these types of injections produce a short-lived response.

Intravenous injections should never be used as primary injections, and they must not contain large particulate matter. Because the injection will introduce the antigen directly into vital organs, harsh chemicals must be avoided. Similarly, adjuvants such as Freund's should not be used (p. 96).

1. Isolate the mouse in a small cage or container. Heat the mouse with an infrared lamp. This will increase the blood supply to the tail, making the veins easier to inject. Be careful of the length of time the mouse is left under the lamp. If it's too hot for your hand, it's too hot for the mouse.

2. Move the mouse to a restraining device as shown in Figure 6.8.

3. Swab a portion of the tail with alcohol about 1.5 cm below the base. The veins on the tail should be easily visible.

4. Use a 1.0-ml syringe fitted with a 26- or 27-gauge needle. Hold the tail firmly with one hand and guide the needle into one of the veins. Gently draw back on the plunger. If the needle is in the vein, there will be very little resistance and blood will appear in the barrel. If there is strong resistance and no blood appears, the needle is not in the vein. Withdraw the needle and move to a second site or to another vein. This technique may require a little practice before you can hit the vein readily. Practice with injections of PBS.

5. After you are sure the needle is in the vein, slowly deliver the injection. Pause a few seconds, remove the needle, and return the mouse to its cage. The antigen should be in solution and no adjuvant should be included. Except for specialized injections, the maximum amount to give a mouse by this method is 0.2 ml.

## NOTES

i. Injection by this route into immunized animals may cause an anaphylactic reaction. This can be prevented by a prior injection of an antihistamine. Contact your local animal committee for guidance.

ii. Solutions for iv injections should not contain high concentrations of denaturing agents and should be free of toxic chemicals such as sodium azide.

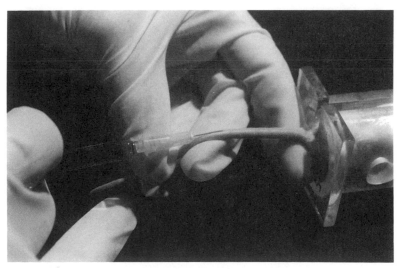

**FIGURE 6.8**
Intravenous injection of a mouse.

## INJECTIONS DIRECTLY INTO LYMPHOID ORGANS

These are injections for more specialized delivery of antigens. They are often appropriate for small amounts of antigen and particularly for secondary or later boosts. In theory, these types of injections may be the best methods for giving a final boost, but because they demand more skill, they are not in common use. The two most frequently used sites of injection are the footpad and the spleen.

In general, footpad and spleen injections should be done only for highly specialized antigens and then only after consulting local authorities for the proper protocols.

## ▓ Identifying Individual Mice

Beginning with the first test bleeds, it is essential to mark the mice so that the immune response can be monitored in individuals. There are a number of methods that are currently used to identify mice. These include ear punches, toe clips, and tail markings. If your animal facility has a standard method, consult them for the proper codes. If not, an acceptable method that is not harmful to the mice is to color the toes of their hindlegs with an indelible marker. This procedure is relatively easy, and, because the marks are on the back legs, the mice don't seem to work as hard to remove the markings as on other sites. Even so, the marks need to be reapplied twice a week.

If there is a large difference between the responses in individual mice, it may be worthwhile to isolate individual mice in separate cages to ensure that the proper mouse is given the final boost.

## ▓ Test Bleeds

Except in unusual circumstances, it is seldom worthwhile to fuse antibody-secreting cells from animals that do not have a usable titer of antibodies in their serum. Periodic test bleeds collected from immunized animals should be checked for the desired antibodies. Tests are run conveniently on small batches of serum prepared from tail bleeds of immunized mice.

The test bleed will yield small samples of polyclonal sera. These sera should be tested in assays that will detect the presence of antibodies specific for the antigen. These tests are discussed in detail in Chapters 10–14.

To make appropriate comparisons in these tests, two practical matters need to be considered. First, the test bleeds should always be titered to monitor the development of the response. The appropriate dilutions will depend on the strength of the response and on the type of assay, but in general 1 in 5 or 1 in 10 dilutions will be satisfactory. Second, the proper negative control should be another polyclonal serum and not an unrelated monoclonal antibody. Most often, this negative control will be serum collected either from another uninjected animal or from an animal that has been boosted with an unrelated antigen. Although it is not always necessary, using serum from a test bleed collected before immunization of the animal is the best negative control. These bleeds are known as preimmune sera.

## COLLECTING SERA FROM A MOUSE BY TAIL BLEED

1. Isolate the mouse in a small cage or container. Heat the mouse under an infrared lamp for a few minutes. This treatment will increase the blood flow to the tail, but be careful not to hurt the mouse. If it's too hot for your hand, then it's too hot for the mouse. Place the mouse in a restricted space as shown in Figure 6.8.

2. Swab a portion of the tail about 1.5–2 inches from the body with alcohol. Using a sterile scalpel, nick the underside of the tail across one of the lower veins that should be visible. Collect several drops of blood in a tube and return the mouse to its cage.

3. Incubate the blood at 37°C for 1 hr. Flick the tube several times to dislodge the blood clot.

4. Transfer to 4°C for 2 hr or overnight.

5. Spin at 10,000g for 10 min at 4°C.

6. Remove the serum from the cell pellet. Discard the cell pellet and spin the supernatant a second time for 10 min. Remove the serum, being careful to avoid the packed cell pellet.

7. Add sodium azide to 0.02% and test. Any remaining serum may be frozen at −20°C. The yield is approximately 100–200 $\mu$l.

## ■ Deciding to Boost Again or to Fuse

Three factors will influence the decision to proceed with the production of monoclonal antibodies. They all are related to the quality and strength of the immune response. First is whether the antibodies recognize the antigen of interest. This is the most straightforward of the factors and the simplest to determine. The second is a complicated set of properties of the antibodies themselves and the strength of the immune response. These properties are manifested as different titers of antibodies and different affinities of the antibody for the antigen. The third factor is the appearance of spurious antibody activities against unrelated antigens.

In many cases the tests will be relatively easy, and the interpretation apparent. First, the sera should be checked for antibodies that bind to the immunogen itself. For example, if a purified antigen is used, sera could easily be tested for activity in a simple antibody capture assay (p. 175), or if whole cells are used then testing for binding to the cell surface should be done first (p. 184). However, if the monoclonal antibodies will be used for tests other than these simple assays, test bleeds should be checked in assays that resemble, as closely as possible, the tests for which the antibodies are being prepared. For many antibodies, the most useful test will be the immunoprecipitation of the antigen (p. 429). This assay is easy when only testing a few samples, and it will identify antibodies that will be useful in a large number of tests that depend on binding to the native antigen. If, however, the antibodies will be used extensively in immunoblot analyses, in immunohistochemical staining, or in other tests in which many antibodies may fail to work, these tests should be run as well.

Second, sera should also be monitored for the concentration of specific antibodies by titering the test bleeds in the appropriate assays. As the immune response matures, higher levels of specific antibodies will be found. However, higher levels of antibodies do not necessarily mean higher affinities. If high affinity is crucial to the intended use of the antibody, the sera should be titered and compared in assays that are sensitive to antibody affinity, such as immunoprecipitation.

The third factor to consider is the appearance of antibody activities against extraneous antigens. This response may be directed against other antigens in your preparations or may be a response to other antigens in the mouse's environment, including invasion by a pathogenic organism. If the mouse is ill, do not proceed with hybridoma construction. Isolate the mouse in a separate cage and allow it to recover before continuing. If a particularly valuable antigen is being used, more care and veterinary help may be needed. If the antibody activities are to contaminating antigens in the immunogen, a decision must be made whether to proceed. In general, making monoclonal antibodies against complex and multicomponent antigens is a very useful way of isolating specific immunochemical probes, particularly when the antigen is difficult to purify further. However, if the response against the other antigens continues to increase without a concomitant strengthening of the response to the desired antigen, other approaches may need to be taken. Either other mice should be tested (other individuals or other strains) or the antigen may need to be purified further before proceeding.

## ◼ DEVELOPING THE SCREENING METHOD

Because most hybridoma cells grow at approximately the same rate, the tissue culture supernatants from all the fusion wells usually are ready to screen within a few days of one another. This means that screening is normally the most labor-intensive segment of hybridoma production. Care in developing the proper screen will help to keep the amount of work needed to identify positive wells to a minimum.

Approximately 1 week after the fusion, colonies of hybrid cells will be ready to screen. During the screening, samples of tissue culture media are removed from wells that have growing hybridomas and are tested for the presence of the desired antibodies. Successful fusions will produce between 200 and 20,000 hybridoma colonies, with 500–1000 colonies being the norm. Depending on the fusion, individual wells will become ready to screen over a 2- to 6-day period. Typically, the first wells would be ready to screen on day 7 or 8, and most of the wells will need to be screened within the next 4 or 5 days.

A good screening procedure must: (1) reduce the number of cultures that must be maintained to a reasonable level (seldom more than 50 cultures at one time), (2) identify potential positives in 48 hr or less (24 hr or less is ideal), and (3) be easy enough to perform for all the needed wells. Positive wells may be as rare as 1 in 500 or as common as 9 out of 10. Several screening steps can be combined to identify the desired clones, as long as the first screen reduces the tissue culture work to a manageable level. After the first round of screens, handling the tissue culture necessary for 100 wells is difficult for one person, 50 wells is reasonable, and less than 20 is relatively simple.

> All screening procedures must be tested and validated before the fusion has begun. After the fusion, there is seldom enough time to try out new ideas or to refine methods. The test bleeds should be used to set up and test the screening assay.

## Screening Strategies

There are three classes of screening methods, antibody capture assays, antigen capture assays, and functional screens. Currently, most screens are done by either antigen or antibody capture, but as functional assays become easier to use, more fusions will be screened by these methods. Table 6.5 depicts several of the more common screening techniques.

In general, the more antigens in the immunizing injections, the more difficult the screen. Researchers with pure or partially pure antigens should use methods for antibody capture. If the subcellular location of an antigen is known, positive tissue culture supernatants can be identified by cell staining. If the immunizations used complex antigen solutions, procedures such as immunoprecipitation or other antigen capture assays may be the only alternatives.

In addition to the tests described below, any of the assays used for analyzing antigens can be adapted for use as a screen (see Chapters 10–14).

## Antibody Capture Assays

Antibody capture assays are often the easiest and most convenient of the screening methods. In an antibody capture assay the following sequence takes place: the antigen is bound to a solid substrate, the antibodies in the hybridoma tissue culture supernatant are allowed to bind to the antigen, the unbound antibodies are removed by washing, and then the bound antibodies are detected by a secondary reagent that specifically recognizes the antibody. In this assay the detection method identifies the presence of the antibody, thus determining a positive reaction.

Most antibody capture assays rely on an indirect method of detecting the antibody. This is commonly done with a secondary reagent such as rabbit anti-mouse immunoglobulin antibodies. These antibodies can be purchased from commercial suppliers or can be prepared by injecting purified mouse immunoglobins into rabbits (see p. 622). The rabbit antibodies can be purified, labeled with an easily detectable tag (pp. 288 and 319), and used to determine the presence of mouse monoclonal antibodies.

Alternatively, positives can be located by other reagents that will bind specifically to antibodies. Two proteins that may be used for these reactions are protein A and protein G (see p. 615). Both of these polypeptides are bacterial cell wall proteins that have high affinities for a site in the Fc portion of some antibodies. Protein A and protein G can be purified and labeled with an appropriate tag.

**TABLE 6.5**
**Methods for Screening Hybridoma Fusions**

| Method | Examples |
|---|---|
| | PVC wells |
| Antibody Capture | Nitrocellulose |
| | Whole cells |
| | Permeabilized cells |
| Antigen Capture | Ab/Ag in solution |
| | Ab/Ag on solid phase |
| Functional | Blocking |
| | Depletion |

| Advantages | Disadvantages |
| --- | --- |
| Easy<br>Rapid | Need pure or partially<br>    pure antigen<br>Doesn't discriminate<br>    between high- and low-affinity Ab |
| Relatively easy<br>Doesn't need pure Ag<br>Learn Ag locale | Ag prep can be tedious<br>Doesn't discriminate<br>    between high- and low-affinity Ab |
| Only detect<br>    high-affinity Ab | Unless you have<br>    pure labeled Ag,<br>    assay is tedious<br>    and slow |
| Rapid | Need pure labeled Ag<br>Setting up solid phase<br>    is tricky |
| Ab immediately<br>    useful | False positives<br>Potentially tedious |
| Ab immediately<br>    useful<br>Only detect high<br>    affinity Ab | Tedious<br>Ag must be<br>    limiting |

## ANTIBODY CAPTURE ON NITROCELLULOSE—DOT BLOTS*

If the antigen is a protein that is available in large amounts, dot blots are one of the assays of choice. The antigen is bound directly to the nitrocellulose sheet. Many assays are performed on a single sheet; therefore, the manipulations are simple.

Assays using polyvinylchloride multiwell plates in place of nitrocellulose sheets are good alternatives to dot blots. They are discussed on pp. 180 or 182.

1. A protein solution of at least 1 $\mu$g/ml is added to a nitrocellulose sheet at 0.1 ml/cm$^2$. Allow the protein to bind to the paper for 1 hr. Higher concentrations of proteins will increase the signal and make screening faster and easier. If the amount of protein is not limiting, concentrations of 10–50 $\mu$g/ml should be used. Nitrocellulose can bind approximately 100 $\mu$g of protein per cm$^2$.

2. Wash the nitrocellulose sheet three times in PBS.

3. Place the sheet in a solution of 3% BSA in PBS with 0.02% sodium azide for 2 hr to overnight. To store the sheet, wash twice in PBS and place at 4°C with 0.02% sodium azide. For long-term storage, shake off excessive moisture from the sheet, cover in plastic wrap, and store at −70°C.

4. Place the wet sheet on a piece of parafilm, and rule with a soft lead pencil in 3-mm squares. Cut off enough paper for the number of assays.

5. Apply 1 $\mu$l of the hybridoma tissue culture supernatant to each square. Incubate the nitrocellulose sheet on the parafilm at room temperature in a humid atmosphere for 30 min.

   Along with dilutions of normal mouse serum, include dilutions of the mouse serum from the last test bleed as controls. Dilutions of the test sera are essential to control correctly for the strength of the positive signals. Mouse sera will often contain numerous antibodies to different regions of the antigen and therefore will give a stronger signal than a monoclonal antibody. Therefore, dilutions need to be used to lower the signal. Good monoclonal antibodies will appear 10-fold less potent than good polyclonal sera.

6. Quickly wash the sheet three times with PBS, then wash two times for 5 min each with PBS.

*Adapted from Sharon et al. (1979); Glenney et al. (1982); Hawkes et al. (1982); Herbrink et al. (1982); Huet et al. (1982); Yen and Webster (1982); and reviewed in Hawkes (1986).

7. Add 50,000 cpm of $^{125}$I-labeled rabbit anti-mouse immunoglobulin per 3-mm square in 3% BSA/PBS with 0.02% sodium azide (about 2.0 ml/cm$^2$).

8. After 30–60 min of incubation with shaking at room temperature, wash extensively with PBS until counts in the wash buffer approach background levels.

9. Cover in plastic wrap and expose to X-ray film with a screen at −70°C.

## NOTES

i. The types of bonds that hold protein to nitrocellulose are not known. However, the binding is blocked by oils or other proteins. Wear gloves and use virgin nitrocellulose sheets. For short- or long-term storage, do not leave the paper in buffers containing other proteins. Slow exchange occurs and this will lower the strength of the signal. For long-term storage, freeze the paper in plastic wrap at −70°C.

ii. In all of the antibody capture assays, the method for detecting the antibody can be substituted with other techniques. Most techniques employ either $^{125}$I-labeled rabbit anti-mouse immunoglobulins or horseradish peroxidase-labeled rabbit anti-mouse immunoglobulins. For nitrocellulose tests using enzyme-linked reagents, only substrates that yield insoluble products should be used (p. 681). Both enzyme- and $^{125}$I-labeled reagents can be purchased from commercial suppliers or prepared in the lab (p. 319). If the detection method uses horseradish peroxidase rather than $^{125}$I-labeled rabbit anti-mouse immunoglobulin, sodium azide will block the development of the color.

iii. In all the assays in which proteins are bound to nitrocellulose or poly-vinylchloride, the remaining binding sites on the solid support must be blocked by adding extraneous proteins. Solutions that are commonly used include 3% BSA/PBS, 1% BSA/PBS, 0.5% casein/PBS, 20% skim dry milk/PBS, or 1% gelatin/PBS. There are conflicting reports in the literature as to the best reagents to use, and this may reflect differences in suppliers or slight differences in techniques. Test several and use whatever works best for you. If the blocking solutions are prepared in advance and stored, sodium azide should be added to 0.02% to prevent bacterial or fungal contamination.

## ANTIBODY CAPTURE IN POLYVINYLCHLORIDE WELLS— $^{125}I$ DETECTION*

Antibody capture assays in polyvinylchloride (PVC) plates are one of the most commonly used assays. Each well serves as a separate assay chamber, but because they are molded together the manipulations are simple. Two variations of these techniques are listed here, one for $^{125}I$ detection and one for enzyme-linked assays (p. 182). Both are easy and accurate. The radioimmune assay is easier to quantitate, but the enzyme assay is adequate for most purposes and avoids the problems of radioactive handling.

1. Prepare a solution of approximately 2 $\mu$g/ml of the antigen in 10 mM sodium phosphate (pH 7.0). Higher concentrations will speed the binding of antigen to the PVC, but the capacity of the plastic is only about 300 ng/cm$^2$, so the extra protein will not bind to the PVC. However, if higher concentrations are used, the protein solution can be saved and used again. Many workers prefer to use 50 mM carbonate buffer (pH 9.0) in place of the phosphate buffer.

2. Add 50 $\mu$l of antigen to each well. Incubate at room temperature for 2 hr or overnight at 4°C.

3. Remove the contents of the well. If the solution will not be saved, it can be removed by aspiration or by flicking the liquid into a suitable waste container.

4. Wash the plate twice with PBS. A 500-ml squirt bottle is convenient.

5. Fill the wells with 3% BSA/PBS with 0.02% sodium azide. Incubate for at least 2 hr at room temperature.

6. Wash the plate twice with PBS.

7. Add 50 $\mu$l of the tissue culture supernatant. Incubate 1 hr at room temperature.

8. Wash the plate twice with PBS.

9. Add 50 $\mu$l of 3% BSA/PBS containing 50,000 cpm of $^{125}I$-labeled rabbit anti-mouse immunoglobulin to each well. Incubate 1 hr at room temperature. ($^{125}I$-Labeled reagents can be purchased or prepared as described on p. 324.)

*Adapted from Catt and Tregear (1967); Salmon et al. (1969).

10. Wash the plate with PBS until there are no more counts in the wash buffer.

11. Either cut the wells apart and count in a gamma counter or expose the entire plate to film.

## NOTES

i. In all the assays in which proteins are bound to nitrocellulose or polyvinylchloride, the remaining binding sites on the solid support must be blocked by adding extraneous proteins. Solutions that are commonly used include 3% BSA/PBS, 1% BSA/PBS, 0.5% casein/PBS, 20% skim dry milk/PBS, or 1% gelatin/PBS. There are conflicting reports in the literature as to the best reagents to use, and this may reflect differences in suppliers or slight differences in techniques. Test several and use those that work best for you. If the blocking solutions are prepared in advance and stored, sodium azide should be added to 0.02% to prevent bacterial or fungal contamination.

ii. Antigen-coated plates can be prepared in advance and stored. After the blocking solution has been removed, store at $-70°C$.

## ANTIBODY CAPTURE IN POLYVINYLCHLORIDE WELLS—
## ENZYME-LINKED DETECTION*

As an alternative to using $^{125}$I-labeled reagents for antibody capture assays in polyvinylchloride (PVC) plates (p. 180), enzyme-linked assays can be employed. The assay is preformed identically to the $^{125}$I-labeled assay except the detection methods are changed.

1. Prepare a solution of approximately 2 $\mu$g/ml of the antigen in 10 mM sodium phosphate (pH 7.0). Higher concentrations will speed the binding of antigen to the PVC, but the capacity of the plastic is only about 300 ng/cm$^2$, so the extra protein will not bind to the PVC. However, if higher concentrations are used, the protein solution can be saved and used again. Many workers prefer to use 50 mM carbonate buffer (pH 9.0) in place of the phosphate buffer.

2. Add 50 $\mu$l of antigen to each well. Incubate at room temperature for 2 hr or overnight at 4°C.

3. Remove the contents of the well. If the solution will not be saved, it can be removed by aspiration or by flicking the liquid into a suitable waste container.

4. Wash the plate twice with PBS. A 500-ml squirt bottle is convenient.

5. Fill the wells with 3% BSA/PBS (no sodium azide). Incubate for at least 2 hr at room temperature.

6. Wash the plate twice with PBS.

7. Add 50 $\mu$l of the tissue culture supernatant. Incubate 1 hr at room temperature.

8. Wash the plate twice with PBS.

9. Add 50 $\mu$l of 3% BSA/PBS (without sodium azide) containing a dilution of horseradish peroxidase-labeled rabbit anti-mouse immunoglobulin antibody to each well. Incubate 1 hr at room temperature. (Horseradish peroxidase-labeled reagents can be purchased or prepared as described on p. 344. Most commercial reagents should be diluted 1 in 1000 to 1 in 5000. Try several dilutions in preliminary tests and choose the best.)

10. Wash the plate with PBS three times.

*Adapted from Catt and Tregear (1967); Salmon et al. (1969); Engvall and Perlmann (1972).

11. During the final washes prepare the TMB substrate solution. Dissolve 0.1 mg of 3',3',5',5'-tetramethylbenzidine (TMB) in 0.1 ml of dimethylsulfoxide. Add 9.9 ml of 0.1 M sodium acetate (pH 6.0). Filter through Whatman No. 1 or equivalent. Add hydrogen peroxide to a final concentration of 0.01%. This is sufficient for two 96-well plates. (Hydrogen peroxide is normally supplied as a 30% solution. After opening, it should be stored at 4°C, where it is stable for 1 month.)

12. After the final wash in PBS, add 50 $\mu$l of the substrate solution to each microtiter well.

13. Incubate for 10–30 min at room temperature. Positives appear pale blue.

14. Add 50 $\mu$l of stop solution, 1 M $H_2SO_4$, to every well. Positives now appear bright yellow. To quantitate the binding, read the results at 450 nm.

## NOTES

i. In all the assays in which proteins are bound to polyvinylchloride, the remaining binding sites on the solid support must be blocked by adding extraneous proteins. Solutions that are commonly used include 3% BSA/PBS, 1% BSA/PBS, 0.5% casein/PBS, 20% skim dry milk/PBS, or 1% gelatin/PBS. There are conflicting reports in the literature as to the best reagents to use, and this may reflect differences in suppliers or slight differences in techniques. Test several and use whatever works best for you.

ii. Antigen-coated plates can be prepared in advance and stored. After the blocking solution has been removed, store at −70°C.

iii. Do not include sodium azide in solutions when horseradish peroxidase is used for detection.

## ANTIBODY CAPTURE ON WHOLE CELLS—
## CELL-SURFACE BINDING*

If the subcellular location of an antigen is known, cell staining assays can be used to screen hybridoma tissue culture supernatants. Two assays are given here, one for cell-surface screening using detection with $^{125}$I-labeled reagents and a second for internal localization using enzyme-labeled reagents. Also, any of the techniques for cell staining in Chapter 10 can be adapted for screening.

1. Prepare a suspension of target cells at $1–2 \times 10^6$ cells/ml in 1% BSA/PBS with 0.1% sodium azide. If the cell pellets are particularly difficult to see, add a drop of a suspension of red blood cells or other colored particle that will not interfere with the assay.

2. Add 100 $\mu$l of the cell suspension to the wells of a V-bottomed polyvinylchloride (PVC) plate.

3. Centrifuge the PVC plate for 5 min at 400g. Many centrifuge manufacturers supply suitable plate carriers for these types of assays.

4. Carefully remove the supernatants by aspiration.

5. Resuspend the cell pellet without adding any buffer by tapping the plate or by using a microshaker. Dispersing the cell pellet is important for the rapid binding of antibody to the surface antigens.

6. Add 50 $\mu$l of tissue culture supernatant. Incubate at 4°C for 1 hr with periodic shaking.

7. Centrifuge the PVC plate for 5 min at 400g. Remove the tissue culture supernatant by aspiration.

8. Wash the cells twice by resuspending in 200 $\mu$l of ice cold 1% BSA/PBS with 0.1% sodium azide and centrifuging the plate.

9. Add 50 $\mu$l of ice-cold 1% BSA/PBS with 0.1% sodium azide containing 50,000 cpm of $^{125}$I-labeled Fab fragment of rabbit anti-mouse immunoglobulin antibodies. Incubate at 4°C for 90 min with periodic shaking. If the target cells do not have a receptor for the Fc portion of immunoglobulins, then $^{125}$I-labeled rabbit anti-mouse immunoglobulin antibody may be used in place of the Fab fragment.

*Goldstein et al. (1973); Jensenius and Williams (1974); Morris and Williams (1975); Welsh et al. (1975); Galfre et al. (1977); Williams (1977).

10. Centrifuge the PVC plate as before. Wash the cells three times with ice-cold 1% BSA/PBS with 0.1% sodium azide.

11. Cut the wells and count in a gamma counter or expose the plate to film.

## NOTES

i. Keeping the cells cold and in the presence of sodium azide throughout this procedure will slow the rate of capping and internalization of the surface antigens.

ii. In all of the antibody capture assays, the method for detecting the antibody can be substituted with other techniques. Most techniques employ either $^{125}$I-labeled rabbit anti-mouse immunoglobulin antibodies or enzyme-linked rabbit anti-mouse immunoglobulin antibodies. Both these reagents can be purchased from commercial suppliers or prepared in the lab (p. 321).

## ANTIBODY CAPTURE ON PERMEABILIZED CELLS— CELL STAINING*

One major advantage of using cell staining in hybridoma screens is that the assays give an extra level of information. Unlike other antibody capture assays that rely on the simple detection of antibody, cell staining also determines the localization. This extra information makes cell staining a good assay when using complex antigens.

Both fluorochrome- and enzyme-labeled reagents can be used to detect the presence of the antibodies (see Chapter 10 for more details), but if the levels are high enough to be detected using enzyme-labeled reagents, enzyme methods should be used. Enzyme-labeled assays can be scored by using the light microscope. Scoring assays using the fluorescent microscope will give more resolution, but long-term observation under this microscope is disorienting for most people.

1. Grow cells in standard tissue culture conditions on standard tissue culture plates with fetal calf serum. Some staining will be more pronounced on subconfluent cells, some on fully confluent cells.

2. Pour off the medium, and flood the plate with PBS.

3. Pour off the PBS. Flood the plate with freshly prepared 50:50 acetone/methanol mixture. Incubate at room temperature for 5 min.

4. Pour off the acetone/methanol, and allow to air dry (approximately 5 min).

5. Score the bottom of the plate with a marking pen to form a grid of small squares to identify the location of the hybridoma tissue culture supernatants. Fifty tests can easily be done on one 100-mm tissue culture dish.

6. Add 2–5 $\mu$l of tissue culture supernatant to the fixed and permeabilized cell sheet above the appropriate mark. Incubate for 1 hr at room temperature in a humid atmosphere.

7. Wash the entire plate by flooding with PBS. Pour off the PBS and repeat three times.

8. For a 100-mm dish, add 3 ml of rabbit anti-mouse immunoglobulin antibody-horseradish peroxidase solution (diluted 1/200 in 3% BSA/PBS) to the plate. Incubate for 1 hr at room temperature.

9. Pour off and wash three times with PBS.

*Lane and Lane (1981); see also Chapter 10 for historical and alternative methods.

10. During the last wash, dissolve 6 mg of 3,3′-diaminobenzidine in 9 ml of 50 mM Tris (pH 7.6). A small precipitate may form. Add 1 ml of 0.3% (wt/vol) $NiCl_2$ or $CoCl_2$. Filter through Whatman No. 1 filter paper (or equivalent). This is sufficient for one 100-mm plate.

11. Add 10 $\mu$l of 30% $H_2O_2$. ($H_2O_2$ is generally supplied as a 30% solution and should be stored at 4°C, where it will last about 1 month.)

12. Add 10 ml of substrate solution per 100-mm dish. Develop at room temperature with agitation until the spots are suitably dark. A typical incubation would be approximately 1–15 min.

13. Pour off the enzyme substrate and wash several times with water. Store in water with 0.02% sodium azide.

14. Look for black/brown spots in the marked squares and examine positives under the microscope.

## NOTE

i. Cells grown in calf serum often show high background staining due to binding of the second antibody to calf immunoglobulins.

## ■ Antigen Capture Assays

In an antigen capture assay, the detection method identifies the presence of the antigen. Often this is done by labeling the antigen directly. These assays require the monoclonal antibody to have a high affinity for antigen since the labeled antigen is normally added at very low concentration in free solution.

There are two types of antigen capture assays, and these assays differ by the order in which the steps are performed. In one variation, the antibodies in the tissue culture supernatant are bound to a solid phase first, and then the antigen is allowed to react with the antibody. In the second variation, the antibody–antigen complex is allowed to form prior to the binding of the antibody to a solid phase. In either case, once the antibody–antigen complexes are bound to the solid support, the unbound antigen is removed by washing, and positives are identified by detecting the antigen.

Detection of the antigen can be performed by a number of techniques. If the antigen is available in pure form, it can be labeled by radiolabeling, fluorescent tagging, or enzyme coupling. If the antigen itself is an enzyme, positives may be determined by the presence of the enzymatic activity. Any property that is unique to the antigen can be used to identify positives.

## ANTIGEN CAPTURE ON NITROCELLULOSE—
## REVERSE DOT BLOT*

Reverse dot blot assays are more complicated to use than many of the other screening assays, but they are particularly valuable if pure or partially pure antigen is available, although only in limited quantities. The monoclonal antibodies in the supernatants are "captured" on an anti-immunoglobulin antibody layer, previously bound to nitrocellulose or PVC (p. 192 for this variation), and then labeled antigen is added. Positives can be determined by the location of the antigen. Because the antigen can be labeled to high specific activity with either $^{125}$I or enzyme, very little antigen is used in the screening procedure. However, the assays are tricky to set up and demand careful use.

1. Prior to the assay, purify the immunoglobulin fraction from rabbit anti-mouse sera using one of the standard methods (p. 288). Purification on protein A beads is probably the easiest for the rabbit antibodies. Alternatively, purchase the purified antibodies from a commercial source.

2. Cut nitrocellulose paper to the size of a dot blot apparatus. Add rabbit anti-mouse immunoglobulin solution (approximately 200 $\mu$g of purified antibody/ml in PBS) to nitrocellulose paper. Use 10 ml/100 cm$^2$. Incubate for 60 min at room temperature.

3. Wash the paper three times with PBS, 5 min for each wash.

4. Incubate in 3% BSA/PBS with 0.02% sodium azide for 1 hr at room temperature.

5. Load into a 96-well dot blot apparatus.

6. Add 50 $\mu$l of hybridoma tissue culture supernatant to each well. Incubate for 1 hr at room temperature.

7. Draw the supernatant through the nitrocellulose paper by applying vacuum to the bottom chamber of the dot blot apparatus.

8. Wash the nitrocellulose paper and wells three times with 3% BSA/PBS.

9. Remove the paper from apparatus and incubate with $^{125}$I-labeled antigen (10 ml/96-well sheet, 50,000 cpm/well in 3% BSA/PBS) at room temperature for 1 hr with shaking. ($^{125}$I-Labeled antigens can be prepared as described on p. 324.)

---

*Adapted from Wide and Porath (1966); Catt and Tregear (1967); Engvall and Perlmann (1971); Van Weeman and Schuurs (1971).

10. Wash the paper with PBS until counts in the wash buffer approach background levels.

11. Cover in plastic wrap and expose to X-ray film at $-70°C$ with a screen.

## NOTES

i. The types of bonds that hold protein to nitrocellulose are not known. However, the binding is blocked by oils or other proteins. Wear gloves and use virgin nitrocellulose sheets. For short- or long-term storage, do not leave the paper in buffers containing other proteins. Slow exchange occurs and this will lower the strength of the signal. For long-term storage, freeze the paper in plastic wrap at $-70°C$.

ii. In all the assays in which proteins are bound to nitrocellulose, the remaining binding sites on the solid support must be blocked by adding extraneous proteins. Solutions that are commonly used include 3% BSA/PBS, 1% BSA/PBS, 0.5% casein/PBS, 20% skim dry milk/PBS, or 1% gelatin/PBS. There are conflicting reports in the literature as to the best reagents to use, and this may reflect differences in suppliers or slight differences in techniques. Test several and use whatever works best. If the blocking solutions are prepared in advance and stored, sodium azide should be added to 0.02% to prevent bacterial or fungal contamination.

## ANTIGEN CAPTURE IN POLYVINYLCHLORIDE WELLS*

This assay is similar to the nitrocellulose reverse dot blot assay described on p. 190. The major difference is the use of polyvinylchloride (PVC) plates in place of nitrocellulose. This makes the handling of the individual assays easier, because each well is used for a separate assay; with the nitrocellulose, this is achieved by using a dot blot apparatus. The major disadvantage of using the PVC is the lower binding capacity of the PVC wells.

1. Add 50 $\mu$l of affinity-purified rabbit anti-mouse immunoglobulin in PBS (20 $\mu$g/ml) to each well. Incubate for 2 hr at room temperature or overnight at 4°C. Many workers prefer to use 50 mM carbonate buffer (pH 9.0) in place of the PBS.

   Because of the low binding capacity of PVC, the signal will be stronger with affinity-purified rabbit anti-mouse immunoglobulin antibodies. These are the subset of antibodies in the anti-mouse immunoglobulin sera that bind to mouse antibodies. Affinity-purified antibodies can be prepared in the laboratory (p. 313) or can be purchased from commercial sources. When using concentrations above about 20 $\mu$g/ml, the solution should be saved for reuse.

2. Wash twice with PBS.

3. Add 200 $\mu$l of 3% BSA/PBS with 0.02% sodium azide to each well. Incubate for at least 2 hr at room temperature.

4. Wash twice with PBS. Add 50 $\mu$l of tissue culture supernatant to each well. Incubate for 2 hr at room temperature.

5. Wash three times with PBS. Add 50 $\mu$l of 3% BSA/PBS with 0.02% sodium azide containing 50,000 cpm of $^{125}$I-labeled antigen per well. Incubate for 1 hr at room temperature. (Procedures for labeling antigens are described on p. 324.)

6. Wash with PBS until counts in the wash buffer approach background levels.

7. Either cut the wells apart and count in a gamma counter or expose the plate to X-ray film at −70°C with a screen.

*Adapted from Wide and Porath (1966); Catt and Tregear (1967); Engvall and Perlmann (1971); Van Weemen and Schuurs (1971).

## NOTE

i. In all the assays in which proteins are bound to polyvinylchloride the remaining binding sites on the solid support must be blocked by adding extraneous proteins. Solutions that are commonly used include 3% BSA/PBS, 1% BSA/PBS, 0.5% casein/PBS, 20% skim dry milk/PBS, or 1% gelatin/PBS. There are conflicting reports in the literature as to the best reagents to use, and this may reflect differences in suppliers or slight differences in techniques. Test several and use whatever works best for you. If the blocking solutions are prepared in advance and stored, sodium azide should be added to 0.02% to prevent bacterial or fungal contamination.

## ANTIGEN CAPTURE IN SOLUTION—IMMUNOPRECIPITATION

Immunoprecipitation is seldom used for screening hybridoma fusions, because the assays are tedious and time consuming. However, because the antigen is normally detected after SDS-polyacrylamide electrophoresis, it is simple to discriminate potential positives from authentic ones. The added information gained about the molecular weight of an antigen makes these assays particularly useful when using complex antigens.

1. Prepare sufficient radiolabeled antigen for overnight detection of 100 samples. Samples can be labeled directly using $^{125}$I (p. 324) or prepared from radiolabeled extracts of cells (p. 429).

2. Either use tissue culture supernatants directly or pool in such a way that no more than 98 samples need to be handled (p. 215).

3. Label microfuge tubes and add samples of radiolabeled antigen to each. Add 50 µl of hybridoma tissue culture supernatants to each tube. Include one positive control (probably from the last test bleed) and one negative control (probably from a nonimmune mouse). Incubate for 1 hr on ice.

   If all classes of immunoglobulins are wanted, 30 min into the incubation add 0.5 µl of rabbit anti-mouse immunoglobulin serum. Normally it is easiest to add this as 10 µl of 1 in 20 dilution of the serum in PBS. Keep on ice.

4. Add 20 µl of a 10% suspension of prewashed SAC (p. 620). Incubate for 30 min on ice.

5. Spin for 1 min at 10,000g. Remove supernatant by aspiration. Resuspend the pellet in 750 µl of PBS.

6. Spin for 1 min at 10,000g and repeat wash.

7. Spin for 1 min at 10,000g and remove supernatant by aspiration. Add 50 µl of Laemmli sample buffer (p. 684). To make the resuspension easier, snap freeze by placing the tubes in a dry ice–ethanol bath. Resuspend and load onto a polyacrylamide gel. Handle as for normal electrophoresis (p. 636).

## Functional Assays

In functional assays, the antibodies in the hybridoma tissue culture supernatants are used either to block a reaction or as a molecular handle to deplete an essential component of a reaction mix. Any antibodies that are identified using these assays form an immediately useful set of reagents. However, the assays are difficult to perform and interpret, and therefore are seldom used.

# ■ PRODUCING HYBRIDOMAS

Although hybridoma production is the most discussed of the stages of monoclonal antibody preparation, most of the steps have been analyzed in enough detail that they are now routine. The ease with which this stage proceeds is dependent on how well the previous stages of immunization and development of the screen have gone. A strong immune response and the use of a good screening method will make the production of the hybridomas an easier task.

Once a good immune response has developed in an animal and an appropriate screening procedure has been developed, the construction of hybridomas is ready to begin. For the actual fusion, antibody-secreting cells are isolated from the appropriate lymphoid tissue, mixed with myeloma cells, centrifuged to generate good cell-to-cell contacts, and fused with polyethylene glycol (PEG). The fused cells are then removed from the PEG solution, diluted into selective medium, and plated in multiwell tissue culture dishes. Beginning approximately 1 week later, samples of the tissue culture supernatants are removed from wells that contain growing hybridomas and tested for the presence of the appropriate antibodies. Cells from positive wells are grown, single-cell cloned, and frozen. Finally, the monoclonal antibodies are collected and used.

Hybridoma production demands good tissue culture facilities and a worker with tissue culture experience. An experienced worker will be able to perform the entire fusion procedure from removal of the lymphoid tissue to the plating of the final fused cells in less than 2 hr. Little work is then required until the screening begins in about 1 week. This step is the most labor intensive of the entire project. Approximately 1 week is needed to complete the screening of the hybridoma wells, and if the fusion has been successful, another 2 weeks of tissue culture work will be needed until a suitable stage for a break has been reached. *Do not begin hybridoma production without the time needed for these operations.*

Although resultant hybridomas are relatively easy to grow, in the first stages following the fusion, they may be particularly fragile and need extra care. Because they are the final result of a long series of operations, and because they are produced as individual clones with no backup, the cells are quite valuable. At the early stages contaminated cultures cannot be recovered.

Chapter 7 (p. 245) contains descriptions of the techniques used for growing and maintaining hybridoma and myeloma cell lines as well as lists of appropriate growth media.

## ■ Preparation for Fusions

Prior to the time of fusion, several solutions must be prepared. In addition, unless you have purchased batches of fetal bovine serum and PEG that have been prescreened by manufacturers for their use in fusions, these solutions should be tested.

## SCREENING FOR GOOD BATCHES OF FETAL
## BOVINE SERUM

Only about one in five lots of fetal bovine serum (FBS) is particularly good at supporting hybridoma growth. The key constituents that distinguish good batches of serum from bad are not known. Order test batches from several suppliers or purchase prescreened serum directly from the distributor.

1. Test each batch of FBS against your present lot. Test the FBS with your most commonly used myeloma line as well as two hybridoma lines (if available). If possible, use one hybridoma that is easy to maintain and one that is more difficult.

2. For each lot of serum to test, prepare 30 ml of 10% FBS in medium (p. 247) without any further additives. Dispense 100 $\mu$l of the test medium in all the wells of a 96-well tissue culture dish using a multiwell pipettor. Prepare three trays per test and place the trays back in a $CO_2$ incubator to adjust the pH.

3. Wash the three test cell lines (one myeloma and two hybridomas) in medium without serum. The cells should be healthy and growing rapidly before the test. Resuspend the cells in medium without serum at a concentration of approximately $10^5$ cells/ml. You will need approximately 2.5 ml for these tests.

4. Add 100 $\mu$l of the cell suspensions to each of the eight wells in the left-hand row of the test plates—one cell line per plate; three plates/sample of FBS.

5. Using an eight-well multipipettor mix the contents of the left hand row. Then remove 100 $\mu$l from the first row and do serial 1 in 2 dilutions across the plate. Incubate at 37°C in a $CO_2$ incubator.

6. Check the wells under a microscope after 7, 10, and 14 days. The wells in the left-hand side of the plate should all grow. Depending on the ability of the individual batches of serum to support growth, you will see growth extending to the wells with smaller number of cells. This assay tests directly for the ability of different serum samples to replace the feeder effects of high-density hybridoma culture and mimics the problems of individual cells attempting to grow out from either single-cell cloning or hybrid fusions.

## NOTES

i. Good batches of serum should support growth of as few as 20 cells per well. Do not purchase batches that support less than 100.

ii. Any method that is used for single-cell cloning can be adapted to screen serum batches.

iii. Serum is stable when stored at −20°C for 1 year.

## PREPARING OPI

OPI is a solution of oxaloacetate, pyruvate, and insulin that helps support the growth of hybridoma and myeloma cells at low densities. It is not required for high-density culture.

1. To prepare 100 ml of 100× OPI, dissolve 1.5 grams of oxaloacetate, 500 mg of sodium pyruvate in 100 ml of $H_2O$ suitable for tissue culture work.

2. Add 2000 IU (international units) of bovine insulin.

3. Filter-sterilize.

4. Dispense in sterile tubes in 2.0-ml aliquots. Freeze at −20°C.

   OPI is stable at −20°C for 6 months to 1 yr.

## PREPARING POLYETHYLENE GLYCOL

1. Melt PEG 1500 in a 50°C water bath. Place a small glass vial on a top-loading balance, and add melted PEG. Add either 0.5 gram or 0.3 gram of PEG, depending on which fusion method will be used (pp. 211 or 212). Most workers use PEG 1500 for fusions, but others use anything from PEG 1000 to PEG 6000 with good results.

2. Cap the vials and autoclave to sterilize.

   PEG is stable at room temperature for many years.

**NOTE**

i. PEG can also be weighed dry and then autoclaved.

## SCREENING FOR GOOD BATCHES OF POLYETHYLENE GLYCOL

Because fusions require so little polyethylene glycol (PEG), good batches will last a long time. Because differences in batches normally are small, and only the odd batch is unusable, most workers do not bother to test different lots of PEG. Bad batches contain trace amounts of toxic chemicals. Buy the highest grade of PEG that is available.

1. Add 100 $\mu$l of medium with 10% serum to each well of a 96-well microtiter dish (one plate per batch of PEG to be tested).

2. Wash myeloma cells by centrifugation at 400$g$ for 5 min (approximately $10^6$ cells per assay). Resuspend the cells in medium without serum and then respin.

3. While washing the myeloma cells, melt one vial (0.5 gram) of each of the PEG samples to be tested at 50°C (p. 201 for PEG preparation). Add 0.5 ml of medium without serum to each vial and place in a 37°C water bath.

4. Resuspend the cells in medium without serum and aliquot samples containing $10^6$ cells into fresh centrifuge tubes (one tube per batch of PEG). Spin at 800$g$ for 5 min. Carefully remove the supernatant from the cell pellet.

5. Resuspend each cell pellet with the 50% PEG solutions by pipetting. Include one control that is resuspended in medium without serum or PEG. Incubate at room temperature for 2 min. Add 10 ml of medium with 10% FBS. Take 100 $\mu$l of this suspension and dilute into a second 10 ml of medium with 10% FBS. Spin the final dilution at 400$g$ for 5 min.

6. Aspirate the supernatant and resuspend the myeloma cells in 1.0 ml of medium with serum. Transfer 100 $\mu$l of the cell suspension into each of the eight wells on the left-hand side of the 96-well tissue culture dish. Using an eight-well multipipettor do 1 in 2 serial dilutions across the plate. Return to the $CO_2$ incubator.

7. Check the plates at day 14. Good batches of PEG will only slightly inhibit growth and will resemble the no PEG controls. Other batches should be discarded.

## NOTE

i. The American Type Culture Collection (ATCC) and Boehringer Mannheim supply high-quality PEG that does not need to be tested prior to use.

## ■ Drug Selections

Hybridoma cell lines are selected by the addition of drugs that block the de novo synthesis of nucleotides (see p. 277 for details of the theory of drug selection). The most commonly used agents are aminopterin, methotrexate, and azaserine. All are effective agents to select against the growth of the myeloma fusion partner. When using aminopterin or methotrexate, de novo purine and pyrimidine synthesis are blocked, whereas azaserine blocks only purine biosynthesis. Consequently, aminopterin and methotrexate are supplemented with hypoxanthine and thymidine. Azaserine solutions are supplemented with hypoxanthine.

## PREPARING HAT SELECTION MEDIUM*

Hypoxanthine, aminopterin, and thymidine selection (HAT) medium is commonly prepared from two stock solutions, 100× HT and 100× A.

1. To prepare 100 ml of 100× HT, dissolve 136 mg of hypoxanthine and 38 mg of thymidine in 100 ml of $H_2O$ suitable for tissue culture. Heat gently (70°C) if they do not dissolve completely. The 100× stock is 10 mM hypoxanthine and 1.6 mM thymidine.

2. To prepare 100 ml of 100× A, add 1.76 mg of aminopterin to 100 ml of $H_2O$ suitable for tissue culture. Add 0.5 ml of 1 N NaOH to dissolve. Titrate with 1 N HCl to neutral pH, being sure not to overshoot to acid pH, as aminopterin is sensitive to acid pH. The concentration of the 100× stock is 0.04 mM aminopterin.

3. Filter-sterilize the two solutions independently.

4. Dispense 2.0-ml aliquots in sterile tubes. Store at −20°C.

   Both 100× HT and 100× A are stable at −20°C for 1 year.

*Littlefield (1964).

## PREPARING HMT SELECTION MEDIUM

HMT selection medium is commonly prepared from two stock solutions, 100× HT and 100× M.

1. To prepare 100 ml of 100× HT, dissolve 136 mg of hypoxanthine and 38 mg of thymidine in 100 ml of $H_2O$ suitable for tissue culture. Heat gently (70°C) if they do not dissolve completely. The 100× stock is 10 mM hypoxanthine and 1.6 mM thymidine.

2. To prepare 100 ml of 100× M, add 49 mg of methotrexate to 100 ml of $H_2O$ suitable for tissue culture. Add 0.5 ml of 1 N NaOH to dissolve. Titrate with 1 N HCl to neutral pH. The 100× stock solution is 1 mM methotrexate.

3. Sterilize the 100× HT and 100× M solutions by filtration.

4. Dispense 2.0-ml aliquots in sterile tubes. Store at −20°C.

   Both 100× HT and 100× M are stable at −20°C for 1 year.

## *PREPARING AH SELECTION MEDIUM**

1. To prepare 100 ml of 100× AH, add 0.136 gram of hypoxanthine in H$_2$O suitable for tissue culture. Heat to 70°C to dissolve, and then add 10 mg of azaserine. The 100× stock is 0.58 mM azaserine and 10 mM hypoxanthine. 100× H will be needed for growing hybridoma cells while removing the azaserine selection. 100× H is prepared as above, but without the addition of azaserine.

2. Filter sterilize.

3. Dispense into sterile tube in 2.0-ml aliquots. Store at −20°C.

   100× AH is stable when stored at −20°C for 1 year.

*Foung et al. (1982).

## Final Boost

Three to five days before the fusion, the immunized mouse is given a final boost. This boost should be done at least 3 weeks after the previous injection. This interval will allow most of the circulating antibodies to be cleared from the blood stream by the mouse. Serum titers in the mouse begin dropping about 14 days after an immunization. If the levels of circulating antibodies are high, they will bind to the antigen and lower the effective strength of the boost.

The final boost is used for two purposes: to induce a good, strong response and to synchronize the maturation of the response. If this synchronization occurs, a large number of antigen-specific lymphocytes will be present in the local lymphoid tissue about 3 or 4 days after the boost. This will allow an increase in the relative concentration of the appropriate B-lymphocyte fusion partners. Consequently, the final boost should be directed to the source of the cell collection. In most cases, the spleen is the best choice for lymphocyte isolation, and, therefore, the final boost should try to localize the response to the spleen. This is best achieved by an iv injection done concurrently with an ip injection. If your source of antigen is limited, a single iv injection should be used. Remember that the antigen solution should be compatible with an iv boost (no Freund's adjuvant, SDS concentration below 0.1%, urea below 1 M, etc., see p. 110). If the antigen cannot be injected directly into the blood stream, an ip injection should be used, and the fusion should be done 5 days after the final boost.

In some specialized cases, for example foot pad injections or preparing IgA monoclonal antibodies, a regional lymph node may be the preferred site of lymphocyte collection. In the two examples given, the B-cell partners would be prepared from the inguinal node or from the Peyers patches, respectively.

## Preparing the Parental Cells for Fusions

Prior to the fusion the myeloma cells that will serve as fusion partners must be removed from frozen stocks and grown. The technique below is designed to place the cells in the best conditions for the fusion, but any tissue culture techniques that keep them healthy and rapidly growing are suitable. On the day of the fusion, the antibody-secreting cells are isolated from the mouse. Both types of cells can be washed together just prior to the fusion (p. 210).

## PREPARING MYELOMA CELLS FOR FUSIONS

Myeloma cells should be thawed from liquid nitrogen stocks at least 6 days prior to the fusion. Longer times may be necessary if the viability of frozen stocks is poor. The myeloma cells should be checked routinely for mycoplasma contamination, particularly if your laboratory or your tissue culture facility has a history of mycoplasma problems (p. 265). Any cells that test positive for mycoplasma should be replaced. The myelomas should be growing rapidly and healthy before the fusion. One day before the fusion, split the cells into fresh medium supplemented with 10% serum at a final concentration of $5 \times 10^5$ cells/ml (see p. 255 for methods for counting cells). On the morning of the fusion, dilute 10 ml of the overnight culture with an equal volume of medium supplemented with 20% FBS and 2× OPI.

Chapter 7 discusses the growth and maintenance of hybridoma cells (p. 245).

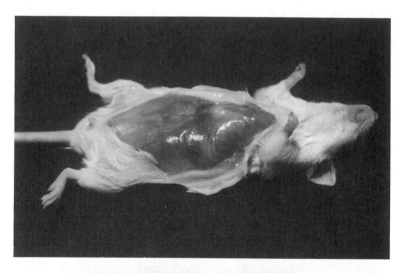

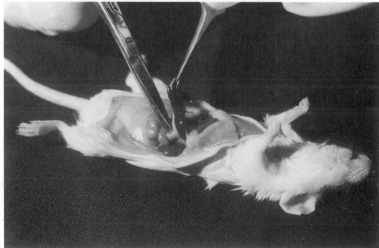

**FIGURE 6.9**
Splenectomy of a mouse.

## PREPARING SPLENOCYTES FOR FUSIONS

If possible, at the time of the fusion you should collect the blood from the immunized animal to use as polyclonal sera specific for your immunogen. Collecting sera from laboratory animals is discussed in Chapter 5.

1. Sacrifice the mouse. See your local authorities on animal handling for advice on the proper humane procedures. Aseptically remove the spleen from an immunized animal (Fig. 6.9) and place in a 100-mm tissue culture dish containing 10 ml of medium without serum (prewarmed to 37°C). Trim off and discard any contaminating tissue from the spleen.

2. Tease apart the spleen using 19-gauge needles on 1.0-ml syringes. Continue to tease until most of the cells have been released and the spleen has been torn into very fine parts. Disrupt any cell clumps by pipetting. Transfer the cells and medium into a sterile centrifuge tube. As you transfer the cells, leave behind the larger pieces of tissue.

3. Wash the tissue culture plate and tissue clumps with 10-ml of medium without serum (prewarmed to 37°C) and combine with the first 10 ml in the tube.

4. Allow the cell suspension to sit for approximately 2 min. This will allow the larger cell clumps to settle to the bottom of the tube. Carefully remove the supernatant from the sediment and transfer to a fresh centrifuge tube.

   A spleen from an immunized mouse contains approximately $5 \times 10^7$ to $2 \times 10^8$ lymphocytes.

## ▓ FUSIONS

Over the last 10 years a number of variations in fusion techniques have evolved. Most are based on the techniques of Galfre et al. (1977) or Gefter et al. (1977) and Kennett (1978). Examples of each are given below. Normally the fusion technique that is used does not make a major difference in the success of the fusion. You should use the method that is easiest and most convenient for you.

---

COMMENTS ▓ Hybridoma Production

- Fusions are easy. If the immunization and the development of the screen have gone well, doing multiple fusions may be very helpful.

- If the drug selection you have chosen to use depends on thymidine in the culture media, mycoplasma contamination will doom the fusion. Mycoplasma are extremely efficient in converting thymidine to thymine (see p. 265 for methods of detecting mycoplasma).

- If the antigen is particularly valuable, it may be prudent to perform a practice fusion prior to using your reagents.

## *FUSION BY STIRRING (50% PEG)**

1. Wash the splenocytes twice by centrifugation at 400g in medium without serum (prewarmed to 37°C). At the second wash also spin 20 ml of the myeloma cells in a separate centrifuge tube. The myeloma cells should also be washed in medium without serum; they get one wash versus the two for the splenocytes. During these washes, melt a vial with 0.5 gram of PEG (p. 201) in a 50°C water bath. Add 0.5 ml of medium without serum and transfer the vial to a 37°C water bath.

2. After the final washes resuspend the two cell pellets in medium without serum (prewarmed to 37°C) and combine the cell suspensions. Centrifuge these cells together at 800g for 5 min. Carefully remove the medium as completely as possible.

3. Remove the 50% PEG solution from its container with a Pasteur pipet and slowly add the PEG to the cell pellet while resuspending the cells by stirring with the end of the pipet. Add the PEG slowly over 1 min. Continue stirring for an additional minute. Fill a 10-ml pipet with 10 ml of medium without serum (prewarmed to 37°C). Add 1.0 ml to the cell suspension over the next minute, while continuing to stir with the end of the 10-ml pipet. Then add the remaining 9.0 ml over the next 2 min with stirring. Centrifuge the cells at 400g for 5 min.

4. Remove the supernatant and resuspend the cells in 10 ml of medium supplemented with 20% prescreened fetal bovine serum (prewarmed to 37°C), 1× OPI, and 1× AH. Transfer the cells to 200 ml of medium with 20% prescreened fetal bovine serum (prewarmed to 37°C), 1× OPI, and 1× AH.

5. Dispense 100 $\mu$l of cells into the wells of 20 96-well microtiter plates using a multiple pipettor (a 96-well pipettor is easiest). Place at 37°C in a $CO_2$ incubator.

6. Clones should be visible by microscopy at about day 4 and by eye starting at about day 7 or 8.

### NOTES

i. Other strategies can be employed for the number of wells into which the fusion is plated. These are discussed on p. 212. The one in the protocol above will yield approximately 2000 wells.

ii. Other drug selection methods can be used to select against the growth of the myeloma cells. These are described on p. 203. AH selection medium has only minor advantages over the other drug selections, but in general the hybridoma clones will appear more quickly in this medium. HAT selection is the most widely used.

*Galfre et al. (1977).

## *FUSION BY SPINNING (30% PEG)\**

1. Melt a vial of 0.3 gram of PEG in a 50°C water bath. Add 0.7 ml of medium without serum and transfer to a 37°C water bath.

2. Centrifuge the spleen cells from the immunized animal at 400g for 5 min. At the same time centrifuge 20 ml of the myeloma cells. Resuspend both cell pellets in 5 ml of medium without serum.

3. Combine the two cell suspensions and transfer to a 15-ml round-bottomed centrifuge tube. Centrifuge for 5 min at 400g. Carefully remove all medium.

4. Add 0.2 ml of PEG solution. Suspend the cells by lightly tapping the tube.

5. Centrifuge for 5 min at 400g. Add 5 ml of medium without serum to disperse the pellet. Flick the tube, if necessary, to resuspend the cells. Do not pipet the cells. Then add 5 ml of medium with 20% fetal bovine serum (prescreened to support hybridoma growth, p. 198).

6. Centrifuge for 5 min at 400g. Remove the supernatant and resuspend the cells in 10 ml of medium supplemented with 20% prescreened fetal bovine serum, 1× OPI, and 1× AH. Add the cells to 200 ml of medium supplemented with 20% prescreened fetal bovine serum, 1× OPI, and 1× AH.

*Adapted from Gefter et al. (1977); Kennett (1978).

## Plating Strategies

The difficulty of the screen will determine the number of clones or pools of clones that can be tested in 1 day. This factor will determine the number of wells into which the fusion is plated. A number of plating strategies have been used successfully to identify positive hybridoma clones. There are no rules to govern the correct choice, but the suggestions given below may serve as a general guide.

• If the screening procedure is easy (300 or more tests per day), dispense the fusion into a large number of wells in an attempt to have no more than one viable hybridoma per well (2000 or more wells). This is true whether the immunogen is very antigenic or not. The time needed in later stages of single-cell cloning will be reduced, and there is a greater chance that unique clones will not be lost by being overgrown by other cells prior to cloning.

• If the screening procedure is moderately difficult (100–200 tests per day), the

7. Dispense 100 $\mu$l of cells into the wells of 20 96-well microtiter plate using a multiple pipettor (a 96-well pipettor is easiest). Place at 37°C in a $CO_2$ incubator.

8. Clones should be visible by microscopy at about day 4 and by eye starting at about day 7 or 8.

i. Other strategies can be employed for the number of wells into which the fusion is plated. These are discussed on p. 212. The one in the protocol above will yield approximately 2000 wells.

ii. Other drug selection methods can be used to select against the growth of the myeloma cells. These are described on p. 203. AH selection medium has only minor advantages over the other drug selections, but in general the hybridoma cells will appear more quickly in this medium. HAT selection is the most widely used.

number of wells to be used is determined by the strength of the immune response. If a strong response has developed, plate the fusion in a large number of wells (2000 or more wells). Because positive wells will arise at high frequency, even if all the wells do not get screened, a good number of positive clones can be identified. If the antigen has not elicited a strong response, and further immunizations have not improved this response, then either plate the fusion into 500 wells and test each well individually or plate into 2000 wells and screen pools of tissue culture supernatants. Determine the pool size by the total number of assays that can be performed in 1 day.

• If the screening procedure is very difficult (50 or less tests per day), the number of wells to be used is determined by the strength of the immune response. If the immune response is good, plate the fusion into approximately 1000 wells, and test pools of these cells. If the immune response is not strong, plate the fusion into 100 wells, and test the wells individually.

## ■ Feeding Hybridomas

Two strategies are used in deciding whether to feed hybridomas prior to screening. Early protocols suggested the removal of medium and then addition of fresh medium. This approach has not proven to be of much advantage, but does lead to more work. The addition of fresh medium to the hybridoma cultures at about day 4 or 5 after the fusion improves the general health of the cultures and will keep them rapidly growing throughout the screening procedure. Some workers prefer not to feed the cultures, allowing any poorly growing clones to die and then to concentrate on the hardiest of the clones that grow up. Feeding is done by adding 100 $\mu$l of fresh medium supplemented with 20% fetal bovine serum, 1× OPI, and 1× AH.

## Pooling Strategies

Pooling of tissue culture supernatants is an effective method to reduce the total number of tests that must be done in 1 day. The major difficulty encountered when supernatants are screened as pools is that the positive well in the pool will normally need to be identified by immediate rescreening.

In some cases the pool size may need to be adjusted during the screening. Also, for some assays the pool size will be limited by the sensitivity of the test. For example, some assays are dependent on the concentration of the positive antibody and dilution of the antibody during pooling may reduce the signal below the detection level.

Three types of pooling strategies are in common use.

- **Simple pools**  The most widely used method of pooling is a simple combination of several tissue culture supernatants. The most important variable is the choice in the number of wells that will form a pool. Because the goal of any screening strategy is to eliminate approximately 90% of the cultures, gauge the size of the pool so that only about 1 in 10 of the pools is positive.

- **Matrices**  If the positive supernatants are likely to be rare, arrange the tissue culture supernatants in a two-dimensional matrix. Because pools of supernatants are prepared from each vertical column and each horizontal row, the correct wells can be identified by the location of intersecting positives.

- **Soft agar**  An effective method of preparing pools of supernatants is to plate the original fusion in a semi-solid medium that will allow the diffusion of antibodies, but will hold the hybridomas in place. This is most commonly done in soft agar (see p. 226 for the preparation of semi-solid media). After the soft agar has solidified, the cells are overlaid with regular tissue culture media. The antibodies can then diffuse into the liquid media and the media can be removed for testing. If a well is scored as positive, the clones within the agar are removed, grown up individually, and retested. One additional advantage of this technique is that the division rate of cells in soft agar is normally slower than in liquid media, so there may be extra time for screening. One disadvantage is that even the hardiest hybridomas do not have high plating efficiencies in soft agar, and therefore the number of clones that arise in these conditions will be reduced.

## ■ Screening

Wells containing hybridomas are ready to start screening approximately 7–14 days after fusions (about day 7 for AH selection and days 10–14 for others). Representative wells at different times after fusion are shown in Figure 6.10. For most assays, clones that are just visible by eye are about the right stage for screening.

Some examples of appropriate screening methods are discussed on pp. 175–195. Other methods based on any of the techniques in Chapters 10–14 can be used as well.

1. When the clone or clones in a well are ready to screen, mark the wells with some convenient numbering system. Aseptically remove about 50 $\mu$l of tissue culture supernatant. This can be done conveniently either by using a pipettor or by removing the approximate amount with a Pasteur pipet. The supernatants are removed from the top of the medium without disturbing the hybridomas on the bottom. After removing the supernatant refeed the well with fresh medium.

2. Transfer the supernatant to a suitable container. If all the supernatant will be used in the screening assay, the hybridoma tissue culture medium can be transferred directly to the assay. However, it is normally more convenient to transfer the supernatants to a microtiter tray. Mark the tray to correspond to the original well number.

3. For most antigens, the supernatants are ready for testing without any special treatments. In some instances lower backgrounds can be achieved by removing any cells in the supernatants by centrifugation before testing.

Any remaining supernatant should be stored at −20°C. This can be used for confirmations, further testing, or identifying individual wells from pooled samples.

### NOTE

i. Although splenocytes do not grow in standard tissue culture medium, they do not die immediately and will continue to secrete antibodies. This may lead to detecting false-positive wells in early screens. All positive cells should be rescreened prior to freezing or single-cell cloning.

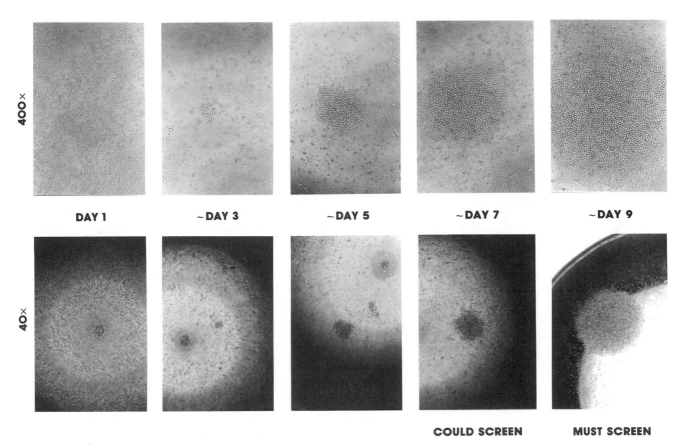

**FIGURE 6.10**
Representative wells at various days after fusion.

## Detection Methods

Antibodies or antigens can be labeled with a number of compounds (see Chapter 9). Most common detection methods depend on reagents that are tagged with a radioactive isotope, a fluorescent compound, or an enzyme.

- **Radioactive isotopes**   Any isotope that is easy to detect may be used to label antigens or antibodies. For example, proteins can be labeled in vitro by iodination or in vivo by growing cells in the presence of radioactive precursors. Radioactivity can be located by using $\beta$- or $\gamma$-counters or by normal X-ray film detection methods.

- **Fluorescence**   Fluorochromes may be bound to either antibodies or antigens in vitro. Measurement of the bound fluorochrome will require an appropriate source of radiation and a detection device.

- **Enzymes**   Antibodies and many antigens can be linked to functional enzymes in vitro. The antibody or antigen can then be detected by the addition of appropriate substrates for the enzyme. This is commonly achieved by using chromogenic substrates.

## ■ Expanding and Freezing Positive Clones

After a positive well has been identified, the cells are transferred from the culture in the 96-well plate to 0.5 ml of medium supplemented with 20% fetal bovine serum, 1× OPI, and 1× AH in a 24-well plate. After the 24-well culture becomes dense, it is transferred into 5.0 ml in a 60-mm dish and then to 10 ml in a 100-mm dish. Once the cells are transferred into the 60-mm dish, the drug selection can begin to be removed. This is done by first growing the cells for several passages in complete medium with hypoxanthine but lacking azaserine, or in complete medium with hypoxanthine and thymidine but lacking aminopterin or methotrexate. In either case, growing the cells with the base but without the drug allows all of the inhibitors to be diluted to a safe level before removing the bases.

At the 100-mm dish stage, the cells should be frozen. This is a convenient stage to collect 10 ml of supernatant, if any further testing of hybridomas needs to be done before concentrating on particular clones. However, if the correct clones have already been identified, the cells should be single-cell cloned as early as possible. This can be begun as early as at the 60-mm dish stage. Techniques for the freezing and storage of hybridoma cell lines are described on p. 257.

Often the transfer of hybridomas from one size of culture dish to the next is a difficult step to maintain cell viability. Presumably this is caused by the dilution of the growth factors in the medium and may be caused in part by overgrowth in the previous stage. If these problems exist, try using feeder cultures at these stages (see pp. 220–221 for the preparation of feeder cultures). Also, adding a sample of the diluted culture back into the original well will serve as a good backup if any problems arise.

---

## COMMENTS ■ More Fusions?

After the screen has been completed, the decision on the appropriate next steps will depend on the number of positives that have been identified. If no positives are found, and the immunization yielded a strong response, the fusion should be repeated, but the choice of screening method should be reevaluated. If the immune response was weak, new approaches to the immunization should be tried. If only a few positive clones were identified, these should be tested as early as possible to determine whether they will perform adequately in the appropriate assays. If a comprehensive set of immunochemical reagents are needed, additional fusions are likely to be needed. If the fusion has been very successful (greater than 50 positives), it is seldom worthwhile and often practically impossible to carry and maintain all the clones. In these situations many of clones are likely to result from fusion of sibling antibody-secreting cells and therefore will not generate new antibody activities. Some sort of secondary screen should be considered to identify particularly valuable clones. This might be based on affinity or perhaps subclass of the resultant antibodies.

## Single-cell Cloning

After a positive tissue culture supernatant has been identified, the next step is to clone the antibody-producing cell. The original positive well will often contain more than one clone of hybridoma cells, and many hybrid cells have an unstable assortment of chromosomes. Both of these problems may lead to the desired cells being outgrown by cells that are not producing the antibody of interest. Single-cell cloning ensures that cells that produce the antibody of interest are truly monoclonal and that the secretion of this antibody can be stably maintained.

Isolating a stable clone of hybridoma cells that all secrete the correct antibody is the most time-consuming step in the production of hybridomas. Depending on the chances of the original positive being derived from a single cell, the easiest and quickest methods to prepare single-cell clones will differ. If the positive well contains multiple clones or if secretion of the antibody is highly unstable, the cloning should be done in two or more stages. In the first cloning, you should try to identify a positive well with only a few clones, and then try to isolate a single-cell clone from this stage. This often can be achieved by a combination of different cloning methods. For example, quick cloning by limiting dilution could be followed by cloning with a single cell pick.

Because hybridoma cells have a very low plating efficiency, single-cell cloning is normally done in the presence of feeder cells or conditioned medium. Good feeder cells should secrete the appropriate growth factors and should have properties that allow them to be selected against during the future growth of the hybridomas. Feeder cell cultures are normally prepared from splenocytes, macrophages, thymocytes, or fibroblasts.

> To ensure that a hybridoma is stable and single-cell cloned, continue repeating the cloning until every well tested is positive.

## PREPARING SPLENOCYTE FEEDER CELL CULTURES

Although splenocyte feeders can be used immediately, they are most effective when they are prepared approximately 1 day before the single-cell cloning. Because spleen cells do not grow in normal tissue culture conditions, they are lost during the subsequent expansion of the hybridoma cells. Use a female mouse of the same genetic background as your hybridoma.

1. Sacrifice the mouse. See your local authorities on animal handling for advice on the proper humane procedures. Remove the spleen aseptically from the mouse and place in a 100-mm tissue culture dish containing 5 ml of medium without serum (see p. 209). Trim off and discard any contaminating tissue from the spleen.

2. Tease apart the spleen using 19-gauge needles on 1.0-ml syringes. Continue to tease until most of the cells have been released and the spleen has been torn into very fine parts. Disrupt any cell clumps by pipetting. Transfer the cells and medium into a conical centrifuge tube leaving behind all of the larger pieces of tissue. Wash these clumps and the plate with an additional 5 ml of medium without serum and combine with the first 5 ml.

3. Allow the cell suspension to sit at room temperature for approximately 2 min. This will allow the larger cell clumps to settle to the bottom of the tube. Carefully remove the medium and cells avoiding the sediment, and transfer to 100 ml of medium with 10% fetal bovine serum prescreened to support hybridoma growth (1 spleen per 100 ml is about $10^8$ cells/100 ml or $10^6$/ml). Either use directly or prepare conditioned medium.

4. **Either:** To use directly for 96-well cloning (pp. 222–224), plate 50 $\mu$l of the spleen cell solution into each of the wells of a 96-well tissue culture dish. Allow to grow for 24 hr at 37°C.

   **Or:** To use directly for soft-agar cloning (p. 226), the medium with the feeder cells is used to dilute the hybridoma cell suspension prior to mixing with the soft-agar.

   **Or:** To prepare conditioned medium, transfer the splenocyte cell suspension to several tissue culture dishes. Place at 37°C in a $CO_2$ incubator for 3 days. Collect the cell suspension and remove the cells by centrifugation at 400$g$ for 10 min. Filter sterilize and dispense in convenient sizes. Freeze at −70°C. Use the conditioned medium mixed 1:1 with medium supplemented with 20% FBS and 2× OPI.

### NOTE

i. To avoid any possible problems with a particular spleen feeder culture, it may be best to combine several batches.

## PREPARING FIBROBLAST FEEDER CELL CULTURES

Certain fibroblast cultures secrete the necessary factors to allow the growth of hybridoma cells at low plating densities. Early studies used fibroblast cultures that had been treated with mitomycin C or lethal doses of irradiation. Both of these treatments made it impossible for the feeder cells to contaminate future cultures of the hybridomas. More recently, this has been shown not to be necessary for fibroblast cultures that adhere strongly to the plastic tissue culture surface. Other studies have compared the ability of different fibroblast cells to support single-cell cultures of hyridoma cells and have found that the human diploid cells MRC 5 are the most effective in this test. These cells are not an established cell line, and so will need to be replaced in the future by another source. The MRC 5 cells are currently available from several sources including the American Type Culture Collection.

1. The MRC 5 cells are grown and maintained in Ham's F12 medium supplemented with 10% fetal bovine serum. They should be used at passages below 40.

2. Trypsinize the cells and count. Prepare a solution of $2 \times 10^5$ cells/ml of medium with 10% fetal bovine serum prescreened to support hybridoma growth.

3. For cloning using 96-well tissue culture dishes (pp. 222–224), add 50 $\mu$l of the cell suspension to the wells. Allow the cultures to grow for 1 day at 37°C.

   **Or:** For soft-agar cloning using 60- or 100-mm tissue culture dishes (p. 226), add 10 ml of the cell suspension to a 100-mm dish or 3 ml to a 60-mm dish. Allow the cells to adhere to the plastic overnight at 37°C. Remove the medium and add the soft agar hybridoma cell suspension to the plate.

## SINGLE-CELL CLONING BY LIMITING DILUTION

Cloning hybridoma cells by limiting dilution is the easiest of the single-cell cloning techniques. Two approaches are given below, one rapid technique for generating cultures that are close to being single-cell cloned and one for single-cell cloning directly.

Even though every attempt is made to ensure that the cells are in a single-cell suspension prior to plating, there is no way to guarantee that the colonies do not arise from two cells that were stuck together. Therefore, limiting dilution cloning should be done at least twice to generate a clonal population.

### Limiting Dilution (Rapid)

1. Using a multiwell pipettor (8-, 12-, or 96-well), add 50 $\mu$l of medium with 20% FBS and 2× OPI to each well of a 96-well plate. The wells should already contain 50 $\mu$l of feeder cells (pp. 220 or 221) giving 100 $\mu$l total volume/well.

2. The hybridomas should be growing rapidly. Remove 100 $\mu$l of the hybridoma cell suspension using a pipetman and transfer to the top left-hand well. Mix by pipetting.

3. Do 1 in 2 doubling dilutions down the left-hand row of the plate (8 wells, 7 dilution steps). Discard tip.

4. Do 1 in 2 doubling dilutions across the plate using an 8-well multi-pipetter.

5. Clones should be visible by microscopy after a few days and normally will be ready to screen after 7–10 days. Score the wells by microscopy. There should be a line running on a 45° diagonal that contains approximately the same number of clones per well. If the cells are nearly cloned when you start, screen only wells with one or two clones. If not, screen a selection of wells with multiple clones as well as all those with only one clone.

6. Select the best wells and either grow up or repeat the cloning procedure directly.

### Limiting Dilution (Slow)

1. The hybridomas should be healthy and rapidly growing at the time of cloning. Prepare four dilution tubes with medium supplemented with 20% fetal bovine serum and 2× OPI for each cell to be cloned. Three tubes should have 2.7 ml and the fourth should have 3.0 ml.

2. Add 10 $\mu$l of the hybridoma cells to the tube containing the 3.0 ml of medium. Do 1 in 10 dilutions of the hybridomas by removing 0.3 ml and transferring into the 2.7-ml tubes.

3. Add 100 $\mu$l of each dilution into 24 of the wells of a 96-well tissue culture plate (24 wells/dilution; 4 dilutions/plate, i.e., one hybridoma/plate). The wells should already contain 50 $\mu$l of feeder cells (pp. 220 or 221), giving 150 $\mu$l total volume/well. If the cells from the highest dilution are plated first, then the pipet does not need to be changed during the plating.

   If many hybridomas are being cloned at the same time, it may be worthwhile to plate the dilutions by using a 10-ml or larger pipet. One drop from these pipets will deliver approximately 100 $\mu$l.

4. Clones will begin to appear in 4 days and should be ready to screen starting about days 7–10.

   Screens can be done from wells containing multiple clones as well as from wells containing only single clones.

## SINGLE-CELL CLONING BY PICKS*

Cloning hybridomas by picking a single cell from a growing culture is the only cloning method that ensures that clones arise from a single cell. During the cloning procedure, the cell is followed under the microscope to be certain that the clone comes from only one cell.

1. Add approximately 100 $\mu$l of medium with 20% FBS and 2$\times$ OPI to the wells of a 96-well plate (approximately 20 wells/hybrid). The wells should already contain 50 $\mu$l of feeder cells (pp. 220 or 221) giving 150 $\mu$l total volume/well.

2. At the time of the cloning all cells should be growing rapidly. Do serial 1 in 5 dilutions of the hybridoma cells in 60-mm dishes. Use about 0.3 ml into 1.2 ml; this will allow enough volume to cover the bottom of the plate, but not so deep as to make the pipetting difficult. Observe the cells under the microscope and choose a plate with well-separated cells.

3. Use a drawn out 50-$\mu$l capillary pipet connected to a mouth pipetting device with a 0.2-$\mu$m filter fitted in the line. Partially fill the pipet with complete medium from a separate plate without cells. While watching under the microscope, draw a single cell into the pipet. Move to an area of the plate without any cells and blow out the cell to make sure you have only one cell. Draw it up again and transfer to one of the wells with feeders. With practice, single-cell picks take about 1 min.

4. The clones should be ready to screen in 7–10 days.

*J. Wyke (pers. comm.).

## NOTE

i. Because this technique demands working under the microscope on the open bench, one might expect contamination to be common. However, the only portion of the tissue culture medium that is exposed to the open air for long is the dish that you are picking from and you only transfer a very small volume at one time. So the chances of contamination are low. Needless to say, this technique should only be done in an area without drafts.

## SINGLE-CELL CLONING BY GROWTH IN SOFT AGAR

Cloning of hybridoma cells in semisolid medium is one of the most commonly used methods for producing single-cell clones. The technique is easy, but, because it is performed in two stages, it does take longer than other methods. Not all cells will grow in soft agar, and there may be a bias on the type of colony that appears. However, most of the commonly used myeloma fusion partners have relatively good cloning efficiencies in soft agar, and consequently, so do most hybridomas.

Even though every attempt is made to ensure that the cells are in a single-cell suspension prior to plating, there is no way to guarantee that the colonies do not arise from two cells that were stuck together. Therefore, single-cell cloning in soft agar should be repeated at least twice before the cells are considered clonal.

1. Prior to cloning prepare 3% agarose (Seaprep 15/45, FML Corporation or equivalent) in $H_2O$ suitable for tissue culture. Sterilize by autoclaving. This is stable for 6 months to 1 year.

   Prepare double-strength medium, normally from powdered medium. Add 100 $\mu$g/ml of gentamicin, and sterilize by filtration. Store at 4°C. Stable for about 1 month at 4°C, but at that time if fresh glutamine is added to 2 mM the shelf life can be extended to 3 months.

2. Melt agarose in a boiling water bath or in a microwave oven and cool to 37°C.

3. To the 2× medium add fetal bovine serum to 20% and OPI to 2×. Warm to 37°C in a water bath.

4. Cells should be healthy and growing rapidly at the time of cloning. The cells should be as free of clumps as possible. Do 1 in 10 dilutions of hybridomas in 1× medium. If not using feeders, the 1× medium is prepared by diluting a sample of the complete 2× medium with sterile $H_2O$. If using feeders that grow in suspension, the medium used for these dilutions should be the cell suspension from the feeder cell preparation (p. 220). If using fibroblast feeders (p. 221), these cells should be plated on the tissue culture dishes to be used for the cloning 24 hr earlier, and the 1× medium should be prepared by diluting the 2× complete medium.

5. Add 150 $\mu$l of cells from the dilutions between $10^5$ and $10^2$ cells/ml to 60-mm tissue culture plates (2 plates/dilution). Do not bother to count cells. If you are uncertain about the exact concentration of cells, it is easier to do an extra dilution than to count the cells.

6. Mix the 3% agarose and the 2× medium 1 : 1. Add 4 ml to each plate, and mix by pipetting.

7. Place the plates at 4°C for 45 min and then transfer to 37°C in a $CO_2$ incubator.

8. Macroscopic clones will appear beginning about day 10. Pick clones from the highest dilution that shows growth. Remove a plug of agarose containing the colony with a sterile Pasteur pipet. Transfer the plug to 1 ml of medium in a 24-well plate. Disperse the clone by pipetting.

9. Supernatants from these wells will normally be ready for screening 48–72 hr later.

## NOTE

i. As an alternative, the cells may be grown in the dilution tubes themselves (Civin and Banquerigo 1983). Add 2 ml of the 1.5% agarose/medium solution to each tube and grow as described above.

## ■ Unstable Lines

If hybridomas continue to produce less than 100% positive wells, even after four or more single-cell cloning steps, the lines probably have an unstable assortment of chromosomes. If the antibodies produced by these cells are particularly valuable, extra work to save these lines may be necessary. Two strategies are used. In the first and most straightforward, the single-cell cloning is continued on a regular basis, trying to isolate a stable subclone. Perhaps surprisingly, this often works. The screening assays should be adjusted to screen not only for the presence of the appropriate antibody, but also for the levels of antibody produced. Wells that contain a stable subclone of the original should produce higher levels of antibodies. If the stable variant is generated early in the proliferation within a well, the differences in antibody production between the well containing the variant and those that do not will be significant. At this stage many workers stop screening with an antigen-specific assay and only screen for the level of mouse antibody produced (see p. 560 for examples). After a stable line is generated, the specificity of the antibody should be reestablished.

A second strategy is to refuse the important line with a myeloma and allow the chromosomes to reassort from the beginning, hoping to isolate the stable variant from this source. To date, most re-fusions have been done by standard techniques and extensive screening. However, the introduction of a selectable drug selection marker into a suitable myeloma cell line should make selection against the parental myelomas easier. The hybridoma would carry a functional HPRT gene, while the myeloma would carry, for example, a neomycin gene. Selection for both genes should yield only successful secondary hybridomas.

## ■ Contamination

During the early stages of the fusion, contamination will mean the loss of the well or the fusion; however in later stages, important hybridomas can sometimes be saved.

## CONTAMINATION IN THE FUSION WELLS—
## A FEW WELLS ONLY

1. Contaminated wells can be identified by their unusual pH or turbidity. Confirm the presence of the contaminating organisms by observing under the microscope. Mark the wells.

2. Move to the tissue culture hood and carefully remove the lid. If the underside of the lid is damp, replace with a new lid. Dry the top and edges of the plate itself by aspiration before replacing. If there is contaminated medium on the lid, autoclave the whole plate without any further work.

3. Remove the medium from the contaminated well by aspiration. Try to avoid generating any aerosols. Add enough 10% bleach to the well to bring the level right to the rim. Allow it to sit for 2 min at room temperature.

4. Remove the bleach from the contaminated well by aspiration. Add enough ethanol to the well to bring the level right to the rim. Remove by aspiration and repeat.

5. Dry the well by aspiration.

## *CONTAMINATION IN THE FUSION WELLS—GROSS*

1. Autoclave the plates.

## *CONTAMINATION OF A CLONED LINE*

1. If the line has been frozen, it is easiest to go back to the most recent freeze down and thaw a fresh vial of the cells.

2. If the line has not been frozen, inject the cells into mice that have been primed for ascites production (p. 274). The animals must be of a compatible genetic background to your hybrids (e.g., BALB/c × BALB/c into BALB/c or BALB/c × C57Bl/B6 into BALB/c × C57B1/B6 $F_1$). If no mice have been primed with 0.5 ml of pristane the required 1 week in advance, inject 0.5 ml of Freund's adjuvant into the peritoneum. Wait 4 hr to 1 day and inject the hybridomas. Inject at least two mice for each contaminated culture.

3. When and if ascites develop, tap the fluid and transfer into a sterile centrifuge tube (see p. 274 for more information on ascites production).

4. Spin the ascites at 400*g* for 5 min at room temperature.

5. Remove the supernatant. Resuspend the cell pellet in 10 ml of medium supplemented with 10% fetal bovine serum and transfer to a tissue culture plate. The supernatant can be checked for production of the appropriate antibody. If positive, save for use.

6. Handle as for normal hybridomas, except keep the cells separate from the other cultures until there is little chance of the contamination reappearing.

The success rate may be as high as 80%.

### NOTE

i. Animals injected with infected cultures should be kept isolated from the main animal colony.

## ▨ Classing and Subclassing of Monoclonal Antibodies

Many techniques for using monoclonal antibodies require antibodies with specific properties. One set of these properties is unique to the individual antibody itself and includes such variables as specificity and affinity for the antigen. These properties all depend on differences in the antigen-combining domain of the antibody and can be assayed by comparing the properties of the monoclonal antibodies in tests that measure antigen binding activity.

A second set of important properties for monoclonal antibodies is determined by the structure of the remainder of the antibody, sequences encoded by the antibody common regions. These properties include the class or subclass of the heavy chain or the light chain. The different classes or subclasses will determine the affinity for important secondary reagents such as protein A (see p. 616). The type of heavy and light chain can be distinguished by simple immunochemical assays that measure the presence of the individual light- and heavy-chain polypeptides. This is normally achieved by raising antibodies specific for the different mouse heavy- and light-chain polypeptides (p. 622). The production of these antibodies is possible because the light- and heavy-chain polypeptides from different species are sufficiently different to allow them to be recognized as foreign antigens. Most often these anti-mouse immunoglobulin antibodies are raised in rabbits as polyclonal sera, and then the antibodies specific for a particular heavy or light chain are purified on immunoaffinity and immuno-depletion columns. Although these chain-specific rabbit anti-mouse immunoglobulin antibodies can be made in the laboratory, it is normally easier to purchase them from commercial sources. There are a large number of different assays used, and some of the more common are listed below.

## *DETERMINING THE CLASS AND SUBCLASS OF A MONOCLONAL ANTIBODY BY OUCHTERLONY DOUBLE-DIFFUSION ASSAYS\**

Originally, the Ouchterlony double-diffusion assays were the most common method for determining class and subclass of a monoclonal antibody. They have been largely superseded by other techniques, but they still are useful, particularly when only a few assays will be performed. In these assays, samples of tissue culture supernatants (often concentrated tenfold) are pipetted into a well in a bed of agar. Class- and subclass-specific antisera are placed in other wells at equal distance from the test antibody. The two groups of antibodies diffuse into the agar. As they meet, immune complexes form, yielding increasing larger complexes as more antibodies combine. When large multimeric complexes form, the immune complexes will precipitate, forming a line of proteins that is either visible to the naked eye or that can be stained to increase the sensitivity. The precipitated proteins form what is referred to as a precipitin line.

1. Prepare a 10-ml sample of tissue culture supernatant from a hybridoma. Grow the cells in medium supplemented with 10% FBS and allow the culture to overgrow and die.

2. Spin the tissue culture supernatant at 1000g for 10 min. Collect the supernatant. If the supernatant is not clear of all debris, either filter it through a 0.45-$\mu$m filter (sterility is not important) or spin at 7000g for 15 min.

3. Concentrate the supernatant 10-fold using an ultrafiltration manifold. This is most easily done with adaptors that are designed to concentrate in the centrifuge. Many of the ultrafiltration specialty companies now supply these devices; follow the manufacturers' instructions. Remove the tissue culture supernatant when the 10 ml sample has been reduced to 1 ml.

   Tissue culture supernatants may also be concentrated by ammonium sulfate precipitation (p. 298).

4. Prepare a 1.4% agarose solution in PBS with 5 mM EDTA. Melt the agarose in a boiling water bath or in a microwave. Cool to 45°C.

   Ouchterlony plates can also be purchased commercially.

5. On a level surface pipet 3 ml of the agarose solution onto the top of a 3 × 5-cm clean glass slide. The agarose should form a layer about 2 mM deep. The surface tension of the agarose should hold the agarose on the slide. Allow to harden at room temperature.

\*Ouchterlony (1949).

6. Using a 200-$\mu$l capillary pipet or a commercial apparatus, carefully core vertical small holes in the agarose in a pattern that looks like this:

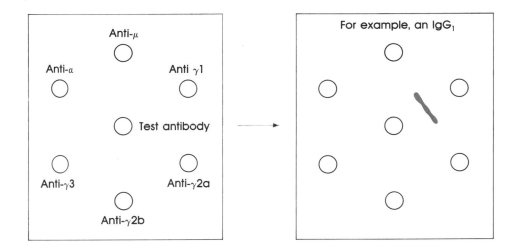

If the capillary pipet is attached to a pipetting device, a light suction while preparing the wells will allow the plugs to be withdrawn easily.

7. Add 5 $\mu$l of rabbit anti-mouse immunoglobulin sera specific for the various classes, subclasses, or light chains to each of the wells in the outer ring.

8. Add 5 $\mu$l of the concentrated tissue culture supernatant to the middle well.

9. Incubate in a humid atmosphere overnight at room temperature.

10. Score positive reactions by the appearance of a precipitin line between the wells with reactive antibodies.

## NOTE

i. The sensitivity of these assays can be increased by staining the bands with Coomassie brilliant blue. Cover the gel with wet filter paper and place in a 50°C oven. Incubate until dry. Wet the paper and remove from the gel. Wash for 30 min in several changes of PBS. Repeat the drying procedure. Stain with Coomassie for 15 min (p. 649). Destain in 7% acetic acid, 25% methanol.

## DETERMINING THE CLASS AND SUBCLASS OF MONOCLONAL ANTIBODIES USING ANTIBODY CAPTURE ON ANTIGEN-COATED PLATES

Any of the assays used to screen hybridoma fusions that detect antibodies with a secondary anti-mouse immunoglobulin antibody can be adapted to screen for class or subclass. For example, if the detection method used $^{125}$I-labeled rabbit anti-mouse immunoglobulin to locate antibodies bound to the antigen, then substituting anti-class or subclass-specific antibodies for the $^{125}$I-reagent will identify the type of heavy chains. An example of these types of reactions is given below using an antigen bound to 96-well PVC plates, but similar tests could be developed for any of the antibody capture assays.

1. Prepare a solution of approximately 2 $\mu$g/ml of the antigen in 10 mM sodium phosphate (pH 7.0). Higher concentrations will speed the binding of antigen to the PVC, but the capacity of the plastic is only about 300 ng/cm$^2$, so the extra protein will not bind to the PVC. However, if higher concentrations are used, the protein solution can be saved and used again. Many workers prefer to use 50 mM carbonate buffer (pH 9.0) in place of the phosphate buffer.

2. Add 50 $\mu$l of antigen to each well. Incubate at room temperature for 2 hr or overnight at 4°C.

3. Remove the contents of the well. If the solution will not be saved, it can be removed by aspiration or by flicking the liquid into a suitable waste container.

4. Wash the plate twice with PBS. A 500-ml squirt bottle is convenient.

5. Fill the wells with 3% BSA/PBS (no sodium azide). Incubate for at least 2 hr at room temperature.

6. Wash the plate twice with PBS.

7. Add 50 $\mu$l of each tissue culture supernatant to be tested to every well of a vertical row (8 wells/test). Incubate 1 hr at room temperature.

8. Wash the plate twice with PBS.

9. Add 50 $\mu$l of 3% BSA/PBS (without sodium azide) containing a dilution of horseradish peroxidase-labeled rabbit anti-mouse immunoglobulin class- or subclass-specific antibody to each well as shown below:

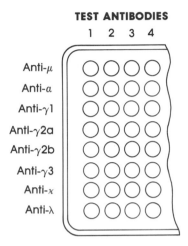

Incubate 1 hr at room temperature. (Horseradish peroxidase-labeled reagents can be purchased or prepared as described on p. 344. Most commercial reagents should be diluted 1 in 1000 to 1 in 5000. Try several dilutions in preliminary tests and choose the best.)

10. Wash the plate with PBS three times.

11. During the final washes prepare the TMB substrate solution. Dissolve 0.1 mg of 3′,3′,5′,5′-tetramethylbenzidine (TMB) in 0.1 ml of dimethylsulfoxide. Add 9.9 ml of 0.1 M sodium acetate (pH 6.0). Filter through Whatman No. 1 or equivalent. Add hydrogen peroxide to a final concentration of 0.01%. This is sufficient for two 96-well plates. (Hydrogen peroxide is normally supplied as a 30% solution. After opening, it should be stored at 4°C, where it is stable for 1 month.)

12. After the final wash in PBS, add 50 $\mu$l of the substrate solution to each microtiter well.

13. Incubate for 10–30 min at room temperature. Positives appear pale blue.

14. Add 50 $\mu$l of stop solution, 1 M $H_2SO_4$, to every well. Positives now appear bright yellow. To quantitate the binding, read the results at 450 nm.

## NOTE

i. Do not include sodium azide in solutions when horseradish peroxidase is used for detection.

## *DETERMINING THE CLASS AND SUBCLASS OF MONOCLONAL ANTIBODIES USING ANTIBODY CAPTURE ON ANTI-Ig ANTIBODIES*

One of the easiest methods for determining the class and subclass of a monoclonal antibody is to bind class- or subclass-specific antibodies to the wells of a polyvinylchloride (PVC) plate. The test monoclonal antibody is added to each well, but will bind only to wells coated with antibodies that are specific for its subclass or class. These bound antibodies are detected using a secondary antibody specific for all mouse antibodies.

1. Purify the antibodies from rabbit anti-mouse immunoglobulin class- or subclass-specific antibodies. Techniques for these purifications are discussed in Chapter 8. For most purposes, protein A beads are probably the easiest to use. (Rabbit anti-mouse immunoglobulin class- and subclass-specific sera can be purchased from several suppliers.)

2. After purification dilute the antibodies to 20 $\mu$g/ml in PBS. Add 50 $\mu$l to the wells of a PVC plate in the pattern below. Each monoclonal antibody being tested will need one row.

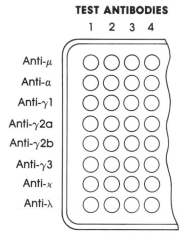

3. Incubate for 2 hr to overnight at room temperature in a humid atmostphere.

4. Remove the antibodies and save for future use. The antibodies can be reused approximately five times.

5. Fill the wells with 3% BSA/PBS with 0.02% sodium azide. Incubate for 2 hr or overnight at room temperature.

6. Wash three times with PBS. Add 50 $\mu$l of tissue culture supernatant from each hybridoma to the appropriate wells.

7. Incubate at room temperature for 2 hr in a humid atmosphere. Shake out the unbound antibody, and wash three times with PBS.

8. Add 50,000 cpm of $^{125}$I-labeled rabbit anti-mouse immunoglobulin antibody to each well (diluted in 3% BSA/PBS with 0.02% sodium azide).

9. Incubate for 2 hr at room temperature in a humid atmosphere. Discard the iodinated antibodies in an appropriate waste container.

10. Wash the wells three times with PBS. Cut the wells from the plate and count in a gamma-counter.

## NOTE

i. Other detection methods can be substituted for the iodinated antibodies. Common alternatives include enzyme-labeled reagents.

## ■ Selecting Class-switch Variants

During the normal development of a humoral response, the predominant class of antibodies that are produced changes, beginning primarily with IgMs and developing into IgGs. These changes and others like them occur by genetic rearrangements that move the coding region for the antigen binding site from just upstream of the IgM-specific region to the IgG region. These events are described in detail in Chapter 2 (p. 7). These rearrangements help the host animal tailor the immune response to the various types of infection. The different classes and subclasses of antibodies also have properties that make them more or less useful in various immunochemical techniques. These differences make the preparation of antibodies of certain classes or subclasses very valuable.

Recently, it has been shown that a process that appears similar to the natural class and subclass switching occurs in vitro, although at a very low frequency. Therefore, any population of hybridomas will have a small proportion of cells secreting antibodies with a different class or subclass of antibody. The antigen binding site will be identical in these antibodies. If these cells can be identified and cloned, then antibodies with the same antigen binding site but with different class or subclass properties can be isolated. These "shift variants" generally are useful in one of two cases, either switching from IgM to IgG or from $IgG_1$ to $IgG_{2a}$. Often these switches are used to produce antibodies that bind with higher affinity to protein A.

When trying to identify any class or subclass switching variants, it is important to remember that the rearrangements that occur will remove and destroy the intervening sequences, so only those heavy-chain constant regions that are found further downstream can be selected for. The order of the heavy-chain constant regions is $\mu$, $\delta$, $\gamma_3$, $\gamma_1$, $\gamma_{2b}$, $\gamma_{2a}$, $\epsilon$, and $\alpha$. Workers should also be certain they need these variants, as the assays are tedious. It may often be more advantageous to set up another fusion rather than isolate switch variants.

The most useful approach for most laboratories has been developed by Scharf and his colleagues (for a summary, see Spira et al. 1985). First, a suitable assay must be developed. Because of the large number of assays that must be performed, enzyme-linked assays are generally more useful. The assay for antibody capture on p. 180 can be easily adopted by changing the detection reagent to an IgG- or $IgG_{2a}$-specific rabbit anti-mouse immunoglobulin antibody. (Not all companies supply reagents that are sufficiently specific for these tests; one useful source is Southern Biotechnical Associates. All sources should be tested carefully before use.)

Hybridoma cells should be washed by centrifugation. Resuspend the cell pellet in medium supplemented with 20% fetal bovine serum at a density of $10^4$ cells/ml and dispense 100 $\mu$l into the wells of 10 96-well microtiter plates. This yields approximately 1000 cells per well with about 1000 wells. Therefore, about $10^6$ cells are being screened per assay. After the cells have grown, remove a sample of the tissue culture supernatant and screen for the presence of the IgG or IgG$_{2a}$ antibodies. Between one and five positive wells may be seen. Choose the strongest positive, and transfer these cells to fresh medium. Continue passaging the cells until they are numerous enough to clone again. In the second round, the cells should be plated at 100 cells per well. The procedure is repeated and then the cells are plated at 10 cells per well. In the last round the cells are single-cell cloned using one of the techniques described on p. 219.

## INTERSPECIES HYBRIDOMAS

Antibody-secreting cells isolated from one species but fused with myelomas from another species yield interspecies hybridomas. These types of fusions were common in the early years of hybridoma production. Often these hybrids would be formed by immunizing rats and fusing with mouse myeloma cells. This was done before good rat myeloma fusion partners were available. These fusions yield hybridomas that secrete rat antibodies, but the hybridoma cells cannot be grown conveniently as ascites tumors. Therefore, antibody production is almost entirely limited to tissue culture sources.

Although some important monoclonal antibodies have been produced using interspecies fusions, there seems little need for using these types of fusions today.

## ■ HUMAN HYBRIDOMAS

One of the most exciting areas for hybridoma research over the last 5 years has been the development of systems for the production of human hybridomas. Human monoclonal antibodies will be used extensively for clinical applications. Although this field has been marked by exciting publications announcing new breakthroughs, the actual progress in setting up the routine production of human hybridomas for laboratory use has been slow. For most research applications, producing human hybridomas still does not offer many, if any, advantages. The two most successful strategies that are used are standard fusions with human myeloma cells and the use of the Epstein-Barr virus (EBV) to transform antibody-secreting cells. One of the major problems in producing human hybridomas has been the lack of a suitable myeloma partner. Several of these lines have been isolated and are now in use.

The use of EBV-transformation to allow antibody-secreting cells to grow in standard tissue culture systems has solved some of the problems in human monoclonal antibody production. One unfortunate drawback of this approach is that the resultant transformants seldom secrete large amounts of antibodies. This has been overcome in some cases by fusing the EBV-transformed cell with a mouse myeloma cell line to allow the secretion of large amounts of antibodies. The combined use of EBV and secondary fusions points out two important aspects in hybridoma research. One is the use of other vectors to deliver important genetic information such as oncogenes. Second, if a particular hybrid does not possess all of the properties that are needed for a particular use, the line may be refused with other hybrids to achieve these properties.

There are several publications that describe progress in the isolation of human antibody-secreting cells, and these types of references should be checked for the details of producing human hybridomas.

## ■ FUTURE TRENDS

Few changes in the techniques used to produce hybridomas have been adopted since the original methods of Köhler and Milstein were reported. However, hybridoma construction is likely to change radically during the next 10 years. In several areas, preliminary work has already been reported that will form the basis for more widespread use of new techniques.

1. **In vitro immunizations**   Although the first in vitro immunization procedures were described in the early 1980s, they have not come into common use. The two major advantages of in vitro immunizations are the small amount of antigen that is required (as low as 1 ng) and the lack of cellular regulation on the developing immune response. Both of these factors make in vitro immunizations a potentially powerful technology. They have not been widely used to date, because so far they do not allow the development of high-affinity antibodies and because many of the antibodies that are produced are from the IgM class.

2. **Electrofusion**   PEG fusions routinely produce one viable hybridoma from $10^5$ starting cells, and this may be below the needed efficiency. One method that is gaining more widespread use is fusing cells by applying high-voltage electrical gradients across cell populations—short bursts fuse adjacent membranes and yield hybrid cells. This method has been applied successfully to hybridoma production, and the higher fusion efficiency allows production of more hybrid cells. In general, this has not been important for most fusions, because hybridoma production is normally limited by the screening method rather than by the frequency of hybridoma production. As more rapid screening procedures are developed, this fusion method will become more important. Also, as techniques are developed that allow the selection of the desired antibody-secreting cell prior to fusion, this and other high-efficiency methods will become increasingly valuable.

3. **Retroviral vectors**   Recombinant retroviral vectors hold the most promise for the efficient transformation of antibody-secreting cells. These vector systems can be engineered to deliver oncogenes into cells. However, the exact gene or combination of genes that will immortalize plasma cells but will not affect antibody secretion has not been determined. Also, because there will be little discrimination between the desired parental cells and undesired ones, this technology will be useful only when other methods of physically isolating the correct antibody-secreting cell are routinely used.

4. **Antigen-directed fusions**  A number of methods are being developed that, prior to fusion, physically couple myeloma cells with cells that are secreting the desired antibodies. These techniques take advantage of the antigen-combining site of surface antibodies found on some secreting cells. This combining site is used as a target for a modified antigen that will also bind to myeloma cells. After fusion, the frequency of appearance of hybridomas secreting the desired antibodies is much higher than in undirected fusions.

5. **Fusion partners**  More sophisticated methods of identifying cells that secrete the desired antibodies are being developed. Most of these methods use a fluorescence-activated cell sorter to identify and purify cells with surface immunoglobulins having the correct specificity. These cells then can either be fused with myeloma cells or transformed by other methods. These technologies will continue to improve, giving better and more refined choices for antibody selection prior to fusion or transformation. In addition, new myeloma fusion partners are constantly being described that have better properties for successful fusions.

6. **Defined medium**  Many of the growth factors that are necessary for the cultivation of hybridomas have been identified, and several defined medium have been developed. These culture conditions allow hybridomas to be grown in medium that do not contain other immunoglobulins (often bovine), and the low levels of proteins in these solutions make purification of antibodies from the tissue culture supernatants easier.

# GROWING HYBRIDOMAS

This chapter is divided into three sections: the first discusses the techniques used to grow and maintain myeloma and hybridoma cell lines, the second covers the production and collection of monoclonal antibodies, and the third deals with methods for drug selection used in hybridoma work.

## GROWING HYBRIDOMAS AND MYELOMAS

Because hybridomas are the result of long hours of work, and at one stage exist only as a single culture, they demand extreme care. Handling these cells requires experience with mammalian tissue culture, but little in the way of specialized equipment. A standard mammalian tissue culture laboratory will have all of the facilities needed to grow these cells.

## Tissue Culture

Myeloma and hybridoma cell lines can be grown in a broad range of standard tissue culture medium. The two most commonly used are Dulbecco's modified Eagle's (DME) and RPMI 1640 medium (Table 7.1). For routine use, these media are supplemented with fetal bovine serum (FBS) to a final concentration of 10%. FBS is used normally, because it contains very low concentrations of IgG molecules that may interfere in some assays or purifications. Other sera may be good supplements where contamination by foreign IgG is unimportant. In particular, both calf and horse sera are often used in place of FBS because of their lower cost. Other growth supplements are not required except when the cells are grown at very low densities (below approximately $10^3$ cells/ml). Tissue culture medium can be purchased either as powder or liquid from a number of commercial suppliers; follow the manufacturer's instructions for their preparation. In addition, most media are supplemented with antibiotics, most commonly penicillin and streptomycin, but note that both of these antibiotics have relatively short half-lifes at 37°C (see Table 7.2). Use penicillin and streptomycin at 100 U/ml and 100 $\mu$g/ml, respectively. Alternatively, use gentamicin at 50 $\mu$g/ml.

The ability to grow hybridomas and myelomas readily in standard tissue culture medium suggests that these cells have fairly simple growth requirements, which has been confirmed by a number of studies and has led to the development of defined media for these cells. The use of these serum-free media has some advantages over serum-containing solutions (see below), and as the costs of serum-free and serum-supplemented media are approaching equality, the use of serum-free medium is increasing. As might be expected, the growth requirements for hybridoma lines are similar to their parental myeloma lines. Table 7.3 lists the components of two defined media that are currently in use. These media or equivalents are available commercially.

When cells are grown at low densities, such as after fusions and during single-cell cloning, media are supplemented to help the cells

**TABLE 7.1**
**Standard Tissue Culture Media**[a]

| Component | DME | RPMI 1640 | Ham's F-12 |
|---|---|---|---|
| $CaCl_2$ | 200 | — | 33 |
| $Ca(NO_3)_2 \cdot 4H_2O$ | — | 100 | — |
| $CuSO_4 \cdot 6H_2O$ | — | — | 0.0023 |
| $Fe(NO_3)_3 \cdot 9H_2O$ | 0.1 | — | — |
| $FeSO_4$ | — | — | 0.46 |
| KCl | 400 | 400 | 220 |
| $MgSO_4 \cdot 7H_2O$ | 200 | 100 | 150 |
| NaCl | 6400 | 6000 | 7600 |
| $NaHCO_3$ | 3700 | 2000 | 1200 |
| $NaH_2PO_4 \cdot H_2O$ | 125 | — | — |
| $Na_2HPO_4$ | — | 800 | 140 |
| $ZnSO_4 \cdot H_2O$ | — | — | 0.86 |
| Glucose | 4500 | 2000 | 1800 |
| Glutathione | — | 1 | — |
| Hypoxanthine | — | — | 4.1 |
| Lipoic Acid | — | — | 0.21 |
| Methyl linoleate | — | — | 0.088 |
| Phenol Red, Na | 15 | 5 | 1.2 |
| Putrescine · 2HCl | — | — | 0.16 |
| Sodium pyruvate | 110 | — | 110 |
| Thymidine | — | — | 0.73 |
| L-Alanine | — | — | 8.9 |
| L-Arginine | — | 200 | — |
| L-Arginine · HCl | 84 | — | 210 |
| L-Asparagine · $H_2O$ | — | 57 | 15 |
| L-Aspartic acid | — | 20 | 13 |
| L-Cystine | 48 | 50 | — |
| L-Cysteine · HCl · $H_2O$ | — | — | 35 |
| L-Glutamic acid | — | 20 | 15 |
| L-Glutamine | 580 | 300 | 150 |
| Glycine | 30 | 10 | 7.5 |
| L-Histidine | — | 15 | — |
| L-Histidine · HCl · $H_2O$ | 42 | — | 21 |
| L-Hydroxyproline | — | 20 | — |
| L-Isoleucine | 105 | 50 | 3.9 |
| L-Leucine | 105 | 50 | 13 |
| L-Lysine · $H_2O$ | 150 | 40 | 37 |
| L-Methionine | 30 | 15 | 4.5 |
| L-Phenylalanine | 66 | 15 | 5.0 |
| L-Proline | — | 20 | 35 |
| L-Serine | 42 | 30 | 11 |
| L-Threonine | 95 | 20 | 12 |
| L-Tryptophan | 16 | 5 | 2.0 |
| L-Tyrosine | 72 | 20 | 5.4 |
| L-Valine | 94 | 20 | 12 |
| Biotin | — | 0.2 | 0.0073 |
| D-Calcium Pantothenate | 4 | 0.25 | 0.24 |
| Choline Chloride | 4 | 3 | 14 |
| Folic Acid | 4 | 1 | 1.3 |

| | | | |
|---|---|---|---|
| *i*-Inositol | 7 | 35 | 18 |
| Nicotinamide | 4 | 1 | 0.037 |
| *Para*-aminobenzoic Acid | — | 1 | — |
| Pyridoxal · HCl | 4 | — | — |
| Pyridoxine · HCl | — | 1 | 0.062 |
| Riboflavin | 0.4 | 0.2 | 0.038 |
| Thiamine · HCl | 4 | 1 | 0.34 |
| Vitamin $B_{12}$ | — | 0.005 | 1.4 |

[a]In mg/1000 ml.

**TABLE 7.2**
**Defined Media for Myelomas and Hybridomas**

| Component | KSLM medium[a] | SFH medium[b] |
|---|---|---|
| **Basal Medium** | RPMI 1640 : DME : F-12 2:1:1 | RPMI 1640 |
| L-Glutamine | add 300 $\mu$g/ml | add 300 $\mu$g/ml |
| Sodium Pyruvate | add 100 $\mu$g/ml | add 110 $\mu$g/ml |
| HEPES (pH 7.6) | 15 mM | 15 mM |
| Insulin (bovine) | 10 $\mu$g/ml | 10 $\mu$g/ml |
| Transferrin (human) | 10 $\mu$g/ml | 5 $\mu$g/ml |
| 2-Aminoethanol | 10 $\mu$M | 20 $\mu$M |
| 2-Mercaptoethanol | 10 $\mu$M | — |
| LDL (human) | 2 $\mu$g/ml | — |
| Oleic Acid with Bovine Serum Albumin Complex | 4 $\mu$g/ml | — |
| Linoleic Acid | — | 5 $\mu$g/ml |
| Bovine Serum Albumin | — | 1 mg/ml |
| Ascorbic Acid | — | 3 $\mu$g/ml |
| Hydrocortisone | — | 2 ng/ml |
| $CdSO_4 \cdot 8/3H_2O$ | — | 50 nM |
| $CoCl_2 \cdot 6H_2O$ | — | 10 nM |
| $CuSO_4 \cdot 2H_2O$ | — | 10 nM |
| $(NH_4)_6Mo_7O_{24} \cdot 4H_2O$ | — | 0.5 nM |
| $MnCl_2 \cdot 6H_2O$ | — | 0.5 nM |
| $NiSO_4 \cdot 6H_2O$ | — | 0.25 nM |
| $Na_2SeO_3$ | 1 nM | 40 nM |
| $Na_2SiO_3$ | — | 200 nM |
| $SnCl_2 \cdot 2H_2O$ | — | 0.25 nM |
| $NH_4VO_3$ | — | 2.5 nM |
| $ZnSO_4 \cdot 7H_2O$ | — | 1 $\mu$M |
| **Myelomas Supported** | NS-1, Sp2/0-Ag14, X63-Ag8.653 | FO, X63-Ag8.653, Sp2/0-Ag14 |
| **Hybridomas Supported** | NS-1 derivatives X63-Ag8.653 derivatives | FO derivatives Sp2/0-Ag14 derivatives |
| **Recommended for** | NS-1 and derivatives X63-Ag8.653 and derivatives | FO and derivatives Sp2/0-Ag14 and derivatives |

[a]Kawamoto et al. (1986).
[b]Kovář and Franěk (1986).

**TABLE 7.3**
**Antibiotics**

| Antibiotic | Effective against | Working concentration |
|---|---|---|
| Benzylpenicillin (Penicillin G) | Gram-positive bacteria | 100 units/ml (approximately 100 $\mu$g/ml) |
| Streptomycin | Gram-negative bacteria | 100 $\mu$g/ml |
| Kanamycin | Gram-negative bacteria | 100 $\mu$g/ml |
| Tetracycline | Broad-spectrum bacteria | 10–50 $\mu$g/ml |
| Gentamicin (Gentamycin) | Broad-spectrum bacteria, mycoplasma | 50 $\mu$g/ml |
| Lincomycin | Mycoplasma, Gram-positive bacteria | 50 $\mu$g/ml |
| Tylosin | Mycoplasma, Gram-positive bacteria | 10 $\mu$g/ml |
| Amphotericin B (Fungizone) | Fungi, yeast | 2.5 $\mu$g/ml |
| Nystatin (Mycostatin) | Fungi, yeast | 20 units/ml |

proliferate. Medium used for these steps should contain 20% FBS that has been prescreened to support good hybridoma growth (p. 198). These media can also be supplemented with other additives such as OPI (p. 200), or the cells can be grown in the presence of feeder cultures (pp. 220–221). Although these extra precautions are not necessary for normal maintenance of hybridomas or myelomas, their use may help poorly growing cells or ones that need a little pampering.

| Stock solution | Half-life in media (37°C) | Structural class | Mechanism of action |
|---|---|---|---|
| 10,000 units/ml | 2 days | Penicillins | Inhibits cell wall synthesis |
| 10 mg/ml | 4 days | Aminoglycosides | Inhibits translation (30S subunit) |
| 10 mg/ml | | Aminoglycosides | Inhibits translation (30S subunit) |
| 5 mg/ml | | Aminoglycosides | Inhibits translation (30S subunit) |
| 5 mg/ml | 15 days | Aminoglycosides | Inhibits translation (30S subunit?) |
| 5 mg/ml | | | Inhibits translation (50S subunit) |
| 5 mg/ml | | Macrolide | Inhibits translation (50S subunit) |
| 250 μg/ml | 4 days | Polyene | Alters membrane permeability |
| 2,000 units/ml | | Polyene | Alters membrane permeability |

## SUBCULTURING HYBRIDOMAS OR MYELOMAS

Most hybridomas and myelomas are easy to culture in vitro. Like other lymphoid cells, they do not attach well to plastic or glass and, therefore, can be transferred by pipetting. Hybridomas do not secrete any noticeably toxic by-products, so the medium can be replenished by diluting growing cultures with fresh medium. These two characteristics make transferring hybridomas and myelomas a simple task.

When the concentration of cells becomes about $10^6$ cells/ml, transfer the hybridomas or myelomas to fresh medium (see Fig. 7.1). With experience, the cells seldom need to be counted and the correct time for passing these cells can be determined by observing them under the microscope. Never allow the plasma membranes to become granulated. Healthy cells have smooth membranes.

2. Hybridomas can be diluted conveniently 1 : 10 or 1 : 20 with fresh medium. Larger dilutions may be fine, but this will depend on the individual cell line. Remember that these cells are somatic cell hybrids, and each line will have properties of its own.

Some hybridoma or myeloma cell lines will adhere lightly to the plastic surface. Even in these situations, the cells can be removed by washing the plastic with a strong stream of medium from a pipet. Check the flask or dish under the microscope after passage to ensure that the cells have been removed.

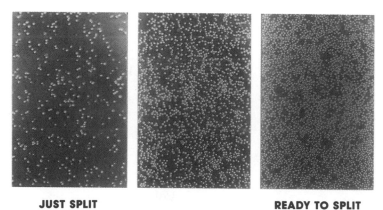

**JUST SPLIT**                    **READY TO SPLIT**

**FIGURE 7.1**
Hybridomas growing in tissue culture.

### NOTES

i. The average doubling time of hybridomas usually will depend on the myeloma fusion partner and normally will be between 12 and 24 hr. Therefore, most hybridomas will need to be subcultured every 2–4 days, depending on the dilution used in the previous passage.

ii. Hybridoma and myeloma cells can be grown in tissue culture dishes or T flasks. Dishes are cheaper and easier to handle than T flasks for subculturing cells. T flasks are useful for mailing cells (see below) or for growing cells that are suspected of contamination.

In addition, because myeloma and hybridoma cells do not adhere well to plastic, it is not necessary to culture these cells in dishes that are made from the more expensive tissue culture-grade plastics. Consequently, many workers grow hybridoma and myeloma cells in bacterial petri dishes.

iii. To move cells from one medium to another or from one type of serum to another, first make sure the hybridomas are growing rapidly. Dilute the cells 1 : 1 with the new medium. Continue to dilute the cell suspension with increasing concentrations of the new medium. Periodically check the viability (p. 256) and the doubling time. The last step is to check the cells for the continued production and yield of the appropriate antibodies.

## Sending and Receiving Hybridomas and Myelomas

• To ensure that hybridomas obtained from outside sources are stable and will continue to produce the needed antibodies, these cells should be single-cell cloned after they are received. These procedures are described on p. 219. Failure to check these cells for stable antibody production may lead to the loss of valuable cell lines.

• Assume that all incoming cells are contaminated with mycoplasma until completing the tests.

• Hybridoma or myeloma cell lines can be sent either as frozen stocks shipped on dry ice or as growing cells in liquid culture. Liquid cultures should be sent at several cell densities in sterile containers filled with medium. Some workers have reported that RPMI 1640 is better for long-term survival under these conditions. (Be sure to determine import and custom regulations for international exchange of cell lines.)

• There are a number of cell repositories that will supply hybridoma and myeloma cultures for a reasonable fee. These include the American Type Culture Collection and the NCI Repository. If a hybridoma that you produce becomes widely requested, you should consider depositing the cell line with an agency such as these.

iv. When large volumes of tissue culture supernatants are needed, it may be easiest to prepare these batches in roller bottles or spinner culture. Cultures for roller bottles are prepared as usual. Most hybridoma cell lines adapt well to spinner cultures, and this will eliminate the cost of plasticware. Start with approximately 50 ml of $5 \times 10^5$ cells from standard plate stocks. Dilute this with 50 ml of fresh medium. Do not use any of the specialized media for spinner cultures, such as Joklik's modified minimal essential medium; use your standard tissue culture medium. Transfer to a spinner apparatus and spin gently. Count the cells daily for the next few days until the cells have adapted. Once the cells have adapted to spinner culture, keep the concentration between $5 \times 10^4$ and $5 \times 10^5$ cells/ml.

Other methods for preparing large batches of antibodies from cells grown in tissue culture include hollow fiber growth chambers and airlift fermenters. Consult the manufacturer's instructions for use of these culture conditions.

v. Rat myelomas and hybridomas often adhere more tightly to plastic than mouse lines. If vigorous agitation does not remove the cells, they can be trypsinized.

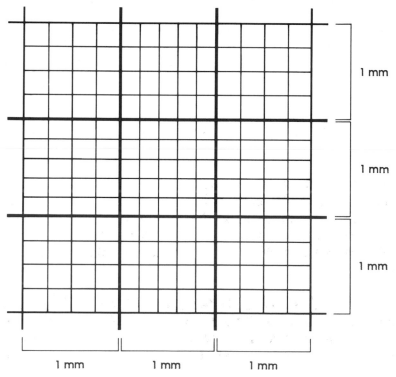

**FIGURE 7.2**
Counting hybridomas with a hemocytometer.

## COUNTING MYELOMA OR HYBRIDOMA CELLS

For most purposes the number of hybridoma or myeloma cells can be estimated simply by observing the cells under the microscope. When an exact cell count is needed, the number can be determined by using a hemocytometer (improved Neubauer counting chambers are the most commonly used). This is a simple device in which a special coverslip rests on supports that hold it 0.1 mm above the base of the slide. The slide is engraved with a series of lines that form $1 \times 1$-mm squares. By counting the number of cells within the 0.1-mm$^3$ chamber formed by the $1 \times 1$-mm square and the height of the coverslip, an accurate quantitation of cells per milliliter can be calculated.

1. Follow the manufacturer's directions for applying the coverslip to the slide. Some are designed to rest on the supports and others need to be secured by using the surface tension of water.

2. Using a sterile Pasteur pipet, transfer a drop of your cell suspension to the edge formed between the coverslip and the slide. In general, most hybridoma cultures will be at a suitable density for counting without dilution. Capillary action will draw the liquid into the chamber. The volume of the drop should be sufficient to draw evenly into the chamber and fill it, but not enough to overflow or to float the coverslip.

3. Examine under the microscope. Figure 7.2 shows what the field will look like under different magnifications. Count the number of cells in several of the $1 \times 1$-mm squares. The total number of cells that are counted will determine the accuracy. To obtain an accurate number, continue counting squares until the total number of cells counted exceeds 200. Divide the total number of cells by the number of squares counted and multiply by $10^4$ (1 ml of $H_2O$ = 1 cm$^3$; 1 cm$^3$ = 1000 mm$^3$; therefore 1 ml = 10,000 0.1 mm$^3$ chambers).

$$\frac{\text{Total number of cells}}{\text{Number of squares counted}} \times 10^4 = \text{number of cells/ml}$$

## VIABILITY CHECKS

To determine the percentage of viable cells within a population, the cell suspension is mixed with a vital dye and observed under the microscope. Vital dyes are excluded from living cells but stain dead cells. The most common dye used for these stains is trypan blue.

1. Prior to staining, prepare a 0.25% (wt/vol) solution of trypan blue in PBS. Either filter-sterilize and handle aseptically, or add sodium azide to 0.02%. (Trypan blue is stable at room temperature for several years.)

2. Mix a drop of your cell suspension with a drop of the trypan blue.

3. **Either:** If only a general impression of the viability of the cells is required, then put the drop on a slide and observe under the microscope.

   **Or:** If an exact percentage is required, transfer the mixture to a Neubauer counting chamber and count (p. 255).

## NOTE

i. If a precise 1 : 1 dilution is made of the cells in the trypan blue solution, then the total number of viable cells can be calculated directly.

$$\frac{\text{Total number of viable cells}}{\text{Number of squares counted}} \times 2 \times 10^4 = \text{number of viable cells/ml}$$

## Long-term Storage of Cell Lines

Hybridoma and myeloma cell lines can be stored by slowly freezing cells in an appropriate solution of nutrients and a cryoprotectant such as dimethylsulfoxide (DMSO). Cells are centrifuged at 4°C, resuspended in the freezing solution, and then transferred to an appropriate freezing vial. The vials are then slowly frozen and stored in liquid nitrogen ($LN_2$).

## FREEZING CELLS FOR LIQUID NITROGEN STORAGE

> **Caution**   $LN_2$ is at $-185°C$. Safety precautions should include gloves and a face mask when working in the $LN_2$ freezer.

1. Use only cells that are healthy and rapidly dividing. On the day before freezing, split the cells $1:10$ into fresh medium.

2. On the day of freezing, transfer the cells to a sterile, chilled centrifuge tube. Spin at $400g$ for 5 min at $4°C$.

3. Remove the supernatant as completely as possible. Gently re-suspend the cell pellet in 8% DMSO/92% fetal bovine serum ($4°C$). Concentrations of DMSO lower than 5% do not allow good recovery of most hybridoma cell lines (N. Williamson, pers. comm.).

   The final cell concentration should be approximately $5 \times 10^6$ to $5 \times 10^7$ cells/ml. If this concentration cannot be achieved conveniently (i.e., $< 10^6$/ml), make up the difference with spleen cells from a mouse of the same genetic background as your cells. The spleen cells will not grow when you thaw the vial, but will give the needed cell volume to produce a good freezedown.

4. Transfer 0.5 ml of the cell suspension to a chilled freezing vial (on ice). Vials that will be stored in $LN_2$ should contain a sealing ring of some compressible rubber that can withstand $-185°C$ (e.g., silicon). This will keep the $LN_2$ from entering the vial and avoid possible contamination. A good seal is also important because vials may explode during thawing if $LN_2$ has entered the vial.

5. Place the vials in a freezing rack at $-70°C$ overnight, or wrap the vials in several layers of paper towels and then place at $-70°C$. Normally it is best not to freeze too many vials at the same time. Prolonged exposure to DMSO at these concentrations is toxic to cells, and trying to handle more than about 10–20 vials at one time will lead to extended exposure to DMSO prior to freezing.

   Cells can be stored temporarily at $-70°C$ and will remain viable for several weeks or months. Long-term storage and high efficiency recovery require storage at $-185°C$ ($LN_2$).

6. Transfer the vials to $LN_2$ the day following the freezedown

Cells stored in LN₂ will remain viable for years. Once every year, a vial should be removed to check for viability. If the viability has dropped significantly, the cells should be expanded, checked for continued antibody production (subclone again, if necessary), and refrozen.

## NOTES

i. Several measures should be employed to ensure the safekeeping of a hybridoma line. At each freezing, prepare enough vials to handle the foreseeable needs. Normally, 10 vials are appropriate.

ii. Prepare at least two batches of frozen stocks for each line (most often on separate days). Variations in the viability of cells will change from one time to the next. To avoid these problems, prepare multiple batches.

iii. Approximately 1 week after each batch has been moved to the LN₂ freezer, remove one vial, thaw, and put in culture (p. 247). After 24 hr, check viability (p. 256). This will determine the percent survival of the cells. Compare this value with future tests (see above).

iv. To avoid the loss of a valuable line by freezer malfunction, "sabotage", or operator error, store samples of each line in at least two different freezers.

## *RECOVERING CELLS FROM LIQUID NITROGEN STORAGE*

**Caution**   $LN_2$ is at $-185°C$. Safety precautions should include gloves and a face mask when working in the $LN_2$ freezer.

1. Remove the frozen vial from the $LN_2$ storage and transfer to a container of dry ice. Move to the tissue culture facility and begin to thaw the cells in a 37°C water bath. Keep the lip of the freezing vial above the surface of the water to lessen the chances of contamination. Occasionally swirl the contents to hasten the thaw.

2. When the cells are almost thawed (i.e., only a little chunk of ice) move to the tissue culture hood. Wipe the outside of the vial with 70% ethanol and remove the top. Carefully remove the cell suspension using a sterile Pasteur pipet. Do not touch the sides of the tube and remove the cells from the bottom of the vial. Transfer the contents to a centrifuge tube containing 10 ml of medium with 10% FBS (room temperature).

3. Spin the cell suspension gently (200g) for 5 min.

4. Remove the medium and resuspend the cells with a second 10 ml of medium with 10% FBS. Transfer the cells to a tissue culture dish, and then to a $CO_2$ incubator. With good recoveries, you need to subculture the cells within 24 hr of plating.

## NOTE

i. If a hybridoma cell line has poor viability when removed from liquid nitrogen storage, supplement the culture medium with 20% FBS, $1\times$ OPI (p. 200) and plate over feeder cells (pp. 220 or 221).

## Contamination by Bacteria or Fungi

The long tissue culture manipulations involved in hybridoma production and maintenance sometimes can lead to contamination. In all cases, if frozen stocks are available, the best course is to autoclave the contaminated cultures and start anew. If uncontaminated stocks are not available, then it may be possible to rescue these cultures. Although mycoplasmas are bacteria, because of their unique biology, they are discussed separately below.

Bacterial or fungal contamination of a myeloma or hybridoma culture normally is first detected by turbidity in the medium or by the unusual pH of the medium. In both cases, this indicates that the growth of the contaminating organism is advanced; consequently, it is seldom worthwhile to try and save the culture. If frozen stocks are available, recover the cell line from the $LN_2$ storage.

If no frozen vials are available and the line cannot be replaced easily, two potential methods for rescuing the lines are given below. If the contamination has been detected early, then the use of drugs to halt the growth of the contaminating organism may succeed. However, if the contamination is gross, then passing the cell line through a mouse is the only method that may succeed. All contaminated cultures must be isolated from other cultures to avoid spreading the contamination. Use a separate incubator and grow the cultures in T flasks.

## RIDDING CELLS LINES OF CONTAMINATING MICROORGANISMS BY DRUG SELECTION

1. Observe the cells under the microscope to determine whether the contaminating microorganism is bacterial or fungal. Transfer approximately $10^5$ cells from the contaminated culture to a sterile centrifuge tube. Spin at 200$g$ for 5 min.

2. Remove the supernatant, being careful not to generate any aerosols. Transfer the supernatant to a suitable container for autoclaving. Resuspend the cell pellet in 10 ml of PBS and spin at 200$g$ for 5 min.

3. Carefully remove the PBS as before and repeat the PBS wash.

4. Resuspend the cells in 10 ml of fresh medium supplemented with 10% FBS and an appropriate drug (see Table 7.3). Normally it will be best to try several different antibiotics. Transfer the cells to a T flask, and close the top securely. Return the cells to the incubator. (Tissue culture medium is normally supplemented with penicillin and streptomycin, so the medium will now contain at least three drugs.)

5. Check the cells under the microscope after 1 day. If the contamination has become worse, autoclave the flask. Try passing the cells through a mouse (p. 264) to eliminate the contamination.

   If the contamination is the same or better, return the flask to the incubator and check the cells over the next 2 days. After several days, if the contamination has not progressed and the cells are growing, split the cells into fresh medium with the same drug selection. Continue to grow the cells in T flasks with tightened caps.

6. After several weeks of growth in these conditions, the cells should have outgrown the contaminating organism. At this stage the hybridoma or myeloma cell should be single-cell cloned in drug-containing medium. Refer to pp. 219–226 for these methods.

   The rapid limiting dilution method on p. 222 is probably the easiest method. Feeder cells should be included, and growing these cells on macrophage feeders has some extra advantages because these phagocytic cells may help clear the infection (see p. 268).

7. Select several clones that appear to arise from single cells and grow these to the 100-mm dish stage in medium supplemented with the same drug that has slowed the growth of the contaminating organism.

8. Remove half of each culture and spin at 400g for 10 min. Resuspend the cells in medium without the antibiotic. Transfer to a T flask and return to the incubator.

9. Grow these cells in T flasks for several passages while checking for reappearance of the contamination. If no growth appears, handle the cells as normal.

Success rate—maybe 20%.

## NOTES

i. An extra step that may help if the contaminating organism adheres to the hybridoma is to treat the cells at step 3 with trypsin for 5 min. Then wash extensively.

ii. Myelomas and hybridomas can be separated from most bacteria by Ficoll density centrifugation. Try this after several washes in PBS and trypsinization (step 3 and note i).

## RIDDING CELLS LINES OF CONTAMINATING MICROORGANISMS BY PASSAGE THROUGH MICE

1. Inject $10^7$ (or less) cells into female mice that have been injected ip about 1 week earlier with 0.5 ml of pristane or incomplete Freund's adjuvant. These types of injections are also used to prime mice for ascites production, and this may serve as a convenient source of appropriate hosts (p. 274). If no mice are available, inject mice with incomplete Freund's adjuvant and wait 4 hr to 1 day before injecting the hybridoma cells. The animals must be of the same genetic background as your cell line.

2. If an ascites develops, tap the fluid and transfer into a sterile centrifuge tube. (More information on ascites is found on p. 274.)

3. Spin the ascites at 400g for 5 min at room temperature.

4. Remove the supernatant. Resuspend the cell pellet in 10 ml of medium supplemented with 10% fetal bovine serum and transfer to a tissue culture plate. The supernatant can be checked for the presence of the antibody and used for further work if needed.

5. Handle as for normal hybridomas, except keep the cells separate from the other cultures until there is little chance of the contamination reappearing.

   Success rate—maybe 80%.

### NOTES

i. This technique has worked with the contents of a single microwell. Therefore, it can be used to rescue a relatively small culture.

ii. It may be necessary to single-cell clone immediately after collecting the ascites.

iii. If the cells form a solid tumor rather than the normal ascitic tumor, it is still possible to rescue the cell line using this technique. Cells quickly grow out of tumor fragments dispersed in tissue culture medium.

iv. When mice are injected with contaminated hybridoma cells, they should be isolated from the remainder of the mouse colony to lessen the chances of infecting the other mice.

## ■ Contamination by Mycoplasma

Mycoplasmas are an unusual group of bacteria. They are particularly small and do not have a cell wall. They are fastidious in their nutritional requirements, and in nature grow in close conjunction with higher eukaryotes. Mycoplasmas have been isolated from a diverse group of organisms ranging from insects to plants to man. Once introduced into mammalian tissue culture, they grow in close association with the plasma membranes of cells. If they contaminate tissue culture supplies, they pose a particularly difficult problem. Mycoplasmas are both difficult to detect and difficult to eradicate.

Unlike contamination with other bacteria or fungi, mycoplasmas normally do not cause any change in the pH or turbidity of the medium. This, plus the fact that they do not grow to high numbers in tissue culture medium, make mycoplasmas impossible to detect by visual observation. Therefore, mycoplasma testing needs to be performed routinely on all tissue culture lines.

Because mycoplasmas are a diverse group of organisms and are difficult to culture, a number of different strategies for detecting mycoplasma contamination have been developed. To date, no one test is suitable for detecting all of the possible mycoplasmas that may contaminate hybridoma or myeloma cultures. Therefore, it is sensible to consider employing several methods. The two most commonly used techniques are described below. These types of tests are available as commercial services, if desired.

Two new methods that have received attention recently are tests that rely on detecting mycoplasma by DNA hybridization or by employing a panel of anti-mycoplasma monoclonal antibodies. Both of these methods appear to be based on solid techniques and should speed the identification of mycoplasma. Both tests are supplied commercially, and workers should consult the manufacturers for instructions.

### Sources of Mycoplasma Contamination

- Because of their size and the flexibility provided by the lack of a cell wall, mycoplasma will pass through the standard 0.22-$\mu$m filters used to sterilize tissue culture media.

- Most mycoplasmas found in tissue culture are either human, bovine, equine, or porcine strains.

The most common sources of mycoplasma contamination include operator error (mouth pipetting or contaminated pipetting devices are the most common) and contaminated lots of serum.

## TESTING FOR MYCOPLASMA CONTAMINATION BY GROWTH ON MICROBIAL MEDIUM

1. Prior to testing, prepare mycoplasma growth medium.

   *Solution I*

   70 ml H$_2$O
   2 grams brain/heart infusion
   2.5 grams yeast extract
   Adjust to pH 7.8
      and autoclave to sterilize

   *Solution II*

   10 ml H$_2$O
   0.5 grams glucose
   0.5 grams L-arginine
   0.2 mg calf thymus
      DNA (sheared)
   0.2 phenol red
   Adjust to pH 7.8
      and filter-sterilize

   Aseptically combine solutions I and II. Add 20 ml of horse serum. Agar medium is prepared similarly, but 1.5 grams of agar is added to solution I before autoclaving. This solution is melted in a microwave oven or by autoclaving. Cool to 45°C, before combining the two solutions and pouring the plates.

2. Prepare two sets of inoculated medium from each cell line that will be tested. Inoculate 0.1 ml and 1.0 ml into 10-ml broth cultures and 0.1 ml on agar plates (six cultures/cell line, four broth and two plates).

3. Incubate one set of cultures at 37°C in a standard 5% CO$_2$ incubator and the other at 37°C in a 5% CO$_2$/95% N$_2$ atmosphere.

4. Check the broth and plates every week for 4 weeks. Growth in the broth cultures can first be detected either as turbidity or as changes in pH. Any potential positive from the broth cultures should be subcultured on agar plates.

   Mycoplasma colonies form a characteristic "fried egg" appearance on agar plates, and this is the diagnostic feature used to confirm mycoplasma contamination. The colonies are small and are most easily seen with an inverted microscope.

## TESTING FOR MYCOPLASMA CONTAMINATION
## BY HOECHST DYE (33258) STAINING

One of the quicker methods for testing for mycoplasma takes advantage of the DNA-intercalating dye Hoechst 33258. Fixed cells are stained with the dye, and contaminated cultures are detected by the bright, punctate cytoplasmic staining of the mycoplasma DNA.

1. Because hybridoma and myeloma cells grow in suspension, the first step in this procedure is to transfer the contamination to a fibroblastic indicator cell. This is achieved by cocultivating the hybridoma or myeloma to be tested with a monolayer of cells such as mouse 3T3 cells. Obviously, these cells must be mycoplasma-free. These cocultures should be grown for 1 week in antibiotic-free medium.

2. Hoechst stain 33258 is made up to 0.05 $\mu$g/ml in tissue culture medium without serum. Add sodium azide to 0.02%. Store in the dark at 4°C.

3. Remove the medium from the plate and wash the monolayer twice with PBS. Add 5 ml of 25% glacial acetic acid/75% methanol. Incubate for 10 min at room temperature.

4. Wash the cell monolayer three times with PBS and add 3 ml of the Hoechst dye solution. After the addition of the dye, incubate at room temperature in the dark for 10 min.

5. Wash the cell monolayer three times with distilled $H_2O$. Observe under a fluorescent microscope (p. 409). Hoechst dye 33258 excites at 360 nm and emits at 490–500 nm.

## *RIDDING CELLS OF MYCOPLASMA CONTAMINATION USING*
## *ANTIBIOTICS AND SINGLE-CELL CLONING*

Mycoplasmas are difficult to eradicate from tissue culture cells. The easiest approach to handle mycoplasma contamination is to autoclave the entire culture and to start with an uncontaminated source. If an uncontaminated line is not available, several methods can be attempted, but successful eradication is not guaranteed. Because mycoplasmas are composed of a wide range of different genera and species, methods that will succeed for one contaminated culture may not work with the next. Two methods are suggested here. The first is selection in antibiotic-containing medium combined with single-cell cloning over activated macrophage feeders. A second method is passage through a mouse (p. 264). Both methods have been used successfully.

1. Remove approximately $10^5$ contaminated cells. Spin the cells at 400g for 10 min.

2. Resuspend the cell pellet in 10 ml of fresh medium supplemented with 50 $\mu$g/ml of lincomycin and 10 $\mu$g/ml of tylosin. Transfer the cells to a fresh tissue culture dish.

3. Culture the cells in medium supplemented with lincomycin and tylosin for approximately 3 weeks.

4. One day before setting up the single-cell cloning, begin the preparation of peritoneal macrophage feeders. Inject 0.5 ml of 0.5% sodium thioglycolate in PBS ip into several mice (p. 160).

5. On the day of the single-cell cloning, inject 5.0 ml of PBS ip into the thioglycolate-primed mice. Massage the mouse's abdomen with your finger for approximately 15 sec and remove as much of the peritoneal fluid as possible with a 19-gauge needle attached to a 5-ml syringe. Combine with the fluid from the other mice in a sterile polypropylene centrifuge tube.

6. Wash the cells once by centrifugation at 400g for 10 min and resuspension in PBS. During the wash, count a sample of the cells (p. 255). After the last wash, prepare a cell suspension of $5 \times 10^5$ cells/ml in standard tissue culture medium supplemented with lincomycin and tylosin. Plate 100 $\mu$l of the cell suspension in each well of a 96-well tissue culture dish. Return the plates to the $CO_2$ incubator to allow the pH to readjust.

7. Single-cell clone the contaminated cells over the macrophage feeder cells (p. 219). The rapid limiting dilution method on p. 222 is one of the better methods for this.

8. Beginning about 7–10 days after the single-cell cloning, colonies will be visible. Pick several from the lowest dilutions that show any growth. Transfer these to 24-well plates and then expand to 60-mm and 100-mm tissue culture dishes. Keep the cells in the standard tissue culture medium supplemented with lincomycin and tylosin.

9. After the cells have been expanded to the 100-mm dish stage, split the cultures in half. One half gets transferred to medium without lincomycin and tylosin; the other stays in the drug selection. Grow the cells for 2 weeks without the drugs and retest for mycoplasma contamination (pp. 266 or 267). If the test proves negative, continue to handle as for normal hybridoma or myeloma cultures but check periodically for mycoplasma. If positive, repeat the procedure but use other drugs or try passing the cells through mice.

Success rate—as high as 70%.

## NOTES

i. Early work suggested that some mycoplasmas are particularly sensitive to heat, thus several workers heat-shock their cells (41°C) for 18 hr during the early stages of drug selection.

ii. Cells can be trypsinized at steps 1 and 5 to help remove mycoplasma from the cell membranes.

iii. Other workers prefer to grow contaminated cells in the presence of anti-mycoplasma antibodies. These sera can be purchased commercially and should be added to medium prepared with non-heat-inactivated sera, as the killing is primarily done by complement-mediated lysis. This treatment should not be used in place of drugs or single-cell cloning over macrophage feeders, but as an addition to these treatments. Grow the contaminated cultures in anti-mycoplasma sera during the 2 or 3 weeks before the single-cell cloning. During single-cell cloning the binding of antibodies to the mycoplasma will also increase phagocytosis by the macrophages.

## RIDDING CELLS OF MYCOPLASMA BY PASSAGE THROUGH MICE

Growing hybridoma cells as ascites tumors is one of the most effective methods of removing mycoplasma contamination. The mycoplasma are removed from the hybridoma cell surface by the immune system of the mouse. Mice must be of the same genetic background as the hybridomas.

This technique is identical to the method for ridding cells of bacterial or fungal infection described on p. 264. Cells recovered from the ascites should be rescreened using one of the techniques on pp. 266 or 267.

# PRODUCING AND STORING MONOCLONAL ANTIBODIES

Monoclonal antibodies normally are collected as either tissue culture supernatants or as ascitic fluid. Table 7.4 summarizes the properties and the uses of these different fluids.

**TABLE 7.4**
**Using Monoclonal Antibodies (Tissue Culture Supernatants vs. Ascites)**

|  | Tissue culture supernatant with 10% FBS | Ascites |
|---|---|---|
| **Properties** | | |
| Specific antibody concentration | 50 $\mu$g/ml | 0.9–9 mg/ml |
| Total antibody concentration | 1 mg/ml | 1–10 mg/ml |
| Total protein concentration | 6 mg/ml | 20 mg/ml |
| Contaminating proteins | Fetal bovine serum proteins | Mouse serum proteins |
| Cost/mg of antibody | High | Low |
| **Technique** | | |
| Cell staining | Usually use neat, but some antigens require more concentrated antibody | Must titer |
| Immunoprecipitation | Use neat | Must titer |
| Immunoblots | Use neat | Must titer |
| Immunoaffinity purification | Need large batches | Good |
| Immune assays | Must titer | Must titer |

## COLLECTING TISSUE CULTURE SUPERNATANTS

For most immunochemical methods, tissue culture supernatants will be the most useful source of monoclonal antibodies. The supernatants are not contaminated with high levels of other antibodies and the concentration is high enough for most assays.

1. When collecting tissue culture supernatants for antibodies, allow the individual cultures to grow until the hybridomas die. This will allow the collection of higher-titer supernatants. In general, antibodies are resistant to the proteases that are released from dying cells, so allowing the cells to die should not affect the quality of the antibodies.

2. Remove the debris by centrifugation. Spin the cultures at 1000g for 10 min.

3. Decant the supernatant from the cell pellet.

4. If there is debris remaining in the supernatant, spin at 7000 rpm for 10 min. Spins harder than 7000 rpm will cause any viable cells remaining in the medium to burst and release debris.

Yield is approximately 20–50 $\mu$g of antibody/ml of supernatant.

### NOTE

i.  If extraneous IgG molecules will alter any of the assays for which the supernatants are being prepared, use medium with fetal bovine serum-or use serum-free medium.

## Preparing Tissue Culture Supernatants with Higher Concentrations of Antibody

Tissue culture supernatants with higher concentrations of antibodies will be helpful or essential for several uses.

- As the cell concentration reaches saturation (approximately $5 \times 10^5$ to $10^6$ cells/ml), remove the cells by centrifugation. Asceptically remove and save the supernatant and then reinoculate the supernatant with a small sample of the cell pellet (approximately 5%). Allow the cells to grow to saturation and death as normal (p. 272).

- As the cell concentration reaches saturation, add additional glucose to 1% and HEPES (pH 7.2) to 25 mM. Grow and collect as normal.

- Short of purifying the antibodies from the supernatant, they can be concentrated conveniently either by ultrafiltration (for small volumes the ultrafiltration devices that run in the centrifuge are ideal) or by ammonium sulfate precipitation (50%, p. 298).

- If extraneous proteins present a problem either in your assays or in concentrating supernatants as above, transfer the hybridomas to serum-free medium (p. 252).

## COLLECTING ASCITIC FLUID*

Ascitic fluid (also called ascites) is an intraperitoneal fluid extracted from mice that have developed a peritoneal tumor. For antibody production, the tumor is induced by injecting hybridoma cells into the peritoneum, which serves as a growth chamber for the cells. The hybridoma cells grow to high densities and continue to secrete the antibody of interest, thus creating a high-titered solution of antibodies for collection. Antibody concentrations will typically be between 1 and 10 mg/ml.

1. Prime adult female mice (at least 6 weeks old) of the same genetic background as your hybridomas by injecting 0.5 ml of pristane (2,6,10,14-tetramethyldecanoic acid, Potter 1972) or incomplete Freund's adjuvant (originally described by Lieberman et al. 1960; see also Mueller et al. 1986; Gillette 1987) into the peritoneum. These solutions will act as irritants to the mice, which respond by secreting nutrients and recruiting monocyte and lymphoid cells into the area. This creates a good environment for the growth of the hybridoma cells.

2. After 7–14 days, inject $5 \times 10^5$ to $5 \times 10^6$ hybridoma cells ip. Prior to injection, the cells should be growing rapidly. Centrifuge the cells and wash once in PBS. Inject the cells in no more than 0.5 ml of PBS.

3. Ascitic fluid may begin to build up within 1–2 weeks following the injection of the cells. Tap the fluid when the mouse is noticeably large, but before the mouse has difficulty moving. Carefully withdraw as much fluid as possible with 18-gauge needle attached to a 5-ml syringe. (Sedating the mouse with $CO_2$ will make the collection of the ascites fluid easier. Consult your local officials for the best methods.)

4. Return the mouse to its cage. Many mice will produce a second or third batch of ascitic fluid. You can also bleed out the mouse and combine the blood with the ascitic fluid.

5. Incubate the fluid at 37°C for 1 hr. Transfer to 4°C overnight.

6. Spin the fluid at 3000g for 10 min. If there is an oil layer, remove this first and discard. Carefully remove the supernatant from the cell pellet. Spin again if necessary.

   A single mouse may yield as much as 10 ml of ascitic fluid per batch. Antibody concentrations may be as high as 10 mg/ml.

*Hoogenraad and Wraight (1986).

## NOTES

i. Between 2 and 10% of the antibodies from ascitic fluid will be from the mouse's current antibody repertoire and not from your hybridoma.

ii. Both pristane and incomplete Freund's adjuvant have been used successfully to produce ascites. Recent test results suggest that the adjuvant may be better. Try both if there is any question.

iii. Some workers have reported that retired breeders (i.e., old males) are particularly good for ascites production. Follow the protocol as suggested.

iv. Some cells will preferentially form solid tumors in the peritoneum rather than the soft-fluid tumors that yield good antibody titers. If these develop, switch the type of priming injection you are using (i.e., pristane to incomplete Freund's adjuvant or reverse) and inject the hybridoma cells into multiple sites ip. Even with these changes, this problem often cannot be overcome.

v. In cases where raising an ascitic tumor is difficult, try the injections in $F_1$ or $F_2$ mice rather than the isogenic strain. For example, a hybridoma prepared by fusing BALB/c parents could be injected into a BALB/c × CBA $F_1$ mouse. Because the $F_1$ (or $F_2$) mouse will carry one set of the BALB/c genes, the hybridoma will not be seen as a foreign cell. These types of inoculations often will allow the production of ascites where none were possible or increase the volume of ascites fluid. One source of mice for these types of injections that is reported to be very useful for large-volume ascites production is a backcross from BALB/c × DBA-1 $F_1$ females with BALB/c males (CD1 × C N1).

vi. Ascites tumors may be passaged serially from one mouse to a second primed mouse or set of mice. Transfer approximately $10^5$ to $5 \times 10^5$ cells directly from the ascites fluid into the second mouse. Such serial passage may be used to convert a line that predominately produces solid tumors to a line that grows as ascites. Wash the peritoneal cavity of a mouse bearing a solid tumor with 2 ml of PBS and transfer to a primed mouse.

## *STORING TISSUE CULTURE SUPERNATANTS AND ASCITIC FLUID*

The most common problem encountered in storing tissue culture supernatants or ascites is contamination of these solutions with bacteria or fungi. This can be prevented by the addition of sodium azide.

> **Caution**   Sodium azide is highly poisonous. It blocks the cytochrome electron transport system, and extreme care should be used.

1. Hybridoma tissue culture supernatants should be buffered by the addition of 1/20 volume of 1 M Tris (pH 8.0). Ascitic fluid is already well buffered.

2. Add sodium azide to 0.02%.

3. Dispense the antibodies in convenient volumes (normally 1 to 20 ml for tissue culture supernatants, 0.1 to 1.0 ml for ascites) and store at $-20°C$.

   Most antibodies are stable for years when stored at $-20°C$.

   Working solutions can be conveniently stored at 4°C where they are stable for at least 6 months.

### NOTES

i. Sodium azide will block a number of biological assays. Do not add sodium azide if the antibodies are to be used for in vivo tests. Instead, filter-sterilize the antibodies and handle aseptically.

ii. Antibody solutions should not be repeatedly frozen and thawed, as this can lead to aggregation of the immunoglobulins.

iii. Long-term storage at 4°C may produce a precipitate of lipids and/or cryoproteins. These can be removed by centrifugation.

iv. Ascitic fluid may also contain proteases. If this becomes a problem with a batch of ascites, either add protease inhibitors (p. 677) or purify the antibodies (p. 288).

## ■ DRUG SELECTION

Cells have two pathways for the synthesis of nucleotides, the de novo and salvage pathways. Since cells in tissue culture can survive by using either of these pathways, mutations in the enzymes responsible for nucleotide synthesis have become common and readily manipulated targets for mutagenesis of mammalian cells. The most common target is the enzyme hypoxanthine-guanine phosphoribosyl transferase (HPRT). HPRT performs one of the essential steps in the salvage pathway, catalyzing the condensation of phosphoribosyl pyrophosphate (PRPP) and a purine base to form a purine monophosphate nucleotide.

## SELECTING FOR HPRT MUTANTS WITH 8-AZAGUANINE

Mutations in the HPRT gene can be selected by growing cells in the presence of purine analogs such as 8-azaguanine (8-AG). HPRT will recognize 8-AG as a substrate and convert it to the monophosphate nucleotide. The 8-AG-containing nucleotide is then processed further and incorporated into DNA and RNA, where it is toxic. Therefore, cells with a functional HPRT enzyme grown in the presence of 8-AG will die. However, because the HPRT enzyme is part of a nonessential pathway (the de novo pathway is capable of supporting good cell growth in tissue culture), cells harboring a mutant HPRT gene can continue to grow. Therefore, selection with 8-AG will kill cells with a wild-type HPRT, but will not affect cells with a mutant HPRT.

The HPRT gene is found on the X chromosome, so the normal rate of mutagenesis in mammalian cells is sufficiently high to produce one cell bearing a nonfunctional HPRT in approximately $10^7$ cells. Therefore, no mutagenesis is necessary to select for the HPRT-negative phenotype, making selection relatively simple. $10^8$ or more cells are treated with 8-AG and surviving cells are tested for the loss of HPRT.

1. Grow approximately $10^8$ cells in standard tissue culture medium.

2. If the cells do not adhere to plastic, centrifuge the cells at 400g for 10 min. Resuspend the cell pellet in 100 ml of medium supplemented with 20 $\mu$g/ml of 8-AG and transfer to 10 $\times$ 100-mm dishes. If the cells adhere to plastic, grow the cells to 80% confluence and remove the medium by aspiration. Feed the cells with fresh medium supplemented with 20 $\mu$g/ml of 8-AG.

3. After 4 days feed the cells by removing one-half of the medium and replacing it with fresh medium containing 20 μg/ml of 8-AG. Continue to feed the cells every fourth day until colonies appear. Transfer the individual resistant clones to fresh medium with 20 μg/ml of 8-AG.

4. Test each clone for sensitivity to azaserine, aminopterin, or methotrexate as described on p. 280. Cells from clones that are sensitive to any of these drugs should be single-cell cloned prior to use (see p. 219).

## CHECKING CELL LINES FOR HPRT DEFICIENCY

Cells can be assayed for the presence of a nonfunctional HPRT enzyme by growing them in the presence of drugs that inhibit the de novo synthesis of nucleotides. In these conditions, cells are forced to rely on the salvage pathway for synthesis of nucleotides, and those with a mutant HPRT enzyme will die.

Several drugs can be used to block the de novo synthesis of nucleotides. The three that are commonly used in hybridoma studies are azaserine, aminopterin, and methotrexate. Azaserine (O-diazoacetyl-L-serine) is an analog of glutamine and will covalently bind to and inactivate two of the essential enzymes in the de novo synthesis of purines, glutamine phosphoribosyl amidotransferase, and phosphoribosyl glycinamindine synthase. Adding azaserine to culture medium will block the de novo synthesis of purines and therefore force cells to synthesize purine nucleotides through the salvage pathway. Therefore, a substrate for the purine salvage pathway must be included in the selection medium to allow cells to survive. For azaserine, the purine that is normally added is hypoxanthine. The preparation of azaserine/hypoxanthine (AH) selection medium is described on p. 206.

Methotrexate (4-amino-10-methylfolate) and aminopterin (4-aminofolate) are analogs of folate and compete for the enzyme dihydrofolate reductase. Competition prevents the production of tetrahydrofolate, a substrate eventually used in deoxythymidine synthesis, thus blocking pyrimidine synthesis. Methotrexate and aminopterin indirectly block purine synthesis by lowering the supply of the folate coenzyme. Because both purine and pyrimidine synthesis are inhibited, both need to be produced by the salvage pathway and precursors for these pathways must be included in the selection medium. The preparation of hypoxanthine/methotrexate/thymidine (HMT) and hypoxanthine/aminopterin/thymidine (HAT) selection medium is described on pp. 205 and 204, respectively.

Cells that have been selected for resistance to 8-AG (p. 278) should be checked for sensitivity to drugs that block the de novo synthesis of DNA. Also, all myeloma cell lines should be checked periodically for reversion of their drug selection markers. Any line that is not killed completely by drug selection should either be reselected or replaced with a new line.

1. Grow approximately $10^7$ cells in standard tissue culture medium.

2. Collect the cells by centrifugation, 400g for 10 min at room temperature.

3. Resuspend the cells in 100 ml of medium supplemented with 10% fetal bovine serum (FBS) and 1× AH, 1× HAT, or 1× HMT (p. 203). Plate the cells in 10 × 100-mm dishes and return to the $CO_2$ incubator at 37°C.

4. After several days the cells should begin to die. After 1 week the cells from the 10 plates should be combined by centrifugation at 400g for 10 min.

5. Resuspend the cell pellet in a total of 10 ml of a 50 : 50 mix of feeder cells and fresh medium. Production of feeder cells is described on pp. 220–221. The 50 : 50 medium mix should be supplemented with 10% FBS and 1× AH, 1× HAT, or 1× HMT as before. Plate in one 100-mm dish and return to the $CO_2$ incubator.

6. Check the cells periodically over the next 2 weeks. If any colonies appear, either discard the line or reselect with 8-AG.

# STORING AND PURIFYING ANTIBODIES

283

This chapter covers techniques that are used to store and purify antibodies. The first portion of this chapter summarizes the common methods of storing antibodies and the second discusses methods used to purify antibodies.

## ■ STORING ANTIBODIES

One of the practical advantages of antibody molecules having compact and stable protein domains (see p. 7 for further information) is that they are resistant to a broad range of mildly denaturing conditions. This makes long-term storage of antibodies relatively easy. Storage buffers seldom need to be supplemented with glycerol or other stabilizing compounds. The only problem commonly encountered in storing antibodies is contamination of these solutions with bacteria or fungi, which can be prevented by the addition of sodium azide to 0.02%. However, sodium azide will block a number of biological assays and will interfere with some coupling methods. For coupling reactions, the sodium azide can be removed by dialysis or gel filtration. If the antibodies are to be used for in vivo tests, it is important not to add sodium azide. In these cases, the antibody solution should be filter-sterilized (filters made with materials that do not bind appreciable amounts of proteins are best), and the sterile antibody solution should be handled aseptically.

Antibodies are stored conveniently at −20°C in the serum, tissue culture supernatant, or ascitic fluid in which they are collected. Methods for collecting these solutions are described on pp. 116, 272, and 274, respectively. Salt concentrations between 0 and 150 mM are suitable for most applications. Antibody solutions should not be frozen and thawed repeatedly, as this can lead to aggregation. Aggregation of antibodies can cause the loss of activity by steric interference of the antigen combining site or by generating insoluble material that is lost during centrifugation or filtration.

## *STORING SERA, TISSUE CULTURE SUPERNATANTS, OR ASCITES*

1. Hybridoma tissue culture supernatants should be buffered by the addition of 1/20 volume of 1 M Tris (pH 8.0). Serum or ascitic fluid are already well buffered.

2. If there is no reason to avoid the use of sodium azide, add to 0.02%.

> **Caution**   Sodium azide is poisonous. It blocks the cytochrome electron transport system, and care should be used.

3. Dispense the antibodies in convenient volumes and store at −20°C. Most antibodies are stable for years when stored at −20°C.

   Working solutions can be stored conveniently at 4°C where they are stable for at least 6 months.

### NOTES

i. Some monoclonal antibodies and some components of serum or ascitic fluid are members of a class of polypeptides known as cryoproteins. As the name implies, these proteins are sensitive to low-temperature storage, where they precipitate. If cryoproteins are simply a contaminant, then their precipitation provides a convenient method to remove them from the antibody solution. Centrifugation at 10,000g is normally sufficient. However, if this treatment also removes the desired antibodies, storage at room temperature may be essential.

ii. Many antibody solutions will generate an insoluble lipid component with prolonged storage. This may look like microbial contamination, but this is unlikely if sodium azide has been included. The precipitate can be removed by centrifugation at 10,000g. In many cases, the lipids will form a layer above the aqueous phase. Remove the aqueous phase and store as described above.

## *STORING PURIFIED ANTIBODIES*

1. Adjust the pH of the antibody solution to neutral. Phosphate-buffered saline (PBS) or similar isotonic solutions are commonly used buffers for storing purified antibodies.

2. Solutions of purified antibodies should be stored at relatively high concentrations (i.e., $\geq 1$ mg/ml) at neutral pH. Concentrations up to 10 mg/ml are commonly used. Antibodies at lower concentrations should be concentrated prior to freezing. Use any of the standard methods such as ultrafiltration or ammonium sulfate precipitation (p. 298).

   If purified antibodies will not be labeled, they can be stored at lower concentrations with the addition of 1% BSA.

3. If there is no reason to avoid the use of sodium azide, add to 0.02%.

   **Caution** Sodium azide is poisonous. It blocks the cytochrome electron transport system, and care should be used.

4. Dispense the antibodies in convenient volumes and store at $-20°C$. Most antibodies are stable for years when stored at $-20°C$.

   Working solutions can be conveniently stored at 4°C where they are stable for at least 6 months.

# ■ PURIFYING ANTIBODIES

Purified antibodies are required for a number of techniques. Table 8.1 lists several techniques that rely on purified antibodies, at least in some steps. In many of the examples listed in this table, purified antibodies are labeled with an easily detected "tag" (p. 319), and these labeled antibodies are then used to determine the presence of an antigen or another antibody. When labeled anti-immunoglobulin antibodies (p. 622) are used to measure the presence of other antibodies, it is seldom worthwhile to prepare and label these reagents yourself. They can be purchased from a number of commercial sources, where they are prepared and tested in large, economic batches. However, when labeled antibodies will be used to detect an antigen directly, the primary antibody must be purified first. Direct labeling also allows two antibodies to be compared in the same assay by marking them with different tags. In other instances, purified antibodies are necessary for different applications. For example, purified antibodies may lower the background in some assays or purification may be the easiest method to concentrate antibody solutions.

There are a wide variety of methods used to purify antibodies. The correct choice of purification method will depend on a number of variables, including the use for which the antibodies are intended, the species in which it was raised, its class and subclass if it is a monoclonal antibody, and the source that will serve as the starting material for the purification. Table 8.2 summarizes the possible sources of antibodies for purification. Also included in this table are the possible sources of antibody contamination and the expected level of purity. Both of these factors may influence the choice of starting material.

**TABLE 8.1**
**Techniques That Require Purified Antibodies**

| Technique | Antibody use | Antibody type | Best sources | Comments |
|---|---|---|---|---|
| **Cell Staining** | Direct localization | Anti-antigen | Polyclonal or monoclonal | Prepare yourself |
| | Indirect localization | Anti-antibody | Polyclonal | Available commercially |
| **Immunoassays** | Direct detection | Anti-antigen | Monoclonal, but polyclonal OK in some cases | Prepare yourself |
| | Indirect detection | Anti-antibody | Polyclonal | Available commercially |
| **Immunoblots** | Direct detection | Anti-antigen | Polyclonal or monoclonal | Prepare yourself |
| | Indirect detection | Anti-antibody | Polyclonal | Available commercially |
| **Immunoaffinity** | Purification | Anti-antigen | Monoclonal | Prepare yourself |

Table 8.3 summarizes the commonly used methods for antibody purification and also lists the advantages and disadvantages of each method. As can be seen from this table, it is often necessary to combine several methods to achieve the desired purification. Finally, Tables 8.4 and 8.5 compare the expected results of the different purification methods when using either polyclonal (Table 8.4) or monoclonal (Table 8.5) antibody sources.

Although no single suggestion can fulfill all of the requirements for different purification needs, most workers have found that purification on protein A beads is the most useful technique. This technique has become the method of choice for antibodies with high affinities for protein A.

## Conventional Methods

Antibodies from serum or ascites can be purified using conventional methods involving precipitation and column chromatography. When similar techniques are used on tissue culture supernatants, the degree of purity achieved will be lower because of the lower concentrations of the specific antibodies. For tissue culture supernatants, purification using protein A beads (p. 309) or anti-immunoglobulin antibody affinity columns (p. 316) is recommended.

## Quantity and Quality of Purified Antibodies

During the purification of antibodies, several variables need to be monitored. These include the purity, the amount, and the antigen binding activity of the antibody. The individual methods to assess these variables are found in other sections, as noted below.

• **Purity**  At any stage, the simplest method to determine the purity of an antibody solution is to run a portion of the sample on an SDS-polyacrylamide gel. The gel can be stained either with Coomassie blue (sensitivity 0.1–0.5 $\mu$g/band) or silver (sensitivity 1–10 ng/band) (see pp. 649–653).

• **Quantitation**  If the antibody is not yet pure, a convenient method for quantitation will be to separate the heavy and light chains by SDS-polyacrylamide gel electrophoresis and compare the staining to known standard controls (p. 659). If many samples need to be analyzed, it is easier to quantitate the amount of antibody in a immunoassay (p. 561).

  If the antibody is pure, either of these methods can be substituted by techniques that measure the total amount of protein. These include the methods described on pp. 670–675. A convenient method is UV absorbance. The amount of antibody can be determined by an absorbance measurement at 280 nm (1 OD = approximately 0.75 mg/ml of purified antibody).

• **Antigen binding activity**  Compare the purified antibody to the starting material in a series of titrations, normalizing to the total amount of antibody in each preparation. This should give a good estimate of the loss of activity through purification.

**TABLE 8.2**
**Sources for Purifying Antibodies**

| Source | Antibody type | Concentration, total antibody (in mg/ml) | Concentration, specific antibody (in mg/ml) | Contaminating antibodies | Possible purity of specific antibody |
|---|---|---|---|---|---|
| **Serum** | Polyclonal | 10 | 1 at best (10% max) | Other serum antibody | 10% at best[a] |
| **Tissue Culture Supernatant with 10% FBS** | Monoclonal | 1 | 0.05 (5%) | Calf antibody | >95% |
| **Tissue culture Supernatant Serum-free Media** | Monoclonal | 0.05 | 0.05 (100%) | None | >95% |
| **Ascites** | Monoclonal | 1–10 | 0.9–9 (90%) | Mouse antibody | 90% |

[a]Except for antigen affinity column.

**TABLE 8.3**
**Techniques for Purification of Antibodies**

| Technique | Appropriate for | Advantages | Disadvantages |
|---|---|---|---|
| **Ammonium Sulfate** | Not recommended as single step<br>Useful for concentration and partial purification of antibody from all sources and species | Cheap<br>Concentrates<br>Convenient for large volumes,<br>Easy | Yields impure antibody<br>Must couple with other techniques |
| **Caprylic Acid** | Not recommended as single step<br>Useful for IgG from all sources and species | Cheap<br>Convenient for large volumes<br>Easy | Yields impure antibody<br>Must couple with other techniques<br>Yields dilute sample<br>Only IgG |
| **DEAE** | Not recommended as single step<br>Useful for concentration and partial purification of antibody from all sources and species | Cheap<br>Concentrates<br>Convenient for large volumes | Yields impure antibody<br>Must couple with other techniques |
| **Hydroxyapatite** | Useful for all antibody from all sources | Concentrates<br>No dialysis<br>Antibody recovered in immediately useful form | Yields impure antibody<br>Moderately expensive |

| Method | Source/Use | Advantages | Disadvantages |
|---|---|---|---|
| **Gel Filtration** | IgM from all sources. Not recommended as a single step | Separates IgM from other antibody in polyclonal sera | Low capacity. Dilutes sample. Yields impure antibody. Not useful for IgG |
| **Ammonium Sulfate, DEAE** | Ascites, polyclonal sera | Nearly pure antibody. Cheap. Convenient for large volume | Multiple steps. Moderate yield |
| **Caprylic acid, Ammonium Sulfate** | Ascites, polyclonal sera or ascites | Nearly pure antibody. Cheap. Convenient for large volumes | Multiple steps. Moderate yield. Only IgG |
| **Protein A Beads** | IgG that bind protein A from all sources | Pure antibody. Easy. Single step. High yield | Expensive. Not suitable for all species and classes |
| **Antigen Affinity Column** | Polyclonal sera | Yields pure antibody. Yields specific antibody | Expensive. Multiple steps. Requires pure antigen. Can inactivate antibody |
| **Anti-Ig Affinity Column** | Rat, sheep, and goat antibody. Class-, subclass-specific antibody from all polyclonal sera, Rat monoclonal antibody | Yields species and class-specific antibody | Expensive. Low to moderate yield. Multiple steps |

**TABLE 8.4**
Purifying Polyclonal Antibodies Using Various Techniques

| Source | $(NH_4)_2SO_4$, DEAE | Caprylic acid, $(NH_4)_2SO_4$ | Hydroxyapatite, $(NH_4)_2SO_4$ | Gel filtration, DEAE |
|---|---|---|---|---|
| **Mouse** | Relatively pure antibody | Moderately pure IgG | Relatively pure antibody | Relatively pure IgM |
| **Rat** | Relatively pure antibody | Moderately pure IgG | Relatively pure antibody | Relatively pure IgM |
| **Hamster** | Relatively pure antibody | Moderately pure IgG | Relatively pure antibody | Relatively pure IgM |
| **Guinea Pig** | Relatively pure antibody | Moderately pure IgG | Relatively pure antibody | Relatively pure IgM |
| **Rabbit** | Relatively pure antibody | Moderately pure IgG | Relatively pure antibody | Relatively pure IgM |
| **Sheep** | Relatively pure antibody | Moderately pure IgG | Relatively pure antibody | Relatively pure IgM |
| **Goat** | Relatively pure antibody | Moderately pure IgG | Relatively pure antibody | Relatively pure IgM |
| **Human** | Relatively pure antibody | Moderately pure IgG | Relatively pure antibody | Relatively pure IgM |

| Protein A low salt | Protein A high salt | Antigen affinity | Anti-Ig affinity | Source |
| --- | --- | --- | --- | --- |
| Pure IgG but lacking IgG$_1$ | Pure IgG | Pure antibody antigen specific | Not recommended | **Mouse** |
| Not recommended | Not recommended | Pure antibody antigen specific | Pure antibody | **Rat** |
| Pure IgG | Not recommended | Pure antibody antigen specific | Not recommended | **Hamster** |
| Pure IgG | Not recommended | Pure antibody antigen specific | Not recommended | **Guinea Pig** |
| Pure IgG | Not recommended | Pure antibody antigen specific | Not recommended | **Rabbit** |
| Not recommended | Not recommended | Pure antibody antigen specific | Pure antibody | **Sheep** |
| Not recommended | Not recommended | Pure antibody antigen specific | Pure antibody | **Goat** |
| Pure IgG but lacking IgG$_3$ | Not recommended | Pure antibody antigen specific | Not recommended | **Human** |

**TABLE 8.5**
**Purifying Monoclonal Antibodies Using Various Techniques**

| Source | $(NH_4)_2SO_4$, DEAE | Caprylic acid, $(NH_4)_2SO_4$ | Hydroxyapatite, $(NH_4)_2SO_4$ | Gel filtration, DEAE |
|---|---|---|---|---|
| **Mouse Ascites** | Relatively pure antibody | Moderately pure IgG | Relatively pure antibody | Relatively pure IgM |
| **Mouse TCSN[1] with FBS** | Not recommended | Moderately pure IgG | Relatively pure antibody | Not recommended |
| **Mouse TCSN Serum Free** | Relatively pure antibody | Moderately pure IgG | Relatively pure antibody | Not recommended |
| **Rat Ascites** | Relatively pure antibody | Moderately pure IgG | Relatively pure antibody | Relatively pure IgM |
| **Rat TCSN with FBS** | Not recommended | Moderately pure IgG | Relatively pure antibody | Not recommended |
| **Rat TCSN Serum Free** | Relatively pure antibody | Moderately pure IgG | Relatively pure antibody | Not recommended |

[1]TCSN, Tissue culture supernatant.

| Protein A low salt | Protein A high salt | Antigen affinity | Anti-Ig affinity | Source |
|---|---|---|---|---|
| Pure IgG low yield IgG$_1$ | Pure IgG | Not recommended | Not recommended, except IgM | **Mouse Acsites** |
| Pure IgG low yield IgG$_1$ | Pure IgG | Not recommended | Not recommended, except IgM | **Mouse TCSN[1] with FBS** |
| Pure IgG low yield IgG$_1$ | Pure IgG | Not recommended | Not recommended except IgM | **Mouse TCSN Serum Free** |
| Not recommended | Not recommended | Not recommended | Pure antibody | **Rat Ascites** |
| Not recommended | Not recommended | Not recommended | Pure antibody | **Rat TCSN with FBS** |
| Not recommended | Not recommended | Not recommended | Pure antibody | **Rat TCSN Serum Free** |

## AMMONIUM SULFATE PRECIPITATION

Ammonium sulfate precipitation is one of the most commonly used methods for removing proteins from solution. Proteins in solution form hydrogen bonds with water through their exposed polar and ionic groups. When high concentrations of small, highly charged ions such as ammonium or sulfate are added, these groups compete with the proteins for binding to water. This removes the water molecules from the protein and decreases its solubility, resulting in precipitation. This can be reversed conveniently by lowering the concentration of ammonium sulfate. The factors that will affect the concentration at which a particular protein will precipitate include the number and position of the polar groups, the molecular weight of the protein, the pH of the solution, and the temperature at which the precipitation is performed.

Although other salts such as sodium sulfate are sometimes used, precipitation of antibodies is commonly done with ammonium sulfate. For most purposes, only the highest purity of ammonium sulfate should be used. The concentration at which antibodies will precipitate varies somewhat from species to species. Most rabbit antibodies can be precipitated with a 40% saturated solution, while mouse antibodies need 45–50% saturation. Because most of the components of serum do not precipitate in this range, it is seldom worthwhile to distinguish between these concentrations, and 50% saturation is a convenient level to use for most applications. One disadvantage of ammonium sulfate precipitation of antibodies is that the resulting antibodies will not be pure. They will be contaminated with other high-molecular-weight proteins, as well as proteins that are trapped in the large flocculant precipitates. Therefore, ammonium sulfate precipitation is not suitable for a single-step purification but must be combined with other methods if pure antibody preparations are needed. The table on p. 658 lists the amount of ammonium sulfate needed to prepare the various concentrations required for most work.

1. Determine the volume of serum, tissue culture supernatant, or ascitic fluid. Centrifuge at 3000g for 30 min.

2. Transfer the supernatant to an appropriate beaker or flask. Add a stirring bar and place on a magnetic stirrer.

3. (**Optional**) To remove large protein aggregates and any proteins that may precipitate with low concentrations of ammonium sulfate, it is often worthwhile to include a preprecipitation step that will remove these proteins prior to collecting the desired antibodies.

While the antibody solution is stirring gently, slowly add 0.5 volume of saturated ammonium sulfate (p. 658; for the conditions described here, saturated ammonium sulfate refers to a solution that has been adjusted to neutral pH with HCl). The ammonium sulfate is added slowly to ensure that the local concentration around the site of addition does not exceed the desired salt concentration. After all the ammonium sulfate is added, move the container to 4°C for 6 hr or overnight.

Centrifuge at 3000g for 30 min. Carefully remove the supernatant and transfer it to a clean container. Add a stir bar to the supernatant and return to a magnetic stirrer (room temperature).

4. While the antibody solution is being stirred gently, slowly add enough saturated ammonium sulfate solution (p. 658) to bring the final concentration to 50% saturation. If step 3 has been used, add 0.5 volume of the starting volume. If step 3 has been skipped, add an equal volume. Transfer to 4°C for 6 hr or overnight.

5. Centrifuge the precipitate at 3000g for 30 min.

6. Carefully remove and discard the supernatant. Drain well. For serum or ascites, resuspend the pellet in 0.3–0.5 volumes of the starting volume in PBS. For monoclonal antibody tissue culture supernatants, resuspend the pellet in 0.1 volume of the starting volume in PBS. Avoid bubbles and frothing.

7. Transfer the antibody solution to dialysis tubing (p. 678) and dialyze versus three changes of PBS overnight. Be sure to allow enough space for expansion of the antibody solution during dialysis. Normally twice the resuspended volume is sufficient.

8. Remove the antibody solution from the tubing. Centrifuge to remove any remaining debris.

Determine the concentration and purity by any of the standard methods (pp. 670–674). Store as for pure antibodies, in the presence of 0.02% sodium azide, if appropriate.

## NOTES

i. If the antibody solution will be immediately transferred to a DEAE column, the dialysate should be the starting buffer for this chromatography step (p. 301).

ii. If trapping of other serum proteins becomes a problem, the ammonium sulfate pellet should be washed in a 50% saturated solution, then redissolved in PBS and reprecipitated.

## ANTIBODY PURIFICATION USING CAPRYLIC ACID*

In mildly acidic conditions, the addition of short-chain fatty acids such as caprylic acid to serum will precipitate most serum proteins with the exception of the IgG molecules. In combination with other purification steps such as binding to a DEAE-matrix or ammonium sulfate precipitation, caprylic acid will yield a relatively pure antibody preparation.

1. Measure the volume of serum, ascites, or tissue culture supernatant (original volume). Transfer to a beaker or other container, add a stir bar, and place on a magnetic stirrer.

2. Add 2 volumes of 60 mM sodium acetate buffer (pH 4.0). Check the pH and adjust to 4.8, if necessary.

3. Slowly (dropwise) add caprylic acid (octanoic acid, best grade available). Continue stirring for 30 min at room temperature.

| *Source* | *Caprylic acid (per 10 ml of original volume)* |
|----------|------------------------------------------------|
| Human    | 0.7 ml   |
| Sheep    | 0.7 ml   |
| Rabbit   | 0.75 ml  |
| Mouse    | 0.4 ml   |

4. Centrifuge at 5000g for 10 min. Carefully decant and save the supernatant.

5. Transfer the supernatant to dialysis tubing (p. 678) and dialyze versus PBS.

### NOTE

i. If necessary, concentrate and purify further by ammonium sulfate precipitation (p. 298).

*Steinbuch and Audran (1969); Russ et al. (1983).

## Antibody Purification Using DEAE

Because antibodies have a more basic isoelectric point than the majority of other serum proteins, ion-exchange chromatography is a useful method for purifying these molecules. Two strategies are commonly used. In the first method listed below, the pH is kept below the isoelectric point for most antibodies, and therefore they will not bind to an anion exchange resin such as DEAE-cellulose. In the second approach the pH is raised above the isoelectric point where the antibodies will bind to the DEAE-groups. As the salt concentration is raised, the antibodies will be the first of the serum molecules to elute. Both methods have been used successfully.

Most commonly, the purification of antibodies using a DEAE-matrix is performed after an ammonium sulfate precipitation (p. 298).

## PURIFICATION OF ANTIBODIES BY USING A DEAE-MATRIX (BATCH)

In this procedure the pH and salt conditions of the antibody solution are adjusted so that the antibodies do not bind to the matrix. Therefore, the antibodies remain in solution. This is a particularly quick method, because the binding is done in bulk. More precise elution can be achieved by following standard chromatography procedures (p. 304). By itself, this procedure will not yield a pure antibody preparation, but must be combined with other techniques.

1. Prior to use, wash the DEAE-matrix extensively with 0.5 N HCl and then 0.5 N NaOH. This can be done conveniently on a sintered glass filter or by gentle centrifugation. Next wash the matrix with 20 volumes of 5 mM sodium phosphate (pH 6.5). Check the pH and repeat the washes until the pH is 6.5.

2. Dialyze the serum, ascites, or ammonium sulfate-purified antibodies against three changes of 5 mM sodium phosphate (pH 6.5).

3. Combine the washed DEAE-matrix with the dialyzed antibodies (for most DEAE-matrices, 2 ml of wet matrix will bind the proteins found in 1 ml of serum). Gently stir or agitate the slurry for 1 hr at room temperature.

4. Either remove the matrtix by centrifugation or by passing the slurry through a suitable filter. The supernatant or flowthrough contains the antibody.

Determine the concentration and purity by any of the standard methods (p. 636). SDS-polyacrylamide gel electrophoresis of antibodies will yield polypeptides of approximately 25,000 and 55,000 daltons.

Store as for pure antibodies (p. 287), in the presence of 0.02% sodium azide if appropriate.

## NOTES

i. If the yield of immunoglobulin is low, try raising or lowering the pH of the binding buffer by 0.5 unit and testing small batches of serum until a better pH is found.

ii. To regenerate the DEAE-matrix, wash sequentially with 0.5 N HCl and 0.5 N NaOH.

## PURIFICATION OF ANTIBODIES BY USING A DEAE-MATRIX (CHROMATOGRAPHY)

When using a DEAE-matrix for column chromatography of antibodies, the pH of the starting antibody solution is raised above the pI of the antibodies. The antibodies are allowed to bind to the DEAE-matrix, washed, and then eluted by increasing the ionic strength of the column buffer. DEAE-chromatography will not yield a pure antibody preparation when used as a single step, and therefore, must be combined with other steps to prepare an antibody solution suitable for most applications.

1. Prior to setting up the column, wash the DEAE-matrix with 0.5 N HCl and then 0.5 N NaOH. This can be done conveniently on a sintered glass filter or by gentle centrifugation. Next wash the matrix with 20 volumes of 10 mM Tris (pH 8.5). Check the pH and repeat the washes until the pH is 8.5. Transfer the matrix into a column. Use approximately 2 ml of wet matrix for each ml of serum.

2. Dialyze the serum, ascites, or ammonium sulfate-purified antibodies against three changes of 10 mM Tris (pH 8.5).

3. Pass the antibody solution down the column. Wash the column with 10 bed volumes of 10 mM Tris (pH 8.5).

4. Sequentially elute the column with increasing NaCl concentrations in the original 10 mM Tris (pH 8.5) buffer. This can be done either with a gradient maker or by step buffers (i.e., three column volumes of 50 mM NaCl in 10 mM Tris, followed by three column volumes of 100 mM NaCl in 10 mM Tris, etc.). Most antibodies will elute with salt concentrations below 500 mM.

5. Determine the presence of the antibody peak by polyacrylamide gel electrophoresis (p. 636) or other antibody-specific quantitative methods. SDS-polyacrylamide gel electrophoresis of antibodies will yield polypeptides of approximately 25,000 and 55,000 daltons.

Combine the peak fractions and store as pure antibody as described on p. 287.

Determine the concentration and purity by any of the standard methods (p. 636).

## NOTES

i. If the antibodies do not bind efficiently to the DEAE-matrix, raise the pH of the starting buffer in 0.5 unit steps, until quantitative binding occurs.

ii. To regenerate the matrix, wash sequentially with 0.5 N HCl and 0.5 N NaOH.

## PURIFICATION OF ANTIBODIES BY HYDROXYAPATITE CHROMATOGRAPHY*

Antibodies can be purified by a simple column chromatography procedure on hydroxyapatite (HA). The procedure is rapid, applicable to large-scale separations, and requires no modification of the sample before loading. Yields are high, but while good purity is achieved from ascites fluid or polyclonal serum, the antibodies isolated from tissue culture supernatants containing fetal bovine serum (FBS) usually are contaminated with albumin. This can be removed by ammonium sulfate precipitation (p. 298). An additional advantage of the HA procedure is that it concentrates the antibodies in the sample and elutes them in a buffer compatible with most labeling reactions. Also, HA chromatography works well for IgM and for subclasses of IgG that do not bind well to protein A.

1. Prepare a column of HA (BioRad grade HT, or equivalent) by resuspending the fully hydrated material in 10 mM sodium phosphate (pH 6.8). Approximately 100 ml of HA will bind the antibody from 500 ml of tissue culture supernatant or from up to 5 ml of ascites or polyclonal serum.

2. Wash the column with 10 column volumes of 10 mM sodium phosphate (pH 6.8).

3. Centrifuge the antibody solution at 4000g for 15 min. Serum or ascites fluid should be diluted 1 : 10 with distilled water prior to centrifugation.

4. Pass the sample down the column. Wash the column with 20 column volumes of 10 mM sodium phosphate (pH 6.8).

5. Elute the antibodies by raising the phosphate concentration. This can be done by applying a linear gradient of 120–300 mM sodium phosphate (pH 6.8). Alternatively, the column can be eluted with a series of steps (50 mM sodium phosphate pH 6.8; 100 mM sodium phosphate, pH 6.8; 150 mM sodium phosphate, etc.).

*Bukovsky and Kennett (1987).

6. Determine the location of the antibodies by SDS-polyacrylamide gel electrophoresis or other antibody-specific assay. SDS-polyacrylamide gel electrophoresis of IgG antibodies will yield polypeptides of approximately 25,000 and 55,000 daltons.

Combine the appropriate fractions and store as for pure antibodies (p. 287).

## NOTES

i. Antibody fractions can be concentrated and further purified by ammonium sulfate precipitation (p. 298).

ii. Because of its low flow rate, DNA-grade HA is not recommended.

iii. The column may be regenerated by washing in 1.0 M NaCl in 10 mM sodium phosphate (pH 6.8).

## GEL FILTRATION CHROMATOGRAPHY

Because IgM molecules are considerably larger than IgG molecules and many of the other molecules found in serum, gel filtration chromatography can be used to separate these proteins. To obtain a pure preparation of IgM, gel filtration must be combined with other techniques such as ammonium sulfate precipitation.

1. Determine the starting volume of the antibody solution. Prepare a column of 20 times the starting volume using gel filtration beads (medium-sized beads, approximately 100 $\mu$m, exclusion limit of ~300,000–500,000 daltons using globular proteins). Follow the manufacturer's instructions for the swelling and preparation of the beads.

2. Transfer the beads into an appropriately sized column. Prewash the column with 20 column volumes of PBS with 0.02% sodium azide. After washing the column, allow the PBS/azide to run just into the top of the column bed. Stop the flow with a valve at the bottom of the column or with modeling clay.

3. Carefully load the antibody solution to the top of the column. Release the flow. Just as the last of the antibody solution runs into the column, carefully top the column up with PBS.

4. Collect fractions of no larger than three times the starting volume. IgM molecules will run with the excluded peak. Check for the presence of IgM by running samples on SDS-polyacrylamide gel. The heavy chain of IgM will migrate with a relative molecular weight of approximately 70,000–80,000 daltons.

### NOTE

i. The IgM fractions can be purified further by ammonium sulfate precipitation without further dialysis.

## ■ Purification on Protein A Beads

Chromatography of antibody solutions over a protein A bead column is one of the most effective and widely used methods for purifying antibodies from many types of crude preparations. The properties of protein A are discussed in detail on pp. 615–622. Two variations on this method are given below. In the first, the antibodies are added to the column in low salt (near physiological levels). This method is applicable when the affinity of the antibody for protein A is sufficient to allow high-capacity binding. In the second method, designed for low-affinity interactions, the affinity of protein A for the antibody is raised by increasing the strength of the hydrophobic bonds that form the basis for the interactions. This is achieved by raising the salt concentration. In both variations the antibodies are eluted by lowering the pH of the buffer.

## ANTIBODY PURIFICATION ON PROTEIN A COLUMNS (LOW SALT)*

Only antibodies with high-affinity binding sites for protein A can be purified conveniently using the low-salt method. These include human, horse, cow, donkey, rabbit, mouse, dog, pig, and guinea pig polyclonal antibodies. Monoclonal antibodies from the mouse $IgG_{2a}$ and $IgG_{2b}$ subclasses will bind well to protein A. Antibodies of the $IgG_3$ subclass bind with intermediate affinity, but $IgG_1$ molecules bind with low affinity (see p. 296 for potential purification methods). Most rat monoclonal antibodies cannot be purified using protein A columns.

In cases where high-affinity binding takes place, a simple one-step purification is possible and will yield an antibody preparation suitable for most applications.

1. Adjust the pH of the crude antibody preparation (serum, tissue culture supernatant, or ascites) to 8.0 by adding 1/10 volume of 1.0 M Tris (pH 8.0).

2. Pass the antibody solution through a protein A bead column. These columns bind approximately 10–20 mg of antibody per milliliter of wet beads (one molecule of protein A binds two molecules of antibody). Serum contains approximately 10 mg/ml of total IgG, tissue culture supernatants contain 20–50 $\mu$g/ml of monoclonal antibody, and ascites between 1 and 10 mg/ml.

3. Wash the beads with 10 column volumes of 100 mM Tris (pH 8.0).

4. Wash the beads with 10 column volumes of 10 mM Tris (pH 8.0).

5. Elute the column with 100 mM glycine (pH 3.0). Add this buffer stepwise, approximately 500 $\mu$l per sample. Collect the eluate in 1.5-ml conical tubes containing 50 $\mu$l of 1 M Tris (pH 8.0). Mix each tube gently to bring the pH back to neutral. As with all protein solutions, avoid bubbling or frothing as this denatures the proteins.

6. Identify the immunoglobulin-containing fractions by absorbance at 280 nm (1 OD = approximately 0.8 mg/ml) or by Bradford dye binding spot test (pp. 673 and 670).

## NOTE

i. For analytical work in which more than one antibody will be purified on the same column, extreme care must be taken. Because different antibodies have such varied affinities for protein A, it is easy to contaminate one batch with the residual from a previous run. Washing columns sequentially with 2 M urea, 1 M LiCl, and 100 mM glycine (pH 2.5) will be sufficient for most needs.

*Adapted from Ey et al. (1978).

## *ANTIBODY PURIFICATION ON PROTEIN A COLUMNS (HIGH SALT)**

Mouse antibodies of the $IgG_1$ subclass do not have a high affinity for protein A. Purification on protein A beads using standard conditions (p. 310) will yield approximately 1/10 the amount of antibody compared with other subclasses. This may be satisfactory for some applications, but by adjusting the pH and salt concentration of the loading and washing buffers, the levels can be increased substantially.

1. Adjust the NaCl concentration of the crude antibody preparation (serum, tissue culture supernatant, or ascites) to 3.3 M. Add 1/10 volume of 1.0 M sodium borate (pH 8.9). Check the pH. (Some authors suggest the addition of 1.5 M glycine in addition to the NaCl.)

2. Pass the antibody solution through a protein A bead column. The capacity of the protein A beads for $IgG_1$ is approximately 5 mg/ml of wet beads. Tissue culture supernatants contain 20–50 µg/ml of monoclonal antibody, and ascites between 1–10 mg/ml.

3. Wash the beads with 10 column volumes of 3.0 M NaCl, 50 mM sodium borate (pH 8.9).

4. Wash the beads with 10 column volumes of 3.0 M NaCl, 10 mM sodium borate (pH 8.9).

5. Elute the column with 100 mM glycine (pH 3.0). Add this buffer stepwise, approximately 500 µl per sample. Collect the eluate in 1.5-ml conical tubes containing 50 µl of 1 M Tris (pH 8.0). Mix each tube gently to bring the pH back to neutral. As with all protein solutions avoid bubbling or frothing as this denatures the proteins.

6. Identify the immunoglobulin-containing fractions by absorbance at 280 nm (1 OD = 0.8 mg/ml) or by Bradford dye binding spot test (pp. 673 and 670).

## NOTE

i. For analytical work in which more than one antibody will be purified on the same column, extreme care must be taken. Because different antibodies have such varied affinities for protein A, it is easy to contaminate one batch with the residual from a previous run. Washing columns sequentially with 2 M urea, 1 M LiCl, and 100 mM glycine (pH 2.5) will be sufficient for most needs.

*Adapted from Ey et al. (1978).

## Immunoaffinity Purification of Antibodies

Immunoaffinity purification is not a commonly used method for isolating antibodies. This technique involves multiple steps and should be used primarily to solve special problems or for special applications. Immunoaffinity purification is discussed in detail in Chapter 13, where it is used to isolate antigens. The principles for both types of affinity purifications are identical.

## IMMUNOAFFINITY PURIFICATION OF ANTIBODIES ON AN ANTIGEN COLUMN*

The only method commonly used to purify antigen-specific antibodies from a preparation of polyclonal antibodies is immunoaffinity purification. In this procedure pure antigen is bound covalently to a solid support. The antibodies within the polyclonal pool that are specific for the antigen are allowed to bind. The unbound antibodies are removed by washing, and the specific antibodies are eluted. This method is unnecessary for monoclonal antibodies, which are homogeneous in their antigen binding activity.

The major strength of this method is its unique ability to isolate specific antibodies from a mixed pool. Its principal disadvantages are that it requires large amounts of antigen and that elution of the specific antibody requires conditions that can lead to some loss of activity.

For protein and peptide antigens, a large number of procedures have been used to prepare affinity columns capable of isolating specific antibodies. The use of activated beads is recommended for most protein antigens. The various types of activated beads are discussed in detail in Chapter 13. The most commonly used activated bead is cyanogen bromide-activated agarose, although this matrix may not be the best choice for many applications (see p. 532). Coupling to beads can also be used for peptide antigens, preferably by first coupling the peptide to a protein carrier (p. 78). For elution from peptide columns, the protein carrier must be distinct from that used for immunization (e.g., BSA peptides for purification of antibodies raised against KLH peptides). The protein carrier–peptide conjugate is then coupled to the activated beads.

*Adapted from Campbell et al. (1951).

1. Covalently attach the antigen to a bead support. A number of different supports are available commercially. The use of activated beads is discussed in detail on p. 528, along with the protocols for coupling.

2. After coupling, carefully drain the resin and retain the supernatant. Measure the protein concentration to confirm that coupling has taken place. In most circumstances, efficient coupling will exceed 90%.

   In some cases, extensive coupling will lead to the loss of the antibody binding sites. If the antibody does not bind well to antigen-coupled beads, try stopping the coupling reaction when only 70% of the antigen is bound.

3. Transfer the beads with the bound antigen to a column. Wash the column with 10 bed-volumes of 10 mM Tris (pH 7.5). Then wash with 10 bed-volumes of 100 mM glycine (pH 2.5), followed by 10 volumes of 10 mM Tris (pH 8.8). Check the pH of the last drops of the Tris wash. If it is not 8.8, continue the wash. Then add 10 bed-volumes of 100 mM triethylamine (pH 11.5, prepared fresh). Wash with 10 mM Tris (pH 7.5) until the pH reaches 7.5.

4. Pass the polyclonal serum through the column to bind the antibody. The serum should be free of any debris prior to loading on the column. If there is any debris, this can be removed by centrifugation or partial purification by ammonium sulfate precipitation (p. 298). If using whole serum, dilute the serum 1 in 10 in 10 mM Tris (pH 7.5) prior to loading. The antibody solution should be passed through the column three times to ensure complete binding, or should be applied at a slow flow rate if using a peristaltic pump.

   If necessary, before loading the antibody on the column, determine the capacity of the beads by titrating the volume needed to clear a sample of the antibody from solution in an immunoprecipitation reaction (p. 429).

5. Wash the column with 20 bed-volumes of 10 mM Tris (pH 7.5), and then with 20 bed volumes of 500 mM NaCl, 10 mM Tris (pH 7.5).

6. Elute the antibodies that are bound by acid-sensitive interactions by passing 10 bed-volumes of 100 mM glycine (pH 2.5), through the column. Collect the eluate in a tube containing 1 bed-volume of 1 M Tris (pH 8.0).

7. Wash the column with 10 mM Tris (pH 8.8) until the pH rises to 8.8.

8. Elute the antibodies that are bound by base-sensitive interactions by passing 10 bed-volumes of 100 mM triethylamine (pH 11.5, prepared fresh) through the column. Collect the eluate in a tube containing 1 bed-volume of 1 M Tris (pH 8.0).

9. Wash the column with 10 mM Tris (pH 7.5), until the pH is 7.5. The column may be reused by storing it in buffers with 0.01% merthiolate.

10. Combine the antibody fractions and dialyze against PBS with 0.02% sodium azide. If necessary, concentrate the antibody solution by ammonium sulfate precipitation (p. 298) or by running on a protein A column (p. 309).

## NOTES

i. High pH and low pH cycles will elute most antibodies in an active state. Some high-affinity antibodies may not be eluted under these conditions, and so these columns eventually will lose antibody binding capacity. If this becomes a serious practical problem and making a new column is prohibited, then elution in harsher reagents such as 1.5 M KSCN, 4 M urea, 1 M propionic acid, or 3.5 M $MgCl_2$ may regenerate the column. These reagents tend to damage the matrix and may restrict flow rates. They may also damage the solid-phase antigen.

ii. Any of the other elution conditions discussed on p. 547 can be used for collecting the antibodies.

iii. In some cases, small amounts of antigen will leach off the antigen columns. If the assays for which the affinity-purified antibodies are being prepared are sensitive to contamination with the antigen (e.g., enzyme assays), then the individual batches of eluted antibodies will need to be tested for the presence of the antigen.

iv. Similar types of purifications can be done from antigens blotted to nitrocellulose or APT paper. See Chapter 12 for more information (p. 498).

## IMMUNOAFFINITY PURIFICATION OF ANTIBODIES ON AN ANTI-IMMUNOGLOBULIN COLUMN*

Antibodies from some species are particularly difficult to purify using conventional means. This is particularly true for rat monoclonal antibodies isolated from tissue culture supernatants. These antibodies often do not bind well to protein A and are present in concentrations that make conventional column chromatography difficult. One useful approach to isolating these antibodies in a single step is to purify them over an anti-rat immunoglobulin antibody column. These columns can be used for many other applications when a certain group of antibodies is needed, for example different species, classes or subclasses.

For this technique, the anti-immunoglobulin antibodies are purified and covalently coupled to activated beads. The test solution containing the antibodies of interest is passed over the column. The beads are washed and the antibodies are eluted by treating the column with a low pH buffer and then a high pH buffer. Any antibodies held by bonds that are broken by these conditions will be eluted and are available for further tests.

1. Purify the antibody fraction from a high-titered antiserum raised against antibodies from the correct species. This anti-immunoglobulin sera can be made using the techniques described on p. 622 or can be purchased commercially. Use any of the techniques described on pp. 291–311 for purification.

2. Couple these antibodies to an activated bead. The various types of activated beads are discussed in detail in Chapter 13 on p. 528. The most commonly used activated bead is cyanogen bromide-activated agarose, although this matrix may not be the best choice for many applications (see p. 532).

   Although more efficient coupling can be achieved at higher pH, do the coupling in 0.1 M sodium phosphate buffer at pH 7.5. This will minimize the number of protonated epsilon-amino groups on lysyl residues and will therefore limit the number of cross-linking sites.

   Other methods for preparing covalently linked antibody matrices are described on p. 519.

3. After coupling, carefully drain the resin and retain the supernatant. Measure the antibody concentration (pp. 670–674) to confirm that coupling has taken place. In most circumstances, efficient coupling will bind approximately 90% of the input antibody.

   If the anti-immunoglobulin antibodies coupled to the column have lost their activity, try reducing the number of cross-linking sites

---

*Adapted from Gurvich and Drizlikh (1964).

by stopping the coupling reaction after only 70% of the antibodies are bound. If this does not restore activity, try one of the other methods described on p. 519

4. Transfer the beads to a column. Wash the column with 10 bed-volumes of 10 mM Tris (pH 7.5). Then wash with 10 bed-volumes of 100 mM glycine (pH 2.5), followed by 10 volumes of 10 mM Tris (pH 8.8). Check the pH of the last drops of the Tris wash. If it is not 8.8, continue the wash. Then add 10 bed-volumes of 100 mM triethylamine (pH 11.5, prepared fresh). Wash with 10 mM Tris (pH 7.5) until the pH reaches 7.5.

5. Pass the antibody preparation through the column to bind the antibody to the anti-immunoglobulins. The antibody preparation should be free of any debris prior to loading the column. If there is any debris, this can be removed by centrifugation. The antibody solution should be passed through the column three times to ensure complete binding.

   To determine the capacity of the column, titrate the volume of beads needed to clear the antibody from solution in an immuno-precipitation reaction (p. 429).

6. Wash the column with 10 bed-volumes of 10 mM Tris (pH 7.5).

7. Elute the antibodies that are bound by acid-sensitive interactions by passing 10 bed-volumes of 100 m glycine (pH 2.5), through the column. Collect the eluate in a tube containing 1 bed-volume of 1 M Tris (pH 8.0).

8. Wash the column with 10 mM Tris (pH 8.8) until the pH rises to 8.8.

9. Elute the antibodies that are bound by base-sensitive interactions by passing 10 bed-volumes of 100 mM triethylamine (pH 11.5, prepared fresh), through the column. Collect the eluate in a tube containing 1 bed-volume of 1 M Tris (pH 8.0).

10. Wash the column with 10 mM Tris (pH 7.5) until the pH is 7.5. The column may be reused by storing it in buffers with 0.01% merthiolate.

11. Combine the antibody fractions and dialyze against PBS with 0.02% sodium azide. If necessary, concentrate the antibody solution by ammonium sulfate precipitation (p. 298) or ultrafiltration.

## NOTES

i. One useful variation on this technique is to preselect the anti-immuno-globulin antibodies on an affinity column prior to setting up this technique. In this variation, a sample of the antibodies that will be purified is bound covalently to a matrix (p. 528). Then the anti-immunoglobulin antibodies are passed over the column and purified by using a known elution buffer. These anti-immunoglobulin antibodies are then used to prepare the actual affinity matrix. The antibodies to be purified are then applied and, because the binding and elution conditions have been preselected, the antibodies are removed easily and efficiently.

ii. Any of the other elution conditions discussed on p. 547 can be used for collecting the antibodies.

# LABELING ANTIBODIES

A wide range of immunological techniques depend on the use of labeled antibodies. The choice of label and the choice between direct and indirect detection methods are discussed below.

## ■ Direct versus Indirect Detection

Detection techniques can be divided into two groups—the direct and indirect methods. In the direct method, the antigen-specific antibody is purified, labeled, and used to bind directly to the antigen. In the indirect method, the antigen-specific antibody is unlabeled and need not be purified. Its binding to the antigen is detected by a secondary reagent, such as labeled anti-immunoglobulin antibodies or labeled protein A (Chapter 15). An important variation that uses aspects of both the direct and indirect methods is to modify the primary antibody by coupling to it a small chemical group such as biotin or dinitrophenol (DNP). The modified primary antibody can then be detected by labeled reagents such as biotin-binding proteins (avidin or streptavidin) or hapten-specific antibodies (such as anti-DNP antibodies).

The choice of the direct or indirect method depends on the circumstances of the experiment. The use of directly labeled antibodies involves fewer steps, is less prone to background problems, but is less sensitive than indirect methods. It also requires a new labeling step for every antibody to be studied or for each labeling method being investigated. In contrast, indirect methods offer the advantages of widely available labeled reagents. The same reagents can be used to detect a large range of antigens and are available commercially. Also, the primary antibody is not modified so the loss of activity is avoided. Therefore, for the majority of applications the indirect methods are the most useful.

## ■ Choice of Label

Regardless of whether the direct or indirect method will be used, the most crucial decision is the choice of label. Table 9.1 lists the most commonly used labels, their advantages and disadvantages, and their preferred use. The choice of label for some techniques is very straightforward. For example, fluorescent labels are essential for cell sorting or high-resolution immunocytochemistry. Routine immunohistology is most satisfactory with enzyme-labeled antibodies. Electron microscopic localization of antigens can only be done with enzyme-labeled or metal-labeled antibodies. For other techniques the choice is less clear, and, for example, both enzyme-labeled and iodinated antibodies perform well in immunoblot and immunoassays. As a very broad generalization, enzyme labels offer the advantage of an instant visual result and great sensitivity but are more difficult to use in a quantitative assay. This is because of the need to measure the rate of the enzyme reaction to get a true measure of the amount of bound enzyme and because the enzyme-labeled reagents are not homogeneous. Iodine-labeled reagents, in particular directly iodinated monoclonal antibodies, can give strikingly accurate quantitative

**TABLE 9.1**
**Choice of Label**

| Label | Detection method | Advantages | Disadvantages | Recommended applications |
|---|---|---|---|---|
| Iodine | Gamma-counter X-ray film | Easy to quantitate Easy to label directly High sensitivity | Short half-life Potential health hazard | Quantitative and qualitative immunoassays Immunoblots |
| Enzyme | Chromogenic substrates detected by eye or absorbance | Long shelf life High sensitivity Direct visualization possible Works at EM level | Multiple steps Some substrates hazardous Endogenous enzymes Poor resolution in cytochemistry | Immunohistochemistry Quantitative and qualitative immunoassays Immunoblots |
| Biotin | Avidin/streptavidin coupled to various labels | Long shelf life High sensitivity Univeral detection Choice of detection methods | Multiple steps Some substrates hazardous Endogenous biotin | Immunohistochemistry, Quantitative and qualitative immunoassays Immunoblots |
| Fluorochromes | UV excitation and visible emission Detected by fluorescence microscope or fluorimeter | Long shelf life, Good resolution in immunocytochemistry | Autofluorescence Quenching Multiple steps Low sensitivity | Immunocytochemistry Cell sorting |
| Biosynthesis | Film Scintillation counter | No damage to antibody Easy | Requires hybridomas Low sensitivity Short shelf life | Immunoassays Immunoblots |

results both in immunoassay and immunoblot procedures. In practical terms, the type of data produced (dpm or X-ray film response) is more easily recorded and communicated than that from the enzyme assays.

Biotinylation of the primary antibody is an underexploited solution to many labeling difficulties. The biotinylation reaction is simple and mild, and rarely inactivates the antibody. Biotinylated antibodies can be stored for years without loss of specific activity. The secondary reagents, avidin and streptavidin, are available commercially in essentially all labeled forms, show very low levels of background binding, and bind so tightly and with such speed that the disadvantages of a multistep procedure are greatly reduced.

## Gold Colloids

One of the major goals of developing ncw immunochemical tools over the last 10 years has been to identify a suitable substrate that is totally inert for all assays. Many of commonly used substrates for binding antibodies or enyzmes are large and/or have a large number of charged groups. One relatively recent development has been the increased use of colloidal gold (for example, see Faulk and Taylor 1971; Brada and Roth 1984; Hsu 1984). Gold is biologically inert and has very good charge distribution. It is becoming more widely available in many useful forms and will probably become one of the substrates of choice during the coming years. Its detection can be enhanced using several silver deposition methods available commercially (p. 412). Colloidal gold also can be detected easily in electron microscopy applications (for example, see Horisberger and Rosset 1977; Horisberger 1979) and can be prepared in discrete and uniform size ranges, permitting double-labeling experiments.

# LABELING ANTIBODIES WITH IODINE

Iodination of antibodies or other proteins is a straightforward and effective method of labeling. Radioactive isotopes of iodine are available commercially in carrier-free solutions, and the use of these sources allows the preparation of proteins with wide ratios of chemical substitution. Although other isotopes of iodine are available, $^{125}I$ is normally used for most immunochemical analyses. The decay of $^{125}I$ yields low-energy gamma and X-ray radiation and, therefore, is easy to detect. The 60-day half-life of $^{125}I$ is a convenient compromise between effective shelf life and waste disposal. Also, $^{125}I$ is cheap.

The major problem when using $^{125}I$ is the potential health hazard. Care must be used to protect anyone handling $^{125}I$. This normally will include protection against potential physical contact with iodide or volatile iodine as well as effective shielding from the $\gamma$-irradiation. Once the iodination of the protein is complete, the major danger is from the radioactive decay itself, and this can be easily handled using shielding and proper waste disposal.

> **Caution**  Before using $^{125}I$, consult officials of your institute about the proper procedures for handling and disposal of this radioactive compound.

Two approaches normally are used to label proteins with iodine. In one, the iodide-125 (normally supplied as NaI) is oxidized to form iodine-125 ($I_2$), which attacks tyrosyl and histidyl side chains. The oxidation can be initiated by chemical or enzymatic means. Chemical oxidants are normally harsher but are less troublesome to manipulate than enzymatic oxidations. The iodine that is produced by oxidation is volatile, so iodinations must be performed in a ventilated hood.

In the second approach, an iodinated reagent containing a reactive coupling group is bound to the protein. The most common compound used for these types of iodinations is the Bolton–Hunter reagent, iodinated *N*-succinimidyl 3-(4-hydroxyphenyl)propionate. Under gentle conditions, this reagent will bind to free amino groups on lysyl or amino-terminal residues. The Bolton–Hunter reagent is prepared by a chemical oxidation reaction similar to the ones discussed above, but as it is available commercially, it is seldom necessary to prepare this reagent.

A third, less commonly used method is the iodine monochloride technique (McFarlane 1958). In this technique nonradioactive iodine is converted to the oxidizing agent iodine monochloride (abbreviated ICl, but actually the hydrochloric acid solution of the trihalide ion $ICl_2^-$). Free $I_2$ is removed and radioactive $^{125}I$ is added. By an exchange reaction, the $^{125}I$ and I reach equilibrium. The solution of iodine monochloride is added to the protein, and the iodine, which is the only oxidizing agent in the solution, attacks the tyrosine side chain, thus iodinating the protein. This technique is seldom used for antibodies, but is commonly used in some fields. See Contreras et al. (1983) for the method and further information.

## Separating Iodinated Antibodies from Unincorporated Label

After an iodination reaction the iodinated antibody is normally separated from the iodotyrosine or iodinated Bolton–Hunter reagent by gel filtration. A simple protocol for this is given below.

1. Use a gel matrix with an exclusion limit of 20,000–50,000 for globular proteins. Use medium-sized beads (approximately 100 $\mu$m in diameter).

2. Prepare a column with 1 ml of bead volume according to the manufacturer's instructions (swelling, etc.). A convenient column for one-time use can be prepared either in a 1-ml or 2-ml syringe barrel with a glass fiber filter cut to fit the bottom of the barrel or in a disposable pipet with a portion of the cotton plug or some glass wool pushed to the bottom of the pipet.

3. To keep the nonspecific binding of the iodinated proteins to a minimum, the column should be prerun with a minimum of 10 column volumes of 1% BSA/PBS with 0.02% sodium azide. Then wash the column with 10 volumes of PBS to remove the BSA. Alternatively swell the beads in buffer containing BSA. Wash with PBS prior to use. Allow the column to run until the buffer level drops just below the top of the bed resin. Stop the flow of the column by either using a valve at the bottom of the column or by plugging the end with modeling clay.

4. After the completion of the iodination reaction, carefully apply the reaction mixture in the stop solution to the column.

5. Release the flow of the column and begin collecting the eluate in a 1.5-ml conical tube. After the iodinated proteins have run into the beads, carefully add 0.3 ml of PBS with 0.02% sodium azide to the column. Continue to collect the eluate in the first tube. When the buffer reaches the top of the beads, change the collection tube to a second 1.5-ml conical tube, and then apply a second 0.3 ml of PBS with sodium azide to the column. Continue to collect the eluate in steps. Monitor the tubes using a minimonitor to identify the peaks of $^{125}$I-labeled antibody and unincorporated label. The antibody should come off in approximately the second to fourth fraction, well ahead of the blue dye which should run with the unincorporated label.

6. Pool the fractions containing the iodinated antibody. It is often convenient to dispose of the unincorporated label by leaving it on the column and putting the entire column into the waste.

Table 9.2 summarizes the properties, advantages, and disadvantages of the various iodination protocols. When labeling a new antibody or other protein, it is normally best to use one of the chemical oxidation methods (iodogen or chloramine T) first. In general, these methods will yield higher specific activities than other methods and are easier to set up. If the protein is damaged by these procedures, other techniques need to be tried. Proteins can be damaged either by oxidation or by distortion due to the introduction of the bulky iodine atom on a critical tyrosine group. In the first case, the oxidative damage may be eliminated altogether by using the Bolton–Hunter reagent or may be limited by using the more gentle oxidation conditions of lactoperoxidase. Because the succinimide ester of the Bolton–Hunter reagent binds to lysyl residues, this technique can be used if the iodination destroys a critical tyrosine residue. In some cases, however, the lysine residue may form an important region for protein function and, thus, modification of these residues may have deleterious effects.

$^{125}$I can be purchased from several commercial suppliers. Buy Na$^{125}$I in alkaline solutions to keep oxidation to a minimum. $^{125}$I can be stored for approximately 6 weeks at room temperature.

**TABLE 9.2**
**Iodination Protocols**

| Technique | Residue | Advantages | Disadvantages |
|---|---|---|---|
| **Chloramine T** | Tyrosine<br>Some histidine | Easy | Generates free I$_2$<br>Possible over substitution<br>Potential oxidation damage |
| **Iodogen** | Tyrosine<br>Some histidine | Easy | Generates free I$_2$<br>Possible over substitution<br>Potential oxidation damage |
| **Bolton–Hunter** | Lysine<br>Free –NH$_2$ | | Lower specific activity |
| **Lactoperoxidase** | Some tyrosine<br>Some histidine | Gentle<br>Detects surface<br>  tyrosine | More steps |

# Iodinations Using Chemical Oxidation

When sodium iodide is exposed to a strong oxidant, the iodide is converted to iodine ($I_2$), which is then free to attack available moieties by a halogenation reaction. For proteins, the most frequent target is tyrosine, but some histidine residues will also be iodinated. These reactions are simple to manipulate and control, and high specific activities can be achieved routinely. The major problems stem from oxidative damage to the proteins or from the modification of a critical tyrosine residue. Oxidative damage can be minimized by using the second technique described for iodogen (p. 331), by switching to the gentler conditions found in lactoperoxidase labeling (p. 334), or by using the Bolton–Hunter reagent which labels free amino groups (p. 338).

## Antibody Purification for Iodinations*

If a large number of antibody preparations need to be iodinated at one time, the purification can be simplified by using the method below.

1. Adjust the pH of the crude antibody prep (serum, tissue culture supernatant, ascites) to 8.9 by adding 0.1 volume of 1.0 M borate (pH 8.9). Add NaCl to 3.0 M. Check the pH; adjust as needed.

2. Add 1.0 ml of tissue culture supernatant or 10 $\mu$l of ascites or serum to a 1.5-ml conical tube. For the ascites or serum, add 1.0 ml of 100 mM borate (pH 8.9). Add 10 $\mu$l of protein A beads. These beads hold approximately 10–20 mg of antibody per milliliter of wet beads.

3. Rock at 4°C for 1 hr.

4. Wash the beads twice with 10 mM borate (pH 8.9), 3.0 M NaCl. The beads can be handled as described for an immunoprecipitation (p. 468).

5. After the second wash, add 100 $\mu$l of 100 mM sodium phosphate (pH 2.8). Incubate at room temperature for 5 min. Spin and remove the supernatant to a fresh 1.5-ml conical tube containing 10 $\mu$l of 1 M sodium phosphate (pH 7.5). Mix gently.

6. Samples are now ready for iodination.

*Adapted from L. Crawford (pers. comm.).

## CHLORAMINE T*

In this technique, the oxidant chloramine T is added to a solution of antibody and iodide. The reaction is terminated by adding a reducing agent and free tyrosine. The labeled antibodies are then separated from the iodotyrosine and reducing agent by gel filtration chromotagraphy.

> **Caution**  This procedure must be carried out in an appropriately ventilated fume hood with protection of the operator from direct radiation and with institute-approved monitoring of radiation exposure and waste disposal.

1. Prior to the iodination, prepare a gel filtration column to separate the labeled antibody from the iodotyrosine after the completion of the iodination. The preparation and use of a suitable disposable column are described on p. 327.

2. Add 10 $\mu$g of protein in a total volume of 25 $\mu$l of 0.5 M sodium phosphate (pH 7.5) to a 1.5-ml conical tube. Other buffers are fine, but do not include any reducing agents (some workers report difficulties in iodinating in the presence of Tris buffers).

3. Add 500 $\mu$Ci of Na$^{125}$I.

4. Add 25 $\mu$l of 2 mg/ml chloramine T (prepared fresh weekly). If oxidation damages the protein, reduce the chloramine T concentration to 0.02 mg/ml (prepared fresh each time).

5. Incubate at room temperature for 60 sec.

6. Add 50 $\mu$l of chloramine T stop buffer [2.4 mg/ml sodium metabisulfite, 10 mg/ml tyrosine (saturated), 10% glycerol, 0.1% xylene cylanol in PBS]. If using 0.02 mg/ml of chloramine T, reduce the sodium metabisulfite concentration to 0.024 mg/ml.

7. Separate the iodinated antibody from the iodotyrosine on the gel filtration column. Pool the fractions containing the iodinated antibody.

*Hunter and Greenwood (1962); Greenwood et al. (1963); McConahey and Dixon (1980).

8. The labeled antibody can be stored in column buffer at 4°C. While the antibody is stable, the iodine is not; the labeled antibody should be used within 6 weeks of preparation. Up to 90% of the input iodine will be incorporated.

## NOTE

i. This labeling procedure will yield specific activities between 5 $\mu$Ci/$\mu$g and 45 $\mu$Ci/$\mu$g. These levels can be adjusted by varying the input iodine and protein concentrations.

## IODOGEN-COATED TUBE—TECHNIQUE I*

In this method, the oxidant iodogen (1,3,4,6-tetrachloro-3$a$,6$a$-diphenyl-glycoluril) is dissolved in a volatile organic solvent such as chloroform or methylene chloride. The solution is added to a test tube, and the solvent is allowed to evaporate. Iodogen is not soluble in water so that aqueous solutions will not remove it from the tube. Next, the reaction mixture of Na$^{125}$I and protein is added. The oxidation is terminated by removing the mixture from the tube, and the iodinated proteins are separated from unincorporated label by gel filtration chromatography.

---

**Caution**  This procedure must be carried out in an appropriately ventilated fume hood with protection of the operator from direct radiation and with institute-approved monitoring of radiation exposure and waste disposal.

---

1. Prior to the iodination, prepare iodogen-coated tubes. Dissolve the iodogen (Pierce Chemical Company, 28600) at 0.5 $\mu$g/ml in chloroform. Dispense into appropriate tubes (2 × 6-mm soda glass or 1.5-ml conical tubes are convenient) at 100 $\mu$l and at 20 $\mu$l per tube. Allow the chloroform to evaporate overnight in a fume hood. Store the tubes in a desiccator at room temperature; this is stable for several years.

   The tubes coated with 20 $\mu$l of iodogen will give a slower rate of iodination, so they can be used if the normal reaction seems to cause damage to the antibody.

2. Prior to the iodination, prepare a gel filtration column to separate the labeled antibody from the iodotyrosine after the completion of the iodination. The preparation and use of a suitable disposable column are described on p. 327.

3. To a 1.5-ml conical tube, add 50 $\mu$l of iodogen-stop buffer [10 mg/ml of tyrosine (saturated), 10% glycerol, 0.1% xylene cylanol in PBS].

4. Behind appropriate shielding in a fume hood, add 50 $\mu$l of antibody to a iodogen-coated tube at room temperature. The antibody should be at a concentration of 0.2–1 mg/ml in 0.5 M sodium phosphate buffer (pH 7.5). Other buffers can be used but no reducing agents should be included.

*Fraker and Speck (1978).

5. Add 500 $\mu$Ci of Na$^{125}$iodide to the iodogen tube with the antibody. Incubate for 2 min. Longer times may be used, but may increase the chances of oxidative damage.

6. Using a Pasteur pipet, transfer contents of the tube to the 50 $\mu$l of iodogen stop. Mix gently.

7. Separate the iodinated antibody from the iodotyrosine on the gel filtration column. Pool the fractions containing the iodinated antibody.

8. The labeled antibody can be stored in the column buffer at 4°C. Although the antibody is stable, the radioactive iodine is not; the labeled antibody should be used within 6 weeks of preparation. Up to 90% of the input iodine can be incorporated.

## NOTE

i. This labeling procedure will yield specific activities between 1 $\mu$Ci/$\mu$g and 45 $\mu$Ci/$\mu$g. These levels can be adjusted by varying the input iodine and protein concentrations.

## *IODOGEN-COATED TUBE—TECHNIQUE II**

In this variation of the iodogen-coated tube technique, the iodide is oxidized to iodine in the absence of the protein. The iodide is then removed from the oxidant and added to the protein solution. The reaction is terminated by adding saturating amounts of free tyrosine, and the iodinated proteins are separated from unincorporated label by gel filtration chromatography.

This technique is designed to reduce the chances of oxidative damage to the protein. In general, antibodies normally are not damaged by oxidation, but this method may be helpful with some antibody preparations or with other more sensitive proteins.

> **Caution** This procedure must be carried out in an appropriately ventilated fume hood with protection of the operator from direct radiation and with institute-approved monitoring of radiation exposure and waste disposal.

1. Prior to the iodination, prepare iodogen-coated tubes. Dissolve the iodogen (Pierce Chemical Company, 28600) at 0.5 $\mu$g/ml in chloroform. Dispense into appropriate tubes ($2 \times 6$ mm soda glass or 1.5-ml conical tubes are convenient) at 100 and at 20 $\mu$l per tube. Allow the chloroform to evaporate overnight in a fume hood. Store the tubes in a desiccator at room temperature; this is stable for several years.

   The tubes coated with only 20 $\mu$l of iodogen will give a slower rate of iodination, so they can be used if the normal reaction seems to cause damage to the antibody.

2. Prior to the iodination, prepare a gel filtration column to separate the labeled antibody from the iodotyrosine after the completion of the iodination. The preparation and use of a suitable disposable column are described on p. 327.

3. To a 1.5-ml conical tube, add 25 $\mu$l of 0.5 M sodium phosphate containing 10 $\mu$g of protein. Other buffers can be used but no reducing agents should be included.

4. Behind appropriate shielding and in a fume hood, add 25 $\mu$l of 0.5 M sodium phosphate to a iodogen-coated tube at room temperature.

5. Add 500 $\mu$Ci of Na$^{125}$iodide to the iodogen tube. Incubate for 2 min.

*Adapted from Fraker and Speck (1978).

6. Using a Pasteur pipet, transfer contents of the iodogen tube to the 25 $\mu$l of antibody. Mix gently. Incubate for 60 sec.

7. Stop the reaction by adding 50 $\mu$l of iodogen-stop buffer [10 mg/ml of tyrosine (saturated), 10% glycerol, 0.1% xylene cylanol in PBS].

8. Separate the iodinated antibody from the iodotyrosine on the gel filtration column. Pool the fractions containing the iodinated antibody.

9. The labeled antibody can be stored in the column buffer at 4°C. While the antibody is stable, the radioactive iodine is not; the labeled antibody should be used within 6 weeks of preparation. Up to 50% of the input iodine can be incorporated.

## NOTES

i. If the protein is damaged during the iodination, try the Bolton–Hunter reagent (p. 338).

ii. This labeling procedure will yield specific activities between 5 $\mu$Ci/$\mu$g and 25 $\mu$Ci/$\mu$g. These levels can be adjusted by varying the input iodine and protein concentrations.

## ◼ Iodinations Using Enzymatic Oxidation

The enzyme lactoperoxidase converts $I^-$ to $I_2$ in the presence of the oxidant $H_2O_2$. The resultant iodine will then attack the side chains of tyrosine or histidine residues yielding an iodinated product (reviewed in Morrison and Schonbaum 1976). Two variations of the lactoperoxidase reaction are commonly used. In the first, the $H_2O_2$ is supplied as a dilute oxidant from a chemical stock. In the second variation, the $H_2O_2$ is produced enzymatically, normally by the enzyme glucose oxidase.

A potentially valuable feature of these techniques is that only amino acid residues found on the surface of the protein will be iodinated efficiently (Morrison 1980). Because the transfer of the iodine to the protein is dependent on the action of the lactoperoxidase enzyme, this technique can be used to identify surface residues.

## LACTOPEROXIDASE LABELING OF ANTIBODIES IN THE PRESENCE OF $H_2O_2$*

> **Caution** This procedure must be carried out in an appropriately ventilated fume hood with protection of the operator from direct radiation and with institute-approved monitoring of radiation exposure and waste disposal.

1. Prior to the iodination, prepare a gel filtration column to separate the labeled antibody from the iodotyrosine after the completion of the iodination. The preparation and use of a suitable disposable column are described on p. 327.

2. Just prior to the iodination, dilute a sample of 30% $H_2O_2$ 1 in 20,000 with PBS. Dilute the lactoperoxidase to 1 unit/ml in PBS (commercial sources of lactoperoxidase are generally about 0.05–0.2 units/$\mu$g). (A stock bottle of 30% $H_2O_2$ will last for approximately 3 weeks at 4°C after it has been opened.)

3. To 10 $\mu$l of protein (approximately 1 mg/ml in PBS), sequentially add 10 $\mu$l of lactoperoxidase and 500 $\mu$Ci of Na$^{125}$I.

4. To initiate the reaction, add 1 $\mu$l of the diluted $H_2O_2$. Mix gently. At 1-min intervals over the next 4 min, add a fresh 1 $\mu$l of diluted $H_2O_2$. Terminate the reaction by adding 25 $\mu$l of lactoperoxidase-stop buffer [10 mg/ml tyrosine (saturated), 10% glycerol, 0.1% xylene cylanol in PBS].

5. Separate the iodinated antibody from the iodotyrosine on the gel filtration column. Pool the fractions containing the iodinated antibody.

6. The labeled antibody can be stored in column buffer at 4°C. Although the antibody is stable, the iodine is not; the labeled antibody should be used within 6 weeks of preparation.

### NOTES

i. Sodium azide will block the action of lactoperoxidase, so it should not be added to the antibodies before the iodination.

ii. If the protein is damaged during the iodination, try the Bolton–Hunter reagent (p. 338).

*Marchalonis (1969).

## LACTOPEROXIDASE LABELING OF ANTIBODIES IN THE PRESENCE OF ENZYMATICALLY GENERATED $H_2O_2$*

> **Caution** This procedure must be carried out in an appropriately ventilated fume hood with protection of the operator from direct radiation and with institute-approved monitoring of radiation exposure and waste disposal.

1. Prior to the iodination, prepare a gel filtration column to separate the labeled antibody from the iodotyrosine after the completion of the iodination. The preparation and use of a suitable disposable column are described on p. 327.

2. Just prior to the iodination, dilute the lactoperoxidase to 1 unit/ml in PBS (commercial sources of lactoperoxidase are generally about 0.05–0.2 units/$\mu$g) and dilute the glucose oxidase to 4 units/ml in PBS (commercial sources of glucose oxidase are generally about 0.1–0.2 units/$\mu$g).

3. To 10 $\mu$l of protein (approximately 1 mg/ml in PBS with 40 mM glucose), sequentially add 10 $\mu$l of lactoperoxidase, and 500 $\mu$Ci of Na$^{125}$I.

4. To initiate the reaction, add 5 $\mu$l of diluted glucose oxidase. Mix gently. Incubate at room temperature for 10 min. Terminate the reaction by adding 25 $\mu$l of lactoperoxidase stop buffer [10 mg/ml tyrosine (saturated), 10% glycerol, 0.1% xylene cylanol in PBS].

5. Separate the iodinated antibody from the iodotyrosine on the gel filtration column. Pool the fractions containing the iodinated antibody.

6. The labeled antibody can be stored in column buffer at 4°C. While the antibody is stable, the iodine is not; the labeled antibody should be used within 6 weeks of preparation.

*Hubbard and Cohn (1972); Schlager (1980).

## NOTES

i. Sodium azide will block the action of lactoperoxidase, so it should not be added to the antibodies before the iodination.

ii. In this technique, the glucose oxidase may also become labeled. To test the level of this labeling, run samples of the iodination on SDS-polyacrylamide gels (p. 636). If the background is a problem, purify the antibodies further (see Chapter 8).

iii. If the protein is damaged during the iodination, try the Bolton–Hunter reagent (p. 338).

## ▓ Iodinations Using Bolton–Hunter Reagent*

Bolton–Hunter reagent [$^{125}$I-labeled *N*-succinimidyl 3-(4-hydroxy-phenylpropionate) is available commercially from a number of suppliers. It is available with either one or two iodine molecules per mole. In aqueous solutions at neutral pH, the hydroxysuccinimide ester will react primarily with free amino groups found on lysyl side chains or the amino-terminal residue. At lower efficiency, the Bolton–Hunter reagent will bind to histidine and even less frequently to tyrosine.

---

**Caution** This procedure must be carried out in an appropriately ventilated fume hood with protection of the operator from direct radiation and with institute-approved monitoring of radiation exposure and waste disposal.

---

1. Prior to the iodination, prepare a gel filtration column to separate the labeled antibody from the iodinated Bolton–Hunter reagent after the completion of the iodination. The preparation and use of a suitable disposable column are described on p. 327.

2. If the Bolton–Hunter reagent comes from the supplier without dimethylformamide (DMF), add DMF to 0.2%.

3. Prepare a protein solution of approximately 0.5–1 mg/ml in 0.1 M sodium borate (pH 8.5). Transfer to 0°C (ice).

   If antibodies have been stored in sodium azide, the azide must be removed prior to coupling. Dialyze extensively against the borate buffer to remove.

4. Remove approximately 500 μCi of Bolton–Hunter reagent and transfer to a 1.5-ml conical tube at 0°C. Dry the reagent in a stream of dry nitrogen gas.

5. Add 10 μl of the protein solution to the dry Bolton–Hunter reagent. Mix gently and return to the ice.

6. Incubate on ice for 15 min. Add 100 μl of a stop solution of 0.5 M ethanolamine, 10% glycerol, 0.1% xylene cylanol, 0.1 M sodium borate (pH 8.5). Incubate for 5 min at room temperature.

*Bolton and Hunter (1973); Langone (1980).

7. Separate the iodinated antibody from the iodinated Bolton–Hunter reagent on the gel filtration column. Pool the fractions containing the iodinated antibody.

8. The labeled antibody can be stored in column buffer at 4°C. Although the antibody is stable, the iodine is not; the labeled antibody should be used within 6 weeks of preparation.

# LABELING ANTIBODIES WITH BIOTIN

Labeling antibodies by covalent coupling of biotinyl groups is a very simple and useful technique (see Becker and Wilchek 1972; Heitzmann and Richards 1974; Heggeness and Ash 1977 for examples of the first uses of biotin and avidin complexes; Green 1963, Green et al. 1972 for a review of avidin and biotin interactions). Biotinylation normally does not have any adverse affect on the antibody, and the coupling conditions are mild. Biotinylated antibodies can be detected using labeled streptavidin or avidin. Both are available commercially, labeled with enzymes, fluorescent dyes, or iodine. Both streptavidin and avidin have exceptionally high affinity for biotin ($K_a$ $10^{14}$ mol$^{-1}$) and therefore bind essentially irreversibly. Streptavidin is preferred for most applications as it has a more favorable pI. Biotin-coupled antibodies are stable under normal storage conditions (p. 287).

The sensitivity of detection with the avidin- or streptavidin-labeled reagents is similar to the sensitivities found when using equivalently labeled anti-immunoglobulin reagents. However, higher sensitivities can be achieved using multimeric avidin or streptavidin complexes. Both avidin and streptavidin are tetravalent and the additional binding sites can be used to increase the size of the detection complex and thus the strength of the signal.

## BIOTINYLATION USING THE SUCCINIMIDE ESTER*

Most biotinylations are performed using a succinimide ester of biotin. The coupling is done through free amino groups on the antibody or other protein, normally lysyl residues. If a free amino group forms a portion of the protein that is essential for activity (for an antibody, normally the antigen combining site), biotinylation with the succinimide ester will lower or destroy the activity of the protein, and other methods of labeling should be tried.

1. Prepare a solution of *N*-hydroxysuccinimide biotin at 10 mg/ml in dimethyl sulfoxide.

2. Prepare an antibody solution of at least 1–3 mg/ml in sodium borate buffer (0.1 M, pH 8.8). If antibodies have been stored in sodium azide, the azide must be removed prior to coupling. Dialyze extensively against the borate buffer to remove.

3. Add the biotin ester to the antibody at a ratio of 25–250 $\mu$g of ester per milligram of antibody. Mix well and incubate at room temperature for 4 hr.

   High concentrations of the biotin ester will lead to multiple biotin groups binding to the antibody and will increase the probability that all of the antibodies will be labeled. Lower ratios will keep biotinylation to a minimum (25 $\mu$g of ester/mg of antibody gives an initial molar ratio of approximately 10 : 1).

4. Add 20 $\mu$l of 1 M $NH_4Cl$ per 250 $\mu$g of ester. Incubate for 10 min at room temperature.

5. Dialyze the antibody solution against PBS or other desired buffer to remove uncoupled biotin. The biotin dialyzes more slowly than would be expected for its size, so extensive dialysis is needed.

6. Store as for pure antibody (p. 287).

### NOTE

i. Many biotinylated succinimide esters are now available. Most of these variations alter the size and characteristics of the spacer arm between the succinimide coupling group and the biotin. For many applications, the spacer arms may be critical. All the esters are handled in a similar manner to the technique listed above.

*Bayer and Wilchek (1980); Bayer et al. (1976); Guesdon et al. (1979).

# LABELING ANTIBODIES WITH ENZYMES

Antibodies can be readily labeled by covalent coupling to enzymes (first reported by Avrameas and Uriel 1966; Nakane and Pierce 1967a,b; and reviewed by Avrameas 1972; Farr and Nakane 1981). The ideal result for any coupling reaction is a 1:1 ratio of antibody to enzyme with no loss of specific activity of either partner. This has not yet been achieved. However, due to the intrinsic amplification of the signal by the enzyme action, even the relatively poor conjugates available can prove exceptionally sensitive. Detection levels in the picomolar range are commonly reported, and the latest developments are producing sensitivities of a few thousand molecules per milliliter.

A large number of enzymes have been used to label antibodies. The most commonly used are horseradish peroxidase, alkaline phosphatase, and $\beta$-galactosidase. Urease and glucose oxidase are also in more limited use. The ideal enzyme needs to be cheap, very stable, small, and easily conjugated. It should have a high catalytic activity and a range of substrates that yield both soluble products (for liquid-phase enzyme immunoassays) and insoluble products (for immunoblotting, immunocytochemical techniques, and solid-phase enzyme immunoassays). Table 9.3 lists the advantages and disadvantages of each enzyme and the most appropriate applications.

In general, it is more convenient to purchase enzyme-linked reagents from commercial sources, but for monoclonal antibodies or affinity-purified antibodies, these reagents will need to be prepared in the laboratory.

**TABLE 9.3**
**Choice of Enzyme**

| Enzyme | Advantages | Disadvantages | Detection method | Recommended application |
|---|---|---|---|---|
| Peroxidase | Good substrates for immunohistochemistry, 1:1 conjugates | High endogenous levels in blood cells Soluble substrates light sensitive | Soluble—TMB | Immunoassays |
| | | | Insoluble—chloronaphthol DAB/cobalt | Immunohistochemistry Immunoblots |
| Alkaline Phosphatase | Excellent substrates for soluble Good substrates for immunoblots Ultrasensitive coupled substrates available | Endogenous activities some in tissues Large conjugates | Soluble—PNPP | Immunoassays |
| | | | Insoluble—BCIP/NBT | Immunoblots |
| β-Galactosidase | Very low endogenous activities Good substrates for soluble and insoluble | More easily inactivated than other enzymes Large conjugates | Soluble—ONPG | Immunoassays |
| | | | Insoluble—BCIG | Immunoblots Immunocytochemistry on blood samples |

## ▦ Coupling Antibodies to Horseradish Peroxidase

Two methods commonly are used for the preparation of antibody–peroxidase conjugates, the two-step glutaraldehyde method and the periodate method. Good batches of horseradish peroxidase (HRP) can be determined by measuring the ratio of the HRP absorbance at 403 and 280 nm (RZ = OD 403 nm/OD 280 nm). This ratio should be at least 3.0.

### Gaining Sensitivity Using Enzyme Assays

One approach to increasing the sensitivity of detection using labeled antibodies is to increase the size of the immune complex. If more labeled antibodies can bind, the signal will be stronger. Three types of approaches have been used successfully. Each method will also increase the background, and a suitable compromise between background and increased sensitivity must be reached by testing the various parameters.

- **Triple layers** Just as a standard indirect detection method will increase the number of labeled molecules bound to an antigen, adding a third layer will yield additional amplification. For example, if, in a direct labeling experiment, one labeled mouse monoclonal antibody molecule can bind to the antigen and using an indirect method allows five labeled rabbit anti-mouse immunoglobulin antibodies to bind, then in a triple-layer experiment 25 labeled goat anti-rabbit immunoglobulin antibodies will bind, increasing the original signal 25 times.

- **Soluble enzyme/anti-enzyme complexes**  In this method, complexes of enzyme and anti-enzyme antibodies are prepared. If the two molecules are added at nearly equal molar ratios, multimeric interactions occur quickly, allowing the formation of large enzyme/anti-enzyme complexes. These complexes are prepared in slight enzyme excess to avoid precipitation. These large complexes are linked to the primary antibody by using a bridging anti-immunoglobulin antibody. For example, a mouse monoclonal antibody bound to an antigen could be detected using a rabbit anti-mouse immunoglobulin antibody to bridge it to a complex of perioxdase/anti-peroxidase, if the anti-peroxidase antibodies are mouse. The most common example of this method is the peroxidase–anti-peroxidase complexes (PAP) (Mason et al. 1969; Sternberger et al. 1970). These complexes can be purchased already in their large aggregated forms.

- **Biotin–streptavidin complexes**  In a third more widely applicable variation, an enzyme is biotinylated and then mixed with streptavidin or avidin (Berman and Basch 1980; Hsu et al. 1981; Kendall et al. 1983). Because both streptavidin and avidin are tetravalent, the biotinylated enzyme and streptavidin or avidin can associate to form large biotin-binding enzymatically active complexes. These complexes can then be used to bind to the antigen either through a directly biotinylated antibody or through a biotinylated anti-immunoglobulin antibody.

## GLUTARALDEHYDE COUPLING*

In the two-step glutaraldehyde method, glutaraldehyde is first coupled to pure horseradish peroxidase (HRP) via the relatively few reactive amino groups available on the enzyme. By performing this step in high glutaraldehyde concentrations, very few HRP–HRP conjugates are formed. The HRP–glutaraldehyde mix is then purified and added to the antibody solution. The method has a relatively low coupling efficiency, and so the HRP–antibody conjugates need to be separated from unconjugated material for optimum sensitivity. The HRP must be pure to minimize cross-linking of enzyme molecules to contaminating proteins during the first step of the procedure.

1. Dissolve 10 mg of HRP in 0.2 ml of 1.25% glutaraldehyde (electron microscopic grade) in 100 mM sodium phosphate (pH 6.8).

> **Caution**   Glutaraldehyde is hazardous. Work in a fume hood.

2. After 18 hr at room temperature, remove excess free glutaraldehyde by gel filtration. To make the column easier to load and run, first add 20 $\mu$l of glycerol and 20 $\mu$l of 1% xylene cylanol.

   Use a gel matrix with an exclusion limit of 20,000–50,000 for globular proteins. Use medium-sized beads (approximately 100 $\mu$m in diameter). Prepare a column with 5 ml of bead volume according to the manufacturer's instructions (swelling, etc.).

   The column should be prerun with a minimum of 10 column volumes of 0.15 M NaCl. Allow the column to run until the buffer level drops just below the top of the bed resin. Stop the flow of the column by either using a valve at the bottom of the column or by plugging the end with modeling clay.

   Carefully load the column with the glutaraldehyde-treated HRP. Release the flow and allow the HRP to run into the column. Just as the level of the HRP solution drops below the top of the column, carefully add 0.15 M NaCl. Run the column with 0.15 M NaCl.

3. Pool the fractions that look brown. These contain the active enzyme.

4. Concentrate the enzyme solution to 10 mg/ml (1 ml final volume) by ultrafiltration or by dialysis against 100 mM sodium carbonate-sodium bicarbonate buffer (pH 9.5) containing 30% sucrose.

*Avrameas and Ternynck (1971); see also Avrameas (1969); Avrameas and Ternynck (1969).

Change the buffer to 100 mM sodium carbonate-bicarbonate (pH 9.5) either by dialysis or by washing on the ultrafiltration membrane.

5. Add 0.1 ml of antibody (5 mg/ml in 0.15 M NaCl) to the enzyme solution and check that the pH >9.0.

6. Incubate at 4°C for 24 hr.

7. Add 0.1 ml of 0.2 M ethanolamine (pH 7.0). Incubate at 4°C for 2 hr.

8. At this stage there will be three species present in the solution, the uncoupled HRP, the uncoupled antibody, and the HRP–antibody conjugate. For some reactions, no further purification is necessary. In these cases, the uncoupled HRP will not bind to any antigen and will be lost during any washes prior to enzyme detection. The uncoupled antibody potentially may block some antibody binding sites, but it will not score in the enzyme reaction.

Further purification will require separation based on the differences between the three species. The easiest separation will be between the uncoupled HRP and the two antibody-containing fractions. If the antibody binds to protein A, the antibodies can be removed simply as described on p. 310. Separation between the two antibody fractions can be achieved by gel filtration (a 50-ml S300, or equivalent, column run in PBS is appropriate) or affinity chromatography on a concanavalin A column (eluted with 0.2 M glucose or $\alpha$-methylmannoside). Alternatively, the whole separation can be achieved on the basis of size by gel filtration. Column eluates can be assayed by enzyme activity, absorbance at 403 nm, or absorbance at 280 nm.

## *PERIODATE COUPLING**

Periodate treatment of carbohydrates opens the ring structure and allows these moieties to bind to free amino groups. Coupling antibodies and horseradish peroxidase with periodate linkage is an efficient method.

1. Resuspend 5 mg of HRP in 1.2 ml of water. Add 0.3 ml of freshly prepared 0.1 M sodium periodate in 10 mM sodium phosphate (pH 7.0).

2. Incubate at room temperature for 20 min.

3. Dialyze the HRP solution versus 1 mM sodium acetate (pH 4.0) at 4°C with several changes overnight.

4. Prepare an antibody solution of 10 mg/ml in 20 mM carbonate (pH 9.5).

5. Remove the HRP from the dialysis tubing and add to 0.5 ml of the antibody solution.

6. Incubate at room temperature for 2 hr.

7. The Schiff's bases that have formed must be reduced by adding 100 $\mu$l of sodium borohydride (4 mg/ml in water). Incubate at 4°C for 2 hr.

8. Dialyze versus several changes of PBS.

   The HRP-antibody conjugates can be used directly or can be purified as described above on p. 345.

## NOTE

i. The concentration of the antibody and the periodate-treated HRP in step 5 can be increased by adding dry Sephadex G-25. Add 0.2× the combined wet weight of the HRP and antibody solutions.

---

* Nakane and Kawaoi (1974); Tijssen and Kurstak (1984).

## ■ Coupling Antibodies to Alkaline Phosphatase*

Satisfactory conjugates of antibodies to alkaline phosphatase can be achieved using a simple, one-step procedure with glutaraldehyde. The resulting conjugates retain good immunological and enzymatic activity but are quite large and heterogeneous. Nevertheless, for most purposes they are satisfactory, and the major drawbacks to the method are the relatively high cost of the enzyme and the need to use very concentrated solutions of enzyme and antibody.

1. Mix 10 mg of antibody with 5 mg of alkaline phosphatase in a final volume of 1 ml. Alkaline phosphatase is usually supplied as a suspension in 65% saturated ammonium sulfate. This should be centrifuged at 4000g for 30 min (or 5 min in microfuge) and the antibody solution can then be used to resuspend the enzyme pellet.

2. Dialyze the mixture against four changes of 0.1 M sodium phosphate buffer (pH 6.8) overnight. This dialysis is essential to remove the free amino groups present in the ammonium sulfate precipitate.

3. Transfer the enzyme–antibody mixture to a container suitable for stirring small volumes (a 5-ml or 10-ml beaker is ideal). In a fume hood, add a small stir bar and place on a magnetic stirrer. Slowly and with gently stirring add 0.05 ml of a 1% solution of EM grade glutaraldehyde.

> **Caution**   Glutaraldehyde is hazardous. Work in a fume hood.

4. After 5 min switch off the stirrer.

5. After 3 hr at room temperature, add 0.1 ml of 1 M ethanolamine (pH 7).

6. After 2 hr further incubation at room temperature, dialyze overnight at 4°C against three changes of PBS.

7. Spin the mixture at 40,000g for 20 min.

8. Store the supernatant at 4°C in the presence of 50% glycerol, 1 mM $ZnCl_2$, 1 mM $MgCl_2$, and 0.02% sodium azide.

### NOTES

i. The procedure can be scaled down to the 1-mg antibody level by reducing the antibody and enzyme concentration by a factor of 10. Under these circumstances the time allowed for coupling should be increased to at least 24 hr, and the yield of conjugate may be reduced.

ii. The dialysis steps can be replaced by gel filtration using medium size beads with exclusion limits of 20,000–50,000 daltons.

*Adapted from Avrameas (1969); Avrameas and Ternynck (1969); Voller et al. (1976).

## ■ Coupling Antibodies to $\beta$-Galactosidase

One advantage of using the $\beta$-galactosidase enzyme conjugates is the large amount of previous research on the enzyme and its activities. Antibodies can be coupled to $\beta$-galactosidase using either a simple one-step glutaraldehyde coupling procedure or maleimidobensoyl-*n*-hydroxysuccinimide ester (MBS) coupling. Glutaraldehyde is more commonly used, but the MBS coupling has many advantages.

## *GLUTARALDEHYDE COUPLING**

The antibody–enzyme conjugates tend to be quite large, but by removing the extremely large conjugates by centrifugation, they can be used for most purposes.

1. Mix 10 mg of antibody with 5 mg of $\beta$-galactosidase in a final volume of 2 ml of PBS.

2. Dialyze the mixture against four changes of 0.1 M sodium phosphate buffer (pH 6.8) overnight. This dialysis is essential to remove the free amino groups.

3. Transfer the enzyme–antibody mixture to a container suitable for stirring small volumes (a 5-ml or 10-ml beaker is ideal). In a fume hood, add a small stir bar and place on a magnetic stirrer. Slowly and with gently stirring add 0.1 ml of a 1% solution of EM grade glutaraldehyde.

> **Caution**   Glutaraldehyde is hazardous. Work in a fume hood.

4. After 5 min switch off the stirrer.

5. After 3 hr at room temperature, add 0.1 ml of 1 M ethanolamine (pH 7).

6. After 2 hr further incubation at room temperature, dialyze overnight at 4°C against three changes of PBS.

7. Spin the mixture at 40,000g for 20 min.

8. Store the supernatant at 4°C in the presence of 50% glycerol and 0.02% sodium azide.

## NOTES

i. The procedure can be scaled down to the 1-mg antibody level by reducing the antibody and enzyme concentration by a factor of 10. Under these circumstances the time allowed for coupling should be increased to at least 24 hr, and the yield of conjugate may be reduced.

ii. The dialysis steps can be replaced by gel filtration using medium-sized beads with exclusion limits of 20,000–50,000 daltons.

*Adapted from Avrameas (1969); Avrameas and Ternynck (1969).

## MALEIMIDOBENSOYL-N-HYDROXYSUCCINIMIDE ESTER COUPLING*

*Meta*-maleimidobensoyl-*N*-hydroxysuccinimide ester (MBS) is a heterobifunctional reagent that links proteins having cysteines with proteins having free amino groups. Because antibodies normally do not contain free cysteines, the reagent is first bound to antibodies through its free amino groups. The excess cross-linker is removed by dialysis and the antibody–MBS added to β-galactosidase.

1. Prepare a solution of antibodies at approximately 1 mg/ml in 50 mM NaCl, 100 mM sodium phosphate (pH 7.0).

2. Dissolve MBS at 20 mg/ml in dioxane. Add 10 μl of MBS solution/ ml of antibody. Mix well.

3. Incubate at 30°C for 1 hr.

4. Dialyze against 10 mM sodium phosphate (pH 7.0) containing 10 mM MgCl₂ and 50 mM NaCl. Several changes for several hours each is appropriate.

5. Dissolve β-galactosidase in 10 mM sodium phosphate to a final concentration of 1.5 mg/ml. Add 1 ml of β-galactosidase per milligram of starting antibody.

6. Incubate at 30°C for 1 hr.

7. Add β-mercaptoethanol to a final concentration of 10 mM.

   The antibody conjugates can be purified from the β-galactosidase by chromatography on protein A beads (pp. 310 or 311) or DEAE ion exchange (p. 304).

---

*Kitagawa and Aikawa (1976); O'Sullivan et al. (1979).

## LABELING ANTIBODIES WITH FLUOROCHROMES

Antibodies can be labeled by direct coupling to fluorochromes. For indirect work, suitable secondary reagents labeled with fluorochromes are available commercially. Direct labeling may be the method of choice where such conjugates are not available or where a direct label is required specifically, for instance, in simultaneously visualizing two antibodies of the same class or subclass.

Four fluorochromes are in common use, fluorescein, rhodamine, Texas red, and phycoerytherin. The properties of these fluorochromes are summarized in Table 10.3 (p. 409). Two labeling procedures are given below, one for the isothiocyanate derivatives of fluoroscein and rhodamine and one for the dichlorotriazinylamine derivative of fluorescein. Methods for labeling antibodies or other proteins with Texas red or the phycoerytherins are not commonly used except for commercial applications.

## *ISOTHIOCYANATE LABELING\**

Both fluorescein and rhodamine isothiocyanate derivatives are available for coupling reactions. The major problem encountered is either over- or undercoupling, but the level of conjugation can be determined by simple absorbance readings (see below).

1. Prior to the coupling, prepare a gel filtration column to separate the labeled antibody from the free fluorochrome after the completion of the reaction. Use a gel matrix with an exclusion limit of 20,000–50,000 for globular proteins. Use fine-sized beads (approximately 50 $\mu$m in diameter).

   To determine the size of the column that is needed, multiply the total volume of the reaction by 20. Prepare a column of this size according to the manufacturer's instructions (swelling, etc.).

   Equilibrate the column in PBS. Allow the column to run until the buffer level drops just below the top of the bed resin. Stop the flow of the column by either using a valve at the bottom of the column or by plugging the end with modeling clay.

2. Prepare a antibody solution of at least 2 mg/ml in 0.1 M sodium carbonate (pH 9.0).

3. Dissolve the fluorescein isothiocyanate (FITC) or tetramethylrhodamine isothiocyanate (TRITC) in dimethyl sulfoxide (best grade available, with no water) at 1 mg/ml. Prepare fresh for each labeling reaction.

4. For each 1 ml of protein solution, add 50 $\mu$l of the dye solution. The dye should be added very slowly in 5-$\mu$l aliquots, and the protein solution should be gently but continuously stirred during the addition.

5. Leave the reaction in the dark for 8 hours at 4°C.

6. Add $NH_4Cl$ to 50 mM. Incubate for 2 hr at 4°C. Add xylene cylanol to 0.1% and glycerol to 5%.

7. Separate the unbound dye from the conjugate by gel filtration. Carefully layer the coupling reaction on the top of the column. Open the block to the column and allow the antibody solution to flow into the column until it just enters the column bed. Carefully add PBS to the top of the column and connect to a buffer supply. The conjugated antibody elutes first and can usually be seen under room light.

*Riggs et al. (1958); The and Feltkamp (1970a,b); Goding (1976).

8. Store the conjugate at 4°C in the column buffer in a lightproof container. If appropriate, add sodium azide to 0.02%. When using low concentrations of antibody (i.e., $<1$ mg/ml), normally it is advantageous to add BSA to a final concentration of 1%.

9. For fluorescein coupling, the ratio of fluorescein to protein can be estimated by measuring the absorbance at 495 nm and 280 nm. For rhodamine, measure at 575 nm and 280 nm. The ratio of absorbance for fluorescein (495–280 nm) should be between 0.3 and 1.0; for rhodamine (575–280 nm), between 0.3 and 0.7. Ratios below these yield low signals, while higher ratios show high backgrounds.

   If the ratios are too low, repeat the conjugation using lower levels of antibody and higher levels of dye. If higher levels are found, either repeat the labeling with appropriate changes or purify the labeled antibodies further on a DEAE column. Equilibrate and load the column with 10 mM potassium phosphate (pH 8.0). Elute with increasing salt concentrations. Measure the ratios (495/280 or 575/280) of each fraction and select and pool the appropriate fractions.

## DICHLOROTRIAZINYLAMINE LABELING*

Fluorescein can be coupled to antibodies or other proteins using a simple one-step conjugation with dichlorotriazinylaminofluorescein (DTAF).

1. Prior to the coupling, prepare a gel filtration column to separate the labeled antibody from the free fluorochrome after the completion of the reaction. Use a gel matrix with an exclusion limit of 20,000–50,000 for globular proteins. Use fine-sized beads (approximately 50 $\mu$m in diameter).

   To determine the size of the column that is needed, multiply the total volume of the reaction by 20. Prepare a column of this size according to the manufacturer's instructions (swelling, etc.).

   Equilibrate the column with PBS. Allow the column to run until the buffer level drops just below the top of the bed resin. Stop the flow of the column by either using a valve at the bottom of the column or by plugging the end with modeling clay.

2. Prepare an antibody solution of at least 2 mg/ml in 0.2 M sodium carbonate (pH 9.0).

3. Dissolve the DTAF in 1.0 M sodium carbonate (pH 9.0) at 2.5 mg/ml.

4. While mixing gently, add 25 $\mu$g of DTAF per milligram of antibody.

5. Continue mixing for 10 min at room temperature.

6. Add $NH_4Cl$ to 50 mM. Incubate for 2 hr at 4°C. Add xylene cylanol to 0.1% and glycerol to 5%.

7. Separate the unbound dye from the DTAF by gel filtration. Carefully layer the coupling reaction on the top of the column. Open the block to the column and allow the antibody solution to flow into the column until it just enters the column bed. Carefully add PBS to the top of the column and connect to a buffer supply. The conjugated antibody elutes first and can usually be seen under room light.

*Blakeslee and Baines (1976); Blakeslee (1977).

8. Store the antibody at 4°C in the column buffer in a lightproof container. If appropriate, add sodium azide to 0.02%. When using low concentrations of antibody (i.e., <1 mg/ml), it is normally advantageous to add BSA to a final concentration of 1%.

9. For fluorescein coupling, the ratio of fluorescein to protein can be estimated by measuring the absorbance at 495 nm and 280 nm. The ratio of absorbance (495 nm–280 nm) should be between 0.3 and 1.0. Ratios below these yield low signals, while higher ratios show high backgrounds.

   If the ratios are too low, repeat the conjugation using lower levels of antibody and higher levels of dye. If higher levels are found, either repeat the labeling with appropriate changes or purify the labeled antibodies further on a DEAE column. Equilibrate and load column with 10 mM potassium phosphate (pH 8.0). Elute with increasing salt concentrations. Measure the ratio (495/280) of each fraction and select and pool the appropriate fractions.

## LABELING MONOCLONAL ANTIBODIES BY BIOSYNTHESIS*

If labeling monoclonal antibodies with any of the conventional techniques causes loss of activity, one convenient method for preparing tagged antibodies is to label the primary amino acid chain by growing the hybridomas in the presence of radioactive precursors. This can only be done with antibodies for which the hybridoma cells are available, but the resultant antibodies have a number of potentially important properties, the most important of which is that the labeled antibodies will be essentially identical to the unlabeled proteins.

1. The hybridoma cells should be healthy and rapidly growing at the time of labeling. Collect approximately $2 \times 10^6$ cells by centrifugation (400g for 10 min).

2. Wash the cells once by resuspending the cell pellet in media without methionine (prewarmed to 37°C) and spinning at 400g for 10 min.

3. Resuspend the cells at approximately $10^6$ cells per milliliter of media without methionine but containing 2% fetal bovine serum. Add [$^{35}$S]methionine (100 $\mu$Ci per labeling will yield approximately $10^5$–$10^6$ cpm of antibody).

4. Incubate overnight at 37°C in a $CO_2$ incubator.

5. Centrifuge the cell suspension at 8000g for 10 min. Remove the supernatant, add 1/20 volume of 1 M Tris (pH 8.0) and sodium azide to 0.02%.

   The antibodies may be purified from the supernatants by standard techniques or used as crude preparations.

### NOTES

i. The [$^{35}$S]methionine used to label these antibodies can be changed to other amino acids by using media lacking these precursors for the labeling. See p. 429 for other labels.

ii. These $^{35}$S-labeled antibodies will be useful for many of the techniques requiring labeled immunoglobulins. However, the metabolically labeled antibodies may not be suitable for some of the techniques that need high-specific-activity labeling.

*Adapted from Cuello et al. (1982).

When labeled antibodies are used to detect antigens in cells or tissues, several characteristics of an antigen can be readily determined. Most importantly, cell staining will demonstrate both the presence and subcellular localization of an antigen. Double-labeling techniques permit the simultaneous detection of two antigens, allowing comparisons of the relative distribution of different antigens. Many cell staining methods can also be used in conjunction with conventional histological stains and autoradiographic methods to compare the localization of the antigen with other markers. Cell staining can also be used in pathology studies for determining such variables as the type of infectious organism, the progenitor of a neoplastic cell, or the presence of an inflammatory response. With certain modifications, the procedure can be used to purify cells on the basis of the antigenic composition of their cell surface.

Cell staining is a versatile technique and, if the antigen is highly localized, can detect as few as a thousand antigen molecules in a cell or tissue. In some circumstances, cell staining may also be used to determine the approximate concentration of an antigen. Recent improvements in antibody labeling methods, microscopes, cameras, and image analyzers are rapidly extending the sensitivity of cell staining procedures and are making these techniques more quantitative. Even without these new improvements, cell staining can yield important qualitative and semi-quantitative data.

Cell staining techniques can be divided into four steps:

(1) cell or tissue preparation
(2) fixation
(3) antibody binding
(4) detection

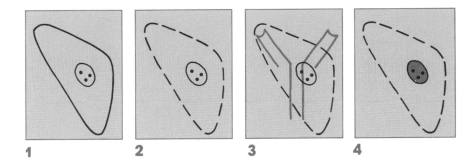

Because the techniques used to prepare cells and tissue sections for staining are different, they are discussed separately here and are considered in individual sections below.

As a first step for most cells, the material to be stained is attached to a solid support to allow easy handling in the subsequent procedures. This can be achieved by a number of methods. Adherent cells may be grown on microscope slides, coverslips, or optically suitable plastic. Suspension cells can be centrifuged onto glass slides, bound to slides using chemical linkers, or in some cases handled in a suspension. For yeast, the cell walls are removed enzymatically, and the cells attached to a solid substrate by a chemical cross-linker. The second step for cell staining usually is to fix and permeabilize the cells to ensure free access of the antibody, although this step may be omitted when examining cell-surface antigens. The cell preparations are then incubated with antibody. Unbound antibody is removed by washing, and the bound antibody detected either directly (if the primary antibody is labeled) or, more commonly, indirectly using a labeled secondary reagent.

For tissue sections, the cells are preserved in one of three methods: (1) the tissue is fixed in situ by perfusing the animal, (2) a dissected tissue sample is frozen, or (3) a sample of tissue is treated with a fixative and embedded in paraffin. Next, sections are cut from the block of tissue with a microtome, and antibodies are applied to the tissue section. Unbound antibodies are removed by washing, and the location of the bound antibodies is determined by using labeled reagents.

# ■ MAJOR CONSTRAINTS

There are three major factors that determine how easily an antigen is detected using cell staining. These are (1) the local antigen concentration, (2) the types of fixative and antibody used, and (3) the type of detection method employed.

The major factor that influences the strength of the signal in cell staining is the local antigen concentration. Diffuse antigens, even when present at high concentrations, are difficult to detect or distinguish from background signals. The antigens that are easiest to detect have a characteristic structure and present a large number of identical antibody binding sites in a small local environment.

The second factor that determines how easy an antigen is to detect is the type of fixative and the characteristics of the antibodies being used. Living cells and tissues are fluid, dynamic structures, and, with the exception of techniques that label live cells, most cell staining methods require that the cells be fixed. Perfect fixation would immobilize the antigens, while retaining authentic cellular and subcellular architecture and permitting unhindered access of antibodies to all cells and subcellular compartments. No fixation method reaches this ideal. Many epitopes will be masked or altered by certain fixatives. Likewise, some antibodies may not be able to recognize epitopes that are altered by fixation or that are hidden within subcellular structures. In some cases, hidden epitopes may be unmasked by limited protease treatment, but many may not.

The resolution and sensitivity of cell staining also depends on the method used for antibody detection. For example, enzyme-labeled antibodies are particularly useful for identifying small amounts of antigen, but do not offer high resolution. On the other hand, high resolution can be obtained using fluorochrome-labeled antibodies, where images of subcellular structures can be studied at magnifications beyond the limit of resolution of the transmitted light microscope. However, these methods are not as sensitive as enzyme-labeled methods.

## CHOICE OF ANTIBODY

Cell staining techniques demand that antibodies bind specifically to antigens in the presence of high concentrations of other macromolecules, and this places exceptional demands on antibody specificity. Antibody preparations that are satisfactory for other techniques such as immunoprecipitation or immunoblotting may show spurious cross-reactions in cell staining. Perhaps surprisingly, antibodies of relatively low affinity often can work well in staining reactions, presumably reflecting bivalent binding to immobilized antigens in high, local concentrations. See Chapter 3 for a more detailed discussion of these interactions.

Three types of antibody preparations can be used for cell staining. These are polyclonal antibodies, monoclonal antibodies, and pooled monoclonal antibodies. Their relative advantages and disadvantages are listed in the summary table on p. 366 and are discussed in detail below.

## Cell Staining With Polyclonal Antibodies

A good polyclonal serum will contain multiple antibodies directed against different epitopes on the antigen. Although steric competition may prevent large numbers of antibodies from binding to the same antigen molecule, several antibodies often will bind, and a strong signal can be expected.

Satisfactory cell staining experiments can be conducted using polyclonal antibodies, but their use requires careful control. In addition to the specific antibodies, polyclonal sera will also contain relatively high concentrations of irrelevant antibodies of unknown specificity. These will include the entire repertoire of antibodies in the animal when the serum was collected. These antibodies create the most troublesome problems for cell staining. Even sera from unimmunized animals will give an intense background when used at high concentrations. Part of this background will be from antibodies binding nonspecifically to the fixed cells, and part will be specific interactions arising from spurious activities in the serum. Because these antibodies will normally not be the major activities in the serum, their binding can often be reduced below the levels of detection by careful titration of the polyclonal antibodies. A second method for lowering these types of background problems is to preadsorb the polyclonal sera with suitable acetone powders (p. 631). A third method used to circumvent these problems is to remove the nonspecific antibodies completely. Immunoaffinity purification of the antigen-specific antibodies (p. 313) can be used to prepare an excellent source of antibodies for cell staining. For some purposes immunoaffinity purification of the antibodies is essential to obtain clear staining patterns.

Because polyclonal sera usually contain antibodies specific for a broad range of epitopes on the antigen, including denaturation-resistant epitopes, they often work well on heavily fixed samples and are the best source of antibodies for techniques such as the staining of paraffin sections of paraformaldehyde-fixed tissues.

## ■ Cell Staining With Monoclonal Antibodies

Monoclonal antibodies often work exceptionally well in cell staining techniques, where their purity and specificity yield low backgrounds over a wide range of antibody concentrations. Most monoclonal antibodies work well on cells or tissue sections that have been fixed in organic solvents or by paraformaldehyde. However, some monoclonal antibodies will not give satisfactory results in cell staining. Presumably, this is due to one of two factors, either the epitope is hidden within a cell structure or is destroyed during fixation. This is particularly true for the harsh conditions encountered during preparation of paraformaldehyde-fixed, wax-embedded tissue blocks. As this is a routine histological fixation procedure, many groups have now started to develop monoclonal antibodies that will perform well on these tissues by immunizing mice with paraformaldehyde-fixed antigens and screening fusions using tissue sections.

Occasionally, monoclonal antibodies will show cross-reactions in cell staining procedures. Some of these cross-reactions are specific and presumably reflect the presence of common epitopes on two cellular components. Cross-reactions with multimeric cellular proteins present in high concentrations, such as cytoskeletal filaments, are the most commonly observed background. This usually results from weak cross-reactions that are enhanced by bivalent binding and by the high concentrations of such antigens. The location of an antigen should be confirmed, whenever possible, by using a panel of monoclonal antibodies directed to discrete epitopes on the same antigen.

Monoclonal antibodies should be tried first as tissue culture supernatants or as purified antibodies isolated from such supernatants. Ascites fluid and antibodies purified from ascites fluid should be used only if necessary, because the presence of contaminating antibodies in these preparations can generate serious and unnecessary background problems.

## ■ Cell Staining With Pooled Monoclonal Antibodies

Pooled monoclonal antibodies will increase the signal strength for a given antigen provided the antibodies in the pool can bind to it noncompetitively. This increase in signal strength can mean the difference between successful detection and failure. Care in pooling must be exercised, however. Each antibody in the pool must be tested individually as part of a panel.

## SUMMARY

### Cell Staining

Detection limit is approximately 1000–10,000 molecules/cell if localized

- Rapid and easy
- Detects antigen presence and localization
- Qualitative to semi-quantitative
- Sensitivity dependent on minimal level and localization of antigen
- High local concentration of antigen, so lower affinity antibody OK

|  | Polyclonal Antibodies | Monoclonal Antibodies | Pooled Monoclonal Antibodies |
|---|---|---|---|
| Signal Strength | Excellent | Fair to good | Excellent |
| Specificity | Good, but some background | Excellent, but some cross-reactions | Excellent |
| Good Features | Signal strength | Specificity Unlimited supply | Signal strength Specificity |
| Bad Features | Nonrenewable Background Need to titer | Lower signal strength | Availability |

## ■ PROTOCOLS FOR CELL STAINING

The protocols for cell staining have been divided into four steps: (1) preparation of cells or tissues, (2) fixation, (3) antibody binding, and (4) detection. See Coons (1956); Goding (1983); Osborn and Weber (1982) for their excellent reviews of immunofluorescence; Avrameas (1972) for enzyme-linked localization.

## ■ Preparation of Cells and Tissues

Cells for staining are usually prepared from one of three sources, adherent cells, suspension cells, or whole tissues. Adherent cells normally are prepared for cell staining by growing on a suitable support. Suspension cells can be fixed directly or can be attached to a solid support by centrifugation, chemical cross-linking, or simple dehydration. Yeast cells are discussed below as a special case of suspension cells. Tissues pose special problems, which are considered separately below.

### Adherent Cells

Adherent cells are easily prepared for cell staining by growing on a suitable microscope slide, coverslip, or plastic tissue culture dish. For high-resolution work and for studies that will be photographed, cells should be grown on coverslips or slides. For lower-resolution work or for large-scale screening procedures, the cells can be grown on tissue culture dishes.

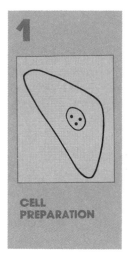

**CELL
PREPARATION**

## *GROWING ADHERENT CELLS ON COVERSLIPS OR MULTIWELL SLIDES*

For high-resolution studies, adherent cells should be grown on the highest available grade glass coverslips, because the controlled thickness, flatness, and good optical properties of a proper coverslip are required to produce the best images. Also, the glass surface is compatible with all fixing and staining solutions. Grade #1 coverslips are the appropriate thickness for most microscopes.

If many antibodies, different dilutions, or various controls are to be tested on the same cell type, plating the cells onto multiwell slides can be helpful. Suitable Teflon-coated slides are available from a number of commercial sources. Select those with white rather than blue Teflon coating, as the latter have been found to detach from the slide in some organic solvents or after washing with detergents.

1. Prior to cell staining, sterilize the glass coverslips or slides. For small numbers of coverslips, you can dip them in 70% ethanol and then flame, while holding loosely between forceps. For large batches, sterilize coverslips or slides by placing them in a glass petri dish and baking in an oven. Do not use the steam cycle of an autoclave, as this will cause them to stick to one another.

2. Place the dry coverslips or slides in suitable tissue culture dishes. Use a sterile Pasteur pipette connected to a vacuum line to pick up the coverslips.

3. Plate the cell suspension into the tissue culture dish and let the cells adhere for at least 24 hr. For greater resolution of most antigens, plate the cells at a low density to avoid overcrowding. If the cells are in short supply, drops of the cell suspension can be placed directly onto the coverslips or slides, left for 4 hr, and then the dish gently filled with medium. This way, most cells will be captured on the glass coverslips or slides. Cells still need approximately 24 hr to spread properly.

Cells are now ready for fixing (p. 384).

**CELL
PREPARATION**

## *GROWING ADHERENT CELLS ON TISSUE CULTURE DISHES*

For low-resolution work, such as crude antigen detection, hybridoma screening, or antibody titration, cells for staining can be grown on regular tissue culture dishes. In this case, the fixative used must not contain more than 50% acetone otherwise the dish will become crazed and optically opaque.

1. Plate the cell suspension into the tissue culture dish and let the cells adhere for at least 24 hr. For easier detection of most antigens, plate the cells at a low density to avoid overcrowding.

2. If the cells are in short supply, mark the bottom of the dish with small circles. Add the cell suspension dropwise above each circle and leave for 4 hr, then gently fill the dish with medium.

**Cells are now ready for fixing (p. 384).**

## Suspension Cells

Suspension cells can be prepared for staining by a number of different methods. For localization of internal antigens, the cells are normally centrifuged onto a glass microscope slide in a cytocentrifuge and then handled as for adherent cells.

For surface staining, the cells can either be maintained as a suspension or can be attached to a slide. Handling cells in suspension is more tedious, because each washing step involves centrifugation, but the cell morphology is seldom distorted. Cells that are stained in suspension can be handled either as live cells or can be fixed using paraformaldehyde. No special preparations are necessary for these cells. The protocol for staining live cells is given on p. 394 and the method for paraformaldehyde is on p. 374. Suspension cells can be attached to slides for surface staining either by cross-linking with poly-L-lysine or by drying.

**CELL PREPARATION**

## ATTACHING SUSPENSION CELLS TO SLIDES USING THE CYTOCENTRIFUGE*

A simple method for detecting intracellular antigens in cells that grow in suspension is to attach the cells to a solid substrate before fixation. This can be achieved by the use of a cytocentrifuge. The centrifugal force also spreads the cells slightly, allowing better resolution of internal structures.

1. (**Optional**) Wash cells by centrifugation at 400g and resuspend in PBS. Resuspend at $1-2 \times 10^6$/ml.

2. Mount microscope slides in cytocentrifuge with card covers. Add 0.1–0.5 ml of cell suspension to reservoir.

3. Accelerate rapidly to 1200g. Spin for 5–10 min.

4. Dry cell monolayer in air for 15–20 min.

Cells are now ready for fixation (p. 384) and can be handled in the same manner as cells that are grown on glass slides or coverslips.

*D. Allen (pers. comm.).

**CELL PREPARATION**

## *ATTACHING SUSPENSION CELLS TO SLIDES USING POLY-L-LYSINE**

Lysine can be polymerized to any desired length, and poly-L-lysine will bind to most solid supports through its charged side chains. The positively charged polymer will provide a site for binding of cells (which carry a overall negative charge). Although this cross-link is not covalent, it is sufficiently strong for most cell staining techniques.

1. Prior to cell staining, prepare a solution of 1 mg/ml poly-L-lysine (average mw 400,000) in distilled water.

2. Coat clean glass slides, coverslips, or tissue culture dishes with poly-L-lysine by dipping or incubating in this solution for 15 min.

3. Wash the solid phase with water. The poly-L-lysine-coated solid phase can be stored at room temperature after drying.

4. Wash the suspension cells by centrifugation at 400g for 5 min and resuspension in PBS. Repeat.

5. Resuspend the cell pellet in PBS at $10^5$ cells/ml. Add to the solid phase. Incubate for 10 min at room temperature.

The cells are now ready for fixing (p. 384), and can be handled in the same manner as cells that are grown on glass slides, coverslips, or dishes.

*Stulting and Berke (1973); Mazia et al. (1975).

**CELL PREPARATION**

## *PREPARING CELL SMEARS*

Increasing use is being made of cell smears for cell staining studies. Cell smears are prepared easily and can be fixed readily by any of the methods used for attached cells.

1. Place a small drop of cells in suspension (including blood and lymph) on one end of a clean glass slide.

2. Touch the edge of a coverslip or another slide to the side of drop nearest the end of the slide. Allow the liquid to be drawn by capillary action to the entire width of the coverslip.

3. Smoothly and evenly push the coverslip across the length of the slide, drawing the liquid in a film over the slide.

4. Allow the samples to air-dry.

**The samples are now ready for fixation, using any of the techniques for attached cells (p. 384).**

## NOTE

i. Some workers find the cells should be smeared in buffers with high protein concentrations (3% BSA for example) to preserve cell morphology.

## Yeast Cells

Staining yeast cells for the presence and location of antigens is particularly challenging. They are small, making the resolution of any antigen difficult; they have a thick cell wall that antibodies cannot penetrate and that is difficult to remove; and they grow in suspension, making handling difficult. Also, background problems can be especially severe, particularly with polyclonal antibodies, because many antisera contain antibodies to yeast cell wall components.

In the method below, adapted from the work of J. Kilmartin (see Kilmartin and Adams 1984), the yeast cells are fixed, the cell wall is removed by enzyme treatment, the spheroplasts are attached to poly-L-lysine-coated slides, and then the cells are stained as per normal. Except in unusual circumstances, the detection reagent should be fluorochrome-labeled.

## Preparing Paraformaldehyde

Standard commercial 37% solutions of formaldehyde are stabilized with 10–15% methanol to inhibit polymerization of the formaldehyde to paraformaldehyde. For most fixatives, paraformaldehyde should be used in place of the commercial formaldehyde. Paraformaldehyde dissolves in water, releasing formaldehyde in the process. To prepare a 4% solution of paraformaldehyde, add 8 grams to 100 ml of water. Heat to 60°C in a fume hood. Add a few drops of 1 N NaOH to help dissolve. When the solid has completely dissolved, let the solution cool to room temperature, add 100 ml of 2× PBS. This stock solution solution should be prepared fresh daily.

**1**

CELL
PREPARATION

## ATTACHING YEAST CELLS TO SLIDES USING POLY-L-LYSINE

1. Prior to cell staining, prepare a solution of 1 mg/ml poly-L-lysine (average mw 400,000) in distilled water. Coat clean glass slides or coverslips with poly-L-lysine by incubating in this solution for 15 min. Wash the solid phase with water. The poly-L-lysine-coated solid phase can be stored at room temperature after drying.

2. Establish a culture of early log-phase yeast cells.

3. **Either:** Add a fresh solution of 4% paraformaldehyde directly to the growing culture (see p. 374 for preparation, but rather than dilute with 2× PBS, use 0.1 M potassium phosphate, pH 6.5). After adding the paraformaldehyde to the yeast culture, adjust the pH to 6.5.

   **Or:** Collect the yeast cells by filtration on 1.2-$\mu$m filters and re-suspend immediately in 4% paraformaldehyde (prepared as above).

4. Incubate for 90 min at room temperature with occasional mixing.

5. Wash the cells by centrifugation 3 times in 0.1 M potassium phosphate (pH 6.5).

6. Resuspend the final pellet in 1.2 M sorbitol, 0.12 M $K_2HPO_4$, 33 mM citric acid (pH 5.9).

7. Add 1/10 volume $\beta$-glucuronidase (crude solution from *Helix pomatia*, approximately 10,000 units/ml final concentration) and 1/100 volume zymolase (or lyticase, 50 units/ml final concentration) to digest the cell wall. Incubate for 90 min at 30°C.

8. Wash three times in the sorbitol buffer.

9. Apply the cell suspension to poly-L-lysine-coated slides or cover-slips by incubating for 15 min at room temperature.

The yeast cells are now ready for lysing as described on p. 389.

## Tissue Sections

Four methods commonly are used to prepare tissue samples for cell staining. These are sectioning of frozen tissues, sectioning of paraformaldehyde fixed, paraffin-embedded tissues, sectioning of tissues prepared from perfused animals, and preparing cytological smears. All these procedures ultimately produce fixed specimens bound to a glass microscope slide ready for the addition of antibody.

**CELL PREPARATION**

## PREPARING FROZEN TISSUE SECTIONS

Cell staining studies on mammalian tissue are often carried out on frozen sections. This is the gentlest method for the preparation of samples and gives good preservation of cell structure and antigens. Its principal disadvantages are that the specimens must be stored frozen and a special microtome, known as a cryostat, is required. Also many clinical specimens are not available in this form and most classical histological descriptions of tissue structure and pathology are based on the use of paraffin-embedded sections of formalin-fixed material.

1. A small sample, no bigger than approximately 1 cm$^2$ × 0.4 cm, of undamaged tissue is dissected.

2. **Either:** If the tissue is very dense and compact (such as skin or muscle), the specimen is placed on one end of a strip of card. The card is labeled, and the end with the sample is submerged in liquid nitrogen. After 60 sec in the liquid nitrogen, the sample is removed, and placed on dry ice. The card is trimmed to just larger than the size of tissue and transferred to a precooled (−70°C), labeled freezing vial. Alternatively, the samples are frozen in isopentane, precooled to liquid nitrogen temperature. Isopentane will not bubble as the sample is immersed, thus shortening the time needed for freezing. Transfer to a precooled (−70°C), labeled freezing vial.

   **Or:** If the tissue has little intrinsic strength (such as spleen or lymph nodes), the specimen is placed in a gelatin capsule containing OCT. (OCT is not o-chlorotoluene, but a mixture of polyvinyl alcohol, polyethylene glycol, dimethyl benzyl ammonium chloride, sold by BDH and other commercial outlets. Another useful freezing solution is Lipshaw Number 1.) The capsule is frozen gradually in liquid nitrogen. Immerse the bottom of the capsule first, allow to freeze, and then submerge the whole capsule. Transfer the capsule to a precooled (−70°C), labeled freezing vial. Alternatively, place a puddle of OCT on a flat piece of dry ice. As the OCT begins to freeze, slide the piece of tissue into the solution. Allow to freeze completely, wrap in foil, and label. The tissue samples can be stored at −70°C for up to 1 year. Long-term storage should be done in air-tight containers.

3. Prior to sectioning, coat clean glass slides with 1% gelatin. Dissolve gelatin in water by heating to 50°C, cool, and add sodium azide to 0.02%. Dip the slides in the solution for 30 sec, remove and allow to air-dry.

4. Prepare sections of frozen tissues by standard techniques. The thickness of the sections will depend on the tissue being studied. In general, thinner sections are better for staining. Sections between 5 and 10 $\mu$m commonly are used. Collect the sections on the coated slides. (Consult the manufacturers' instructions on the proper use of a cryostat.)

5. Depending on the type of tissue and the nature of the antigen, several different steps can be used next.

   **Either:** Allow the section to air-dry.

   **Or:** Fix the section in acetone at room temperature for 30 sec.

   **Or:** Allow the section to air-dry. Dip in freshly prepared paraformaldehyde (see p. 374) for 2 min. Wash with several changes of PBS, and place in 1% NP40, PBS, for 5 min.

6. Rinse in several changes of PBS.

**Antibodies can now be applied to the samples (p. 390).**

## NOTE

i.  If peroxidase- or alkaline phosphatase-labeled detection methods are to be used, it may be necessary to block or inhibit endogenous enzyme activity within the specimen before the application of antibody. Because the blocking procedures may harm some antigens, it may be easier to change the detection reagent rather than block endogenous enzyme activities.

To block endogenous peroxidase activity, incubate the specimen with a solution of 4 parts methanol to 1 part of 3% hydrogen peroxide for 20 min. $H_2O_2$ is generally supplied as a 30% solution and should be stored at 4°C, at which it will last about 1 month. For specimens such as spleen or bone marrow containing high peroxidase activities, better results may be obtained by using a solution of 0.1% phenylhydrazine hydrochloride in PBS. While some sources suggest the simple application of 3% $H_2O_2$ to the specimen; this should not be done, as violent reactions and bubble formation can destroy the specimen.

To block endogenous alkaline phosphatase activity include 0.1 mM levamisole in the substrate solution. This inhibitor does not diminish the activity of intestinal alkaline phosphatase used to prepare the labeled antibody but does inhibit the activity of other tissue phosphatases. Alkaline phosphatase-labeled antibodies should not be used to stain specimens containing endogenous intestinal alkaline phosphatase.

If endogenous activities remain a problem after these blocking steps, try antibodies coupled to $\beta$-galactosidase.

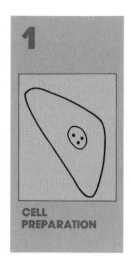

**CELL
PREPARATION**

## PREPARING PARAFFIN TISSUE SECTIONS

Most histological studies are carried out on paraformaldehyde-fixed, paraffin-embedded tissue samples. Therefore, there is an extensive atlas of most tissues and organs prepared from these sources, and comparing the location of antigens to these data is immediately informative. However, the fixation and embedding procedures are harsh and many antigens are not well preserved.

1. Cut small blocks of tissue approximately 1 cm$^2$ × 0.4 cm.

2. Place in

    **Either:** Freshly prepared 4% paraformaldehyde (see p. 374).

    **Or:** Bouin's fixative (To prepare 1 liter, dissolve 2 grams of picric acid in 500 ml of deionized water. Filter through Whatman No. 1, or equivalent. Add 20 grams of paraformaldehyde, and heat to 60°C in a fume hood. Add a few drops of 1 N NaOH to dissolve. Cool and add 500 ml of 2× PBS).

    > **Caution**  Picric acid is harmful. Consult your local safety officer for correct handling procedure.

3. Incubate for 2 hr to overnight.

4. Follow standard paraffin embedding procedures.

5. Collect 4-μm sections onto clean glass slides.

6. Place in 60°C oven for 30 min.

7. Dewax in xylene. Change two times, 3 min each.

8. Rehydrate by passing through graded alcohols (two changes, absolute ethanol, 3 min each, followed by two changes, 95% ethanol, 3 min each).

9. Rinse in water.

Antibodies can now be applied to the specimen (p. 390).

## NOTES

i. If peroxidase- or alkaline phosphatase-labeled detection methods are to be used, it may be necessary to block or inhibit endogenous enzyme activity within the specimen before the application of antibody. Because the blocking procedures may harm some antigens, it may be easier to change the detection reagent rather than block endogenous enzyme activities.

To block endogenous peroxidase activity, incubate the specimen with a solution of 4 parts methanol to 1 part of 3% hydrogen peroxide for 20 min. $H_2O_2$ is generally supplied as a 30% solution and should be stored at 4°C, at which it will last about 1 month. For specimens containing high peroxidase activities, such as spleen or bone marrow, better results may be obtained by using a solution of 0.1% phenylhydrazine hydrochloride in PBS. While some sources suggest the simple application of 3% $H_2O_2$ to the specimen, this should not be done, as violent reactions and bubble formation can destroy the specimen.

To block endogenous alkaline phosphatase activity, include 0.1 mM levamisole in the substrate solution. This inhibitor does not diminish the activity of intestinal alkaline phosphatase used to prepare the labeled antibody but does inhibit the activity of other tissue phosphatases. Alkaline phosphatase-labeled antibodies should not be used to stain specimens containing endogenous intestinal alkaline phosphatase.

If endogenous activities remain a problem after these blocking steps, try antibodies coupled to β-galactosidase.

ii. Because of the harsh fixation, embedding, and preparation conditions, cell staining of paraffin-embedded tissue sections usually requires sensitive detection methods and may need amplification using multiple-layer techniques.

## PERFUSED TISSUES

For laboratory animals, perfusion of tissue or organs with fixative solutions in situ is often used to preserve tissue and cell architecture. This technique is particularly useful for soft tissues. Once preserved, tissue samples can be prepared for cell staining using either of the methods listed above (frozen sections or paraffin-embedded sections). A wide range of different fixative solutions and perfusion methods are used. The best choice of fixative and technique depends on the tissue, antigen, and animal. See Meek (1976); Sternberger (1979); Bullock and Petrusz (1982); Polak and Van Noorden (1983) for suggestions.

Regulations governing this process vary. Consult your local authorities for the proper procedures.

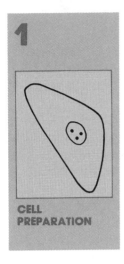

**1**

CELL
PREPARATION

## *PREPARING CELL SMEARS FROM TISSUE SAMPLES*

Cell staining can be carried out on cell smears prepared from biopsy samples, such as needle aspirates, tissue scrapings, or freshly dissected tissues. In these protocols, a thin layer of cells is deposited on a dry slide by physical methods. The most important factor in obtaining good staining patterns is that the smear be only a single cell thick. Tissue smears do not preserve tissue architecture, but are useful for identifying pathological changes and infectious organisms in tissue samples.

1. For tissues or organs that have internal fluid spaces (e.g., spleen or liver), cut a small block of unfixed tissue, and drop directly onto a clean, dry glass slide.

   For tissue scrapings and needle aspirates, deposit the tissue sample at one end of the slide. For well-dispersed samples, touch a second slide to the sample and use it to push the sample across the first slide. For denser samples, use the second slide to pull the sample across the first slide, helping to break up the tissue.

2. Allow to air-dry.

3. Dip the slide in any organic fixative. Most often this will be acetone, but methanol or 50% acetone/50% methanol are also used.

4. Incubate for 2 min at room temperature.

5. Rinse twice in PBS.

The sample is now ready for the addition of antibodies (p. 390).

## ■ Fixation

All fixation protocols must (1) prevent antigen leakage, (2) permeabilize the cell to allow access of the antibody, (3) keep the antigen in such a form that it can be recognized efficiently by the antibody, and (4) maintain the cell structure.

A wide range of fixatives are in common use, and the correct choice of method will depend on the nature of the antigen being examined and on the properties of the antibody preparation. Fixation methods fall generally into two classes, organic solvents and cross-linking reagents. Organic solvents such as alcohols and acetone remove lipids and dehydrate the cells, precipitating the proteins on the cellular architecture. Cross-linking reagents form intermolecular bridges, normally through free amino groups, thus creating a network of linked antigens. Both methods may denature protein antigens, and for this reason, antibodies prepared against denatured proteins may be more useful for cell staining. In some instances, anti-denatured protein antibodies are the only ones that can work.

## Unmasking Hidden Epitopes with Proteases

Fixation in formaldehyde or glutaraldehyde may mask or change some epitopes. These epitopes can often be reexposed by a gentle incubation of the sample in proteases. Trypsin works well. Incubate the specimen in a 0.1% trypsin, 0.1% $CaCl_2$, 20 mM Tris (pH 7.8) solution for 2–20 min at room temperature. Stop the digestion by rinsing the specimen under the cold tap for 5 min.

## *FIXING ATTACHED CELLS IN ORGANIC SOLVENTS*

**2**

**FIXATION**

Cells attached to a solid support are simple to manipulate, because the slides, coverslips, or dishes are easily washed and placed in different fixatives or staining solutions. When using slides or coverslips, they may be easier to handle by placing them in racks or staining jars.

Choosing between fixation in organic solvents or cross-linking agents is empirical. There are no general rules to decide between the two. Try both and choose the one that works best.

1. Rinse the coverslip, slide, or plate once with PBS. Dipping in a beaker of PBS is satisfactory.

2. Drain well, but do not allow the specimen to dry.

3. For coverslips or slides, plunge into a large volume of solvent. Possible solvents include methanol, acetone, or a freshly prepared mixture of 50% acetone/50% methanol (all at room temperature). For tissue culture dishes, simply fill with a freshly prepared mixture of 50% acetone/50% methanol at room temperature.

4. Agitate gently and incubate for 2 min.

5. Drain off or remove from solvent and rinse in PBS.

**Antibodies can now be applied to the samples (p. 390).**

## NOTES

i. Coverslips, slides, and plates fixed in organic solvents can be stored at −70°C. At step 4 in the above protocol, the samples are air-dried without a PBS wash. The dry samples are then stored in a sealed container at −70°C.

For use, the samples must be carefully thawed to avoid damage. Transfer the specimen to a solution of 50% acetone/50% methanol on dry ice. Transfer the container with the samples to room temperature. When the acetone/methanol solution has reached room temperature remove the specimen and rinse in PBS. Antibodies can now be applied to the specimen.

ii. Some antigens give much better reactions after fixation in strongly acidified alcohol solutions, and this fixation method is favored by some laboratories. This technique has become the method of choice for detecting intracellular antibody in lymphoid cells. However, acid alcohol fixation can destroy the antigenicity of many antigens and should not be used as a first choice when examining a new antigen or antibody preparation. In place of the solvents listed above, incubate the sample in 5% glacial acetic acid, 95% ethanol at −20°C for 5 min. Wash as usual.

iii. Some workers use the organic solvents at −20°C.

**2**

**FIXATION**

## FIXING ATTACHED CELLS IN PARAFORMALDEHYDE OR GLUTARALDEHYDE

Fixation in protein cross-linking reagents such as paraformaldehyde or glutaraldehyde preserves cell structure better than organic solvents but may reduce the antigenicity of some cell components. Simple fixation with paraformaldehyde or glutaraldehyde does not allow access of the antibody to the specimen and therefore is followed by a permeabilization step using an organic solvent or nonionic detergent. Using the organic solvent is easy, but it can destroy certain elements of the cell architecture (see p. 385), although the prior fixation with paraformaldehyde does help to preserve the cellular structure. If preservation of cell structure is important, the best first choice would be to use a nonionic detergent.

1. Prior to the staining, prepare either the paraformaldehyde or glutaraldehyde solution. For paraformaldehyde, prepare a 4% solution as described on p. 374. For glutaraldehyde, prepare a 1% solution (electron microscopic grade) in PBS.

> **Caution**   Glutaraldehyde is toxic. Work in a fume hood.

2. Wash the coverslip, slide, or plate gently in PBS.

3. For paraformaldehyde, incubate in a 4% solution for 10 min at room temperature. For glutaraldehyde, incubate in a 1% solution for 1 hr at room temperature in a fume hood.

4. Wash the cells twice with PBS. At this stage, cells fixed with glutaraldehyde can be removed from the fume hood.

5. Permeabilize the fixed cells by incubating in

   **Either:** 0.2% Triton X-100 in PBS for 2 min at room temperature. Some antigens may need as long as 15 min. Check this for each antigen.

   **Or:** Methanol for 2 min at room temperature.

   **Or:** Acetone for 30 sec at room temperature (for cells grown on tissue culture plates, use 50% acetone/50% methanol).

6. (**Optional**) For glutaraldehyde, block free reactive aldehyde groups by incubating with 0.2 M ethanolamine (pH 7.5) for 2 hr at room temperature or by incubating with three changes of 5 min each with 0.5 mg/ml sodium borohydride in PBS. In some cases this may also help paraformaldehyde-fixed cells, but in general is not necessary.

7. Rinse gently in PBS with four changes over 5 min.

**The sample is now ready for the application of antibodies (p. 390).**

## NOTES

i. Paraformaldehyde fixation is not stable. If the paraformaldehyde fixation is followed by nonionic detergent lysis, long incubations in aqueous buffers will reverse the cross-linking and thus should be avoided. Therefore, the use of nonionic detergent permeabilization is not recommended for the detection of soluble antigens.

ii. Sodium borohydride is a strong reducing agent and may change the antigenicity of the sample.

iii. Glutaraldehyde fixation is compatible with electron microscopic work, potentially allowing the specimen to be studied at the levels of both the light and electron microscopy.

iv. Some workers use the organic solvents at $-20°C$.

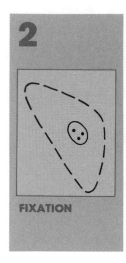

**FIXATION**

## *FIXING SUSPENSION CELLS WITH PARAFORMALDEHYDE*

Staining of suspension cells after fixation is normally used only to detect cell-surface antigens. Because the cells are in suspension, the washing steps are tedious and care should be taken not to centrifuge for long durations or at high speeds.

1. Prior to the staining, prepare a 4% solution of paraformaldehyde (see p. 374).

2. Wash the suspension cells twice with PBS by centrifugation and resuspension.

3. Carefully resuspend the cell pellet in 4% paraformaldehyde/PBS at approximately $10^6$ cells/ml. Incubate at room temperature with occasional mixing for 15 min.

4. Wash the cells by centrifugation at 200g for 5 min and resuspend in PBS. Repeat twice.

**The samples are now ready for the addition of antibodies (p. 390).**

**FIXATION**

## *LYSING YEAST*

After yeast cells have been treated with paraformaldehyde, the cell wall removed by enzymatic digestion, and the spheroplasts attached to poly-L-lysine coated slides, they are ready for methanol fixation.

1. Immerse the slides in methanol for 6 min at −20°C.

2. Transfer the slides to acetone for 30 sec at −20°C.

3. Rinse the slides in four changes of 5 min each of PBS containing 1% BSA.

**The samples are now ready for the addition of antibodies (p. 390).**

## ■ Antibody Binding

Once cells or tissues are fixed and permeabilized, the antibodies are added. Here, as in many other immunochemical techniques, the antibodies can be labeled directly or they can be detected by using a labeled secondary reagent that will bind specifically to the primary antibody. Both the direct and indirect detection methods are in common use, and the choice of method will depend on the experimental design. In general, direct labeling of the primary antibody will produce cleaner signals with lower background. The major disadvantage of direct detection is the time needed to purify and label each preparation of primary antibody. Another potential disadvantage to using direct detection is that the signal will not be as strong as for indirect methods, where the number of labeled molecules will be higher. The purification and labeling of antibodies are discussed in Chapters 8 and 9 (pp. 283 and 319).

For indirect detection, any reagent that will bind specifically to the primary antibody can be "tagged" and used to locate the antibody. The possible reagents include anti-immunoglobulin antibodies (p. 624), protein A or G (p. 615), or, if the first antibody is labeled with biotin, streptavidin (p. 340). The major advantage of indirect detection is that one set of labeled reagents can be used for a number of primary antibodies. Indirect methods will normally give stronger signals, but the backgrounds may be worse.

Detection reagents for cell staining can be labeled with fluorochromes, enzymes, gold, or iodine. Each of the labels has advantages and disadvantages that will vary between different experimental designs. The choice of label is discussed in the detection section below (p. 396). Methods for labeling these reagents are listed in Chapter 9.

Because the antigen in cell staining will be fixed to a solid phase, the time needed for the antibody to find the antigen will be longer than if both molecules are in solution. The incubation times can be adjusted for the experimental design, but seldom will times less than 30 min yield efficient binding. The incubation times can be lowered by increasing the concentration of the antibodies, but this will also increase the background. Usually, some compromise needs to be reached between using enough to achieve a good signal and keeping the background to an acceptable level. In all cases, the antibodies should be diluted in buffers containing high concentrations of nonspecific proteins. Proteins that are commonly used are bovine serum albumin (BSA), fetal bovine serum (FBS), nonfat dry milk, or serum from the same species as the labeled antibody.

## COMMENTS ■ Controls

Specific cell staining reactions should always be compared with control reactions. To assess a staining pattern accurately, two antibodies from the same species should be compared. The best control will be a prebleed from the same animal used to prepare the specific antibodies, but nonimmune sera from the same species is acceptable in most cases. For monoclonal antibodies, the control must be from the same source as the specific antibody, i.e., supernatant versus supernatant, ascites versus ascites, or pure antibody versus pure antibody. If possible, the control antibodies should be of the same class and subclass as the specific antibody. Tissue culture supernatants from the parental myeloma are never appropriate controls, because they do not contain antibodies. Suitable control hybridoma cell lines are available from ATCC.

In addition, if using indirect detection, the secondary reagent should be tested on its own. Finally, when using enzyme-linked detection, the enzyme reaction should be done on the specimen without the addition of any antibodies. This will demonstrate the presence and location of any endogenous enzyme activities.

**3**

ANTIBODY
BINDING

## BINDING ANTIBODIES TO ATTACHED CELLS

Antibodies generally are applied directly to the area of the cells or tissues that is being studied.

1. Cells are fixed and washed as described on p. 384. Place coverslips, slides, or plates in a humidified chamber. Slides or coverslips can be placed in a petri dish containing a water-saturated filter. Coverslips are best placed on a layer of parafilm; this helps to stop the antibody solution from rolling off the edge of the coverslip and makes it easy to pick up the coverslips with fine forceps, as the parafilm is compressible.

2. Add the first antibody solution. All dilutions must be carried out in protein-containing solutions. For example, use PBS containing 3% BSA.

   **For unlabeled primary antibodies:** Monoclonal antibodies are best applied as tissue culture supernatants (specific antibody concentration of 20–50 $\mu$g/ml, use neat). Ascites fluids, purified monoclonal and polyclonal antibodies, and crude polyclonal sera should be tested at a range of dilutions aimed at producing specific antibody concentrations between 0.1–10 $\mu$g/ml. If the specific antibody concentration of the antibody sample is unknown, prepare and test 1/10, 1/100, 1/1000, and 1/10,000 dilutions of the starting material.

   **For labeled primary antibodies:** Primary antibodies can be labeled with enzymes, fluorochromes, or iodine as described on p. 319. They should be assayed at several dilutions in preliminary tests to determine the correct working range. Too-high concentrations will yield high backgrounds; too-low concentrations will make detection difficult. The correct concentration will depend on both the abundance of the antigen under study and the specificity of the antibody.

3. Incubate the coverslips, slides, or plates for a minimum of 30 min at room temperature in the humidified chamber. For some reactions, prolonged incubations of up to 24 hr can increase sensitivity.

4. Wash in three changes of PBS over 5 min. This buffer may be supplemented with 1% Triton X-100 or NP-40 to help with any background problems.

If the first antibody is labeled, the specimen is now ready for the detection step (p. 396).

5. Apply the labeled secondary reagent. It is essential to carry out all dilutions in a protein-containing solution such as 3% BSA/PBS or 1% immunoglobulin/PBS (prepared from the same species as the detection reagent). Useful secondary reagents include anti-immunoglobulin antibodies, protein A, or protein G (see Chapter 15). They can be labeled with enzymes, fluorochromes, gold, or iodine. Labeled secondary reagents can be purchased from several suppliers or can be prepared as described on p. 319.

**For enzyme-labeled reagents:** If using a commercial preparation, test dilutions of the secondary antibodies 1/50 to 1/1000. Alkaline phosphatase-labeled reagents should be handled using Tris-buffered saline, not PBS.

**For fluorochrome-labeled reagents:** If using commercial preparations, test dilutions between 1/10 to 1/300.

**For gold-labeled reagents:** Wash the gold particles once in PBS. Dilute in PBS containing 1% gelatin and add to the specimen.

**For iodine-labeled reagents:** Add the iodinated antibody at approximately 0.1 $\mu$g/ml. Usually, specific activities between 10 and 100 $\mu$Ci/$\mu$g are used.

6. Incubate with the labeled secondary reagent for a minimum of 20 mins at room temperature in the humidified chamber. For gold-labeled reagents, observe periodically under the microscope until a satisfactory signal is obtained.

7. Wash in three changes of PBS (or Tris saline) over 5 min.

The specimen is now ready for the detection step (p. 396).

## NOTES

i. When using horseradish peroxidase-labeled reagents, the buffers used for dilution and washing should not contain sodium azide.

ii. If background problems are seen, the nonspecific binding can often be inhibited by preincubating the specimen with protein. Commonly used proteins are BSA at 3%, fetal bovine serum at 10% (use fetal and not calf, as fetal bovine serum has lower amounts of IgGs), 10% dry milk, or purified antibodies (used at 1%) from the same species as the detection reagent. The blocking protein can be added to each antibody preparation and/or can be used to incubate the samples before the addition of antibody.

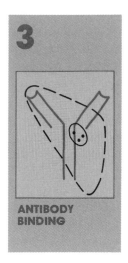

**3**

ANTIBODY
BINDING

## BINDING ANTIBODIES TO CELLS IN SUSPENSION

Cells in suspension normally are stained only for cell-surface antigens. For internal antigens, it is easier to attach the cells to slides using one of the techniques described on p. 370. Surface staining can be done on either fixed or live cells. To minimize capping when using live cells, all buffers should be used cold and sodium azide should be added to a concentration of 0.02%.

1. Resuspend at a density of approximately $5 \times 10^6$ cells/ml in a protein-containing buffer such as 3% BSA/PBS.

2. Add the first antibody solution. All dilutions must be carried out in protein-containing solutions. For example, use PBS containing 3% BSA.

   **For unlabeled primary antibodies:** Monoclonal antibodies are best applied as tissue culture supernatants (specific antibody concentration of 20–50 $\mu$g/ml). Ascites fluids, purified monoclonal and polyclonal antibodies, and crude polyclonal sera should be tested at a range of dilutions aimed at producing specific antibody concentrations between 0.1 and 10 $\mu$g/ml. If the specific antibody concentration of the antibody sample is unknown, prepare and test 1/10, 1/100, 1/1000, and 1/10,000 dilutions of the starting material.

   **For labeled primary antibodies:** Primary antibodies can be labeled with enzymes, fluorochromes, or iodine as described on p. 319. They should be assayed at several dilutions in preliminary tests to determine the correct working range. Too-high concentrations will yield high backgrounds; too-low concentrations will make detection difficult. The correct concentration will depend on both the abundance of the antigen under study and the specificity of the antibody.

3. Incubate for at least 30 min at room temperature.

4. Wash three times in PBS by careful centrifugation and resuspension.

   If the primary antibodies are labeled, the cells are now ready for detection (p. 396).

5. Resuspend the final pellet in the appropriate dilution of the labeled, detecting reagent. All dilutions should be done in a protein-containing solution; PBS with 3% BSA or 1% immunoglobulin (from the same species as the detection reagent) is suitable. Useful secondary reagents include anti-immunoglobulin antibodies, protein A, or protein G (see Chapter 15). They can be labeled with enzymes, fluorochromes, gold, or iodine. Labeled secondary reagents can be purchased from several suppliers or can be prepared as described on p. 319.

**For enzyme-labeled reagents:** If using a commercial preparation, test dilutions of the secondary antibodies from 1/50 to 1/1000. Alkaline phosphatase-labeled reagents should be used with Tris-buffered saline, not PBS.

**For fluorochrome-labeled reagents:** If using commercial preparations, test dilutions between 1/10 to 1/300.

**For gold-labeled reagents:** Wash the gold particles once in PBS. Dilute in PBS containing 1% gelatin and add to the specimen.

**For iodine-labeled reagents:** Add the iodinated antibody at approximately 0.1 $\mu$g/ml. Usually, specific activities between 10 and 100 $\mu$Ci/$\mu$g are used.

6. Incubate for at least 20 min.

7. Wash three times in PBS by careful centrifugation and resuspension.

The specimen is now ready for the detection step (p. 396).

## NOTES

i. If fluorochrome-labeled detecting reagents are used, cell populations can be analyzed in a fluorescence-activated cell sorter as well as by microscopy.

ii. When using horseradish peroxidase-labeled reagents, the buffers used for dilution and washing should not contain sodium azide.

■ **Detection**

Four types of methods are used in cell staining to label the detecting reagent. These are fluorochromes, enzymes, gold, and iodine. The properties of each detection method are summarized in Table 10.1. Fluorochrome and enzyme labels are by far the most common. The decision between enzyme- and fluorochrome-labeled reagents should be made on the basis of the needed sensitivity and resolution. Higher sensitivity, but lower resolution, is found when using enzyme labels, while the opposite is true for fluorescent labels.

Enzyme-labeled antibodies provide extremely sensitive antigen detection and require only a suitable substrate and a light microscope for their detection. Because the signal is detected by differential absorption rather than emission of light and because the insoluble colored product of the chromogenic substrate will be deposited on a small area around the site of enzyme localization, this detection method can never approach the resolution of fluorescent techniques. However, they are compatible with the counterstains commonly used in classical histology.

**Biotin and Streptavidin**

An alternative labeling and detection method that is becoming more popular for cell staining is the biotin/streptavidin system. The primary antibody is purified and conjugated with biotin. For detection, labeled streptavidin or avidin is added and allowed to bind. Streptavidin or avidin can be labeled with enzymes, fluorochromes, gold, or iodine. These labeled reagents can be purchased commercially or can be prepared as described on p. 319. In general, streptavidin reagents are recommended over avidin because of their more favorable pI. The biotin/streptavidin detection methods are recommended when large numbers of samples must be tested with the same first antibody. When many different antibodies are to be tested, the purification and labeling steps for each antibody will make this variation too tedious for general use. The purification and labeling of antibodies is discussed in Chapters 8 and 9.

The resolution of subcellular structures using fluorochrome-labeled antibodies exceeds that of the transmitted light microscope due to the visualization of an expanding cone of emitted light from the excited fluorochrome in the specimen. For example, excellent information on the cellular localization of filamentous structures of 5-nm diameter can be obtained by this method. The major limitation on the sensitivity of this method has been the relatively weak emission from the fluorochrome coupled with the effect of quenching of the fluorescence emission after relatively short periods of exposure to exciting radiation. Thus, good images have only been obtained for antigens present in relatively high local concentrations. While this situation is still broadly true, the field is progressing rapidly with the introduction of more efficient fluorochromes (the phycobiliprotetins), confocal scanning microscopes, and low-light-level camera systems coupled to image analyzers.

Gold-labeled reagents are relatively new for detection at the level of light microscopy, but have been used widely for electron microscopy. When gold-labeled reagents are used with silver enhancement, they offer some excellent advantages. Iodinated reagents yield excellent sensitivity, but are somewhat troublesome to use.

**TABLE 10.1**
**Cell Staining Detection Methods**

| Method | Advantages | Disadvantages | Recommended for |
|---|---|---|---|
| **Fluorescence** | High resolution<br>Double labeling possible<br>Staining live cells possible<br>Cell sorting possible | Requires special equipment<br>Not compatible with most<br>histological stains<br>Loss of signal with time<br>Not compatible with electron<br>microscopy | High-resolution studies<br>Double labeling |
| **Enzyme** | High sensitivity<br>Only need light microscopy<br>Compatible with many<br>histological stains<br>Permanent | Low resolution<br>Endogenous enzyme activities<br>Double staining difficult<br>Some potentially carcinogenic<br>substrates | Low-resolution studies of<br>antigen distribution,<br>compatible with histology |
| **Gold** | High resolution<br>Only need light microscope<br>Compatible with electron<br>microscopy<br>Permanent | Needs silver enhancement<br>for high sensitivity | Staining tissues that<br>autofluoresce<br>or have high endogenous<br>enzyme activities |
| **Iodine** | High sensitivity<br>Easy control of exposure time<br>Only need light microscopy<br>Low background | Low resolution<br>Complex manipulations<br>Slow | Staining tissues that<br>autofluoresce or have high<br>endogenous enzyme activities |

## Double Labeling

Double-labeling experiments are used to detect two different antigens in the same specimen. These types of experiments fall into two classes—those in which the locations of both antigens are known and do not overlap and those in which the antigen locations are not known or are known to overlap. In the first, many possible combinations of labels and detection systems are possible, and these applications are obvious. The second case presents a more difficult task, and only certain combinations of detection systems are possible. For most circumstances, the use of two fluorochromes is the best solution to this problem.

To use two fluorochromes on the same specimen, each antigen must be recognized by only one of the labeled reagents. There are three ways to achieve this distinction. The easiest and the surest is to label each primary antibody directly. For example, when using two purified monoclonal antibodies specific for different antigens, one is labeled with fluorescein and the other with rhodamine. A second method to locate two antigens in the same specimen is to use two detection reagents that are species, class, or subclass specific. For example, a polyclonal rabbit antibody and a mouse monoclonal antibody specific for two different antigens could be studied using a fluorescein-conjugated goat anti-mouse immunoglobulin antibody and a Texas red-conjugated goat anti-rabbit immunoglobulin antibody. In these protocols the species specificity of the anti-immunoglobin antibodies must be checked rigorously. A third method is to perform the staining reactions sequentially. Particular care must be used in this protocol to ensure that no dissociation and reassociation are possible. Fixing between the steps will stop this problem, but adds extra time to the protocol. Another useful way to gain information about the relative location of antigens in a tissue is to stain alternate serial sections cut from the same tissue block with the different antibody preparations.

Care must be used in choosing double-labeling reagents to ensure that no cross-reaction between the labeled detection reagents is possible. This should always be tested in control reactions. To record the results of double-labeling experiments, double exposures are helpful.

## *Detection Using Enzyme-Labeled Reagents*

Enzyme-labeled reagents are detected using soluble chromogenic substrates that precipitate following enzyme action, yielding an insoluble colored product at the site of enzyme localization (Avrameas and Uriel 1966; Nakane and Pierce 1967a,b; Avrameas 1972). A range of substrates is available for each enzyme, and the following protocols represent some of the most useful alternatives. A wide range of conjugated reagents are available commercially or can be prepared as described in Chapter 9 on p. 319. Enzymes can be coupled to anti-immunoglobulin antibodies, protein A, protein G, avidin, or streptavidin.

Table 10.2 summarizes the various enzymes that are in common use and the advantages and disadvantages of each. Extra sensitivity can be found using enzyme-labeled reagents by observing the precipitated enzyme products by interference reflection microscopy.

---

**Caution**   Some chromogenic substrates may be carcinogenic. Consult your local authorities for handling and disposal procedures.

---

**TABLE 10.2**
**Enzymes Used in Cell Staining**

| Method | Advantages | Disadvantages* | Recommended for | Recommended substrates |
|---|---|---|---|---|
| **Horseradish Peroxidase** | Sensitive<br>Many substrates available | Endogenous activities, particularly in macrophages, RBCs, and bone marrow | Tissue culture cells<br>Tissue sections, but endogenous activity gives problems | DAB/metal |
| **Alkaline Phosphatase** | Sensitive<br>Easily conjugated | Endogenous activities, particularly in placenta and intestine | Good alternative to HRP<br>Endogenous activity frequent in cells and tissues | NABP/new fuchsin<br>NBT/BCIP |
| **β-Galactosidase** | Sensitive<br>No endogenous activity in mammalian cells | Not in common use<br>Fewer substrates available | Use when endogenous enzyme activities are persistent problem | BCIG |

*See Pearson et al. (1963); Druget and Pepys (1977); Nanba et al. (1977) for examples of endogenous activities.

**DETECTION**

## *HORSERADISH PEROXIDASE-LABELED REAGENTS*

A range of substrates are useful, including diaminobenzidine, chloronaphthol, and aminoethylcarbazole.

### Diaminobenzidine

Diaminobenzidine (DAB) is the most commonly used substrate and one of the most sensitive for horseradish peroxidase. It yields an intense brown product that is insoluble in both water and alcohol. DAB staining is compatible with a wide range of common histological stains.

1. Dissolve 6 mg of DAB (use DAB tetrahydrochloride) in 10 ml of 0.05 M Tris buffer (pH 7.6).

2. Add 0.1 ml of a 3% solution of $H_2O_2$ in $H_2O$. $H_2O_2$ generally is supplied as a 30% solution and should be stored at 4°C, at which it will last about 1 month.

3. If a precipitate appears, filter through Whatman No. 1 filter paper (or equivalent).

4. Apply to specimen, incubate for 1–20 min. Stop the reaction by washing in water.

5. (**Optional**) Counterstain if necessary (see p. 415).

6. Mount in DPX (p. 417).

### Diaminobenzidine/Metal

The diaminobenzidine substrate for horseradish peroxidase can be made more sensitive by adding metal salts such as cobalt or nickel to the substrate solution. The reaction product is slate gray to black, and the products are stable in both water and alcohol. DAB/metal staining is compatible with a wide range of common histological stains.

1. Dissolve 6 mg of DAB (use DAB tetrahydrochloride) in 9 ml of 0.05 M Tris buffer (pH 7.6).

2. Add 1 ml of a 0.3% wt/vol stock solution of nickel chloride in $H_2O$ (the same amount of cobalt chloride can be used as an alternative).

3. Add 0.1 ml of a 3% solution of $H_2O_2$ in $H_2O$. $H_2O_2$ generally is supplied as a 30% solution and should be stored at 4°C, at which it will last about 1 month.

4. If a precipitate appears, filter through Whatman No. 1 filter paper (or equivalent).

5. Apply to specimen, incubate for 1–20 min. Stop the reaction by washing in $H_2O$.

6. (**Optional**) Counterstain if necessary (see p. 415).

7. Mount in DPX (p. 417).

### Chloronaphthol

Chloronaphthol gives a blue-black product. It is less sensitive than DAB, and the products are soluble in alcohol. It can be used when the DAB reaction gives too high a background, or if the alternate product colors are required.

1. Prepare a stock solution of chloronaphthol by dissolving 0.3 gram of chloronaphthol in 10 ml of absolute ethanol and storing at $-20°C$.

2. Add 100 $\mu$l of chloronaphthol stock with stirring to 10 ml of 0.05 M Tris (pH 7.6).

3. Add 0.1 ml of 3% $H_2O_2$ in $H_2O$. $H_2O_2$ generally is supplied as a 30% solution and should be stored at 4°C, at which it will last about 1 month.

4. A white precipitate forms that is removed by filtering through Whatman No. 1 filter paper (or equivalent).

5. Apply to specimen, incubate for 10–40 min at room temperature. Stop the reaction by washing in $H_2O$.

6. (**Optional**) Counterstain if necessary. Only aqueous stains can be used (see p. 415).

7. Mount in Gelvatol or Mowiol (p. 418).

**Aminoethylcarbazole**

Aminoethylcarbazole (AEC) yields a red product. It is less sensitive than DAB, but can be used if the DAB reaction gives too high a background or if the alternate product colors are required. The products are soluble in alcohol, but not in water.

1. Dissolve 4 mg of AEC in 1 ml of *n,n*-dimethyl formamide (DMF). AEC is stable in DMF, so a stock can be prepared by dissolving 0.4 gram in 100 ml of DMF.

2. Add 1.0 ml of the AEC solution to 15 ml of 0.1 M sodium acetate buffer (pH 5.2) with stirring.

3. Add 0.15 ml of 3% $H_2O_2$ in $H_2O$. $H_2O_2$ generally is supplied as a 30% solution and should be stored at 4°C, at which it will last about 1 month.

4. Filter through Whatman No. 1 (or equivalent).

5. Apply to specimen, incubate for 10–40 min at room temperature. Stop the reaction by washing in $H_2O$.

6. (**Optional**) Counterstain if necessary. Only aqueous stains can be used (see p. 415).

7. Mount in Gelvatol or Mowiol (p. 418).

## NOTE

i. The AEC product fades if oxidized. Take special care to avoid air bubbles on mounting.

**DETECTION**

## ALKALINE PHOSPHATASE-LABELED REAGENTS

BCIP/NBT is the most commonly used of the chromogenic substrates for alkaline phosphatase. It is somewhat more sensitive than NABP/New Fuchsin, but gives less contrast during microscopic observation.

Coupled reagents should be prepared with eukaryotic alkaline phosphatase, as this enzyme is readily inactivated with EDTA. The bacterial enzyme is difficult to stop, causing overdevelopment and leading to high background.

### Naphthol-AS-BI-Phosphate/New Fuchsin (NABP/NF)

Naphthol-AS-BI-phosphate (NABP)/New Fuchsin produces an intense red product that is insoluble in alcohols as well as aqueous solutions and is compatible with a range of histochemical stains.

1. Dissolve 1 mg of New Fuchsin in 0.25 ml of 2 N HCl.

2. Dissolve 1 mg of sodium nitrate in 0.25 ml of $H_2O$.

3. Dissolve 10 mg of naphthol AS-TR phosphate (sodium salt) in 0.2 ml of dimethylformamide.

4. Mix the New Fuchsin solution with the sodium nitrate solution. Shake for 1 min.

5. Add the mixture to 40 ml of 0.2 M Tris (pH 9.0).

6. Add the naphthol AS-TR solution to the above mixture.

7. Apply to specimen, incubate for 10–40 min at room temperature. Stop the reaction by washing with 20 mM EDTA.

8. (**Optional**) Counterstain if necessary (see p. 415).

9. Mount in DPX (p. 417).

### Bromochloroindolyl Phosphate/nitro Blue Tetrazolium (BCIP/NPT)

The bromochloroindolyl phosphate/nitro blue tetrazolium (BCIP/NBT) substrate generates an intense black-purple precipitate at the site of enzyme binding. The reaction proceeds at a steady rate, thus allowing accurate control of the development of the reaction. This allows the relative sensitivity to be controlled by the length of incubation.

1. Prior to developing the cell staining, prepare the three stock solutions. (1) NBT: Dissolve 0.5 gram of NBT in 10 ml of 70% dimethylformamide. (2) BCIP: Dissolve 0.5 gram of BCIP (disodium salt) in 10 ml of 100% dimethylformamide. (3) Alkaline phosphatase buffer: 100 mM NaCl, 5 mM $MgCl_2$, 100 mM Tris (pH 9.5). All stocks are stable at 4°C for at least 1 year.

2. Just prior to developing, prepare fresh substrate solution. Add 66 $\mu$l of NBT stock to 10 ml of alkaline phosphatase buffer. Mix well and add 33 $\mu$l of BCIP stock. Use within 1 hr.

3. Place the washed cells in a suitable container. Add enough substrate solution to cover the cells (for a 100-mm dish, 3–5 ml is appropriate). Develop at room temperature with agitation until the stain is suitably dark. Periodic monitoring under the microscope may be necessary for some antigens. A typical incubation would be approximately 30 min.

4. To stop the reaction, rinse with PBS containing 20 mM EDTA, chelating the $Mg^{2+}$ ions.

5. (**Optional**) Counterstain if necessary (see p. 415).

6. Wash in $H_2O$ and mount in DPX (p. 417).

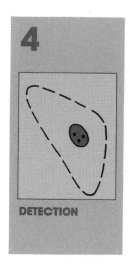

**DETECTION**

## β-GALACTOSIDASE-LABELED REAGENTS

β-Galactosidase has been used extensively as a label in enzyme immunoassays, but has only recently become popular for immunocytochemistry. One good substrate is 5-bromo-4-chloro-3-indolyl-β-D-galactopyranoside (BCIG), which gives an intense blue product. The product is stable and insoluble in alcohol as well as $H_2O$.

### Bromochloroindolyl-β-D-Galactopyranoside (BCIG)

1. Dissolve 4.9 mg of BCIG in 0.1 ml of dimethylformamide.

2. Add the 0.1 ml of the BCIG solution to 10 ml of PBS containing 1 mM $MgCl_2$, 3 mM potassium ferrocyanide.

3. Filter through Whatman No. 1 filter paper (or equivalent).

4. Apply to specimen, and incubate for 10–40 min at room temperature. Stop the reaction by washing with water.

5. (**Optional**) Counterstain if necessary (see p. 415).

6. Mount in DPX (p. 417).

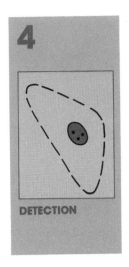

**DETECTION**

## DETECTING FLUOROCHROME-LABELED REAGENTS*

To detect fluorochrome-labeled reagents, a specially equipped microscope is required. The low levels of fluorescence produced in cell staining experiments mean that the microscope must be equipped for epifluorescence in which the exciting radiation is transmitted through the objective lens onto the surface of the specimen. Absorbing radiation of the appropriate wavelength causes the electrons of the fluorochrome to be raised to a higher energy level. As these electrons return to their ground state, light of a characteristic wavelength is emitted. This emitted light forms the fluorescent image seen in the microscope. Individual fluorochromes have discrete and characteristic excitation and emission spectra. Filters are used to ensure that the specimen is irradiated only with light at the correct wavelength for excitation. By placing a second set of filters in the viewing light path that only transmit light of the wavelength emitted by the fluorochrome, images are formed only by the emitted light. This produces a black background and a high-resolution image.

Because some fluorochromes have emission spectra that do not overlap, two fluorochromes can be observed on the same sample. This allows the study of two different antigens in the same specimen even when they have identical subcellular distributions.

The most commonly used fluorochromes are fluorescein and rhodamine. They can be conjugated to anti-immunoglobulin antibodies, protein A, protein G, avidin, or streptavidin. These conjugates are available from many commercial sources or can be prepared as described on p. 353. Filter sets are commonly available that will permit independent observation of these two fluorochromes in the same sample. The properties of the commonly used fluorochromes are listed in Table 10.3.

*Coons et al. (1941); Coons and Kaplan (1950); reviewed in Osborn and Weber (1982).

**TABLE 10.3**
**Properties of Fluorochromes Used in Cell Staining**

| Fluorochrome | Excitation[a] | Emission[a] | Color |
|---|---|---|---|
| DAPI | 365 | >420 | Blue |
| Fluorescein | 495 | 525 | Green |
| Hoechst 33258 | 360 | 470 | Blue |
| R-phycocyanin | 555, 618 | 634 | Red |
| B-phycoerythrin | 545, 565 | 575 | Orange, red |
| R-phycoerythrin | 480, 545, 565 | 578 | Orange, red |
| Rhodamine | 552 | 570 | Red |
| Texas red | 596 | 620 | Red |

Data from Hansen (1967) and various commercial technical brochures.
[a]Values given in nm.

The fluorochrome, Texas red, is also used occasionally for immunofluorescence. It can be detected using the same filter sets as rhodamine.

1. After antibody binding, the fluorochrome-labeled samples are ready for observation. Specimens that are already bound to a solid support can be used directly. Drain the buffer from the slide. Dry the slide around the sample with a paper towel.

   Suspension cells need to be transferred to a clean glass slide at a concentration of approximately $10^5$ cells/ml. Carefully remove as much of the wash buffer as possible. In some cases, using the corner of a paper towel may be gentle enough to remove the buffer. If not, allow to air-dry partially. Do not allow the specimen to dry out.

2. Mount the specimens in Gelvatol or Mowiol as described on p. 418. Observe and photograph under the fluorescent microscope.

## NOTES

i. Fluorescence detection is not compatible with most histochemical stains, because the components of most of these stains autofluoresce strongly. Fluorescence detection is not compatible with enzyme detection systems, because the deposition of insoluble compounds after enzyme detection will block the emission of light from the fluorochrome.

ii. To limit the amount of irradiation of the fluorochrome, it is a good idea to focus and scan the specimen initially using phase-contrast optics with transmission from the UV source blocked. Another approach is to use a second fluorescent probe that emits light at another wavelength. Some of the DNA-intercalating agents such as Hoechst 33258 or DAPI (4',6-diamidino-2-phenylindole) are useful in this respect. Nuclei can be stained with these dyes by including a few microliters of dye (to 0.0005%) in the second antibody incubation or in a subsequent wash buffer. The absorption and emission curves of these dyes do not overlap the narrow band filters for fluorescein or rhodamine. Filters for DAPI and Hoechst 33258 are relatively inexpensive.

## Choosing the Correct Fluorochrome

The choice of fluorochromes is limited primarily by filter sets that are available for microscopy. Most filter sets are best matched to the properties of rhodamine or fluorescein. Texas red can be used with rhodamine filter sets, but its emission spectra is not matched exactly by these filters. The increasing availability of phycobiliproteins, which are theoretically about 50 times brighter, is going to have substantial influence on the design of immunofluorescence experiments over the next few years as other filter sets become available (Oi et al. 1982).

Fluorescein and rhodamine are the most frequently used. Fluorescein emits a yellow-green light that is detected well by the human eye and by most films. However, fluorescein is prone to rapid photobleaching, and bleaching retardants such as DABCO should be added to the mounting medium (see pp. 418 and 419). Rhodamine emits a red color and is not as prone to fading as fluorescein, but the rhodamine conjugates are more hydrophobic and therefore yield higher backgrounds than fluorescein. Texas red also emits a strong red light, and its emission spectra is further from the emissions of fluorescein than rhodamine. It is the least likely to produce problems of fading, but it is not as widely available as either fluorescein or rhodamine.

When using double labels, fluorescein can be combined with either Texas red or rhodamine conjugates.

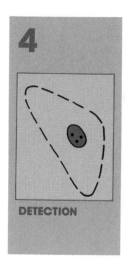

**DETECTION**

## DETECTING GOLD-LABELED REAGENTS*

Colloidal gold particles bind tightly but not covalently to proteins at pH values around the protein's pI. Colloidal gold particles conjugated with a wide range of anti-immunoglobulin antibodies, protein A, or streptavidin are available commercially. Because some of the bound protein may slowly dissociate from the gold particles, the colloid should be washed before use to remove free protein.

Gold labels were developed originally for electron microscopic studies (for example, see Roth et al. 1978), but they also work well at the level of the light microscope. They give higher resolution than enzyme-based methods and avoid the problems of substrate preparation and endogenous enzyme activity. Until recently the gold labels lacked sensitivity at the level of light microscopy, but the recent development of the photochemical silver method of amplification has overcome this problem.

Unamplified gold labels can be detected under the light microscope using bright-field illumination where the label ranges from pale pink to deep red, depending on the strength of reaction. Nomarski differential interference contrast microscopy makes the label appear dark red to black. With the silver enhancement method, the gold particles become coated in metallic silver and yield a black-brown label, best visualized by bright-field optics. Gold labeling methods are compatible with many histochemical stains. Gold labeling reactions are very readily controlled, as the appearance of staining can be monitored directly and continuously under the microscope.

1. Rinse the specimen in distilled water briefly. The samples can be examined directly or can be processed for silver enhancement. For direct observation mount in Gelvatol or Mowiol (p. 418).

2. For silver enhancement, transfer the sample to a darkroom equipped with a safe light. To prepare the developer, mix two parts of 0.5 M sodium citrate (pH 3.5) with three parts of 5.6% hydroxyquinone and 12 parts water. Immediately before use, add three parts of 0.73% silver lactate.

3. Immerse the slides with the gold-labeled specimens into this developer. Incubate for 2–3 min at room temperature.

4. Rinse briefly in 1% acetic acid and fix in a standard photographic fixative for several minutes.

*Danscher (1981); Danscher and Nörgaard (1983); Holgate et al. (1983).

5. (**Optional**) Counterstain if necessary (see p. 415).

6. Samples are now ready for mounting in DPX (p. 417).

## NOTE

i. Recently, several commercial companies have introduced silver enhancement kits that do not require using a darkroom, permitting development to be monitored under the microscope.

**DETECTION**

## DETECTING IODINE-LABELED REAGENTS*

The binding of radiolabeled reagents to cells is detected by coating the specimen with a photographic emulsion. When the emulsion is developed, silver grains can be seen at the location of radiolabel. The silver grains appear as black dots in bright-field microscopy but are most readily detected by dark-field illumination where they are seen as intense silver reflections on a black background. The detection of iodine-labeled reagents is extremely sensitive, and the reaction is readily controlled by setting up duplicate samples and varying the exposure time.

The method is compatible with histological stains and can be used in double-labeling procedures with enzyme- or fluorochrome-labeled reagents. When iodine-labeled and enzyme-labeled reagents are used together for double labeling, the enzyme-labeled reagents must be developed before the photographic emulsion is added.

1. After the samples have been washed following antibody addition, rinse the specimen briefly in water.

2. (**Optional**) Counterstain or develop enzyme label, if necessary.

3. Allow to air-dry.

4. Transfer the specimen to the dark room. Dip the sample in photographic emulsion (Ilford L-4 or equivalent, diluted 1 : 1 in water). Avoid bubbles.

5. Dry under a gentle stream of air. After the emulsion is dry, transfer to a light-tight container. Expose at 4°C.

6. After suitable exposure times (2 days, 10 days, and 3 weeks will usually bracket the times), return the samples to the darkroom for developing (3 min in Kodak D-19 or equivalent at 15°C, rinse in water, 10 min in Kodak Ektaflo or equivalent at 15°C). Rinse in water.

7. Mount in DPX (p. 417).

*Adapted from Berenbaum (1959).

## Counterstains

Counterstains are used to help differentiate the various cell types or subcellular structures seen in cell staining. It is essential for tissue sections, allowing the identification of the cell types, but also may be helpful in other staining reactions. Counterstains should not be used with fluorochromes, as the commonly used counterstains autofluoresce. To choose an appropriate counterstain, first determine whether the cell staining detection reagent (often a chromogenic substrate) is soluble in alcohol.

*Alcohol insoluble*

1. Wash the slide gently with water.

2. Add a few drops of Harris' Hematoxylin (available commercially) to the specimen. Incubate for approximately 5 min. The length of time will determine the intensity of the stain.

3. Wash the slide gently in water.

4. Dip the slide in 0.5% glacial acetic acid/99.5% ethanol for 10 sec.

5. Wash the slide gently in water.

6. Mount in DPX (p. 417).

*Alcohol soluble*

1. Wash the slide gently with water.

2. Add a few drops of Mayer's Hematoxylin (available commercially) to the specimen. Incubate for approximately 5 min. The length of time will determine the intensity of the stain.

3. Wash the slide gently in water.

4. Dip the slide repeatedly into 30 mM $NH_4OH$. Continue until the stain turns blue.

5. Wash the slide gently in water.

6. Mount in Gelvatol or Mowiol (p. 418).

## *Mounting*

Mounting media for immunohistology must be compatible with the detection method used. There are two classes of mounting media, the aqueous and nonaqueous. Suitable aqueous media are made from Gelvatol or Mowiol. If these are not available, glycerol can be substituted, but permanent mounts are not possible. A suitable nonaqueous mounting medium is DPX which is available commercially. DPX is named for its components, 10 grams of distrene 80, 5 ml of dibutyl phthalate, and 35 ml of xylene.

## DPX

1. Samples for mounting in DPX must be dehydrated by passing through graded alcohols. Incubate twice for 3 min each in 75% ethanol, twice for 3 min each in 95% ethanol, and twice for 3 min each in absolute ethanol. Air-dry. Alternately, air-dry directly without passing through the ethanol washes.

2. Add a small drop of DPX to the specimen. If the sample is on a slide or tissue culture dish, carefully place a coverslip (#1) on the drop, avoiding air bubbles. If the sample is on a coverslip, invert the coverslip on a clean glass slide (sample side down).

3. Remove any excess mount with a paper towel.

DPX will set almost immediately and the samples can be observed or photographed as described on p. 419.

## GELVATOL OR MOWIOL*

1. To prepare Gelvatol or Mowiol: Add 2.4 grams of Gelvatol 20-30 (Monsanto Chemicals) or Mowiol 4-88 (Hoechst) to 6 grams of glycerol. Stir to mix. Add 6 ml of water and leave for several hours at room temperature. Add 12 ml of 0.2 M Tris (pH 8.5), and heat to 50°C for 10 min with occasional mixing. After the Mowiol or Gelvatol dissolves, clarify by centrifugation at 5000g for 15 min. For fluorescence detection, add 1,4-diazobicyclo-[2.2.2]-octane (DABCO) to 2.5% to reduce fading. Aliquot in air-tight containers and store at −20°C. Stocks of these mounts are stable at room temperature for several weeks after thawing.

2. Gelvatol and Mowiol can be used directly on the washed specimen.

3. Add a small drop of mounting media to the specimen. If the sample is on a slide or tissue culture dish, carefully place a coverslip (No. 1) on the drop, avoiding air bubbles. If the sample is on a coverslip, invert the coverslip on a clean glass slide (sample side down).

4. Remove any excess mount with a paper towel.

   The mounting media will set overnight. To observe immediately, secure the coverslip to the slide by placing a small drop of nail polish at the edges of the coverslip. The samples can be observed or photographed as described on p. 419.

---

*Heimer and Taylor (1974); Osborn and Weber (1982).

## *Photographing the Samples*

The results of cell staining experiments can be stored on film or videotape. Photography of enzyme-, gold-, or iodine-labeled samples presents no special difficulties. The low light levels emitted by fluorochromes make photographing these specimens more difficult. Long exposure times lead to fading of the emissions, increased chances of vibration, and high backgrounds. Any measures that will shorten the exposure times will improve the quality of the photograph. Objectives should be chosen with the highest available numerical apertures, the microscope should be adjusted to allow 100% of the emitted light to reach the film plane, and use fast films (400 ASA is suitable for many applications, but these films can also be exposed at 800 or 1600 and push-processed in high contrast developer). Black-and-white film has a finer grain and is suitable for all photography except for double labels and color slides for presentations.

## Fluorescence Quenching

Although other fluorochromes such as rhodamine or Texas red are less susceptible to fading, quenching of fluorescein is a serious problem. The problems caused by fading can be minimized in two ways: first by limiting the exposure of the specimen to exciting radiation, and second by including specific anti-fade reagents in the mounting media.

The specimen should always be located under phase-contrast illumination with the UV light source shielded. An initial assessment of the staining reaction should be made and a photograph taken as soon as possible.

A wide range of anti-fade reagents can be used. They appear to work by scavenging free radicals liberated by excitation of the fluorochromes. The free radicals attack unexcited fluorochromes and damage them, thus producing exponential fading. The most useful anti-fade compound is 1,4-diazobicyclo-[2.2.2]-octane (DABCO), because of its solubility and chemical stability (see also Johnson et al. 1981, 1982 for use of *p*-phenylenediamine; Giloh and Sedat 1982 for use of *n*-propyl gallate).

## COMMENTS ■ Troubleshooting for Bad Backgrounds

Background problems in cell staining come primarily from two sources, nonspecific sticking and specific cross-reactions. These problems are discussed separately below. Other problems are encountered with particular detection methods and include autofluorescence and endogenous enzyme activities. Resolving these problems is covered for each detection method within its particular section.

To identify the source of background problems, compare control samples that omit each of the various stages of the staining protocol. Controls must include a nonimmune antibody, no primary antibodies, and no antibodies at all (see p. 391).

*Nonspecific sticking*   Nonspecific background problems are not caused by antigen binding. They are due to either the primary antibodies or secondary reagents binding to the specimen through interactions that do not involve the antigen combining site.

• Spin all antibody preparations or secondary reagents at 100,000g for 30 min.

• Titrate the concentration of the primary antibody and detection reagents to the minimum needed to produce a suitable signal.

• After fixation, soak the specimen in saturating amounts of nonspecific proteins that will not be recognized by the detecting agent. Useful blocking solutions include 5% serum derived from the same species as the labeled antibody, 3% BSA, and 10% nonfat dry milk.

• Dilute the antibody and detection reagents in one of the above blocking solutions.

• After fixation, add 0.2% Tween 20 to all buffers and wash solutions.

• Reduce the incubation time of the primary antibody or the labeled reagent.

• Increase number and duration of washes.

• Switch detection methods.

*Specific backgrounds*   Specific background problems arise from three sources. These are contaminating antibodies with spurious reactions, cross-reacting antibodies, and immunoglobulin binding proteins in the sample.

• If using polyclonal antibodies, titrate the antibody (spurious reactions).

• If using polyclonal antibodies, affinity-purify the antigen-specific antibodies (p. 313, spurious reactions).

• If using polyclonal antibodies, block the spurious activities with an appropriate acetone powder (p. 631).

• Try a different antibody preparation (cross-reaction or spurious).

• Incubate the specimen with normal serum derived from the same species as the labeled antibody. Also dilute specific antibody in 1% normal serum from the same species as the labeled antibody (immunoglobulin binding proteins in the sample).

# 11 IMMUNOPRECIPITATION

Immunoprecipitation is one of the most useful of the common immuno-chemical techniques. When immunoprecipitations are coupled with SDS-polyacrylamide gel electrophoresis (p. 636), a number of important characteristics of the antigen can be determined readily. These assays can determine the presence and quantity of the antigen, relative molecular weight of the polypeptide chain, rate of synthesis or degradation, presence of certain post-translational modifications, and interactions with proteins, nucleic acids, or other ligands. These types of characteristics are difficult to determine using other techniques.

The immunoprecipitation procedure can be divided into four steps:

(1) labeling the antigen (optional)
(2) lysis of the cells to release the antigen
(3) formation of the antibody–antigen complexes
(4) purification of the immune complexes

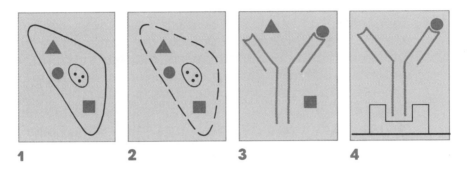

1     2     3     4

In most cases the antigen is labeled by incubating viable cells in medium supplemented with a radioactive precursor, although many antigens can be detected by methods that do not require direct labeling. The antigen is extracted from the cell in an appropriate lysis buffer, and antibodies are added to the lysate to allow formation of the immune complex. A solid-phase matrix containing protein A is added, and the immune complexes are allowed to bind by adsorption of the antibody to protein A. After the protein A–antibody interaction occurs, the unbound proteins are removed by washing the solid phase, leaving the purified antibody–antigen complexes bound to the matrix.

Purified immune complexes usually are analyzed by gel electrophoresis, but they also can be used in a variety of other techniques including enzymatic studies, ligand binding, further immunizations, immunoblotting, or other immunochemical methods. These other techniques, when used in conjunction with immunoprecipitation, can greatly extend the amount of information gathered about an antigen.

## ■ MAJOR CONSTRAINTS

The ease of preparing highly purified antigens by immunoprecipitation depends on two major factors: (1) the abundance of the antigen in the original preparation and (2) the affinity of the antibody for the antigen. Using immunoprecipitations, any soluble polypeptide within a cell can be purified to a sufficient degree that it can be seen as a specific band on an SDS-polyacrylamide gel. Insoluble or highly polymerized antigens may not be able to be studied using immunoprecipitations. Obviously, abundant soluble proteins are easy to detect and analyze by immunoprecipitation. When rare cellular proteins are studied, the immunoprecipitation procedure must achieve purifications in the order of 100,000- to 1,000,000-fold. Although this level of purification would be difficult using other techniques, when immunoprecipitation is coupled to a technique such as SDS-polyacrylamide gel electrophoresis, such rare polypeptide chains can be identified.

The second factor that affects the ease of purification is the affinity of the antibody for the antigen. Comparing the affinities of different monoclonal antibodies provides an excellent example of this problem. Unlike other assays that present the antigen in a highly concentrated local environment, immunoprecipitations rely on the formation of the antigen–antibody complex in solution at relatively low concentrations of the antigen. In practice, this means that quantitative immunoprecipitations normally require antibody affinities of $10^8$ mol$^{-1}$ or higher. Affinities of $10^7$ mol$^{-1}$ may allow the antigen to be detected, but it will not be possible to remove the antigen quantitatively from the solution. Antibodies with affinities of $10^6$ mol$^{-1}$ will score as negative in these assays. These observations mean that some monoclonal antibodies that are positive in other tests may not be usable in immunoprecipitations. The effect of antibody affinity on immunoprecipitation is discussed in detail in Chapter 3 (p. 23).

# CHOICE OF ANTIBODY

Three types of antibody preparations can be used for immunoprecipitations. These are polyclonal antibodies, monoclonal antibodies, and pooled monoclonal antibodies. Their relative advantages and disadvantages are discussed in detail below.

## Immunoprecipitations Using Polyclonal Antibodies

Polyclonal antibodies are the most commonly used reagents for immunoprecipitations. Normally they will contain antibodies that bind to multiple sites on the antigen and therefore will have a much higher avidity for the antigen (see Chapter 3). Having more than one antibody bound to an antigen also has other important advantages. If the immune complexes are collected on any of the solid-phase matrices such as *Staphylococcus aureus* Cowan I (SAC) or protein A beads, the availability of multiple binding sites for the protein A molecules provides a more stable antigen–antibody–protein A complex. Together, multiple antibody–antigen interactions and multiple antibody–protein A interactions provide a multivalent complex that is easy to prepare, stable, and can be treated relatively harshly during the washing procedure.

Although using polyclonal antibodies for immunoprecipitations will often produce stable multivalent interactions, their use also yields higher nonspecific backgrounds than the use of other types of antibodies. Multiple interactions that lead to forming large complexes are more apt to trap or bind nonspecific proteins.

Because polyclonal antibodies normally are used as whole sera, they will contain the entire repertoire of circulating antibodies found in the immunized animal at the time the serum was collected. Therefore, the serum may contain antibodies that specifically recognize spurious antigens. For example, this is particularly troublesome in the analysis of microbial antigens. Sera may contain antibodies against bacteria or fungi that have recently infected the host animal. If the organism under study has proteins that are identical or similar to the ones found in these microorganisms, these antigen–antibody interactions will contaminate the immunoprecipitations. Because this type of contamination is specific, it cannot be removed by methods that are designed to lower nonspecific background (e.g., adding BSA, etc.). In these cases the easiest method to remove these activities is to block the specific antibodies by preincubating the serum with a solution that contains the contaminating proteins (e.g., an acetone powder from a microorganism that does not express the antigen being studied, p. 631). Similar types of specific antibody contaminations are found as autoantibodies to a range of tissue antigens, and can be handled using analogous adsorption techniques.

Therefore, because of contaminating activities and increased nonspecific interactions, immunoprecipitations using polyclonal antibodies normally will have higher backgrounds than other antibody preparations. Many of these problems are inherent in this technique, but some of the background can be effectively removed by titrating the amount of antisera needed to immunoprecipitate the antigen. By providing the smallest amount of serum necessary for the quantitative recovery of the antigen, the background can be kept to a minimum. In addition, because of the stability of the complexes, nonspecific background problems may be lessened by more stringent washing.

## ■ Immunoprecipitations Using Monoclonal Antibodies

The biggest advantage of using monoclonal antibodies for immunoprecipitations is the specificity of their interactions. Because monoclonal antibodies bind to only one epitope, they provide an excellent tool to identify a particular structure on an antigen. Given the right antibody, they can be used not only to detect an antigen but also to distinguish between different forms of the antigen, including conformational changes or post-translational modifications. In addition, because the immune complexes formed using monoclonal antibodies are not usually multimeric and are smaller than those formed when using polyclonal antibodies, there is less of a problem with nonspecific binding. Therefore, the backgrounds are normally cleaner.

Although using monoclonal antibodies for immunoprecipitations may solve or lessen some of the problems found when using polyclonal antibodies, their use also creates another set of difficulties. The most worrisome problem is affinity. Because the antigen is held only by one antibody–antigen interaction (except when the antigen is multimeric), the affinity of the antibody for the antigen is critically important (see discussions of affinity on p. 424). Monoclonal antibodies with affinities lower than about $10^7 \, \text{mol}^{-1}$ are difficult to use in immunoprecipitations. As many screening techniques for hybridoma fusions will detect antibodies with affinities as low as $10^6 \, \text{mol}^{-1}$, not all monoclonal antibodies work well in immunoprecipitations.

Another problem that is commonly encountered when using monoclonal antibodies is their varied affinities for protein A. Most immunoprecipitations use solid-phase immunoadsorbants such as SAC or protein A beads to collect the immune complexes. Unless the antigen is multimeric, there will be only one antibody bound to the antigen, so the antigen will be linked to the immunoadsorbant only through this single antibody molecule. This makes the affinity of the antibody–protein A interaction as critical for the detection of the antigen as the antibody–antigen interaction. Unfortunately, not all subclasses of IgG molecules bind well to protein A. Among the rat monoclonal antibodies, only those from the $IgG_{2c}$ subclass will bind well to protein A. Mouse monoclonal antibodies are somewhat better. $IgG_{2a}$ antibodies bind with high enough affinity at neutral pH that, for all practical purposes, the interaction can be considered stable. $IgG_{2b}$ molecules bind less well, but with high enough affinity to allow easy detection of the antigen. They do not bind tightly enough to protein A to ensure the quantitative removal of all the antibodies from the solution. Immunoglobulins of the $IgG_3$ subclass have a slightly lower affinity for protein A than the $IgG_{2b}$ molecules, and like the $IgG_{2b}$ antibodies, the interactions are sufficiently strong to allow the detection, but not the quantitation, of an antigen. The lowest affinities are found with antibodies from the $IgG_1$ subclass. Here, the affinities are low enough that the immune complexes may not be able to be purified by binding to protein A immunoadsorbants. Any of these protein A affinity problems may be circumvented by using intermediate antibodies such as rabbit anti-mouse or anti-rat immunoglobulin antibodies (p. 624). Because rabbit antibodies have a very high affinity for protein A, this bridging layer will make the interactions stronger both by providing a higher affinity for protein A and by making the antibody–protein A interaction multivalent. However, the addition of rabbit anti-immunoglobulin antibodies to the immune complexes now adds some of the problems of using polyclonal antibodies, including dirtier backgrounds.

A third problem with using monoclonal antibodies is the possibility of detecting spurious cross-reactions with other polypeptides. Because an epitope can be a relatively small protein structure, often composed of only 4 or 5 amino acids (p. 23), there is a reasonable chance that a similar epitope can be found on another polypeptide. In some cases, the common epitopes will form part of an important structural similarity between antigens, and monoclonal antibodies can be used to detect related antigens. Alternatively, the antibodies may detect small structural similarities confined only to the antibody combining site. This is particularly true for antibodies that recognize denaturation-resistant epitopes. Presumably this occurs because these antibodies recognize features found in the primary structure of the polypeptides. Depending on the set of hybridomas, as many as one in five monoclonal antibodies will display these types of cross-reactions. Because of the frequency of these cross-reactions, the precipitation of an unexpected polypeptide should be treated as a contaminant until proven otherwise.

## Immunoprecipitations Using Pooled Monoclonal Antibodies

Using pools of monoclonal antibodies in immunoprecipitation takes advantage of the best properties of both polyclonal and individual monoclonal antibodies. The monoclonal antibodies provide specificity, and the use of multiple antibodies allows the formation of stable multivalent complexes. Consequently, pooled monoclonal antibodies are the best choice of reagents for most immunoprecipitations. Unfortunately, not all antigens have been studied in enough detail to have a set of monoclonal antibodies available for pooling. However, even the use of two antibodies specific for two separate epitopes will greatly increase the avidity for the antigen as well as for protein A. Therefore, whenever possible, pooled monoclonal antibodies should be used for immunoprecipitations.

## SUMMARY

# Immunoprecipitation of Radiolabeled Protein Antigens
Detection limit is approximately 100 cpm/protein

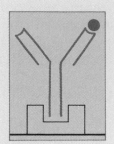

- Multiple steps over several days
- Determines antigen presence, quantity, size, turnover, modifications, complexes
- Semi-quantitative to quantitative
- Sensitivity dependent on specific activity of antigen and affinity of antibody
- Needs high affinity/avidity antibody

|  | Polyclonal Antibodies | Monoclonal Antibodies | Pooled Monoclonal Antibodies |
|---|---|---|---|
| **Signal Strength** | Excellent | Antibody dependent (poor to excellent) | Excellent |
| **Specificity** | Usually good, but some background | Excellent, but some cross-reactions | Excellent |
| **Good Features** | Stable, multivalent interactions | Specificity Unlimited supply | Stable, multivalent interactions Specificity Unlimited supply |
| **Bad Features** | Nonrenewable Background | Need high-affinity Affinity for solid phase | Availability |

# IMMUNOPRECIPITATION PROTOCOLS

Most immunoprecipitations of proteins involve four steps: (1) labeling the antigens (optional), (2) lysing cells, (3) forming the immune complexes, and (4) purifying the immune complexes. The different techniques for each of the four steps are described below.

## Labeling Protein Antigens

Protein antigens can be labeled by several different techniques. They can be labeled by growing cells in the presence of a radioactive amino acid or a radioactive precursor of an amino acid, by iodination of surface proteins, by treatment of surface proteins with radioactive sodium borohydride, or by iodination of the proteins after they have been purified by immunoprecipitation. Most commonly, proteins are labeled by growing cells in the presence of a radioactive precursor.

For some purposes the proteins from the target cells need not be labeled prior to extraction. These include experiments that detect proteins by staining of SDS-polyacrylamide gels, by immunoblotting, by checking for enzymatic activities in the immunoprecipitated proteins, or by testing for ligand binding in the immune complexes.

**LABELING**

## METABOLIC LABELING OF TISSUE CULTURE CELLS

Most often, proteins from cells in tissue culture are labeled by growing cells in the presence of radioactive amino acids. This is convenient for most tissue cultures cells and for most proteins.

For metabolic labeling studies, a protein may label poorly if it only has a few amino acid residues of the radioactive precursor or because it has a low turnover rate. If this becomes a problem, try labeling the cells for longer times (see p. 437) or switching the radioactive precursor.

1. Cells should be healthy and growing rapidly at the time of labeling.

   For *monolayer cultures*, cells should be approximately 75% confluent. Remove the medium and wash the monolayer once with medium without methionine (prewarmed to 37°C). After removing the medium, add approximately 2.0 ml of medium without methionine per 100-mm dish (prewarmed to 37°C, use 0.5 ml/60-mm dish; 0.25 ml/35-mm dish; 0.2 ml/24 well, 0.02 ml/96 well). The amount of medium that is added to the dish may need to be adjusted depending on whether the dish will be rocked occasionally. Slightly larger volumes may be needed for longer labeling or for unattended labeling. Smaller volumes of medium without methionine can be used, but the plates should be rocked during the labeling period. Plates should be observed for drying or changes in pH.

   For *suspension cultures*, collect the cells by centrifugation at 400g for 5 min. Resuspend the cell pellet in medium without methionine (prewarmed to 37°C) and centrifuge as before. Remove the wash buffer as completely as possible. Resuspend the cells at approximately $10^7$ cells/ml in medium without methionine and transfer to a small tissue culture plate.

2. Add the [$^{35}$S]methionine. The amount of [$^{35}$S]methionine to be added depends on a number of variables (see Table 11.1 for some examples).

   For most cells incubated in methionine-free medium supplemented only with [$^{35}$S]methionine, incorporation of radiolabel into total protein will be linear for approximately 6–8 hr. For labeling periods less than this, no additional methionine is needed. By comparing the amount of TCA-precipitable counts (p. 678 or 679) that are incorporated versus time, the exact parameters for a particular cell can be determined. For labeling times beyond the linear portion of this curve, additional cold methionine should be added. See p. 437 for more details on longer labeling times.

3. Incubate at 37°C in a $CO_2$ incubator for the desired length of time. As a starting point for new antigens, 2–4 hr is appropriate.

4. For *monolayer cultures*, remove the medium and wash the monolayer once with PBS.

   For *suspension cultures*, transfer the cells to a test tube and centrifuge at 400g for 5 min. Remove the medium and wash the cells once by resuspending the cells in PBS and centrifuging at 400g.

   The spent medium and the wash should be combined and disposed of in accordance with your institute's procedures for handling radioactive waste.

5. Drain well and process for the appropriate assay.

Cells can be lysed by the techniques listed on p. 448.

## Metabolic Labeling with [$^{35}$S]Methionine

The most commonly used radioactive precursor for labeling proteins is [$^{35}$S]methionine. The $^{35}$S decay is easier to detect than either $^3$H or $^{14}$C, and the $^{35}$S signal is easily enhanced using fluorography (p. 647). In addition, the intracellular pool of methionine is smaller than many other amino acids; therefore, exogenous methionine is easily incorporated into proteins. One other advantage to the use of [$^{35}$S]methionine is that its incorporation into proteins is linear over a wide range of added label. For mammalian tissue culture cells, this is true at least to 5 mCi/10$^6$ cells. This means that the number of counts that are incorporated is dependent primarily on the amount of [$^{35}$S]methionine added to the culture. Table 11.1 suggests some appropriate amounts for the radiolabeling of representative mammalian proteins.

[$^{35}$S]Methionine is available commercially from several distributors at relatively high specific activities (1000 Ci/mmole) and at several levels of purity. For most labeling purposes the impure [$^{35}$S]methionine is fine (not for in vitro translations, however). At the time of going to press, only ICN sells impure [$^{35}$S]methionine (mixed with [$^{35}$S]cysteine). It is available at approximately one-half the best quoted price for purified [$^{35}$S]methionine.

## NOTES

i. For secreted proteins save the labeling media and use for immuno-precipitations. Proceed directly to addition of antibody (p. 464). No lysis is necessary. Dispose of the cell pellet using standard radioactive waste procedures.

ii. This technique can be modified to label proteins with any amino acids by changing both the radioactive precursor and the medium. Other amino acids that are commonly used to label proteins include [$^{35}$S]cysteine, [$^{3}$H]proline, and [$^{3}$H]leucine. In each case, the medium should be changed to one that lacks the amino acid that is used for labeling.

For some purposes, such as sequencing of radioactively labeled proteins, polypeptides can be labeled with two or more amino acids by incubating cells in medium supplemented with two or more radioactive precursors. For high-specific-activity labeling of many proteins, combining [$^{35}$S]methionine with [$^{35}$S]cysteine is a good choice.

Several companies offer amino acid kits for preparing medium without an individual amino acid.

iii. Similar studies can be used for the detection of certain post-transcriptional modifications by changing the radioactive precursor to one that is specific for the modifying group. These types of techniques are described on p. 435.

iv. By using short labeling times, followed by the removal of the label and addition of saturating amounts of cold amino acids, the rates of turnover of the protein can be determined. These pulse-chase experiments can be used with several of the different amino acid precursors and are described on p. 437. Pulse-chase experiments can also be used to analyze other post-translational processing events such as proteolytic cleavage.

v. With the appropriate changes in conditions, long-term labeling can be used to estimate the relative amount of a protein (see p. 437).

vi. Immunoprecipitation is one of the easiest methods to look for potential interactions between the antigen and other macromolecules such as proteins (p. 443).

vii. In addition, the amount of radioactivity incorporated can be used to determine the levels of total protein synthesis and as a general measure of the relative metabolic rate of different cells. The incorporated counts can be determined by TCA precipitation (p. 678 or 679).

**TABLE 11.1**
**Suggested Amounts of [$^{35}$S]Methionine to Use**

| Abundance | Molecules per cell | Examples | $\mu$Ci for 3 hr label[a] (autoradiography/fluorography) |
|---|---|---|---|
| Major | $>10^7$ | Cytoskeleton Nucleoplasmin Heat shock Histones | 5/1 |
| Mid | $10^5$–$10^7$ | Topoisomerases Glycolysis enzymes Cytochromes | 50/10 |
| Minor | $10^3$–$10^5$ | Cellular oncoproteins Polymerases Transcription factors Plasma membrane receptors | 500/100 |

This table is designed only to give a general starting point to determine the amount of label to use in an immunoprecipitation. The relative abundance categories are based on the classes of mRNA determined by $R_0t$ analysis. The major factors not taken into account here are the relative rates of protein turnover, ease of extraction of a given protein, and the number of methionine residues per mole of protein. The figures given above are designed to achieve a readable result after an overnight exposure when the lysate is divided into two immunoprecipitation reactions (control and immune) and exposed to X-ray film. Exposures for both autoradiography and fluorography are given.

[a]For longer labeling times, the amount of methionine can be lowered.

## LACTOPEROXIDASE IODINATION OF CELL-SURFACE PROTEINS*

As an alternative to metabolic labeling, many workers label proteins with $^{125}$I. This is most commonly used for cell-surface proteins that are present at low levels and are difficult to detect using metabolic labeling. Although many of the techniques described for labeling antibodies on p. 324 can be modified for iodinating cell-surface proteins, lactoperoxidase labeling offers the most advantages. It is relatively efficient and is gentle enough not to destroy most plasma membranes.

> **Caution**  This procedure must be carried out in an appropriately ventilated fume hood with protection of the operator from direct radiation and with institute-approved monitoring of radiation exposure and waste disposal.

1. Cells should be rapidly growing and healthy at the time of iodination.

   For *monolayer cultures*, remove the medium and wash the monolayer twice with PBS. Drain well and then add 0.5 ml of PBS per 100-mm dish.

   For *suspension cultures*, spin approximately $5 \times 10^6$ to $10^7$ cells at 400$g$. Remove and discard the supernatant. Resuspend the cells in PBS and wash twice by centrifugation and resuspension in PBS. After the last wash, resuspend the cells in 0.5 ml of PBS.

2. Just prior to the iodination, dilute a sample of 30% $H_2O_2$ 1 in 20,000 with PBS. Dilute the lactoperoxidase to 1 unit/ml in PBS (commercial sources of lactoperoxidase are generally about 0.05–0.2 units/$\mu$g). (A stock bottle of 30% $H_2O_2$ will last for approximately 3 weeks at 4°C after it has been opened.)

3. To the cells, sequentially add 100 $\mu$l of the diluted lactoperoxidase and 500 $\mu$Ci of Na$^{125}$I. Mix gently.

4. To initiate the reaction, add 1 $\mu$l of diluted $H_2O_2$. Mix gently. At 1-min intervals over the next 4 min, add a fresh 1 $\mu$l of diluted $H_2O_2$. Terminate the reaction by adding an equal volume of 1 mM dithiothreitol, 1 mg/ml tyrosine in PBS.

5. Wash the cells twice with PBS. The washes should be combined and disposed of in accordance with your institute's procedures for handling radioactive waste.

Cells can be lysed by the techniques listed on p. 446.

*Marchalonis (1969).

## NOTES

i. Sodium azide will block the action of lactoperoxidase, so it should not be added to any of the solutions prior to the iodination.

ii. $^{125}$I can be purchased from several commercial suppliers. Buy Na$^{125}$I in alkaline solutions to keep the oxidation to a minimum. $^{125}$I can be stored for approximately 6 weeks at room temperature.

iii. Occasionally, internal proteins are labeled during cell surface iodinations. This is caused by the lysis of a small percentage of the cells in the preparation. This can be avoided by purifying whole cells on Ficoll gradients.

## Checking for Post-Translational Modifications

By growing cells in the presence of a radioactive precursor specific for a modifying group, immunoprecipitations can be used to identify proteins that undergo post-translational modifications. There are several hundred different post-translational modifications, and many modifications cannot be studied easily using specific radioactive precursors. This is because either the correct precursors are not available, they cannot be taken up by cells, or the intracellular pools are too large to achieve high enough specific activities to identify the group. However, there are a number of modifications that can be studied in this manner. Several of these are summarized below.

| Modification | Label | Radioactive media | Labeling time | References |
|---|---|---|---|---|
| Phosphorylation | [$^{32}$P]ortho-phosphate | phosphate-free medium | 2 hr | |
| O-linked glycoslyation | [$^{3}$H]fucose, [$^{3}$H]mannose, or [$^{3}$H]galactose | standard | varies | |
| N-linked glycosylation | [$^{3}$H]fucose, [$^{3}$H]mannose, or [$^{3}$H]galactose | standard | varies | |
| Myristylation | [9,10-$^{3}$H]myristic acid | standard plus 10% tryptose phosphate,[1] 1% DMSO | 4 hr | Cross et al. (1984) |
| Palmitylation | [9,10-$^{3}$H]palmitic acid | standard plus 10% tryptose phosphate, 1% DMSO, 5 mM sodium pyruvate | 4 hr | Sefton et al. (1982) |

[1]Tryptose phosphate broth is 2% tryptose, 0.2% glucose, 0.5% NaCl, and 0.25% disodium phosphate.

**LABELING**

## *BOROHYDRIDE LABELING OF SURFACE PROTEINS*

1. Cells should be rapidly growing and healthy at the time of labeling.

   For *monolayer cultures*, remove the medium and wash the monolayer twice with PBS. Drain well and then add 1.0 ml of 1 mM sodium periodate in PBS (0°C) per 100-mm dish. Place the dish on ice.

   For *suspension cultures*, spin approximately $5 \times 10^6$ to $10^7$ cells at 400g. Remove and discard the supernatant. Resuspend the cells in PBS and wash twice by centrifugation and resuspension in PBS. After the last wash, resuspend the cells in 1.0 ml of 1 mM sodium periodate in PBS (0°C). Place the cells on ice.

2. Incubate for 10 min on ice.

3. Add glycerol to a final concentration of 25 mM. Mix gently. Wash the cells twice with PBS.

4. Add 1.0 ml of PBS at room temperature containing 1 mCi of [$^3$H]borohydride. Incubate at room temperature for 30 min.

5. Wash the cells twice with PBS. The washes should be combined and disposed of in accordance with your institute's procedures for handling radioactive waste.

Cells can be lysed by the techniques listed on p. 446.

### NOTE

i. Proteins labeled with $^3$H must be detected using fluorographic techniques (p. 647).

## Determining the Half-life of a Protein Antigen

By combining other techniques with immunoprecipitation, the rate of degradation of a protein antigen can be determined. Cells are radiolabeled with an amino acid precursor, the radioactive precursor is removed from the medium, and then large quantities of unlabeled amino acid are added. At different time points after the pulse, representative samples of the labeled cultures are lysed and the antigen purified by immunoprecipitation. The proteins from each chase time point are immunoprecipitated, and the immune complexes are resolved by SDS-poly-acrylamide gel electrophoresis. The amount of radioactivity in the protein band is determined either by densitometric tracing or, preferably, by counting (p. 648). With careful manipulation of the length of the chases, three points that fall on a straight line can be found. These data points can be used to determine the half-life of the protein.

## Steady-State Levels of Proteins

Three variations of the immunoprecipitation technique can be used to determine the steady-state levels of protein antigens. For all three methods, quantitative precipitation of the antigen is essential to determine the approximate levels.

In the first method, the immunoprecipitated proteins are separated on SDS-poly-acrylamide gels and stained with Coomassie blue or silver stain (pp. 636 and 649–653). When compared to protein markers of known amounts, the approximate steady-state level of the protein can be determined. Since different proteins are stained to different extents, these comparisons can be used only to gain an approximation of protein levels.

For the second method, cells are labeled with [$^{35}$S]methionine for extended periods, and the relative level of the protein is determined by radioactive incorporation. Methionine-free medium is supplemented with a small amount of complete medium. For mammalian cells incubated for 8–12 hr use methionine-free medium supplemented with 5% complete medium. For 12 hr or longer incubations use 10% complete medium. Immunoprecipitated proteins are separated on SDS-poly-acrylamide gels and compared to known proteins or to the total level of incorporation as judged by TCA precipitation (p. 678 or 679). This method can be used only to obtain an approximation of the protein level, as there are no controls for the relative rates of protein turnover nor for the number of methionine residues.

A third method for determining the steady-state levels of proteins is to analyze immunoprecipitated proteins on immunoblots. These techniques are described in Chapter 12.

**LABELING**

## *METABOLIC LABELING OF YEAST CELLS WITH RADIOACTIVE AMINO ACIDS*

Many yeast strains can be labeled in complex media such as YPD. Here the highest specific activities of proteins cannot be reached, but the labeling conditions are suitable for most proteins. Alternatively, yeast cells can be labeled in synthetic media lacking methionine.

1. Grow yeast cells overnight in either YPD or the SD synthetic media.

*YPD*

| | |
|---|---|
| Yeast extract | 10 grams |
| Peptone | 20 grams |
| Glucose | 20 grams |

Add water to 1000 ml; sterilize by autoclaving.

*SD (synthetic media)**

| | |
|---|---|
| Bacto-yeast nitrogen base (without amino acids) | 6.7 grams |
| Glucose | 20 grams |
| Adenine sulfate | 20 mg |
| Uracil | 20 mg |
| L-Tryptophan | 20 mg |
| L-Histidine-HCl | 20 mg |
| L-Arginine-HCl | 20 mg |
| L-Methionine | 20 mg |
| L-Tyrosine | 30 mg |
| L-Leucine | 30 mg |
| L-Isoleucine | 30 mg |
| L-Lysine-HCl | 30 mg |
| L-Phenylalanine | 50 mg |
| L-Glutamic acid | 100 mg |
| L-Aspartic acid[1] | 100 mg |
| L-Valine | 150 mg |
| L-Threonine[1] | 200 mg |
| L-Serine | 400 mg |

Add water to 1000 ml; sterilize by autoclaving.

[1]Add after autoclaving media.

2. Dilute the cells in the appropriate medium and grow to a density of $10^7$ cells/ml.

3. Centrifuge the cells at 5000g for 5 min. Resuspend the cells in PBS and centrifuge again. Resuspend the cells at $10^7$ cells/ml in either YPD or synthetic medium without methionine. Incubate at 32°C for 20 min.

*Sherman et al. (1986).

4. Add 10–50 $\mu$Ci of [$^{35}$S]methionine per ml (5-ml cultures will yield enough counts for several immunoprecipitations).

5. Incubate at 32°C with shaking for 1–3 hr. Collect the cells by centrifugation. Wash once in PBS. The spent medium and the wash should be combined and disposed of in accordance with your institute's procedures for handling radioactive waste.

**The cells are ready for lysis. The suggested methods for extraction are described on p. 446.**

## NOTES

i. For secreted proteins save the labeling media and use for immunoprecipitations. Proceed directly to addition of antibody (p. 464). No lysis is necessary. Dispose of the cell pellet using standard radioactive waste procedures.

ii. For most labeling procedures with yeast, the cells should never be allowed to reach stationary phase during the labeling period. For many antigens the length of labeling time must be optimized. Test different incubation times to determine the best for the antigen being studied.

iii. This technique can be modified to label proteins with any amino acids by changing both the radioactive precursor and the medium. Amino acids that are commonly used to label proteins include [$^{35}$S]methionine, [$^{35}$S]cysteine, [$^{3}$H]proline, and [$^{3}$H]leucine. In each case, the medium should be changed to one that lacks the amino acid used for labeling.

iv. Similar studies can be used for the detection of certain post-transcriptional modifications by changing the radioactive precursor to one that is specific for the modifying group. These types of techniques are described on p. 435.

v. By using short labeling times, followed by the removal of the label and addition of saturating amounts of cold amino acids, the rates of turnover of the protein can be determined. These pulse-chase experiments can be used with several of the different amino acid precursors and are described on p. 437.

vi. With the appropriate changes in conditions, long-term labeling can be used to estimate the relative amount of a protein. See p. 437.

vii. Immunoprecipitation is one of the easiest methods to look for potential interactions between the antigen and macromolecules such as proteins (p. 443).

viii. Not all yeast cells will grow in synthetic media. Check the phenotype of your strain for the particular requirements.

## METABOLIC LABELING OF YEAST CELLS WITH RADIOACTIVE SULFATE

Because $^{35}$sulfate is quickly assimilated into the methionine and cysteine pools in yeast cells, this source of radiolabel can be substituted for the more expensive radiolabeled amino acids.

This method is used primarily for *Schizosaccharomyces pombe*, but can be adapted for other species.

1. Yeast cells should be grown in synthetic medium overnight. Complete synthetic media is:

| | |
|---|---|
| KH phthalate | 3.0 grams |
| $Na_2HPO_4$ | 1.8 grams |
| $NH_4Cl$ | 5.0 grams |
| Glucose | 20.0 grams |
| Vitamin stock | 1.0 ml |
| Mineral stock | 0.1 ml |
| Salt stock | 20.0 ml |

pH to 5.6 with NaOH; add water to 1000 ml; sterilize by filtration.

*Vitamin stock (1000×)*

| | |
|---|---|
| Pantothenic acid | 1.0 grams |
| Nicotinic acid | 10.0 grams |
| Inositol | 10.0 grams |
| Biotin | 10 mg |

Add water to 1000 ml; sterilize by filtration.

*Mineral stock (10,000×)*

| | |
|---|---|
| Boric acid | 5.0 grams |
| $MnSO_4$ | 4.0 grams |
| $ZnSO_4$ | 4.0 grams |
| $FeCl_2 \cdot 6H_2O$ | 2.0 grams |
| $H_2MoO_4$ | 1.6 grams |
| KI | 1.0 grams |
| $CuSO_4 \cdot 5H_2O$ | 0.4 grams |
| Citric acid | 10.0 grams |

Add water to 1000 ml; sterilize by autoclaving.

*Salt stock (50×)*

| | |
|---|---|
| $MgCl_2 \cdot 6H_2O$ | 53.5 grams |
| $CaCl_2 \cdot 2H_2O$ | 0.75 grams |
| KCl | 50 grams |
| $Na_2SO_4$ | 2.0 grams |

Add water to 1000 ml; sterilize by autoclaving.

2. Dilute the cells in synthetic medium and grow to a density of $10^7$ cells/ml.

3. Centrifuge the cells at 5000g for 10 min. Resuspend the cells in PBS and centrifuge again. Resuspend the cells at $10^7$ cells/ml in synthetic medium without $Na_2SO_4$. Incubate at 32°C for 20 min.

4. Add 20 $\mu$Ci of inorganic $^{35}$sulfate per milliliter.

5. Incubate at 32°C with shaking for 1–3 hr. Collect the cells by centrifugation. Wash once in PBS. The spent medium and the wash should be combined and disposed of in accordance with your institute's procedures for handling radioactive waste.

**Cells are ready for lysis. The suggested methods for extraction are described on p. 446.**

## NOTES

i. For secreted proteins save the labeling media and use for immunoprecipitations. Proceed directly to addition of antibody (p. 464). No lysis is necessary. Dispose of the cell pellet using standard radioactive waste procedures.

ii. For most labeling procedures with yeast, the cells should never be allowed to reach stationary phase during the labeling period. For many antigens the length of labeling time must be optimized. Test different incubation times to determine the best for the antigen being studied.

iii. Not all yeast cells will grow in synthetic media. Check the phenotype of your strain for the particular requirements.

## METABOLIC LABELING OF BACTERIA WITH RADIOACTIVE AMINO ACIDS

1. Grow bacteria overnight in rich medium such as LB broth (Luria-Bertani, 1% tryptone, 0.5% yeast extract, and 1% NaCl, pH 7.5).

2. Dilute the culture to 0.1–0.2 OD units at 600 nm (approximately a 1/10 dilution) in 5 ml of medium.

   For high-specific-activity labeling, use M9 medium (0.6% $Na_2HPO_4$, 0.3% $KH_2PO_4$, 0.5% NaCl, 0.1% $NH_4Cl$, pH 7.4), sterilized by autoclaving; after cooling add for each liter, 2 ml of 1 M $MgSO_4$, 10 ml of 20% glucose, and 0.1 ml of 1 M $CaCl_2$ from sterile stocks.

   For low-specific-activity labeling, use LB broth.

3. Grow at 37°C with aeration until the OD reaches 0.4. Add 100 $\mu$Ci of [$^{35}$S]methionine.

4. Incubate for 15 min at 37°C with aeration.

5. Collect the cells by centrifugation. Wash once in PBS. The spent medium and the wash should be combined and disposed of in accordance with your institute's procedures for handling radioactive waste.

**Cells are ready for lysis. Cells can be lysed by the techniques listed on p. 446.**

### NOTES

i. For secreted proteins save the labeling media and use for immunoprecipitations. Proceed directly to addition of antibody (p. 464). No lysis is necessary. Dispose of the cell pellet using standard radioactive waste procedures.

ii. This technique can be modified to label proteins with any amino acids by changing both the radioactive precursor and the medium. Amino acids that are commonly used to label proteins include [$^{35}$S]methionine, [$^{35}$S]cysteine, [$^{3}$H]proline, and [$^{3}$H]leucine.

iii. Many strains of bacteria will require additional supplements to permit growth in M9 medium; check before labeling.

## Protein–Protein Complexes

When lysates are prepared under gentle conditions, antibodies will immunoprecipitate not only the specific antigen, but also any other macromolecules that are bound to it. Immunoprecipitations are often the most useful method for studying these types of interactions, and therefore allow workers the possibility of studying heteropolymeric complexes. However, because immunoprecipitations have many points at which specific and nonspecific contaminants can be detected, extreme care must be used before any band is classified as an associated protein. Minimal requirements should include such other criteria as copurification, coprecipitation with several monoclonal antibodies that bind to different regions of the antigen, or coprecipitation with antibodies specific for both members of a complex.

In many cases the choice of buffers or detergents used in the lysis step will determine whether a multimeric complex remains intact. Try various combinations to test for weaker interactions.

LABELING

## METABOLIC LABELING OF BACTERIA WITH RADIOACTIVE SULFATE

1. Grow bacteria overnight in rich medium such as LB broth (Luria-Bertani, 1% tryptone, 0.5% yeast extract, and 1% NaCl, pH 7.5).

2. Dilute the culture to 0.1–0.2 OD units at 600 nm (approximately at 1/10 dilution) in 5 ml of medium.

   For high-specific-activity labeling, use M9 medium (0.6% $Na_2HPO_4$, 0.3% $KH_2PO_4$, 0.5% NaCl, 0.1% $NH_4Cl$, pH 7.4) sterilized by autoclaving; after cooling add for each liter, 2 ml of 1 M $MgSO_4$, 10 ml of 20% glucose, and 0.1 ml of 1 M $CaCl_2$ from sterile stocks.

   For lower-specific-activity labeling, use LB broth.

3. Grow at 37°C with aeration until the OD reaches 0.4.

   For M9 cultures, spin at 5000g for 5 min. Resuspend cells in M9 without sulfate (use 2 ml of 1 M $MgCl_2$ in place of the $MgSO_4$). Add 20 $\mu$Ci of inorganic $^{35}$sulfate per milliliter.

4. Incubate for 15 min at 37°C with aeration.

5. Collect the cells by centrifugation. Wash once in PBS. The spent medium and the wash should be combined and disposed of in accordance with your institute's procedures for handling radioactive waste.

**Cells are ready for lysis. Cells can be lysed by the techniques listed on p. 446.**

## NOTES

i. For secreted proteins save the labeling media and use for immunoprecipitations. Proceed directly to addition of antibody (p. 464). No lysis is necessary. Dispose of the cell pellet using standard radioactive waste procedures.

ii. Many strains of bacteria will require additional supplements to permit growth in M9 medium; check before labeling.

## Iodinating Immunoprecipitated Proteins

An option that is rarely used, but has many potential advantages for detecting minor proteins, is to immunoprecipitate and concentrate proteins from unlabeled cells and then to iodinate the proteins while they are still on the protein A immunoadsorbant. Any of the techniques described on p. 324 would be appropriate for these types of experiments. When using this variation, it is essential to cross-link the antibody to the protein A matrix prior to the immunoprecipitation (see p. 521). This will ensure that the iodinated antibody heavy and light chains do not interfere with the subsequent detection of the antigen. One disadvantage of this procedure is all background proteins, no matter what their source, will be seen in this assay. Therefore steps to remove background problems (pp. 461 and 469) will be helpful.

## ■ Lysing Cells

Cells can be lysed by several different techniques. Cells without a cell wall can be lysed by treating with detergents, by mechanically disrupting the plasma membranes, or by freezing and thawing. If they have a cell wall, the cells can be lysed by removing the cell wall enzymatically and then using detergents to release the antigens, by mechanical shearing of the cells, or by treating the whole cell with harsh denaturing solutions.

Many extraction conditions will release proteases in the lysis buffer. If protease digestion becomes a problem during immunoprecipitations, two approaches can be used to lessen its effects. First, care should be taken to keep the samples cold. Temperature has a profound effect on the rate of degradation by most proteases. Second, the lysis buffers can be supplemented with protease inhibitors. The table on p. 677 lists many of the commonly used protease inhibitors. In immunoprecipitations the two most commonly used inhibitors are aprotinin and phenylmethylsulfonyl fluoride (PMSF); however, a mixture of several of the different protease inhibitors is better, particularly as a starting point for further tests.

## Lysis Buffers

A number of different lysis buffers can be used to release protein antigens from cells, but no one buffer is sufficient for all purposes. In choosing a buffer, there are two important considerations. The antigen must be released efficiently and must still be recognizable by the antibody. When beginning the analysis of a new antigen, it is often necessary to test a number of different extraction buffers to identify the most efficient conditions for the release of the antigen. A good strategy is to lyse cells with a relatively strong buffer. If the antigen is released, then begin to remove and alter the composition of the lysis buffer until the optimal conditions are achieved. Quantitative release should be judged both by testing the amount of antigen in the lysate and by testing the amount of antigen remaining in the cell debris. This can be done most easily by analysis on immunoblots.

In general, the conditions used for lysis should be as gentle as possible to retain the antibody binding sites and to avoid solubilizing background proteins, but harsh enough to ensure quantitative release of the antigen. For detergents this normally will mean choosing nonionic detergents over ionic, lower concentrations over higher, and single detergents over mixes.

Variables that can drastically affect the release of polypeptide antigens include salt concentration, type of detergent, presence of divalent cations, and pH. Salt concentrations between 0 and 1 M, nonionic detergent concentrations between 0.1 and 2%, ionic detergent concentrations between 0.01 and 0.5%, divalent cation concentrations between 0 and 10 mM, EDTA concentrations between 0 and 5 mM, and pHs between 6 and 9 should all be monitored to determine the optimal conditions for extraction. Other possible additions to lysis buffers that may affect some antigens include RNases or DNases. Appendix III contains a short discussion on various detergents that may be choosen.

1. **NP-40 Lysis**   150 mM NaCl, 1.0% NP-40, 50 mM Tris (pH 8.0).

2. **High Salt Lysis**   500 mM NaCl, 1.0% NP-40, 50 mM Tris (pH 8.0).

3. **Low Salt Lysis**   1.0% NP-40, 50 mM Tris (pH 8.0).

4. **RIPA**   150 mM NaCl, 1.0% NP-40, 0.5% DOC, 0.1% SDS, 50 mM Tris (pH 8.0).

## Lysis of Tissue Culture Cells

For most cells grown in tissue culture, the most useful method of lysis is treating with detergents. The note on p. 447 lists several commonly used lysis buffers. Cell membranes can also be broken by physical shearing. Common methods include using potters, blenders, ultrasound generators, and dounce homogenizers. These methods are particularly useful when large volumes are being used or when detergents must be avoided. Of these, the gentlest method is dounce homogenization. However, dounce homogenization is efficient only with larger volumes (several milliliters or more). Of the methods for lysis of tissue culture cells, the freeze-thaw lysis is the least desirable. For reasons that are not particularly clear, repeated freezing and thawing of lysates leads to excessive proteolytic degradation. Unless your antigen is particularly resistant to proteolytic cleavage or you have previously tested this lysis method, it should not be used as a first choice. In general, detergent lysis is the best method for immunoprecipitations.

**2**

**LYSIS**

## DETERGENT LYSIS OF MONOLAYER CULTURES

1. Wash the monolayer once with room temperature PBS. Drain well.

2. Place the tissue culture dish on ice. Add 1.0 ml of lysis buffer (prechilled to 4°C) per 100-mm dish. The choice of lysis buffer is discussed on p. 447. For new antigens, RIPA buffer (150 mM NaCl, 1% NP-40, 0.5% DOC, 0.1% SDS, 50 mM Tris, pH 7.5) or NP-40 lysis buffer (150 mM NaCl, 1.0% NP-40, 50 mM Tris, pH 8.0) are good first choices.

3. Incubate the plate on ice for 30 min with occasional rocking.

4. **Either:** Tilt the plate on the bed of ice and allow the buffer to drain to one side. Remove the lysate and transfer to a 1.5 ml conical tube.

   **Or:** Scrape the cells and debris from the plate with a rubber policeman and transfer the lysis buffer and cell debris to a 1.5-ml conical tube.

5. Spin for 10 min at 10,000g at 4°C. Remove the lysate to a fresh tube.

6. (**Optional**) Spin the supernatant at 100,000g for 30 min.

**The supernatant is now ready for preclearing (p. 461) or for the addition of antibodies (p. 464).**

### NOTES

i. Some antigens that are particularly sensitive to proteases should not be centrifuged in benchtop microcentrifuges, as the 10-min spin will raise the temperature significantly, even if the centrifuge is kept in the cold.

ii. Rubber policeman can be made by cutting silicon rubber stoppers in quarters. Using a heavy-gauge scalpel, first cut the stopper in half lengthwise. Then cut each piece crosswise. The result will be four pieces with a flat edge that can be used to scrape the plate. The silicon rubber pieces can be mounted on the end of a Pasteur pipet for easy handling.

**LYSIS**

## DETERGENT LYSIS OF SUSPENSION CULTURES

1. Collect the cells by centrifugation at 400g for 5 min. Remove the supernatant.

2. Place the tube containing the washed cell pellet on ice. Resuspend the pellet in 1.0 ml of lysis buffer (prechilled to 4°C) per $10^7$ cells. Choice of lysis buffer is discussed on p. 447. For new antigens, RIPA buffer (150 mM NaCl, 1% NP-40, 0.5% DOC, 0.1% SDS, 50 mM Tris, pH 7.5) or NP-40 lysis buffer (150 mM NaCl, 1.0% NP-40, 50 mM Tris, pH 8.0) are good first choices.

3. Leave on ice for 30 min with occasional mixing.

4. Spin for 10 min at 10,000g at 4°C.

5. Carefully remove the lysate to a fresh tube.

6. (**Optional**)   Spin the supernatant at 100,000g for 30 min.

**The supernatant is now ready for preclearing (p. 461) or for the addition of antibodies (p. 464).**

### NOTE

i. Some antigens that are particularly sensitive to proteases should not be centrifuged in benchtop microcentrifuges, as the 10-min spin will raise the temperature significantly, even if the centrifuge is kept in the cold.

**2**

LYSIS

## DOUNCE HOMOGENIZATION OF SUSPENSION CULTURES

Because of the extra steps of trypsinizing and collecting adherent cells, dounce homogenization is normally only used for suspension cells. This technique is particularly valuable when detergents must be avoided and when cell organelles must remain intact.

1. Collect the cells by centrifugation at 400g for 5 min. Remove the supernatant.

2. Resuspend the cell pellet in 1–3 cell volumes of ice cold hypotonic dounce buffer (1.5 mM MgCl$_2$, 5 mM KCl, 10 mM HEPES, pH 7.5). Incubate on ice for 10 min.

3. Transfer the cell suspension to a dounce homogenizer (prechilled on ice). When choosing the proper size dounce, be sure the total volume of the cell suspension does not exceed one-third the volume of the lower chamber.

4. Using a type-B pestle, lyse the cells by repeated strokes (approximately 10). Even pressure will result in a uniform lysis. Monitor the process by phase microscopy. Vital dyes may be included in the samples used for microscopy to help determine the extent of lysis.

5. Centrifuge the broken cells at 10,000g for 10 min at 4°C.

6. (**Optional**)  Spin the supernatant at 100,000g for 30 min. This new supernatant is often referred to as an S100 fraction.

**The supernatant is now ready for preclearing (p. 461) or for the addition of antibodies (p. 464).**

## Lysis of Yeast Cells

For yeast, the preferred method of lysis is mechanical shearing either by vortexing the cells in the presence of glass beads or by sonication. Both techniques shatter the cell wall and burst the cells, releasing the antigens. Enzymatic degradation of the cell wall is an effective method to produce spheroplasts that can then be lysed with detergents, but the high cost of the enzymes makes this approach unfeasible for routine use.

**2**

**LYSIS**

## *LYSING YEAST CELLS USING GLASS BEADS*

1. Collect the cells by centrifugation at 5000g for 5 min. Remove the supernatant. Resuspend the cells in PBS and spin again. Remove the PBS.

2. Resuspend the yeast cell pellet in a small volume of ice-cold lysis buffer (RIPA, 150 mM NaCl, 1% NP-40, 0.5% DOC, 0.1% SDS, 50 mM Tris, pH 8.0, is commonly used. Use approximately three cell volumes). In many cases protease inhibitors should be included (p. 677). Put on ice.

3. Measure the total volume and add an equal volume of chilled glass beads (500 $\mu$m).

   Glass beads are prepared by washing twice in 1 N HCl and twice in lysis buffer. Store at 4°C in a small volume of lysis buffer.

4. Vortex vigorously for at least 2 min or longer.

5. Spin the lysate with the beads at 10,000g for 5 min at 4°C.

6. (**Optional**)   Spin the supernatant at 100,000g for 30 min.

The supernatant is now ready for preclearing (p. 461) or for the addition of antibodies (p. 464).

### NOTE

i. The extent of lysis can be monitored after step 4 by observation under the microscope either directly or with the addition of a vital dye.

**LYSIS**

## *LYSING YEAST CELLS BY SONICATION*

1. Collect the cells by centrifugation at 4,000g for 5 min. Resuspend in PBS containing 1% NP-40, 10 mM dithiothreitol, and 100 $\mu$g/ml of PMSF.

2. Transfer the tube to ice water. Insert a suitably sized sonicator probe into the yeast suspension.

3. Sonicate for 30 sec at maximum intensity. Return to the ice for 15 sec. Repeat three times.

4. Centrifuge at 10,000g for 10 min at 4°C.

5. (**Optional**)   Spin the supernatant at 100,000g for 30 min.

**The supernatant is now ready for preclearing (p. 461) or for the addition of antibodies (p. 464).**

**LYSIS**

## LYSING YEAST CELLS BY ENZYMATIC DEGRADATION OF THE CELL WALL

1. Collect the cells by centrifugation at 5000g for 5 min. Remove the supernatant.

2. Resuspend the yeast cell pellet in a volume equal to the original culture with 1 M sorbitol, 10 mM $MgCl_2$, 30 mM DTT, 100 $\mu$g/ml of PMSF, 50 mM potassium phosphate (pH 7.8). Incubate at 30°C for 10 min.

3. Centrifuge at 5000g for 5 min at room temperature. Resuspend the cell pellet in a volume equal to the original culture with 1 M sorbitol, 10 mM $MgCl_2$, 2 mM DTT, 100 $\mu$g/ml PMSF, 25 mM potassium phosphate, 25 mM sodium succinate (pH 5.5). Incubate for 2 min at 30°C.

4. Add 0.25 volume of 10 mg/ml of zymolyase (60,000).

5. Incubate for 20–30 min at 30°C. Monitor for spheroplast formation by comparing two samples, one untreated and one in 1% NP-40. Spheroplasts lyse in the NP-40 but whole cells will not.

6. Collect the spheroplasts by centrifugation at 3500g for 5 min. Wash the spheroplasts twice in PBS.

7. Place the tube containing the spheroplasts on ice.

8. **Either:** Resuspend the pellet in 1.0 ml of lysis buffer (prechilled to 4°C) per $10^9$ cells. Choice of lysis buffer is discussed on p. 447. For new antigens, RIPA buffer (150 mM NaCl, 1% NP-40, 0.5% DOC, 0.1% SDS, 50 mM Tris, pH 8.0) is an appropriate choice. Leave on ice for 30 min with occasional mixing.

   **Or:** Sonicate 3 times for 30 sec each. Keep cold during the lysis by returning the tube to the ice after each burst.

   **Or:** Add an equal volume of glass beads and vortex (see p. 453).

9. Spin for 10 min at 10,000g at 4°C.

10. Carefully remove the lysate to a fresh tube.

11. (**Optional**)   Spin the supernatant at 100,000g for 30 min.

The supernatant is now ready for preclearing (p. 461) or for the addition of antibodies (p. 464).

## NOTE

i. This method may lead to extensive protease degradation. Be careful to keep the temperature low, and use protease inhibitors (p. 677).

## Lysis of Bacteria

Several different methods commonly are used for the preparation of bacterial lysates. Here, enzymatically removing the cell wall and then lysing the cells is a common, although tedious, method for Gram-negative bacteria. Another common method is lysing the cells using ultrasound. The sonic impact breaks the cell wall and membranes. Here as well, the technique is fine for handling a low number of cultures, but is less feasible when numerous cultures are being tested.

**2**

LYSIS

## *LYSING BACTERIA BY SONICATION*

1. Collect the cells by centrifugation at 6000*g* for 5 min. Remove the supernatant. Resuspend the cells in PBS and spin again. Remove the PBS.

2. Resuspend the bacterial cell pellet in at least 10 cell volumes of ice-cold lysis buffer. RIPA (150 mM NaCl, 1% NP-40, 0.5% DOC, 0.1% SDS, 50 mM Tris, pH 8.0) is usually suitable. Other lysis buffers are discussed on p. 447.

3. Using a probe- or cup-type sonicator, disrupt the bacterial cells with multiple short bursts of maximum intensity (e.g., 30 sec × 4). Between each burst, return the cells to an ice bath to keep the temperature from rising.

4. Spin at 10,000*g* for 10 min at 4°C.

5. Carefully remove the lysate to a fresh tube.

6. (**Optional**)   Spin the supernatant at 100,000*g* for 30 min.

The supernatant is now ready for preclearing (p. 461) or for the addition of antibodies (p. 464).

**LYSIS**

## LYSING BACTERIA BY LYSOZYME DEGRADATION
## OF THE CELL WALL

1. Collect the cells by centrifugation at 6000$g$ for 5 min. Remove the supernatant. Resuspend the cells in PBS and spin again. Remove the PBS.

2. Resuspend the bacterial cell pellet in 10 cell volumes of 50 mM glucose, 10 mM EDTA, 25 mM Tris (pH 8.0) containing 4 mg/ml of lysozyme. Lysozyme should be added just prior to use.

3. Incubate at room temperature for 5 min.

4. Place the tube containing the spheroplasts on ice. Resuspend the pellet in 1.0 ml of lysis buffer (prechilled to 4°C) per $10^9$ cells. Choice of lysis buffer is discussed on p. 447. For new antigens, RIPA buffer (150 mM NaCl, 1% NP-40, 0.5% DOC, 0.1% SDS, 50 mM Tris, pH 8.0) is an appropriate choice.

5. Leave on ice for 30 min with occasional mixing.

6. Spin for 10 min at 10,000$g$ at 4°C.

7. Carefully remove the lysate to a fresh tube.

8. (**Optional**)  Spin the supernatant at 100,000$g$ for 30 min.

The supernatant is now ready for preclearing (p. 461) or for the addition of antibodies (p. 464).

**2**

LYSIS

## *DENATURING LYSIS*

For many sources of antigens, one useful method of lysis is treating cells with harsh, denaturing solutions to release most of the protein antigens. The lysates are then diluted or dialyzed to reduce the denaturing conditions to levels that are suitable for the formation of antibody–antigen complexes. In these cases, only antibodies that recognize "denaturation resistant" epitopes can be used, unless the proteins renature following the dilution of the denaturant.

1. Collect the cells by centrifugation. Remove the supernatant.

2. Lyse the cells by adding 10 cell volumes of 2% SDS, 50 mM Tris (pH 7.5). Vortex vigorously.

3. Place the tube in a boiling water bath for 10 min.

4. Shear the DNA by sonication. Either a probe- or cup-type sonicator can be used. Several short bursts at maximum intensity will be sufficient.

5. Spin at 10,000g for 10 min. Remove the supernatant and dilute 20-fold in PBS with 2% BSA.

6. Incubate on ice for 10 min. Spin at 10,000g for 10 min.

7. (**Optional**)   Spin the supernatant at 100,000g for 30 min.

**The supernatant is now ready for preclearing (p. 461) or for the addition of antibodies (p. 464).**

### NOTE

i. Urea (8 M) can be used in place of the SDS. Do not heat, but incubate for 30 min on ice. Dilute and process as above.

## ▨ Preclearing of the Lysate

When immunoprecipitating minor cellular polypeptides, it is often necessary to remove proteins in the lysates that bind nonspecifically to the immune complexes or the solid phase prior to adding the specific antibodies themselves.

## *PRECLEARING WITH NORMAL SERUM*

One method of preclearing is to subject the lysates to a series of steps that mimic the actual immunoprecipitation techniques. An antibody that does not recognize the antigen being studied is added to the lysate and processed as normal; however, in this case the supernatant is saved and used for the immunoprecipitation. It is essential that all of the antibody used in the preclearing step be removed quantitatively or it will compete for binding in the immunoprecipitation step itself. Any antibody can be used for preclearing, but rabbit serum is often the best choice. It is available in large amounts and binds tightly and quantitatively to *S. aureus* protein A.

1. Add 50 $\mu$l of normal rabbit serum per 1.0 ml of lysate.

2. Incubate 1 hr on ice.

3. Use the lysate to resuspend a washed pellet of fixed *S. aureus* Cowan I (SAC, 500 $\mu$l of 10% suspension per 50 $\mu$l of serum, washed once in lysis buffer prior to centrifuging to make the pellet). Resuspend the pellet using a Pasteur pipet. Fixed SAC are available commercially or can be prepared as described on p. 620.

4. Incubate 30 min on ice.

5. (**Optional**)  Spin 2 min at 4°C. Carefully remove supernatant and use to resuspend a second SAC pellet. Incubate 30 min on ice.

6. Spin 15 min at 4°C. Carefully remove and save the supernatant. Respin if necessary.

Techniques for forming the immune complexes are described on p. 464.

## PRECLEARING WITH PRELOADED S. AUREUS*

An alternative method used to preclear lysates prior to immuno-precipitations is to incubate the lysates with fixed *S. aureus* Cowan I (SAC) that has been saturated with nonspecific antibodies. This method is quicker than the method above; however, more preparation is needed prior to the immunoprecipitation.

1. Prior to the immunoprecipitation, the antibodies should be bound to the SAC. The preparation of SAC is described on p. 620 or can be purchased commercially from several suppliers. SAC is normally stored at −70°C in a 10% (vol/vol) suspension.

2. Thaw a large volume of SAC, and centrifuge at 6,000g for 5 min. Resuspend the pellet in the lysis buffer used for the antigen under study (see above). Bring the volume to a level equal with the original volume. Add 0.1 volume of normal rabbit sera, and then incubate on ice for 30 min. Wash three times with lysis buffer. Resuspend in one-half the original volume, dispense in convenient volumes and refreeze at −70°C. Final concentration is 20%.

3. Thaw a sample of preloaded SAC equal to the volume of lysate. Add to the lysate.

4. Incubate on ice for 2 hr.

5. Spin for 15 min at 10,000g at 4°C. Carefully remove and save the supernatant.

**Techniques for forming the immune complexes are described on p. 464.**

*Knowles (1987).

## ■ Forming the Immune Complexes

Forming an antibody–antigen complex is the simplest step in an immunoprecipitation. The variables to be considered are the amount of antibody to be added, the final volume, and whether or not an anti-immunoglobulin antibody will be used.

The amount of antibody to be added will depend on how much antigen will be precipitated. This can be determined by titrating the antibody versus a constant amount of antigen. With polyclonal antibodies, titrating the amount of antibody that is required to precipitate the antigen will often help to lower the nonspecific background. With monoclonal antibodies this will often not be necessary. For radiolabeled material, a general starting point will be 0.5–5 $\mu$l of a polyclonal sera, 10–100 $\mu$l of a hybridoma tissue culture supernatant, or 0.1–1 $\mu$l of an ascites fluid. At the midpoint of these three suggested ranges, the amount of heavy chain will be approximately 1 $\mu$g.

When using a polyclonal antibody, the volume of the reaction is seldom a concern. In fact, the avidity is normally high enough to allow efficient binding and eventually removal of the antigen from solution even with dilute samples. This feature means that the antibody–antigen interaction can be diluted with a solution containing nonspecific proteins. With monoclonal antibodies of high affinity, similar approaches are possible. However, with low-affinity monoclonal antibodies, using high concentrations of the antibody in low volumes will help drive the reaction.

Consult Tables 15.1 and 15.2 (pp. 617 and 618) to determine whether anti-immunoglobulin antibodies should be added. Polyclonal antibodies from human, rabbit, guinea pig, or pig never need to be supplemented; those from horse, cow, mouse, or hamster seldom; rat or sheep occasionally; whereas chicken or goat antibodies should always be supplemented. With monoclonal antibodies, rat subclasses should always be supplemented, and mouse subclasses other than $IgG_{2a}$ should be supplemented for quantitative removal of the antibody. For quantitative precipitations, adding anti-immunoglobulin antibodies is recommended, but under some circumstances will increase the nonspecific background. When using anti-immunoglobulin antibodies, it is important to titrate the amount of the anti-immunoglobulin antibody that is needed to bind all of the primary antibody. This should be determined for each batch of antisera.

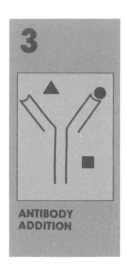

**3**

**ANTIBODY ADDITION**

## ADDING ANTIBODY TO LYSATES

1. Add serum (0.5–5 $\mu$l), hybridoma tissue culture supernatant (10–100 $\mu$l), or ascites fluid (0.1–1.0 $\mu$l) to a sample of lysate.

2. Incubate on ice for 1 hr. High-affinity reactions will be substantially complete considerably sooner than 1 hr; however, the 1-hr incubation time provides a reasonable compromise for most antibodies. Overnight reactions, although suggested in some protocols, are seldom any advantage, but will increase the background.

3. (**Optional**)   If necessary add 0.5 $\mu$l of anti-immunoglobulin antibodies after 30 min. This can be conveniently added as 20 $\mu$l of a 1 in 40 dilution of whole sera.

Techniques for collecting the immune complexes are described on p. 466.

## COMMENTS ■ Controls

The correct controls for immunoprecipitation reactions should be antibody preparations that are as close to the specific antibody as possible. For example, polyclonal serum should be compared to other polyclonal serum from the same species. The best control will be a prebleed from the same animal used for immunization. For monoclonal antibodies, the control must be from the same source as the specific antibody, i.e., supernatant versus supernatant, ascites versus ascites, or pure antibody versus pure antibody. If possible, the control antibodies should be of the same class and subclass as the specific antibody. Tissue culture supernatants from the parental myeloma are never appropriate controls because they do not contain any antibodies. Suitable control hybridoma cell lines are available from ATCC.

EMBARASSING AND AVOIDABLE MISTAKES ARISE FROM THE IMPROPER CHOICE OF CONTROL ANTIBODIES (references available on request).

# ■ Purifying the Immune Complexes

All of the methods for purifying immune complexes rely on secondary reagents that bind to the antibody. The antibody is removed from the solution, and the antigen remains associated with the antibody throughout the purification. The earliest techniques used anti-immunoglobulin antibodies to form a large complex of antibody–anti-immunoglobulin antibodies. This multicomponent complex of molecules is known as a lattice. When the lattice is suitably large, it can be removed from solution by centrifugation. This procedure is still one of the best at yielding clean immunoprecipitations. However, the ability to form large enough complexes to be collected by centrifugation is critically dependent on the molar ratio of anti-immunoglobulin antibodies to primary antibodies. For quantitative removal of the immune complexes, the ratio must be determined empirically for each primary antibody that is used. To some degree, the size of the lattice will also depend on the amount and type of antigen. A large quantity of a multimeric antigen will change the potential size of any latticework compared with a rare, monomeric antigen. A monoclonal antibody will give a different set of problems to forming the lattice.

As a solution to these problems, Kessler (1975) suggested the use of protein A-bearing *S. aureus* Cowan I (SAC) as a solid phase to collect the antibody–antigen complexes. Protein A forms a portion of the cell wall of these bacteria. The cell wall proteins are fixed by treating with formaldehyde, the bacteria killed by heat treatment, and the resulting particles form an excellent solid-phase matrix to bind to the antibodies. The protein A binds to the Fc domain of the antibody, thus its attachment to the antibody does not affect the interaction with the antigen. The use of protein A and other cell wall proteins that bind antibodies is discussed in detail on p. 615. An alternative solid phase that can be used to lower some of the background problems encountered with SAC is a bead with purified protein A covalently attached. Techniques for both of these reagents are described below. Protein A beads give a cleaner background, but they are more expensive than SAC.

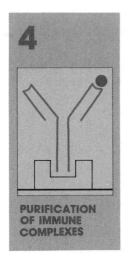

**PURIFICATION OF IMMUNE COMPLEXES**

## COLLECTING IMMUNE COMPLEXES ON FIXED S. AUREUS*

1. To the antibody–antigen reaction, add 50 $\mu$l of 10% fixed *S. aureus* Cowan I (SAC, wash the SAC twice in lysis buffer to remove any free protein A molecules from the solution and to transfer the SAC to the proper buffer). SAC is available commercially or can be prepared as described on p. 620.

2. Incubate on ice for 30 min.

3. Spin at 10,000$g$ for 1 min at 4°C.

4. Remove the supernatant by aspiration and add 0.5 ml of lysis buffer.

5. Resuspend by vortexing. The SAC pellet will be difficult to resuspend, but placing two tubes in the vortex so they violently strike one another during vortexing will facilitate the resuspension.

6. Wash three times in lysis buffer. Remove the last wash as completely as possible. As some buffer will adhere to the sides of tubes, a quick spin to remove the last few microliters may lower backgrounds for critical applications.

7. Use the immune complexes for the appropriate assay.

   For SDS-polyacrylamide gel electrophoresis, add 50 $\mu$l of Laemmli sample buffer (2% SDS, 10% glycerol, 100 mM DTT, 60 mM Tris [pH 6.8], and 0.001% bromphenol blue). Heat to 85°C for 10 min. Spin, remove supernatant, and load supernatant onto gel (p. 636). Samples can be stored at −20°C.

## NOTE

i. If the SAC pellet is difficult to resuspend, several methods can help resuspension. Adding only 50 $\mu$l of lysis buffer prior to vortexing will help speed the resuspension. Vortex and then add the remainder of the 0.5 ml of lysis buffer, vortex briefly, and proceed. Freezing the pellet in a dry ice–alcohol bath prior to adding the wash buffer will also allow the cells to be resuspended quickly.

*Adapted from Kessler (1975, 1981).

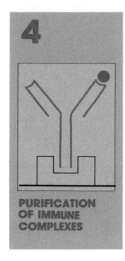

**PURIFICATION
OF IMMUNE
COMPLEXES**

## COLLECTING THE IMMUNE COMPLEXES ON PROTEIN A BEADS

1. To the antibody–antigen reaction, add 100 $\mu$l of protein A beads (10% vol/vol in lysis buffer). Incubate 1 hr at 4°C with rocking.

2. Collect the beads by centrifugation at 10,000g for 15 sec at 4°C. Wash the immune complexes three times with lysis buffer. The lysate and wash buffers are easily removed by aspiration with a 23-gauge needle bent like this:

→ **TO ASPIRATOR**

3. Remove the final wash as completely as possible. The needle should be inserted directly into the beads to remove the remaining wash buffer. Any beads that adhere to the needle can be removed by gently flicking it on the lip of the tube.

4. Use the immune complexes for the appropriate assay.

   For SDS-polyacrylamide gel electrophoresis, add 50 $\mu$l of Laemmli sample buffer (2% SDS, 10% glycerol, 100 mM DTT, 60 mM Tris [pH 6.8], and 0.001% bromphenol blue). Heat to 85°C for 10 min. Spin, remove supernatant, and load supernatant onto gel (p. 636) Samples can be stored at −20°C.

## COMMENTS ■ Troubleshooting—Bad Backgrounds

Probably the most frequently asked question about immunoprecipitations concerns methods for lowering the number and type of background proteins that contaminate the washed immune complexes. Background problems come from many different sources and can be either specific or nonspecific. Antigens that are recognized directly by spurious antibodies in the antibody preparation give rise to specific background bands. There are only two methods to deal with these problems. First, the contaminating antibodies can be removed by purifying the antigen-specific antibodies on an antigen affinity column (p. 313). Second, the activities can be blocked by adding a saturating amount of the cold proteins that are bound by the contaminating antibodies (p. 631).

Nonspecific background proteins result from a number of sources. They can form noncovalent bonds with the immune complexes or with the SAC or protein A beads. They can bind to particulate matter in the preparation. Some contaminating proteins aggregate or polymerize to a large enough size that they are removed by centrifugation. Proteins can be trapped within the latticework of the antigen–antibody–protein A complex. The contaminating proteins can bind to the plasticware that is used to centrifuge the immune complexes. Below are some suggestions to deal with nonspecific background problems.

• Preclear the lysate prior to adding the specific antibodies (see p. 461).

• Add saturating amounts of competitor proteins. BSA, gelatin, acetone powders (p. 631), or blotto (p. 496) are commonly used. Store SAC or protein-A beads in buffers containing 2% BSA.

• Spin the lysate at 100,000$g$ for 30 min prior to the addition of the antibody.

• Try protein A beads for collecting the immune complexes if you have been using SAC.

• Try a different antibody.

• Centrifuge the antibody at 100,000$g$ for 30 min and titrate.

• If using rabbit anti-mouse immunoglobulin, check the background due to this reagent alone. Titrate if necessary.

• Use more stringent washes. Try 1.0 M NaCl, 0.5 M LiCl, 1 M KSCN, 0.2% SDS, or 1% Tween 20. Also, alternating wash buffers from high salt to low salt or different detergents may help. Try washing with distilled water for one wash.

- Increase the number of washes. Leave the solid phase in the wash buffers for 10 min at each wash.

- Transfer the SAC or protein A beads to a fresh tube prior to the last wash and elution of the antibody–antigen complexes.

- Lower the number of cpm of the radiolabel to the minimum needed for antigen detection.

- Make certain the lysates are not frozen before use.

- Spin the antibody–antigen binding reaction for 10 min at 10,000g immediately before adding the solid phase. This may remove aggregates and polymerizing proteins.

- For immunoprecipitations using yeast, the background may be lowered by spinning the antibody–antigen solution for 10 min at 10,000g immediately before the addition of the solid phase. When collecting the solid phase, use very low speed centrifugation. Much of the background for yeast immunoprecipitations arises from continuous polymerization and precipitation of proteins in the extracts.

- If background persists, analyze proteins on two-dimensional gels.

- If specific proteins remain, remember that your antigen may consist of more than one polypeptide chain.

# IMMUNOBLOTTING

Immunoblotting combines the resolution of gel electrophoresis with the specificity of immunochemical detection. Immunoblotting can be used to determine a number of important characteristics of protein antigens—the presence and quantity of an antigen, the relative molecular weight of the polypeptide chain, and the efficiency of extraction of the antigen. It is particularly useful when dealing with antigens that are insoluble, difficult to label, or easily degraded, and thus less amenable to analysis by immunoprecipitation. Immunoblotting can be combined with immunoprecipitation to allow very sensitive detection of minor antigens and to study specific interactions between antigens. It is also a particularly powerful technique for assaying the presence, quantity, and specificity of antibodies from different samples of polyclonal sera. Finally, it can be used to purify specific antibodies from polyclonal sera. Because the antigen is not labeled, only the steady-state level can be determined and further analysis of its biochemical properties, modifications, or half-life is not possible.

The immunoblotting procedure can be divided into six steps:

(1) Preparation of the antigen sample
(2) Resolution of the sample by gel electrophoresis
(3) Transfer of the separated polypeptides to a membrane support
(4) Blocking nonspecific binding sites on the membrane
(5) Addition of the antibody
(6) Detection

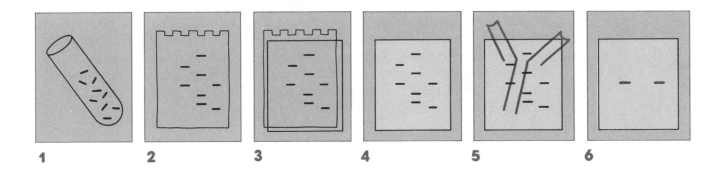

First, an unlabeled solution of proteins, often an extract of cells or tissues, is prepared in a gel electrophoresis sample buffer. The proteins are separated by gel electrophoresis, and transferred to a membrane that binds the proteins nonspecifically. Transfer usually is achieved by placing the membrane in direct contact with the gel and then placing this sandwich in an electric field to drive the proteins from the gel onto the membrane. The remaining binding sites on the membrane are blocked to eliminate any further reaction with the membrane. Finally, the location of specific antigens is determined using a labeled primary antibody or an unlabeled primary antibody, followed by a labeled secondary antibody.

## ■ MAJOR CONSTRAINTS

The major factor that will determine the success of an immunoblotting procedure is the nature of the epitopes recognized by the antibodies. Most high-resolution gel electrophoresis techniques involve denaturation of the antigen sample, so only antibodies that recognize denaturation-resistant epitopes will bind. Most polyclonal sera contain at least some antibodies of this type, but many monoclonal antibodies will not react with denatured antigens.

Immunoblotting does not require particularly high-affinity antibodies. Some antibodies that work poorly in immunoprecipitation will be satisfactory for blots. This is due to both the exposure of different epitopes on denaturation and the high local concentration of antigens on the membrane. The high local concentration provides a better chance for the antibody to bind through both Fab arms, thus greatly raising the avidity of the interaction (see Chapter 3 for a more detailed discussion). The antigen density also favors the retention of low-affinity antibodies on the membrane by increasing the frequency with which antibodies leaving the membrane bind to adjacent sites.

The sensitivity of immunoblotting is also determined by the detection method. For most detection systems, this limit will be about 20 femtomoles. For a 50,000 mw protein, this is approximately 1 ng. Because the loading capacities of gels are limited, an antigen usually cannot be detected when its concentration falls below 1 ng/sample. For a typical SDS-polyacrylamide gel, the capacity of a lane is approximately 150 $\mu$g (see p. 485). Distortion is seen if larger amounts are used. Therefore, a protein antigen must be present at 1 part in 150,000. When analyzing total proteins from mammalian cells, this limit represents approximately 10,000 molecules of a 50,000 mw protein per cell. To detect proteins present at lower levels, partial purification of the antigen is required. This may be done conveniently by cell fractionation, chromatography, or immunoprecipitation. Unless the molecular weight of your antigen is similar to an abundant protein, the capacity of the transfer membrane will not be limiting.

# CHOICE OF ANTIBODY

Three types of antibody preparations can be used for immuno-blotting—polyclonal antibodies, monoclonal antibodies, and pooled monoclonal antibodies. Their relative advantages and disadvantages are discussed in detail below.

## Immunoblots Using Polyclonal Antibodies

Polyclonal antibodies probably are the most widely used type of antibody preparation for immunoblotting. Many sera will contain anti-bodies that will bind to a number of denaturation-resistant epitopes on the antigen. This will result in multiple antibody molecules binding to a given polypeptide and will generate a relatively stronger signal than a monoclonal antibody. As polyclonal sera will usually contain high concentrations of specific antibodies, they can often be diluted exten-sively. Therefore, dilution can be used to reduce nonspecific back-ground problems without reducing sensitivity.

Because polyclonal antibodies normally are used as whole sera, they will contain the entire repertoire of circulating antibodies found in the immunized animal at the time the serum was collected. Therefore, the serum may contain high titered antibodies that specifically recognize spurious antigens. This is particularly troublesome in the analysis of microbial antigens. Sera may contain antibodies against bacteria or fungi that have infected the host animal recently. If the antigen preparation has proteins that are identical or similar to the ones found in these microorganisms, these antigens will be detected specifically, generating additional bands and potentially obscuring the bands of interest. Because this type of contamination is specific, it cannot be removed by methods that are designed to lower nonspecific back-grounds (e.g., more extensive blocking of the membrane or vigorous washing). In these cases, the easiest solution to removing these spuri-ous activities is to block the specific antibodies by preincubating the serum with a preparation that contains the contaminating protein (e.g., an acetone powder for a microorganism that does not express the antigen being studied). Similar types of specific antibody contamina-tion are found as autoantibodies to a range of tissue antigens but can be removed by analogous adsorption techniques.

## ■ Immunoblots Using Monoclonal Antibodies

The major advantage of using monoclonal antibodies for immuno-blotting is the specificity of their interactions. Because monoclonal antibodies bind to only one epitope, they provide an elegant tool for identifying a particular region of the antigen. Monoclonal antibodies that function well in immunoblotting usually interact with epitopes that are defined by short stretches of the protein's primary sequence. As such, they can be used to examine the presence and origin of quite small regions of the antigen, and this could be useful in studies of polypeptide processing.

The principal difficulty in using monoclonal antibodies in immuno-blotting is that many will not recognize epitopes that are destroyed by the denaturing reagents used in the preparation of the sample. However, some monoclonal antibodies work well and can give extremely clean and sensitive results.

A second problem with using monoclonal antibodies in immuno-blotting is the possibility of detecting cross-reactions with other poly-peptides. Because an epitope is a relatively small protein structure which can be composed of only four or five amino acids, there is a chance that a similar epitope can be found on another polypeptide. In some cases, the common epitopes will form part of an important structural similarity between antigens, and in these cases monoclonal antibodies can be used to detect related antigens. Alternatively, the antibodies may detect small structural similarities confined only to the antibody combining site. Depending on the set of hybridomas, as many as one in five monoclonal antibodies will display these types of cross-reactions. Unless there are other reasons to suggest that extensive and physiologically significant similarities accompany a cross-reaction (such as multiple sites in common), an unexpected band on an immunoblot should be treated as a spurious cross-reaction until proven otherwise.

## ■ Immunoblots Using Pooled Monoclonal Antibodies

Using pools of monoclonal antibodies in immunoblotting takes advantage of the best properties of both polyclonal and individual monoclonal antibodies, combining specificity and sensitivity. Obviously, the pool must be composed of monoclonal antibodies that react with distinct denaturation-resistant epitopes and that do not show spurious cross-reactions. Only a few antigens have been studied in enough detail to have such sets of antibodies available. However, whenever possible, a well chosen pool of monoclonal antibodies should be used for immunoblotting.

## Immunoblots

Detection limit is 10–100 fmoles, about 1–10 ng of protein

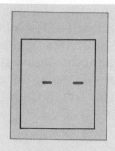

- Multiple steps over several days
- Detects antigen presence, quantity, size
- Semi-quantitative to quantitative
- Antigen detection dependent on denaturation-resistant epitope
- In general, lower affinity antibody may work

| | Polyclonal Antibodies | Monoclonal Antibodies | Pooled Monoclonal Antibodies |
|---|---|---|---|
| Signal Strength | Usually good | Antibody dependent (zero to excellent) | Excellent |
| Specificity | Good, but some background | Excellent, but some cross-reactions | Excellent |
| Good Features | Most recognize denatured antigen | Specificity Unlimited supply | Signal strength Specificity Unlimited supply |
| Bad Features | Nonrenewable Background Need to titer | Many do not recognize denatured antigen | Availability |

## ■ IMMUNOBLOTTING PROTOCOLS ▬▬▬▬▬▬▬▬▬▬▬▬

The immunoblotting method has evolved from early stages where antibodies were used to "stain" polyacrylamide gels directly (Burridge 1976; Showe et al. 1976) to more versatile approaches using replica techniques, where the separated polypeptides are transferred to nitrocellulose membranes, chemically activated paper, or nylon sheets (reviewed in Gershoni and Palade 1983; Towbin and Gordon 1984; Beisiegel 1986). Although there are a number of variations on this basic theme, the most common and easiest method involves electrophoretic transfer to nitrocellulose sheets (Towbin et al. 1979; Burnette 1981).

The immunoblotting techniques described below are separated into six steps: sample preparation, gel electrophoresis, transfer, blocking, addition of antibody, and detection.

## ■ Sample Preparation

Any protein sample can be used as a source for immunoblotting. The protein sample is diluted in sample buffer and separated by gel electrophoresis. Most often, the electrophoresis system of choice will be a Tris/glycine discontinuous SDS-polyacrylamide gel (see p. 636 and Appendix I). Other systems can be substituted, and the only change will be the selection of the appropriate sample buffer. To analyze a protein sample directly, two criteria must be met. (1) If the protein sample is in solution, it must be in a buffer that is compatible with the gel electrophoresis system. (2) The concentration of proteins in the sample must not exceed the loading capacity for the particular gel system. Representative values for both of these variables are given below.

One of the strengths of the immunoblotting technique is that it permits the analysis of proteins from sources that cannot be studied using other immunochemical techniques. For example, samples that cannot be labeled or are insoluble in gentle extraction buffers often can be analyzed by immunoblotting. The most common sources of proteins for immunoblotting are whole-cell lysates prepared from cells, tissues, or organs.

In most cases, crude samples may be run directly on a gel without further purification. However, because the total amount of protein that can be resolved adequately on a gel is limited, some minor proteins will be present at amounts below the level of detection. In these cases, the protein of interest must be partially purified by standard chromatographic and extraction protocols or by immunoaffinity techniques. Immunoaffinity purification is particularly useful, because it greatly extends the range of the immunoblotting technique. Most often the purification will be done by immunoprecipitating the protein of interest (p. 429) prior to being run on a gel. When immunoprecipitation will be used, the antigen is often located using a direct detection technique, because the heavy and light chains from the antibody used for the immunoprecipitation will also be present on the membrane and will interfere with many indirect detection reagents.

No steps specific to immunoblotting are required prior to electrophoresis, so standard extraction and electrophoresis conditions can be used (p. 636). For SDS-polyacrylamide electrophoresis, samples can be prepared by a number of methods. The two most common methods of preparing samples for immunoblots are either to lyse the entire cell or tissue mass in sample buffer and load the total cell lysate on a gel or to mix any protein sample 1 : 1 with double-strength sample buffer and load.

**SAMPLE
PREPARATION**

## PREPARING TOTAL CELL LYSATES FOR IMMUNOBLOTTING

In this procedure, samples of cells or tissues are disrupted in an electrophoresis sample buffer. After the chromosomal DNA is sheared to lower viscosity, the samples are processed and run as usual for the particular gel system. In the example given here, the samples are prepared for electrophoresis in a standard Tris/glycine SDS-polyacrylamide gel.

1. Measure the volume of sample to be lysed. For tissue culture cells $10^9$ cells $\cong 1$ ml $\cong 1$ gram. For tissues or organs, weighing the sample will be sufficient, 1 gram $\cong 1$ ml. For bacterial or yeast cells, spin the samples and estimate the volume.

2. Add 10 volumes of sample buffer (for Tris/glycine SDS-polyacrylamide gel electrophoresis, the 1× sample buffer is 2% SDS, 100 mM dithiothreitol, 60 mM Tris, pH 6.8, 0.01% bromphenol blue). Mix vigorously.

   For lysing yeast cells, the buffer should be supplemented with glass beads before vortexing. Measure the total volume and add an equal volume of chilled glass beads (500 $\mu$m). Glass beads are prepared by washing twice in 1 N HCl and twice in sample buffer. Store in a small volume of sample buffer. Alternatively, the yeast cells can be lysed by sonication. The yeast are resuspended in sample buffer and sonicated at full intensity for four bursts of 30 sec each. Between each sonication step, they are kept on ice for 15 sec.

3. Most samples should be boiled for 5 min. Some antigens will come out of solution under these conditions; if no signal is seen, check different temperatures.

4. Shear the chromosomal DNA. Either pass the sample repeatedly through a 20-gauge needle and then through a 26-gauge needle or sonicate the sample to shear the DNA. Sonicators with either an immersible tip or cuphorn are both suitable, approximately 15–30 sec at full force is sufficient to shear the DNA.

5. Spin the sample at 10,000g for 10 min. Recover the supernatant and discard any pellet. If necessary, determine the relative protein concentration. After the final spin, samples may be frozen at −20°C. They are stable at this step for months.

**The samples are now ready for electrophoresis (p. 485).**

## NOTE

i. When comparing the levels of antigens in multiple samples, be certain to correct for the total amount of protein in the sample. This can be done by many of the commonly used protein concentration determination methods (pp. 670–674). Protein quantitation will be easier without the addition of bromphenol blue, which can be left out of the sample buffer with no detrimental effects.

**SAMPLE PREPARATION**

## *PREPARING PROTEIN SOLUTIONS FOR IMMUNOBLOTTING*

Most protein solutions can be mixed with sample buffer and analyzed directly using immunoblots. This approach can be used on detergent lysates of cells, samples from column eluates, and most other sources of antigens. However, the buffer of the protein solution must be compatible with that of the gel separation technique. For Tris/glycine SDS-polyacrylamide gels, the pH of the protein solution should be approximately neutral, salt concentrations should be below approximately 200 mM unless a large dilution into the sample buffer will be used, but most other buffer components will not affect the running of the gel.

1. Mix a portion of the protein solution with sample buffer. For Tris/glycine SDS-polyacrylamide gel electrophoresis, the final sample buffer conditions are 2% SDS, 100 mM dithiothreitol, 60 mM Tris (pH 6.8), and 0.01% bromphenol blue. It is often convenient to add the sample buffer from a 2× stock an equal volume of protein solution.

2. Place the sample in a boiling water bath for 5 min. After the final spin, samples may be frozen at −20°C. They are stable at this step for months.

**The samples are now ready for electrophoresis (p. 485).**

### NOTE

i. When comparing samples, be certain to correct for the total amount of protein in the sample. This can be done by many of the commonly used protein concentration determination methods (pp. 670–674). Protein quantitation will be easier without the addition of bromphenol blue. This can be left out of the sample buffer with no detrimental effects.

## ■ Gel Electrophoresis

Although Tris/glycine SDS-polyacrylamide gels (p. 636) are the most common method for separating proteins for immunoblotting, any gel separation technique can be used. Other techniques for high-resolution separation of proteins that have been used successfully for immunoblotting include isoelectric focusing, isoelectric focusing followed by SDS-polyacrylamide gels, SDS-polyacrylamide gels using buffers other than Tris/glycine, urea-polyacrylamide gels, and native low-salt polyacrylamide gels. When using these or other types of separation systems, it is important that the gel buffers are compatible with the type of coupling being used to bind the proteins to the transfer membrane. For example, glycine buffers cannot be used with diazotized paper, as the binding to this paper will be blocked by the primary amines on glycine.

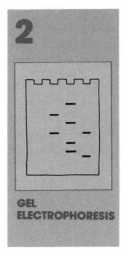

**GEL ELECTROPHORESIS**

## GEL ELECTROPHORESIS

Two decisions concerning the conditions for electrophoresis need to be made for immunoblotting: first, the total amount of protein that can be loaded on a gel without distortion and, second, the physical dimensions and acrylamide concentration of the separating gel. If an extract from whole cells will be used, the total amount of protein should not exceed approximately 30 $\mu$g/mm$^2$ of loading surface (150 $\mu$g/5 $\times$ 1-mm slot). For individual bands, loading more than 0.3 $\mu$g/mm$^2$ will lead to distortion (1.5 $\mu$g/5 $\times$ 1-mm slot); however, because immunoblots are so sensitive, this upper limit for individual proteins will never be used.

Since all blotting techniques require a successful transfer of proteins from the gel, some consideration as to the kind of gel to be used is helpful. Crudely, the lower the percentage of the acrylamide and cross-linker, the easier the transfer will be. Therefore, running the softest gel that yields the desired resolution is best. For efficient transfer of high-molecular-weight proteins, the concentration of the acrylamide should be lowered to a level where the protein of interest will migrate at least one-third of the distance from the top of the separating gel to the bottom within the time it takes the dye front to reach the bottom. In general, proteins that run slower than this will be difficult to elute.

Transfer will be faster and more complete from thinner gels, although ultra-thin gels may cause handling problems. In most cases, a 0.4 mm thickness represents the lower practical limit. Remember that during transfer, all the proteins in the width of a band become concentrated at the membrane surface, so running larger samples on thicker gels will increase sensitivity. Transfers from gels over 2 mm in width are inefficient, however.

**Gels are run under standard conditions (p. 636) and can be transferred by the methods on p. 486.**

## ■ Transfer of Proteins from Gels to Membranes

Nitrocellulose, activated paper, and activated nylon have all been used successfully to bind the transferred proteins (Towbin et al. 1979; Renart et al. 1979; Gershoni and Palade 1982). Nitrocellulose is the most commonly used, but it has certain disadvantages. The proteins are not covalently bound to it, and nitrocellulose is brittle, especially when dry. Activated paper (diazo groups) can be purchased commercially and will covalently bind proteins. It has good mechanical strength, but unfortunately the coupling method is incompatible with many gel electrophoresis systems. The linkage is through primary amines, so systems that use gel buffers without free amino groups must be used with this paper. (See Jovin, Dante, and Chrambach [1971] for alternative electrophoresis buffers.) Also, the paper is expensive and the reactive groups have a limited half-life once the paper is activated. Consult commercial suppliers' instructions for using this paper. Nylon itself can bind only a small amount of protein and is not suitable for most applications. However, it has excellent mechanical strength and does not change size during washing or drying. Activated nylon (positively charged) solves some of these problems, but has higher backgrounds. Because of these problems, the best compromise for most situations is nitrocellulose membranes, particularly during early stages of an experimental series.

Transfer of proteins from the gel onto the nitrocellulose membrane is achieved in three ways: simple diffusion (Renart et al. 1979; Bowen et al. 1980), vacuum-assisted solvent flow (Peferoen et al. 1982), and electrophoretic elution (Towbin et al. 1979; Bittner et al. 1980; Reiser and Wardale 1981; Kyhse-Andersen 1984). Simple diffusion has few advantages and is seldom used. Vacuum-assisted solvent flow has a number of good properties, and it is easy and quick. However, with the advent of semi-dry electrophoretic transfer, there are few reasons to use vacuum transfers. Except in unusual circumstances, the two electroblotting techniques given below are the methods of choice.

Electrophoretic elution is the most commonly used method and is highly recommended. Its principal advantage lies in the speed and completeness of transfer. Electrophoretic elution can be achieved either by complete immersion of a gel–membrane sandwich in a buffer (wet transfer) or by placing the gel–membrane sandwich between absorbent paper soaked in transfer buffer (semi-dry transfer). For the wet transfer, the sandwich is placed in a buffer tank with platinum wire electrodes. For the semi-dry transfer, the gel–membrane sandwich is placed between carbon plate electrodes. Apparatuses for both types of transfers are available commercially. Both work well, although semi-dry transfer is quicker and often more complete.

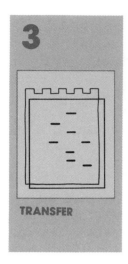

TRANSFER

## SEMI-DRY ELECTROPHORETIC TRANSFER*

Transfer of proteins from a gel to nitrocellulose using a semi-dry method gives even and rapid transfer and does not require a large power source. It can also be adapted to handle stacks of gel–membrane sandwiches (up to six). In the original description, different buffers were described for the anode and cathode, but recently it has been demonstrated that a simple buffer system using diluted Tris/glycine SDS-polyacrylamide gel running buffer is just as effective and much easier.

1. Rinse electrode plates of the semi-dry apparatus with distilled $H_2O$.

2. Cut six sheets of absorbent paper (Whatman 3MM or equivalent) and one sheet of nitrocellulose to the size of the gel. If the paper overlaps the edge of the gel, the current will short-circuit the transfer and bypass the gel, preventing efficient transfer.

3. Soak the nitrocellulose membrane in distilled $H_2O$. Nitrocellulose should be wetted by carefully laying it on the surface of the water. Allow the nitrocellulose to wet by capillary action from the bottom (several minutes), then submerge the sheet for 2 min. Wet the absorbent paper by soaking in transfer buffer. Transfer buffer is:

|  | Concentration | for 1000 ml |
|---|---|---|
| Tris base | 48 mM | 5.8 grams |
| Glycine | 39 mM | 2.9 grams |
| SDS | 0.037% (vol/vol) | 0.37 grams |
| Methanol | 20% | 200 ml |
| Distilled water |  | Make up to 1000 ml |

4. On the bottom plate of the apparatus (the anode), assemble the gel, nitrocellulose, and paper in this order:
   - bottom electrode
   - three layers absorbent paper soaked in transfer buffer
   - one nitrocellulose membrane soaked in $H_2O$
   - polyacrylamide gel slightly wetted with $H_2O$
   - three layers absorbent paper soaked in transfer buffer

5. Check carefully for air bubbles and gently remove them either by using a gloved hand or by rolling a pipet over the sandwich. Dry any buffer that may surround the gel–paper sandwich. Many apparatuses have a gasket that surrounds the gel–paper sandwich to lessen the chance of allowing a short-circuit of the transfer. Place this in plate, if available.

*Kyhse-Andersen (1984).

6. Carefully place the upper electrode (the cathode) on top of the stack. Connect the electrodes (positive or red lead to the bottom) and commence transfer. Running time is 45 min to 1.5 hr with a current of 0.8 mA/cm$^2$ of gel. Avoid extending the running time, as this can cause drying out.

7. After transfer, disconnect the power source. Carefully disassemble the apparatus. Mark membrane to follow orientation (usually by snipping off lower left-hand corner, the number one lane).

8. The polyacrylamide gel can now be stained with Coomassie (pp. 648–650) or silver stain (p. 651 or 653) to verify transfer. Some workers run prestained or radioactive markers on their gels to serve as internal markers for transfer and molecular weight.

**Process the nitrocellulose sheet for staining (p. 493) or blocking (p. 497) as appropriate.**

## NOTES

i. The type of bonds that hold proteins to nitrocellulose are not known; however, the binding is blocked by oils or other proteins. Wear gloves and use virgin nitrocellulose sheets.

ii. Several workers recommend allowing the membrane to dry completely before processing the nitrocellulose for blocking or staining. In some cases, this seems to lessen the chance of the transferred proteins being lost during subsequent steps. After drying, wet the membrane as in step 3. An alternative to drying is to treat the membrane with isopropanol for 1 min.

iii. Multiple gels can be transferred on the same apparatus by separating complete stacks with dialysis membranes (presoaked in transfer buffer). The top of the final sandwich should not have a membrane, because direct contact between the electrode and the absorbent paper is required. For a two-gel stack:
- Bottom electrode
- Three layers absorbent paper soaked in transfer buffer
- One nitrocellulose membrane soaked in H$_2$O
- Polyacrylamide gel slightly wetted with H$_2$O
- Three layers absorbent paper soaked in transfer buffer
- Dialysis membrane
- Three layers absorbent paper soaked in transfer buffer
- One nitrocellulose membrane soaked in H$_2$O
- Polyacrylamide gel slightly wetted with H$_2$O
- Three layers absorbent paper soaked in transfer buffer
- Top electrode

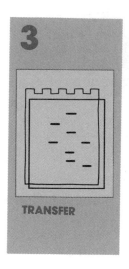

**3**

TRANSFER

## WET ELECTROPHORETIC TRANSFER*

For submerged or wet transfers, no one set of transfer conditions offers complete and even transfer of proteins with good retention on the membrane. However, most of the common transfer techniques are adequate.

1. Cut one sheet of nitrocellulose paper and four sheets of absorbent filter paper (Whatman 3MM or equivalent) to the size of the gel.

2. Soak the nitrocellulose membrane in distilled $H_2O$. Nitrocellulose should be wetted by carefully laying it on the surface of water. Allow the nitrocellulose to wet by capillary action from the bottom (several minutes), then submerge the sheet for 2 min. Move the membrane to soak in transfer buffer for 5 min. Wet the absorbent paper by soaking in transfer buffer.

Transfer buffer 1 (proteins 20,000–400,000 mw):

|  | Concentration | for 1000 ml |
|---|---|---|
| Tris | 50 mM | 5.8 grams |
| Glycine | 380 mM | 29 grams |
| SDS | 0.1% (wt/vol) | 1.0 gram |
| Methanol | 20% | 200 ml |
| Distilled water |  | Make up to 1000 ml |

Transfer buffer 2 (proteins <80,000 mw):

|  | Concentration | for 1000 ml |
|---|---|---|
| Tris | 25 mM | 2.9 grams |
| Glycine | 190 mM | 14.5 grams |
| Methanol | 20% | 200 ml |
| Distilled water |  | Make up to 1000 ml |

3. Immerse gel, membrane, filter papers, and support pads in transfer buffer to ensure they are thoroughly soaked. Be careful to exclude air bubbles from the support pads.

*Towbin et al. (1979).

4. Assemble the transfer sandwich as shown below. Keep all the components wet and make sure the sandwich is tightly assembled.

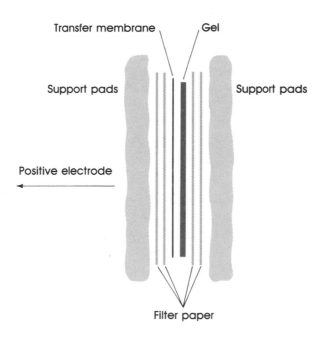

Transfer membrane    Gel

Support pads    Support pads

Positive electrode

Filter paper

5. Place the complete sandwich in the transfer tank with the membrane closest to the positive electrode (anode, red electrode).

6. Transfer for 4–18 hr. Thicker gels and higher-molecular-weight proteins require longer transfers. The following conditions have been optimized to permit transfer and retention of a very wide range of molecular weight proteins.

   For proteins over 100,000 mw on a 15 × 15 × 0.1-cm gel, transfer at 28 volts for 1 hr, then at 84 volts for 14–16 hrs.

   For proteins under 100,000 mw on a 15 × 15 × 0.1-cm gel, transfer at 63 volts for 4–16 hrs.

   In all transfers, the temperature will rise substantially during the run, and it is essential to use a cooling coil or to run the transfer in a cold room to avoid the generation of gas bubbles in the sandwich.

7. After transfer, disconnect the power supply. Disassemble the sandwich and mark the nitrocellulose membrane by clipping one corner (normally the bottom left hand corner, the number one lane).

8. The polyacrylamide gel can now be stained with Coomassie (pp. 648–650) or silver stain (pp. 651 or 653) to verify transfer. Some workers run prestained or radioactive markers on their gels to serve as internal markers for transfer and molecular weight.

**Process the nitrocellulose sheet for staining (p. 493) or blocking (p. 497) as appropriate.**

## NOTES

i. The types of bonds that hold proteins to nitrocellulose are not known. However, the binding is blocked by oils or other proteins. Wear gloves and use virgin nitrocellulose sheets.

ii. Several workers recommend allowing the membrane to dry completely before processing the nitrocellulose prior to blocking or staining. In some cases, this appears to lessen the chance of the transferred proteins being lost during subsequent steps. After drying, rewet the membrane as in step 2. An alternative to drying is to treat the membrane with isopropanol for 1 min.

**TABLE 12.1**
**Stains Used for Immunoblots**

| Stain | Advantages | Disadvantages |
|-------|-----------|---------------|
| **India Ink** | Cheap<br>Sensitive<br>Compatible with<br>    radiolabel detection | For enzyme<br>    detection<br>    block in Tween,<br>    detect then stain |
| **Ponceau S** | Cheap<br>Compatible with all<br>    antigen detection methods | Insensitive<br>Not permanent |
| **Amido black** | Cheap | Insensitive<br>No antigen detection possible<br>Distorts membrane |

## ■ Staining the Blot for Total Protein (Optional)

Prior to probing blots for the presence of an antigen, the total composition of the proteins transferred to nitrocellulose can be determined by staining the nitrocellulose sheet. Staining for proteins can be used to determine the position of molecular weight markers and to ensure that efficient transfer has occurred. It also gives a remarkably good picture of the complexity of the transferred proteins. The conventional procedures used for staining polyacrylamide gels are unsuitable for staining immunoblots. Both Coomasie blue and silver staining methods give high backgrounds due to nonspecific dye adsorption, and when using nitrocellulose, these methods damage and distort the membrane. A number of suitable staining methods have now been developed that give a sensitive and high-quality stain of nitrocellulose replicas.

Four basic approaches can be adopted: (1) the set of samples to be analyzed is run in duplicate, and one-half of the sheet stained for protein and the other half for antigen. This avoids any problems of interference between the visualization of the antigen and total protein but creates problems of alignment; (2) the same sheet is stained for total protein and used for antigen detection using protocols that allow a distinction between the two images; (3) the blots are blocked with Tween 20, antigen is detected immunochemically, and then the blots are stained for total protein; or (4) the proteins are labeled before the transfer, creating a direct image on the sheet that is probed for the antigen using nonradioactive detection. For simple detection of molecular weight markers, this last approach can be conveniently employed by including a lane of prestained or $^{14}$C-labeled molecular weight markers on the gel.

Protocols 2, 3, and 4 avoid problems of alignment, but the staining step may alter the immunological detection by modifying or obscuring the antigen. The choice of procedure depends on the information required from the experiment. In many cases, all that is required of the protein stain is that it identifies molecular weight markers and the position of individual tracks. In other experiments, for instance in two-dimensional gel analysis or when assessing a protein purification, the protein stain must permit the unambiguous assignment of antigenic activity to a particular stained spot or band.

Table 12.1 lists the commonly used stains and summarizes their advantages and disadvantages. The starting point for all staining protocols is the completion of the transfer step, when stains are generally applied. It is somewhat surprising that blocked and immunologically probed filters can sometimes be stained to allow a useful rescue or verification of old blots.

## STAINING THE BLOT WITH PONCEAU S

Ponceau S is applied in an acidic aqueous solution. Staining is rapid but not permanent; the red stain will wash away in subsequent processing. Because the binding is reversible, the stain is compatible with most antigen visualization techniques. Therefore, Ponceau S can be used routinely to verify transfer and to locate molecular weight markers. However, the stain is not very sensitive and is rather difficult to record photographically. The stain can be used effectively by marking the position of molecular weight markers and lane positions with pencil or indelible pen.

1. Prior to staining, prepare a concentrated stock of Ponceau S. The concentrated solution is 2% Ponceau S (3-hydroxy-4-[2-sulfo-4-(sulfo-phenylazo)phenylazo]-2,7-naphthalene disulfonic acid) in 30% trichloroacetic acid, 30% sulfosalicylic acid. The concentrated dye solution is diluted 1 in 10 with water to prepare the working solution. (The stock solution is stable at room temperature for over 1 year.)

2. Wash the nitrocellulose sheet once in Ponceau S solution.

3. Add freshly diluted Ponceau S and incubate for 5–10 min at room temperature with agitation.

4. Transfer the nitrocellulose sheet to PBS and rinse for 1–2 min with several changes of PBS.

5. Mark position of transfer and molecular weight markers as required.

The nitrocellulose sheet is now ready for blocking (p. 497).

## STAINING THE BLOT WITH INDIA INK*

This method of staining blots is highly recommended. It is cheap, reliable, sensitive, and yields a permanent record. Also, it appears not to interfere with subsequent binding of antibody to the antigen (Glenney 1986). Because of a lack of contrast between the colored reaction product, India ink is not compatible with photography of enzyme-based detection methods. However, it is ideally suited for staining blots, if radioisotopic detection is used. It is also an excellent choice, if staining a duplicate blot.

The staining depends on the preferential adherence of the colloidal carbon particles in India ink to the immobilized protein on the filter. It is essential that the correct India ink be used. Pelikan Fount India drawing ink (original recipe, black) or equivalent is suitable.

1. Wash blot in 0.4% Tween 20/PBS, with two changes of 100 ml at 5 min each.

2. Place blot in ink solution (100 $\mu$l of ink in 100 ml of 0.3% Tween 20/PBS).

3. Incubate at room temperature for 15 min to 18 hr. Longer incubations increase sensitivity.

4. Destain by washing the blot in multiple changes of PBS.

The nitrocellulose sheet is now ready for blocking (p. 497).

### NOTE

i. India ink staining can be used after immunoenzymatic detection is complete. In this case, the remaining sites on the nitrocellulose must be blocked with Tween 20 prior to antigen detection (see Table 12.2, p. 496).

*Hancock and Tsang (1983).

## STAINING THE BLOT WITH AMIDO BLACK*

Under certain circumstances, the sensitivity of India ink may be inappropriate and, therefore, a less sensitive procedure is needed. Amido black is suitable for a number of staining needs but is not readily compatible with immunological detection. Also, this method can cause distortion of the nitrocellulose sheet.

1. Wash nitrocellulose blot in 0.3% Tween/PBS, with three changes of 15 min each at room temperature.

2. Immerse the sheet in amido black stain (0.1% amido black 10-B, 45% methanol, 10% acetic acid). Incubate for 5 min at room temperature with agitation.

3. Destain in 90% methanol/2% acetic acid/8% water.

4. Rinse in water and dry.

As described above, no further steps are possible.

*Kuno and Kihara (1967).

**TABLE 12.2**
**Blocking Solutions**

| Blocking buffer | Composition |
| --- | --- |
| Blotto | 5% wt/vol nonfat dry milk<br>0.02% sodium azide<br>    in PBS |
| Blotto/Tween | 5% wt/vol nonfat dry milk<br>0.2% Tween 20<br>0.02% sodium azide<br>    in PBS |
| Tween | 0.2% Tween 20<br>0.02% sodium azide<br>    in PBS |
| BSA | 3% BSA (IgG free, e.g., Cohn fraction V)<br>0.02% sodium azide<br>    in PBS |
| Horse Serum | 10% horse serum<br>0.02% sodium azide<br>    in PBS |

## Blocking Nonspecific Binding Sites on the Blot

Before the blot can be processed for antigen detection, it is essential to block the membrane to prevent nonspecific adsorption of the immunological reagents. Blocking can be achieved with a range of protein or detergent solutions. Table 12.2 summarizes the commonly used blocking solutions and their advantages and disadvantages. Although all of these work satisfactorily in some circumstances, a careful choice is necessary to ensure compatibility with the detection reagent. For example, using whole serum to block is inappropriate, if protein A is used as a detection reagent. The compatibility of the blocking reagent with the detection system can be tested by taking a fresh piece·of membrane, blocking it and then treating with the detection system.

Two blocking solutions that are compatible with nearly all detection systems are nonfat dried milk (Johnson et al. 1984) and BSA (Towbin et al. 1979). If protein blocking solutions cause problems with excessive background, then blocking in Tween 20 should be tried (Batteiger et al. 1982).

| Advantages | Disadvantages | Blocking buffer |
|---|---|---|
| Cheap<br>Clean background | Deteriorates rapidly<br>Disguises some antigen | **Blotto** |
| Cheap<br>Clean background | Deteriorates rapidly<br>Disguises some antigen | **Blotto/Tween** |
| Allows staining<br>    after antigen detection<br>Cheap | May get some<br>    residual background | **Tween** |
| Good signal strength | Relatively expensive | **BSA** |
| Clean backgrounds | Moderately expensive<br>Incompatible with some<br>    anti-immunoglobulin<br>    antibody detection | **Horse Serum** |

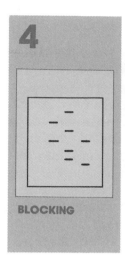

**BLOCKING**

## BLOCKING THE BLOT

1. Rinse the filter several times with PBS.

2. Add one of the blocking solutions (see Table 12.2).

3. Incubate at room temperature with agitation for 2 hr.

**The nitrocellulose is now ready for the addition of antibody (p. 499).**

### NOTE

i. Different antigens or applications may require different blocking buffers. For example, both BSA (fraction V) and milk contain phosphotyrosine, and this will confuse the interpretation of certain tests with antibodies specific for anti-phosphotyrosine antibodies. Alkaline phosphate is inhibited by some preparations of milk. If no signal or a high background is detected, check different blocking buffers.

---

## Purifying Antibodies from Immunoblots

Antibodies that bind to a given polypeptide band can be purified from polyclonal sera by a simple modification of the immunoblotting technique (Olmsted 1981; Smith and Fisher 1984). These band-specific antibodies, though recovered in small amounts, can be useful in studies of complex antigens, often allowing the confirmation of antigen identity. With antigens that are available in larger amounts, such as those isolated from bacterial overexpression systems, this procedure can be scaled up to produce a larger amount of antigen-specific antibodies.

Load the antigen into all of the wells of the gel or into one large preparative well. The blots are run as normal, but after the incubation with the primary antibody and washing, the location of the antibodies bound to the antigen of interest is determined by cutting strips of the blot from the sides. These side strips are processed as normal for antigen detection, while the main body of the blot is stored in PBS supplemented with 0.02% sodium azide at 4°C. After the localization of the antigen–primary antibody complexes on the side strips, align the strips with the untreated portion of the blot. Excise the area of the blot that contains the antigen and primary antibody using a sharp scalpel. Cut this into small pieces and transfer to a test tube. Incubate with 100 mM glycine (pH 2.5) for 10 min or with 1 M KSCN for 10 min. Remove the buffer and neutralize with one-tenth volume of 1 M Tris (pH 8.0) for the glycine elution or dilute with 10 volumes of PBS for the KSCN elution. The antibodies are now ready for further tests.

For larger-scale purifications, covalent coupling of the antigen to the solid support using APT paper will increase the capacity of the blot and will allow the use of harsher conditions.

# ■ Addition of Antibody

Detecting the antigen in an immunoblot is achieved by binding antibodies to the proteins immobilized on the membrane. The antibodies can be labeled themselves and, thus, can detect the location of the antigen directly, or the antibodies can be used unlabeled and located by labeled secondary reagents. These two methods are known as direct and indirect detection.

Direct detection of antigens has several advantages. It normally results in a lower background, is quicker and easier, and decreases the chances of detecting spurious cross-reactions. However, direct detection is less sensitive than indirect techniques. The major disadvantage of direct detection is the tedium of purifying and labeling each antibody before use. Employing indirect detection methods allows the use of one batch of reagents to detect a wide range of antibodies. Except when the specificity of direct detection is needed, most immunoblots should be analyzed using indirect methods. For both direct and indirect techniques, the antibodies can be labeled with iodine or enzymes. The procedures for labeling these reagents are discussed on p. 319 in Chapter 9 or they can be purchased from any of a number of commercial suppliers. Choosing the type of label is discussed below in the detection section (p. 502).

An alternative to labeling primary antibodies with iodine or enzymes is biotinylation (p. 340). Biotin-labeled antibodies are not detected directly, but need the addition of a secondary reagent, usually avidin or streptavidin coupled with an enzyme or iodinated. Using biotinylated antibodies takes advantage of a wide range of secondary reagents that are available commercially. Because the detection is performed using secondary reagents, a single batch of avidin or streptavidin can be used with a variety of primary antibodies. However, each primary antibody needs to be purified and labeled before it can be used. For immunoblots, biotinylation of the primary antibodies is a good choice, when many samples need to be screened with the same antibodies. Biotinylated antibodies are treated in the same manner as unlabeled antibodies.

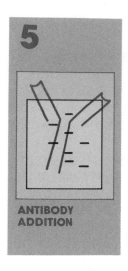

**5**

**ANTIBODY
ADDITION**

## ANTIBODY BINDING

1. Remove the blot from the blocking solution and wash twice for 5 min each in PBS.

2. Add the primary antibody solution. Use 10 ml per 15 × 15-cm blot or 0.5 ml for a 15 × 0.5-cm strip. Primary antibodies can be labeled directly or can be detected using labeled secondary reagents. All dilutions should be done in protein containing solutions such as 3% BSA/PBS. Incubations can be performed in heat-sealed bags or shallow trays. In heat-sealed bags, avoid air bubbles to ensure even contact.

   *For unlabeled primary antibodies*: Use tissue culture supernatants either undiluted or diluted up to 1 in 10. This yields an antibody concentration between 1 and 50 $\mu$g/ml. For polyclonal sera, ascites fluid, or purified antibodies, the final antibody concentration should also be approximately in this range. If the concentration of an antibody solution is unknown, try several dilutions to determine the correct range.

   *For labeled primary antibodies*: The correct concentration of a labeled primary antibody solution must be determined in preliminary tests. As a starting point with $^{125}$I-labeled antibodies, add approximately $10^6$ cpm per 15 × 15-cm$^2$ blot. For enzyme-labeled primary antibodies, a good starting point will be to add the equivalent of 0.1 to 1 $\mu$g of antibody per 15 × 15-cm$^2$ blot.

3. Incubate the blot with antibody for at least 1 hr at room temperature with agitation. Some workers have reported increased sensitivity by using overnight (12–18 hr) incubation.

4. Wash the blot with four changes of PBS for 5 min each.

**If the primary antibodies are labeled, the blot is now ready for detection (see p. 502).**

5. The blot is now ready for the addition of the labeled secondary reagent. Incubation can be done in a shallow tray when probing a full blot. For probing strips of blots, use heat-sealable bags or test tubes. All dilutions should be done in protein containing buffers such as 3% BSA/PBS.

*For iodine-labeled secondary reagents*: Remove the blot from the final wash and add approximately $10^6$ cpm of $^{125}$I-labeled secondary reagent (protein A, protein G, or anti-immunoglobulin antibodies) in approximately 10 ml of blocking buffer per 15 × 15-cm blot. These reagents are available commercially or can be prepared as described on p. 324.

*For enzyme-labeled secondary reagents*: In general, conjugates should be used at concentrations of 5–0.5 $\mu$g/ml (usually a dilution of 1/200 to 1/2000 of commercial stocks). Reagents labeled with horseradish peroxidase should be diluted in blocking buffer without sodium azide. Reagents labeled with alkaline phosphatase should be diluted and washed with Tris-buffered saline and not PBS. Enzyme-labeled protein A, anti-immunoglobulin antibodies, avidin, and streptavidin are available commercially or can be prepared using the protocols described on p. 342.

6. Incubate for 1 hr at room temperature with agitation.

7. Wash the blot with 4 changes of PBS for 5 min each.

**The blot is now ready for detection (see p. 502).**

## NOTES

i. When probing multiple strips with different primary antibodies, the strips should always be handled separately.

ii. When using alkaline phosphatase, the blot should be washed twice in 150 mM NaCl, 50 mM Tris (pH 7.5) before adding the enzyme conjugate.

## ■ Detection

Two methods used for detection are radioactive- or enzyme-linked reagents. Radioactive detection methods offer two distinct advantages. First, they are easy to quantitate by timed exposure to film and densitometry or by excising labeled bands from the membrane and counting them directly. Second, the autoradiographic image is reproduced easily and accurately for publication.

Detection with enzyme-labeled antibodies is the most popular method, normally using horseradish peroxidase- or alkaline phosphatase-coupled antibodies, and a range of soluble substrates that yield insoluble colored products. These methods are simple and do not require special equipment. They also provide an immediate result. However, due to the nature of the enzymatic reactions involved, it is not easy to obtain quantitative results. There can also be technical problems with photography of enzyme-developed blots, due to poor contrast between the color of the reaction product and the membrane itself.

## Increasing the Sensitivity

Four methods commonly are used to increase the sensitivity of immunoblots. The first relies on the purification of the antigen by immunoprecipitation prior to electrophoresis. This is discussed on p. 480. The other three increase the strength of the detection method. The signal can be intensified by using a triple-layer detection system, i.e., primary, secondary, and tertiary antibodies are all complexed to the antigen. This increases the number of potential labeled antibodies that can be bound. For example, if five secondary antibodies can bind to the primary antibody and five tertiary antibodies can bind to each secondary antibody, a 25-fold increase in the signal over a labeled primary antibody can be achieved. Another approach is to use the avidin or streptavidin molecules as intermediates. These reagents are tetravalent and therefore when combined with biotin can be used to form large multimeric complexes. A further approach to increase the sensitivity is to use enzyme/anti-enzyme complexes in the detection. These last three methods are discussed on p. 344.

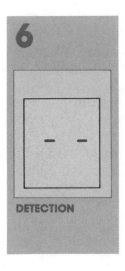

**DETECTION**

## DETECTION WITH RADIOLABELED REAGENTS

Most radioactive detection methods use $^{125}$I.

1. After the last wash, drain the blot by placing it on a piece of absorbant paper for a few minutes.

2. Wrap the blot in plastic wrap. To allow alignment of the film and blot, put labels on three of the corners of the plastic wrap. Mark each label with a small symbol using radioactive ink. Radioactive ink can be prepared by adding 100 $\mu$l of ink to an empty [$^{35}$S]methionine container. The small amount of radioactivity that remains on the side of the container will be enough for most applications. Allow the ink to dry thoroughly.

3. Expose the blot to X-ray film at $-70°$C with an intensifying screen.

### NOTE

i. The plastic wrap will block the exposure of $^{14}$C-labeled markers. If radioactive markers have been included, the blot must be dried before exposure to film and plastic wrap is not used. However, the nitrocellulose must be dry. Even small amounts of buffer from the last wash will cause the film to stick to the blot. If the nitrocellulose adheres to the film, do not attempt to peel it off; instead, immerse the entire sandwich in PBS. Allow the nitrocellulose to rehydrate completely and then remove it from the film.

### Detection with Enzyme-labeled Reagents

Two enzymes have been used extensively to label antibodies for immunoblot detection. Alkaline phosphatase-labeled antibodies developed with the bromochloroindolyl phosphate/nitro blue tetrazolium substrate (BCIP/NBT) are probably the best choice for immunoblots, because they yield an intense purple-black color with little background. Satisfactory results can be obtained with horseradish peroxidase-labeled antibodies developed with 4-chloro-1-naphthol, 3-amino-9-ethylcarbazole (AEC), or diaminobenzidine (DAB). These combinations are particularly useful, because of the wide commercial availability of horseradish peroxidase-labeled conjugates.

---

**Caution** Some chromogenic substrates may be carcinogenic. Consult your local authorities for handling and disposal procedures.

---

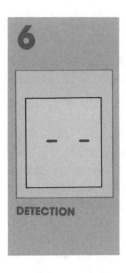

**DETECTION**

## ALKALINE PHOSPHATASE

The bromochloroindolyl phosphate/nitro blue tetrazolium (BCIP/NBT) substrate generates an intense black-purple precipitate at the site of enzyme binding. The substrate solution is stable in the absence of enzyme. The reaction proceeds at a steady rate, thus allowing accurate control of the development of the reaction. This allows the relative sensitivity to be controlled by the length of incubation. The BCIP/NBT substrate characteristically produces sharp bands with very little background coloring of the membrane. There are a number of other chromogenic substrates for alkaline phosphatase (see p. 681), but for immunoblots, the BCIP/NBT substrate is best.

Coupled reagents should be prepared with eukaryotic alkaline phosphatase, as this enzyme is readily inactivated with EDTA. The bacterial enzyme is difficult to stop, causing overdevelopment and leading to high background.

1. Prior to developing the blot, prepare the three stock solutions. (1) NBT: Dissolve 0.5 gram of NBT in 10 ml of 70% dimethylformamide. (2) BCIP: Dissolve 0.5 gram of BCIP (disodium salt) in 10 ml of 100% dimethylformamide. (3) Alkaline phosphatase buffer: 100 mM NaCl, 5 mM $MgCl_2$, 100 mM Tris (pH 9.5). All stocks are stable at 4°C for at least 1 year.

2. Just prior to developing the blot, prepare fresh substrate solution. Add 66 $\mu$l of NBT stock to 10 ml of alkaline phosphatase buffer. Mix well and add 33 $\mu$l of BCIP stock. Use within 1 hr.

3. Place the washed immunoblot in a suitable container. Add 10 ml of substrate solution per 15 × 15-$cm^2$ membrane. Develop the blot at room temperature with agitation until the bands are suitably dark. A typical incubation with the blot would be approximately 30 min.

4. To stop the reaction, rinse with PBS containing 20 mM EDTA, chelating the $Mg^{2+}$ ions.

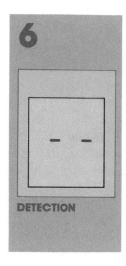

**6**

DETECTION

## HORSERADISH PEROXIDASE

Three substrates are suitable for developing immunoblots stained with horseradish peroxidase-coupled antibodies: chloronaphthol, amino-ethylcarbazole, and diaminobenzidine.

### Chloronaphthol

4-Chloro-1-naphthol (chloronaphthol) is relatively insensitive but gives an intense blue-black reaction product in an easily controlled reaction. Blots developed with this reagent are readily photographed, but the color fades noticeably on storage. Background binding of the product to the filter is low.

1. Prior to developing the blot, prepare the stock solution of 4-chloro-1-naphthol. Dissolve 0.3 gram of chloronaphthol in 10 ml of absolute ethanol. The chloronaphthol stock is stable at −20°C for at least 1 year.

2. Just prior to developing the blot, add 0.1 ml of chloronaphthol stock to 10 ml of 50 mM Tris (pH 7.6). A copious white precipitate will form.

3. Remove the precipitate by filtering through Whitman No. 1 filter paper (or equivalent).

4. Add 10 $\mu$l of 30% $H_2O_2$. $H_2O_2$ is generally supplied as a 30% solution and should be stored at 4°C, where it will last about 1 month.

5. Place the washed immunoblot in a suitable container. Add 10 ml of substrate solution per $15 \times 15$-cm$^2$ membrane. Develop the blot at room temperature with agitation until the bands are suitably dark. A typical incubation with the blot would be approximately 30 min.

6. To stop the reaction, remove the $H_2O_2$ by rinsing with PBS.

### Aminoethylcarbazole

3-Amino-9-ethylcarbazole (AEC) yields an aesthetically pleasing red reaction product. It is slightly more sensitive than chloronaphthol but harder to photograph.

1. Prior to developing the blot, prepare the stock solution of AEC. Dissolve 0.4 gram of 3-amino-9-ethylcarbazole in 100 ml of dimethylformamide. The AEC stock is stable at room temperature for at least 1 year.

2. Just prior to developing the blot, add 0.67 ml of AEC stock to 10 ml of 0.1 M sodium acetate (pH 5.2) with stirring. A small precipitate may form.

3. Filter through Whatman No. 1 filter paper (or equivalent).

4. Add 10 $\mu$l of 30% $H_2O_2$. $H_2O_2$ is generally supplied as a 30% solution and should be stored at 4°C, where it will last about 1 month.

5. Place the washed immunoblot in a suitable container. Add 10 ml of substrate solution per $15 \times 15$-cm$^2$ membrane. Develop the blot at room temperature with agitation until the bands are suitably dark. A typical incubation with the blot would be approximately 30 min.

6. To stop the reaction, remove the $H_2O_2$ by rinsing with PBS.

### Diaminobenzidine

Diaminobenzidine (DAB, 3,3',4,4'-tetraaminobiphenyl) is an exceptionally sensitive substrate, yielding a brown reaction product. Its sensitivity can be further enhanced by adding cobalt or nickel ions to the substrate solution. In the presence of these ions, the product is slate black in color and more easily photographed than the unenhanced product. The drawback to DAB is that it reacts so fast that it is very easy to overdevelop the blot giving a high background.

1. Just prior to developing the immunoblot, dissolve 6 mg of 3,3'-diaminobenzidine (use DAB tetrahydrochloride) in 10 ml of 50 mM Tris (pH 7.6). A small precipitate may form.

2. Filter through Whatman No. 1 filter paper (or equivalent).

3. Add 10 $\mu$l of 30% $H_2O_2$. $H_2O_2$ is generally supplied as a 30% solution and should be stored at 4°C, where it will last about 1 month.

4. Place the washed immunoblot in a suitable container. Add 10 ml of substrate solution per $15 \times 15$-cm$^2$ membrane. Develop the blot at room temperature with agitation until the bands are suitably dark. A typical incubation with the blot would be approximately 1–5 min.

5. To stop the reaction, remove the $H_2O_2$ by rinsing with PBS.

### Diaminobenzidine with Metal Ion Enhancement (DAB/Metal)

1. Just prior to developing the immunoblot, dissolve 6 mg of 3,3′-diaminobenzidine (use DAB tetrahydrochloride) in 9 ml of 50 mM Tris (pH 7.6). A small precipitate may form. Add 1 ml of 0.3% (wt/vol) $NiCl_2$ or $CoCl_2$.

2. Filter through Whatman No. 1 filter paper (or equivalent).

3. Add 10 $\mu$l of 30% $H_2O_2$. $H_2O_2$ is generally supplied as a 30% solution and should be stored at 4°C, where it will last about 1 month.

4. Place the washed immunoblot in a suitable container. Add 10 ml of substrate solution per 15 × 15-cm$^2$ membrane. Develop the blot at room temperature with agitation until the bands are suitably dark. A typical incubation with the blot would be approximately 1–5 min.

5. To stop the reaction, remove the $H_2O_2$ by rinsing with PBS.

## COMMENTS ■ Troubleshooting—Bad Backgrounds

The most frequently encountered problem in immunoblots is the presence of a high background consisting either of extra, discrete bands or of a general diffuse signal covering the entire membrane.

Specific background bands are generated either by contaminating antibodies present in polyclonal sera or by specific cross-reactions between different antigens. The latter problem is especially prevalent with monoclonal antibodies. Nonspecific diffuse backgrounds result either from insufficient blocking of the membrane or from a specific reaction of the detecting reagent with a component of the blocking buffer.

### Diffuse Backgrounds

- If using an indirect technique, first determine whether the secondary reagent is contributing to the background. Omit the primary antibody and observe the background generated by the secondary reagent alone. If the background is generated by the secondary reagent: (1) use an alternative label; (2) try adsorbing the secondary reagent with a sample of the starting antigen preparation (an acetone powder may be suitable); (3) reduce the time in substrate or use a less sensitive substrate.

- Use a different blocking buffer. Try Blotto/Tween, if presently using anything else.

- Titrate the concentration of the primary and secondary antibodies.

- Reduce the incubation time of the primary and secondary antibodies.

- Wash the filter at each step in 1% NP-40, 0.5% DOC, 0.1% SDS, 150 mM NaCl, 50 mM Tris (pH 7.5).

- Add 1% NP-40 to the primary and secondary antibodies.

- Increase the duration of each washing step.

### Specific Background Bands

- If using an indirect technique, first determine whether the secondary detecting reagent is contributing to the background. Omit the primary antibody and observe the background generated by the secondary reagent alone. If the background bands are generated by the secondary reagent, first use an alternative label. Second, try adsorbing the secondary reagent with a sample of the starting antigen preparation. An acetone powder may be suitable.

- If using a monoclonal antibody, use another monoclonal antibody if available. Continued presence of the same band may suggest an amino acid homology between your antigen and the other band.

- If using a polyclonal serum as a source of primary antibodies, try adsorbing the serum with a protein preparation that does not contain the antigen of interest.

# 13

# IMMUNOAFFINITY PURIFICATION

Immunoaffinity purification is one of the most powerful techniques for the isolation of proteins. Under the proper conditions, purifications of 1,000- to 10,000-fold can be achieved routinely in a single step. Purifications of greater than 10,000-fold have been reported, but are possible only with particularly good antibodies or under particularly unique elution conditions.

Immunoaffinity purification can be divided into three steps:

(1) preparation of the antibody–matrix
(2) binding the antigen to the antibody–matrix
(3) elution of the antigen

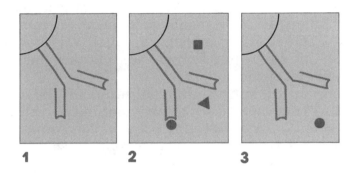

**1**     **2**     **3**

In the first step, either monoclonal antibodies or affinity-purified polyclonal antibodies are covalently attached to a solid-phase matrix. There are a large number of different protocols for covalently binding antibodies to a solid phase, but probably the easiest is linking antibodies to protein A beads. After the preparation of the antibody–bead matrix, the antigen is bound to the antibodies, and contaminating macromolecules are removed by washing. In the third step, the antibody–antigen interaction is broken by treating the immune complexes with strong elution conditions, and the antigen is released into the eluate.

## ▨ MAJOR CONSTRAINTS ▨▨▨▨▨▨▨▨▨▨▨▨▨▨▨▨▨▨▨▨▨▨

There are several factors that contribute to the success of an immuno-affinity purification. The three most crucial are the starting purity of the antigen, the affinity of the antibody for the antigen, and the ease with which the antibody–antigen bond can be broken.

The relative purity of the starting antigen is the single most important factor in determining the purity of the final product using immunoaffinity techniques. Because of the unique properties of the antigen–antibody interaction, no other type of chromatographic technique is likely to yield greater purification in a single step. However, the degree of purification is not unlimited, as the use of affinity columns has certain inherent background problems. Commonly, 1000-fold purifications are routine, and in most applications 10,000-fold purifications are possible. If the protein antigen of interest is rarer than this, immunoaffinity purification must be combined with other methods to achieve a homogeneous product. This may be accomplished by the use of other purification steps either before or after the immunoaffinity column.

The affinity of the antibody for the antigen is the most worrisome of the problems encountered with immunoaffinity purification. The affinity of the antibody will determine both the total amount of antigen that can be removed from the antigen-containing solution. For antibodies with high affinities ($>10^8$ $mol^{-1}$), quantitative removal can be achieved in less than 1 hr. Even at high antibody concentrations, low-affinity antibodies ($10^6$ $mol^{-1}$) will never bind all of the antigen in the solution. Note that all of the times for reaching equilibrium will be considerably longer for immunoaffinity purifications than for techniques such as immunoprecipitation, because the antibodies are bound to a solid support, thus drastically slowing the kinetics of binding. Tricks that can be used to trap antigen more efficiently, such as using multiple antibodies (p. 31), are not appropriate for immunoaffinity purification. Increasing the avidity by these methods will lead to higher antigen binding, but will also increase the problems encountered when eluting the antigen.

The third factor that will influence the success of an immunoaffinity purification is the relative ease with which the antigen can be eluted. This is determined solely by the type and number of bonds that form the antibody–antigen interaction, and is, therefore, related to the antibody affinity. However, the affinity does not determine whether the antigen will be easy to elute. The ideal antibody for an immunoaffinity purification is one that has a high affinity for the antigen and whose binding can be reversed by a simple but gentle change in an easily manipulatable variable such as pH. The methods for designing these manipulations are described on p. 551. A common mistake in designing immunoaffinity purification experiments is to equate low affinity with easy elution. Although this may prove to be the case for some low-affinity antibodies, often it is not true. The bonds that hold the antigen to the antibody are of the same basic types for both high-affinity and low-affinity antibodies (see Chapter 3). For example, both high-affinity and low-affinity antibodies may bind an antigen through a salt bridge and a hydrophobic interaction. Eluting the antigen in both cases requires both bonds to be disrupted and, hence, poses similar problems for the choice of elution conditions. In addition, low-affinity antibodies introduce new problems, including difficult binding (see above), continual leaching from the column during washing, and gradual, rather than sharp, elution profiles.

## ■ CHOICE OF ANTIBODY

Most immunoaffinity purifications are done with monoclonal antibodies or affinity-purified polyclonal antibodies. In some cases, total polyclonal antibodies may be appropriate, but almost never will pooled monoclonal antibodies be useful.

## ■ Immunoaffinity Purification Using Polyclonal Antibodies

Immunoaffinity purification using polyclonal antibodies has limited applications. Because polyclonal antibodies usually bind to numerous sites on an antigen and therefore bind with high avidity, they are difficult to elute. The harsh conditions needed to elute an antigen when it is bound by several different antibodies usually will damage the antibody column and at least partially denature the antigen. Even when the number of binding sites between the antigen and a polyclonal antibody–matrix is kept low by using saturating amounts of antigen, successful elution will be difficult. In this case, the elution conditions for each antibody–antigen interaction will be different, so efficient elution cannot be established. Another disadvantage when using polyclonal antibodies is that they will contain a number of spurious activities against unrelated antigens. These will include all the antibodies in the animal's serum at the time of collection.

Two types of polyclonal antibodies that are useful for immunoaffinity columns are those that have been raised against synthetic peptides or defined regions of an antigen. In both of these cases, because the epitopes for binding the antigen to the column are located in one small region, it may be possible to achieve efficient elution.

One method to make immunoaffinity purifications possible using polyclonal antibodies is to select specific antibodies on an antigen-affinity column (p. 313). This use is limited to cases where sufficient quantities of antigen are available to prepare these columns. However, in some cases, such as purifying a protein from one species with an antibody against an analogous protein from another species, this may be helpful. Affinity-purifying the polyclonal antibodies solves several problems. Only the antibodies that bind to the antigen are collected, and because they have been eluted from the antigen, the exact conditions for releasing the antigen from the immunoaffinity column already have been determined.

## Immunoaffinity Purification Using Monoclonal Antibodies

Using monoclonal antibodies for immunoaffinity purifications has a number of advantages compared with other sources of antibodies. Monoclonal antibodies are available in an essentially unlimited supply, and high-affinity monoclonal antibodies can bind to a large proportion of the available antigen. Because all the antibodies are identical and bind to the same epitope, all of the antigen interactions can be broken under similar conditions. Because there are only a limited number of bonds that will hold the antigen–antibody complexes together, the conditions to release the antigen will normally be gentler than those needed for polyclonal interactions.

The problems that are found when using monoclonal antibodies for immunoaffinity purification normally concern the properties of antigen interaction, commonly low-affinity reactions or cross-reactions. Low-affinity interactions are a problem common to all immunoaffinity purifications and are discussed above. Cross-reactions are particular to monoclonal antibody purifications and are only seen for a subset of antibodies. In these cases, the antibodies bind to other antigens through shared epitopes. These epitopes may be part of a more extensive homology, in which case the monoclonal antibodies may be useful in studying the related proteins but not for purifying the original antigen. On the other hand, the cross-reactions may be limited to the epitope itself. In this case, using other monoclonal antibodies may eliminate the cross-reaction. Another solution would be to purify the antigen away from the spurious protein by some other chromatographic or extraction technique.

## Immunoaffinity Purification Using Pooled Monoclonal Antibodies

Except in unusual circumstances, there is no reason to pool monoclonal antibodies for preparing immunoaffinity columns. All of the problems of elution discussed in the section on polyclonal antibody columns apply to pooled monoclonal antibodies. The only common use of these reagents is in the purification of antigens that will be denatured before use or in preparing an antigen–antibody–bead complex for immunizations (p. 135).

## SUMMARY

### Immunoaffinity Purification

Purification factor is 1000- to 10,000-fold

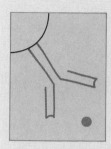

- Rapid antigen purification, columns often reusable
- Yields purified antigen
- Not useful for quantitation
- Efficiency depends on antigen concentration and antibody affinity
- Needs moderate affinity antibody

|  | Polyclonal Antibodies | Monoclonal Antibodies | Pooled Monoclonal Antibodies |
|---|---|---|---|
| Signal Strength | Good | Antibody dependent, (poor to excellent) | Excellent |
| Specificity | Good, but high background | Excellent, but some cross-reactions | Excellent |
| Good Features | Availability | Precise and often gentle elution conditions Specificity Unlimited supply | No advantages |
| Bad Features | Elution often impossible | Finding suitable antibody | Elution often impossible |

## PROTOCOLS FOR IMMUNOAFFINITY PURIFICATION

Immunoaffinity purification can be divided conveniently into three steps: (1) the preparation of the antibody column, (2) the binding of the antigen to the antibody–bead matrix, and (3) the elution of the antigen from the column (affinity chromatography was first introduced by Cuatrecases 1968. See *Methods Enzymol.* vol. 34 and Wilchek et al. 1984, for general considerations).

## Preparing Antibody Affinity Columns

There are a number of methods that can be used for covalent attachment of antibodies to solid-phase matrices. Their advantages and disadvantages are summarized in Table 13.1 along with possible coupling methods. The methods for coupling antibodies to solid-phase matrices are divided into three classes: (1) protein A/G beads, (2) activated beads, and (3) activated antibodies. The most useful of these coupling methods makes use of previously prepared matrices that have been modified to contain secondary reagents binding specifically to antibodies. The most commonly used of these matrices is the protein A beads. Protein A binds specifically to the Fc domain of antibodies, and after the antibody is bound the interaction is stabilized by cross-linking with a bifunctional coupling reagent. The second class of coupling method directly couples the antibody to an activated bead. Beads are activated chemically to contain reactive groups. The beads are mixed and coupled with purified antibodies. The third method of coupling antibodies to beads is to activate the antibody first, placing the reactive group on the soluble antibody. The activated antibody is then mixed and coupled with beads. A variation of this third technique that also combines portions of the activated-bead methods is the biotin/streptavidin approach. For this method, antibodies are modified to contain biotin groups. The antibodies are then bound to streptavidin beads.

For most purposes the matrices that require the least preparation are the best ones to try first. If no background information argues against their use, the protein A bead columns should be tried first. If these present problems, either for reasons of cost or difficulties in the purification, then other methods can be substituted.

**TABLE 13.1**
**Methods for Coupling Antibodies to Beads**

| Method | Variations | Antibody oriented? | Coupling group on antibody | Advantages | Disadvantages |
|---|---|---|---|---|---|
| **Protein A** | Direct coupling | Yes | $-NH_2$ | Good antibody orientation | Expensive |
| | Coupling through anti-Ig antibody | Yes | $-NH_2$ | Easy | Protein A still available for antibody binding |
| **Activated Beads** | Carbonyldiimidazole | No | $-NH_2$ | Cheap | Antibody often damaged by coupling |
| | Cyanogen bromide | No | $-NH_2$ | Activated beads often commercially available | Multiple steps for activating and coupling |
| | Glutaraldehyde | No | $-NH_2$ | | |
| | Hydroxysuccinimide | No | $-NH_2$ | | |
| | Tosyl chloride | No | $-NH_2, -SH$ | | |
| **Activated Antibody** | Carbodiimides | No | $-NH_2, -COOH$ | Cheap | Antibody may be damaged by coupling |
| | Condensing agents | No | $-NH_2$ | | Multiple steps for activating and coupling |
| | Glutaraldehyde | No | $-NH_2$ | | |
| | Periodate | Yes | Sugar | | |
| **Biotin Antibody Streptavidin Beads** | — | No | $-NH_2$ | Bind in solution | Expensive |

## Coupling Antibodies to Protein A Beads

Protein A bead–antibody columns are one of the most versatile column matrices used for affinity purification. The columns are easy to prepare, and because the antibody molecules are bound to the matrix via the Fc domain, the antigen binding site is oriented correctly for maximal interaction with the antigens.

Antibodies can be bound directly to protein A or can be linked via an intermediate anti-immunoglobulin antibody. Antibodies that have an Fc domain with high affinity for protein A can be linked directly; those with a lower affinity may need the intermediate layer (pp. 615 and 622). Once the antibodies are bound to the protein A beads, they are cross-linked to the protein A via a bifunctional coupling reagent. Any bifunctional reagent can be used, but most workers now use dimethylpimelimidate (DMP), which is cheap and easy to handle. Both binding groups of DMP bind to free amino groups. Because the carbon backbone has a great deal of flexibility, most antibody–protein A pairs will have reactive sites within a suitable distance to allow efficient coupling. In the rare cases where this is not true, other cross-linkers with carbon spacers of different lengths can be used. Antibodies that have free amino groups as a key portion of the antigen binding site cannot be linked using these reagents. Other bifunctional reagents can be used and can be tested easily for coupling antibodies to protein A in small-scale procedures that resemble the techniques discussed below.

Because protein A has different affinities for antibodies from different species, classes, and subclasses (see Table 15.1, p. 617, and Table 15.2, p. 618), it may not be possible to couple a high concentration of every antibody to protein A beads. This can be circumvented by three different approaches. First, the antibodies can be bound using protein G beads. Protein G has a different spectrum of binding affinities than protein A (see Table 15.1, p. 617, and Table 15.2, p. 618), and protein G beads are now available commercially. Second, antibodies with a low affinity for protein A can be bound to an intermediate layer of anti-immunoglobulin antibodies that do have a high affinity for protein A. The most commonly used intermediate antibodies are prepared in rabbits. Third, with some antibodies, particularly mouse $IgG_1$s, the affinity with protein A can be increased by adjusting the binding conditions. Normally this is done using high salt concentrations that favor the hydrophobic bonds found in the protein A–Fc binding. Protein G affinity columns are prepared identically to protein A bead columns. The other two methods are discussed below.

One caution that should be included when considering the use of these types of affinity columns is that the protein A molecules on the columns that are not coupled to antibodies will be available for interaction with proteins in the antigen preparation. Therefore, when these columns are used with mammalian antigen sources, the columns will also purify antibodies that are found in the antigen preparation.

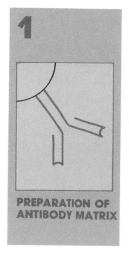

**1**

**PREPARATION OF ANTIBODY MATRIX**

## PREPARING PROTEIN A BEAD–ANTIBODY AFFINITY COLUMNS—DIRECT COUPLING*

This method can be used for coupling mouse monoclonal antibodies from the $IgG_{2a}$, $IgG_{2b}$, and $IgG_3$ subclasses and polyclonal antibodies from mouse, rabbit, human, horse, donkey, pig, guinea pig, dog, or cow.

1. Bind the antibody to protein A beads. Antibodies with high affinity for protein A can be added from any source including serum, tissue culture supernatant, ascites, or purified solutions. If it is essential to know the exact concentration of antibodies on the beads, the antibody should be purified prior to binding to the column (p. 288).

   For general-purpose columns, bind approximately 2 mg of antibody per milliliter of wet beads. When using low-affinity antibodies or in other specialized cases, higher concentrations of antibodies may be appropriate.

   Mix the antibodies and protein A beads. Incubate at room temperature for 1 hr with gentle rocking. Passing the antibody through a column containing the beads will make a gradient of antibody concentrations. The top of the column will have a high concentration, while the bottom of the column will have a low concentration. To avoid this, the antibodies should be bound by mixing in a slurry.

2. Wash the beads twice with 10 volumes 0.2 M sodium borate (pH 9.0) by centrifugation at 3000g for 5 min or 10,000g for 30 sec.

3. Resuspend the beads in 10 volumes of 0.2 M sodium borate (pH 9.0) and remove the equivalent of 10 $\mu$l of beads. Add enough dimethylpimelimidate (solid) to bring the final concentration to 20 mM.

4. Mix for 30 min at room temperature on a rocker or shaker. Remove the equivalent of 10 $\mu$l of the coupled beads.

5. Stop the reaction by washing the beads once in 0.2 M ethanolamine (pH 8.0) and then incubate for 2 hr at room temperature in 0.2 M ethanolamine with gentle mixing.

*After Gersten and Marchalonis (1978); Schneider et al. (1982); Simanis and Lane (1985).

6. After the final wash, resuspend the beads in PBS with 0.01% merthiolate.

7. Check the efficiency of coupling by boiling samples of beads taken before and after coupling in Laemmli sample buffer (p. 684). Run the equivalent of 1 $\mu$l and 9 $\mu$l of both samples on a 10% SDS-polyacrylamide gel (p. 636) and stain with Coomassie blue (p. 649). Good coupling is indicated by heavy-chain bands (55,000 mw) in the "before" but not in the "after" lanes.

The affinity columns are stable for over 1 year when stored at 4°C in buffer containing merthiolate.

**The beads are now ready for binding of the antigen (p. 541).**

## NOTES

i. If there are small amounts of heavy chains in the coupling check (step 7), prewash the coupled beads with 100 mM glycine (pH 3.0) to remove any antibodies that are bound only by noncovalent binding to the protein A molecules.

ii. Coupling with dimethylpimelimidate must be performed above pH 8.3. If coupling is inefficient, check that the pH is above 8.3 after the addition of the dimethylpimelimidate.

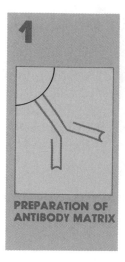

**PREPARATION OF ANTIBODY MATRIX**

## PREPARING PROTEIN A BEAD–ANTIBODY AFFINITY COLUMNS—DIRECT COUPLING IN HIGH SALT*

This method can be used for coupling mouse monoclonal antibodies from the $IgG_1$ subclass.

1. Adjust the pH of the antibody solution to pH 9.0. Check the volume and add NaCl to raise the concentration to 3 M. Check the pH and readjust to pH 9.0, if necessary.

2. Bind the antibody to protein A beads. This coupling procedure is normally used for mouse $IgG_1$ antibodies, and under these salt and pH conditions, protein A beads will bind approximately 2–5 mg of antibody per milliliter of wet beads. Therefore, the binding can be done in antibody excess without worrying about high concentrations of antibody.

3. Mix the antibodies and protein A beads. Incubate at room temperature for 1 hr with gentle rocking. Passing the antibody through a column containing the beads will make a gradient of antibody concentrations. The top of the column will have a high concentration, while the bottom of the column will have a low concentration. To avoid this, the antibodies should be bound by mixing in a slurry.

4. Wash the beads twice with 10 volumes of 3 M NaCl, 50 mM sodium borate (pH 9.0) by centrifugation and aspiration.

5. Resuspend the beads in 10 volumes of 3 M NaCl, 0.2 M sodium borate (pH 9.0) and remove the equivalent of 10 μl of beads. Add enough dimethylpimelimidate (solid) to bring the final concentration to 20 mM.

6. Mix for 30 min at room temperature on a rocker or shaker. Remove the equivalent of 10 μl of the coupled beads.

7. Stop the reaction by washing the beads once in 0.2 M ethanolamine (pH 8.0) and incubating in 0.2 M ethanolamine (pH 8.0) at room temperature for 2 hr with gentle mixing.

*After Gersten and Marchalonis (1978); Schneider et al. (1982); Simanis and Lane (1985).

8. After the final wash resuspend the beads in PBS with 0.01% merthiolate.

9. Check the efficiency of coupling by boiling samples of beads taken before and after coupling in Laemmli sample buffer. Run the equivalent of 1 $\mu$l and 9 $\mu$l of both samples on a 10% SDS-polyacrylamide gel (p. 636) and stain with Coomassie blue (p. 649). Good coupling is indicated by heavy-chain bands (55,000 mw) in the "before" but not in the "after" lanes.

The affinity columns are stable for over 1 year when stored at 4°C in buffer containing merthiolate.

**The beads are now ready for binding of the antigen (p. 541).**

## NOTES

i. If there are small amounts of heavy chains in the coupling check (step 9), prewash the coupled beads with 100 mM glycine (pH 3.0) to remove any antibodies that are bound only by the noncovalent binding to the protein A molecules.

ii. Coupling with dimethylpimelimidate must be performed above pH 8.3. If coupling is inefficient, check that the pH is above 8.3 after the addition of the dimethylpimelimidate.

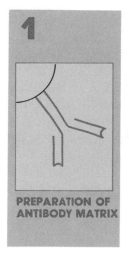

**PREPARATION OF ANTIBODY MATRIX**

## PREPARING PROTEIN A BEAD–ANTIBODY AFFINITY COLUMNS—COUPLING THROUGH AN ANTI-IMMUNOGLOBULIN ANTIBODY LAYER*

Antibodies from all sources can be coupled to protein A using this technique.

1. Bind the anti-immunoglobulin antibodies to the protein A column (p. 310). Normally these antibodies will be prepared in rabbits. If unfractionated polyclonal antibodies will be used, saturate the protein A beads with antibodies (approximately 20 mg/ml of beads). If affinity purified anti-immunoglobulin antibodies are used, bind approximately 2 mg/ml.

2. Bind the antibody to be used for the immunoaffinity column to the anti-immunoglobulin antibody–protein A beads. Antibodies can be added from any source including serum, tissue culture supernatant, ascites, or purified solutions. The concentration of antibodies on the beads will be determined by the amount of anti-immunoglobulin antibodies, so an excess of monoclonal antibodies can be added to the anti-immunoglobulin antibody–protein A beads.

3. Mix the antibodies and anti-immunoglobulin antibody–protein A beads. Incubate at room temperature for 1 hr with gentle rocking. Passing the antibody through a column containing the beads is fine as long as a saturating amount of antibodies is used (approximately 2 mg/ml of beads).

4. Wash the beads twice with 10 volumes 0.2 M sodium borate (pH 9.0) by centrifugation at 3000g for 5 min.

5. Resuspend the beads in 10 volumes of 0.2 M sodium borate (pH 9.0) and remove the equivalent of 10 $\mu$l of beads. Add enough dimethylpimelimidate (solid) to bring the final concentration to 20 mM.

6. Mix for 30 min at room temperature on a rocker or shaker. Remove the equivalent of 10 $\mu$l of the coupled beads.

7. Stop the reaction by washing the beads in 0.2 M ethanolamine (pH 8.0).

*After Gersten and Marchalonis (1978); Schneider et al. (1982); Simanis and Lane (1985).

8. After final wash, resuspend the beads in PBS with 0.01% merthiolate.

9. Check the efficiency of coupling by boiling samples of beads taken before and after coupling in Laemmli sample buffer. Run the equivalent of 1 $\mu$l and 9 $\mu$l of both samples on a 10% SDS-polyacrylamide gel (p. 636) and stain with Coomassie blue (p. 649). Good coupling is indicated by heavy-chain bands (55,000 mw) in the "before" but not in the "after" lanes.

The affinity columns are stable for over 1 year when stored at 4°C in buffer containing merthiolate.

The beads are now ready for binding of the antigen (p. 541).

## NOTES

i. If there are small amounts of heavy chains in the coupling check (step 9), prewash the coupled beads with 100 mM glycine (pH 3.0) to remove any antibodies that are bound only by the noncovalent binding to the protein A molecules. Check coupling again.

ii. Coupling with dimethylpimelimidate must be performed above pH 8.3. If coupling is inefficient, check that the pH is above 8.3 after the addition of the dimethylpimelimidate.

## *Coupling Antibodies to Activated Beads*

The most common method of covalently binding antibodies to a solid-phase matrix is to activate beads using any of a number of chemical agents and then bind purified antibodies to the beads (for general reviews, see Porath and Axén 1976; Scouten 1987). This approach offers several advantages. The beads normally can be activated in much harsher conditions than proteins can sustain, thus allowing the use of a range of activating protocols. Preparing activated beads is relatively cheap, and many of the coupling methods yield a linkage that is stable to a wide range of denaturing conditions. Another advantage is that there are a number of activated beads available commercially. Table 13.2 lists five of the commonly used activating agents and recommended commercial products. In general, the commercial beads provide a good source of activated beads for the initial stages of an immunoaffinity purification. For large-scale preparations or for limited budgets, the activation procedures for several of the common methods are given below.

There are a wide variety of beads available for preparing an activated support. Table 13.3 summarizes the properties of several of the most common. Important variations exist in resistance to pH and temperature or nonspecific binding between the various beads. Most beads are more resistant to pH and temperature changes after coupling with antibodies, as this acts to cross-link the beads and may greatly stabilize them.

**TABLE 13.2**
**Activated Beads**

| Reagent | Binding group on matrix | Ligand attachment | Stability of final linkage | Recommended matrix | Commercial examples |
|---|---|---|---|---|---|
| Carbonyldiimidozole | –OH | –NH$_2$ | Avoid pH > 10 | Agarose<br>Cross-linked agarose<br>Mixed agarose/polyacrylamide<br>Polyacrylic | Reacti-Gel 6X (Pierce)<br><br>Reacti-Gel GF-2000 (Pierce) |
| Cyanogen Bromide | –OH | –NH$_2$ | Good, some leaching | Agarose<br>Cross-linked agarose<br>Mixed agarose/polyacrylamide<br>Polyacrylic | CNBr-activated Sepharose (Pharmacia) |
| Glutaraldehyde | –NH$_2$ | –NH$_2$ | Excellent | Polyacrylamide<br>Mixed agarose/polyacrylamide | Act-Ultragel AcA 22 (IBF) |
| Hydroxysuccinimide | –COOH | –NH$_2$ | Excellent | Cross-linked agarose<br>Modified to contain –COOH | Affigel 10 (BioRad) |
| Tosyl Chloride | –OH | –NH$_2$<br>–SH | Excellent | Cross-linked agarose<br>Polyacrylic | Activated Microspheres (KPL) |

**TABLE 13.3**
**Beads for Activation**

| Matrix | Chemical stability | pH range | Temperature range | Groups available for coupling | Activators | Commercial examples |
|---|---|---|---|---|---|---|
| Agarose Beads | Fair<br>Avoid chaotropic agents, extended exposure to urea, or guanidine | 4–10 | 4–30°C | –OH | Cyanogen bromide<br>Carbonyldiimidazole | Bio-Gel A (BioRad)<br>Sepharose (Pharmacia)<br>Ultragel A (IBF) |
| Cross-linked Agarose Beads | Good<br>Avoid extended exposure to chaotropic agents | 3–14 | 4–120°C | –OH | Cyanogen bromide<br>Carbonyldiimidazole<br>Tosyl chloride | Sepharose CL (Pharmacia) |
| Polyacrylamide Beads | Excellent<br>Strong alkali on uncoupled beads converts amide to carboxylic | 2–10 | 4–120°C | –NH$_2$ | Glutaraldehyde | Bio-Gel P (BioRad) |
| Copolymers of Polyacrylamide and Agarose | Fair<br>Avoid chaotropic agents, extended exposure to urea or guanidine | 4–10 | 4–30°C | –OH<br>–NH$_2$ | Cyanogen bromide<br>Carbonyldiimidazole<br>Glutaraldehyde | Ultragel AcA (IBF) |
| Polyacrylic Beads | Excellent<br>Strong alkali on uncoupled beads converts amide to carboxylic | <1–11 | –20–120°C | –OH | Cyanogen bromide<br>Carbonyldiimidazole<br>Tosyl chloride | Trisacryl (IBF) |

## PREPARING CARBONYLDIIMIDAZOLE-ACTIVATED BEADS*

Antibodies bound to carbonyldiimidazole-activated beads provide a stable complex that is useful for a wide range of elution procedures, but coupled beads should not be treated with strong alkali. Because the activation is done in anhydrous conditions, only beads that are resistant to these conditions should be used. Both cross-linked agarose and polyacrylic beads are good supports. The activating reaction is easy, and the activated beads are stable for several weeks when stored in dry acetone or dioxane.

1. Transfer 10 ml of wet beads to a sintered glass filter. Sequentially wash with 30% dioxane in water, 70% dioxane in water, and 100% dioxane.

2. Resuspend the beads in 10 ml of dioxane and transfer to a dry beaker.

3. Add 0.35 gram of carbonyldiimidazole. Incubate at room temperature for 15 min with mixing on a shaker or rocker.

4. Return the beads to the sintered glass filter. Wash with dioxane, and then remove the dioxane by suction.

   The activated beads can be stored for several days in dry dioxane. Store at 4°C.

Use for coupling (p. 536).

### NOTE

i. Carbonyldiimidazole-activated cross-linked agarose and polyacrylic beads are available commercially from Pierce.

*After Bethell et al. (1979); Hearn et al. (1981); Hearn (1987).

## *PREPARING CYANOGEN BROMIDE-ACTIVATED BEADS\**

Cyanogen bromide-activation is the most commonly used method for preparing antibody-affinity columns. Two advantages of cyanogen bromide (CNBr) coupling are the high capacity and the extensive literature on these types of preparations. Two disadvantages of this method are the highly toxic activating reagent and the isourea bond that couples the antibody to the solid phase. This bond is not as stable as other linkages (Lasch and Koelsch 1978) and adds an extra charged group to the column. The stability will be a major problem only when repeated use of the columns or extended exposure to high temperature, extremes of pH, or nucleophiles (no Tris, azide, etc.) is planned. Because of the additional charge group added by the isourea linkage, all buffers for antigen binding and elution should contain at least 100 mM salt. Agarose, cross-linked agarose, and polyacrylic beads can be activated by CNBr.

> **Caution** Cyanogen bromide is extremely toxic. It is volatile and should only be used in a fume hood.

1. Transfer 10 ml of wet beads to a sintered glass filter. Wash with distilled water. Wash with 1 M sodium carbonate buffer (pH 11.0).

2. Add 10 ml of 1 M sodium bicarbonate (pH 11.0) and transfer the beads to a suitable beaker.

3. Move to a fume hood. Weigh out 1 gram of cyanogen bromide (CNBr) and dissolve in 1 ml of acetonitrile. Add the CNBr to the beads.

4. Incubate at room temperature for 10 min with constant agitation. Monitor the pH. Adjust as necessary to keep between 10.5 and 11.0 by adding 4 N NaOH.

5. Transfer the beads to a sintered glass filter. Use suction to draw the CNBr buffer into a vacuum flask containing 100 mM ferrous sulfate to inactivate the CNBr.

6. Sequentially wash the beads with water, several milliliters of 95% acetone, and several changes of 100 mM sodium phosphate (pH 7.5) (or an alternative binding buffer).

Use immediately for coupling (p. 536).

## NOTE

i. CNBr-activated beads are commercially available from Pharmacia.

*Axén et al. (1967); March et al. (1974); Kohn and Wilchek (1984).

## PREPARING GLUTARALDEHYDE-ACTIVATED BEADS*

Glutaraldehyde can be used to activate beads through $-NH_2$ groups. The glutaraldehyde linkage is very stable with the glutaraldehyde backbone forming a spacer arm between the bead and the antibody in the final product. Polyacrylamide and polyacrylamide–agarose beads provide a suitable support for glutaraldehyde activation.

1. Transfer 10 ml of wet beads to a sintered glass filter. Wash with distilled water and then with several volumes of 0.5 M sodium phosphate (pH 7.5).

2. Transfer the gel to a suitable flask or beaker. In a fume hood add an equal volume of 25% glutaraldehyde (electron microscope grade). Check the pH. It should be between 7.2 and 7.6; adjust if necessary.

3. Gently mix the slurry overnight at 37°C using a shaker or rocker.

4. Transfer the beads to a sintered glass filter and wash extensively with 0.5 M sodium phosphate buffer.

   If the gel will not be used immediately, store in phosphate buffer at 4°C. It is stable for several weeks to months.

The gel is now ready for coupling (p. 536).

### NOTE

i. Glutaraldehyde-activated beads are commercially available from IBF.

*Ternynck and Avrameas (1972); Korn et al. (1972); Guesdon and Avrameas (1976).

## HYDROXYSUCCINIMIDE-ACTIVATED BEADS

Of the methods for activating beads, hydroxysuccinimide is one of the least often used, because the beads must be modified to contain carboxylic acids prior to reacting with the N-hydroxysuccinimide. Although these carboxylic acid beads can be purchased from several suppliers, there is little or no savings over purchasing the final activated beads. Therefore, a protocol is not included for preparing hydroxysuccinimide-activated beads. Two commercial sources offer hydroxysuccinimide-activated beads. Pharmacia offers a resin activated CH-Sepharose 4B, in which the spacer arm is linked to the beads via an isourea bond (p. 532). Because the isourea bond is less stable than other bonds, BioRad's Affi-Gel, which has linkage through an ester, is recommended.

## *PREPARING TOSYL CHLORIDE-ACTIVATED BEADS**

Tosyl chloride-activated beads are not frequently used to bind antibodies to solid-phase matrices for immunoaffinity purification; however, they offer several advantages over other methods. The principle advantage is the stability of the linkages, which are stable to all elution conditions except the strongest alkalis. The activation is done in organic solvents, and therefore only beads that are resistant to these conditions should be used for preparing the activated beads. These include polyacrylic and cross-linked agarose.

1. Transfer 10 ml of wet beads to a sintered glass filter.

2. Wash sequentially with 30% acetone in water, 60% acetone in water, 80% acetone in water, and then dry acetone.

3. Transfer the beads in 10 ml of acetone to a dry beaker in a fume hood. Add 2 ml of pyridine (dry).

4. Dissolve 0.8 gram of tosyl chloride (*p*-toluenesulfonyl chloride) in 5 ml of acetone. While mixing the beads, add tosyl chloride dropwise.

5. Mix for 15 min on a rocker or shaker. Return to the sintered glass filter. Wash with 10 volumes of acetone. Then wash sequentially with 1 mM HCl/80% acetone in water, 1 mM HCl/60% acetone in water, 1 mM HCl/30% acetone in water. Finally wash with 1 mM HCl.

6. If the beads will not be used immediately, wash with acetone after the activation (step 4) and store in 1 mM HCl/100% acetone before the final washes. Store at 4°C in the dark. Stable for several days.

The beads are now ready for coupling (p. 536).

## NOTE

i. Higher rates of coupling can be achieved using 250 $\mu$l of tresyl chloride in place of tosyl chloride, and lowering the pyridine concentration to 250 $\mu$l. The activation is done as normal.

*After Nilsson and Mossbach (1980, 1981, 1984); Nilsson et al. (1981).

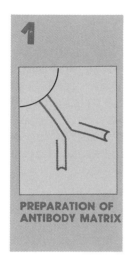

**PREPARATION OF ANTIBODY MATRIX**

## COUPLING WITH ACTIVATED BEADS

Achieving efficient coupling of antibodies to activated beads is a simple task. However, retaining antibody activity is sometimes difficult. All of the coupling reactions bind through amino groups, and presumably most of the loss of activity is due to overcoupling of the proteins. Two variables should be checked to protect against overcoupling. First, if the antibodies are bound below the pK of the amino groups, most of the sites will not be available for coupling, thus limiting the possibility of overcoupling. For this option, approximately neutral pHs are satisfactory. Second, the reaction should be monitored and stopped when approximately 80% of the antibodies are bound. Together, these two steps should limit the loss of antibody activity by the cross-linking process.

Antibodies with amino groups within the antigen-combining site will be particularly sensitive to inactivation following binding to activated beads. Alternatives to this problem include coupling on protein A beads with coupling reagents that do not use amino groups or using activated antibodies (p. 538).

1. The antibody preparations must not contain extraneous compounds with amino groups (or thiol groups, if using tosyl chloride). If these compounds have been used during the purification, the antibody preparation should be extensively dialyzed against the binding buffer of 0.5 M sodium phosphate (pH 7.5).

2. Prepare a solution of antibody at the desired concentration in 0.5 M sodium phosphate (pH 7.5). Save a small sample to determine the binding efficiency. The amount of antibody to be coupled will vary depending on the individual experiment. For most purposes, 5–10 mg of antibody per milliliter of beads will yield a high-capacity column.

3. Add the activated beads (either prepared as described by the manufacturer or prepared as on pp. 531–536).

4. Mix gently overnight at room temperature on a rocker.

5. Wash the beads twice with 0.5 M sodium phosphate (pH 7.5). Save the wash from the overnight binding. Compare the amount of protein from the "before" and "after" samples of the binding buffers.

6. Wash the beads once with 1 M NaCl, 0.05 M sodium phosphate (pH 7.5).

7. Add 10 volumes of 100 mM ethanolamine (pH 7.5). Incubate at room temperature for 4 hr to overnight with gentle mixing.

8. Wash twice with PBS. Add merthiolate to 0.01%.

The beads can be stored at 4°C where they will be stable for several months. Some of the linkages are suitable for prolonged storage (see above).

The beads are now ready for binding of the antigen (p. 541).

## NOTES

i. If the activity of the antibodies has been lowered drastically during coupling (10–50% activity is common), stop the reaction before extensive coupling has occurred. Remove samples at different times during coupling and assay the buffer for the presence of antibodies. Coupling efficiencies of 70–80% are normally a good compromise between good activity and high concentrations. If problems continue, use one of the alternative methods for coupling antibodies.

ii. Low levels of coupling can be increased by raising the pH. For pHs above 8.2 use 0.5 M carbonate buffer in place of the phosphate.

## Preparing Antibody-Affinity Columns with Activated Antibodies

Purified antibodies can be activated by treating with a bifunctional reagent, one group binding to an appropriate group on the antibody and the other remaining free to bind to the matrix. Reagents that are commonly used for indirect coupling are: (1) the water-soluble carbodiimides such as 1-ethyl-3-(3-dimethylaminopropyl)-carbodiimide HCl (EDAC or EDCI) or 1-cyclohexyl-3-(2-morpholoinoethyl)-carbodiimide-metho-*p*-toluene-sulfonate (CMIC or CMCI), (2) the condensing agents for peptide synthesis such as *N*-ethoxycarbonyl-2-ethoxy-1,2-dihydroquinoline (EEDQ), (3) glutaraldehyde, or (4) periodate. After the antibodies are activated, they are bound to beads. Proteins activated by the carbodiimides and condensing reagents will bind to carboxylic acid, while the glutaraldehyde and periodate-activated antibodies will bind to amines.

Methods for activating the antibody are useful alternatives to using activated beads, but in general offer no particular advantages during the initial stages of affinity purification. Their primary advantage is that they allow the choice of a wide range of beads with different spacer arms, combinations that may be important or essential for more specialized purifications. The exception to this is periodate-activated antibodies.

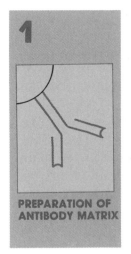

**PREPARATION OF ANTIBODY MATRIX**

## PREPARING COLUMNS USING PERIODATE-ACTIVATED ANTIBODIES

If the orientation of the antibody on the bead is important, i.e., all of the antigen combining sites away from the matrix, and covalent coupling of antibodies to protein A beads is inappropriate, one possible compromise is to bind the antibodies to the matrix through the carbohydrates on the heavy chains. Treating these sugars with periodate will break the ring and allow the antibodies to be coupled to column matrices with $-NH_2$ groups. Possible supports include polyacrylamide beads that already have $-NH_2$ groups or other supports that have been modified by the addition of a spacer arm with a terminal amine. Polyacrylamide beads are available from BioRad as Bio-Gel P, and modified beads with a spacer arm bearing the reactive $-NH_2$ group are available from BioRad (Affi-Gel 102, Aminoethyl Bio-Gel P-2, and Aminoethyl Bio-Gel P-100) and Pharmacia (AH-Sepharose 4B and agarose-adipic acid hydrazide). Polysaccharide beads can also be activated with periodate, yielding beads that will bind to proteins with free amino groups (see Sanderson and Wilson 1971).

1. Dialyze the purified antibody against several changes of 50 mM sodium phosphate (pH 7.2). Remove a sample to check for coupling efficiency. The antibody concentration should be approximately 5 mg/ml. Concentrate if necessary.

2. Add sodium periodate to a final concentration of 50 mM. Check the pH and readjust, if necessary.

3. Incubate for 30 min at room temperature.

4. Remove the periodate by dialysis against 50 mM sodium phosphate (pH 7.2), several changes for several hours each. After the dialysis remove a sample for the coupling check.

5. Add beads (with $-NH_2$ groups) to the antibody solution, 1 ml/5 mg of antibody. Incubate at 4°C overnight with gentle mixing.

6. Transfer the beads to a sintered glass filter. Wash extensively with 50 mM sodium phosphate (pH 7.2). Save a sample of the first wash and compare the protein concentration to the earlier two samples.

7. The Schiff's bases that have formed are reduced by treating with 20 mg/ml of *tert*-butylamine borane or 0.1 mg/ml of $NaBH_3CN$ for 2–4 hr in 20 mM sodium phosphate (pH 6.5).

The beads are now ready for the binding of antigen (p. 541)

**PREPARATION OF ANTIBODY MATRIX**

## *PREPARING COLUMNS USING BIOTIN STREPTAVIDIN–AGAROSE*

One of the most attractive, but little used, techniques for immuno-affinity purification is the use of biotinylated antibodies in conjunction with streptavidin–agarose (Hofmann et al. 1978; Updyke and Nicholson 1984; 1986). Purified biotinylated antibodies are bound to the antigen in solution. The antibody–antigen complexes are then purified on a streptavidin–agarose column, and eluted as normal. The advantages include using solution binding to control better the level and speed of binding. (Procedures for biotinylating antibodies are described on p. 340.) Streptavidin–agarose can be prepared by substituting streptavidin for the antibodies in one of the techniques using activated beads as described above (p. 519) or can be purchased from a commercial source.

## ■ Binding Antigens to Immunoaffinity Columns

Several different approaches can be taken to allow efficient binding of antigens to immunoaffinity columns. Except when using biotinylated antibodies, antigens must bind to antibodies on a solid-phase matrix. Because the antibody is not in solution, the time required for the antibody–bead/antigen interaction will have different kinetics than soluble interactions. It will take considerably longer for equilibrium to be reached than for solution assays such as immunoprecipitation. Therefore, the binding protocol should maximize the degree of interaction. The recommended methods are binding in a constantly mixing slurry of antibody–beads with soluble antigen or by passing the antigen solution down an antibody–bead column, keeping the antigen in contact with the antibody for as long as possible. In either case, high-affinity antibodies will be significantly more efficient at removing the antigen from solution than low-affinity antibodies. Several small-scale columns can be used to determine the best conditions for binding and collecting the antigen.

When using a high-affinity antibody, the amount of matrix that is used should be adjusted to a level just above saturation. This will keep the nonspecific background to a minimum. When using a low-affinity reaction, this will not be possible, because the reaction will need to be driven by a high concentration of antibody. In this case, either use an excess of antibody–beads or the biotin streptavidin–agarose system discussed above.

### Low- Versus High-Affinity Antibodies

Although the exact affinity of an antibody for an antigen can be calculated, for most work the crucial criterion is whether the antibodies will remove the antigen from solution quantitatively. The easiest method to test this is to set up small-scale reactions and examine the first wash buffer for the presence of the antigen. If the antibody–bead matrix does not remove most of the antigen within 1–2 hr, several changes in the experimental design may help. The amount of bound antigen may be increased by using higher amounts of antibodies on the beads, by increasing the number of beads, or by increasing the amount of time for binding. Unfortunately, all of these conditions will raise the nonspecific background, so a compromise normally will result in the highest yields with the lowest acceptable background. On the other hand, using high-affinity antibodies solves the problem of efficiently collecting the antigen. Consequently, they can be used in dilute solutions, at relatively lower concentrations, and for shorter times—all conditions that lead to lower backgrounds.

**BINDING OF ANTIGEN**

## BINDING IN SUSPENSION—ANTIBODY—BEAD SLURRIES

1. Mix the antigen solution with the antibody–beads.

2. Incubate at 4°C with constant agitation for 1 hr to overnight. The agitation should be sufficient to keep the beads in suspension, but not enough to break the beads. Rocking is preferred to the use of a stir bar.

3. Stop the agitation.

4. (**Optional**) An effective method to remove protein aggregates is to allow the beads to settle by gravity at 4°C. Remove the majority of the binding buffer by aspiration, being careful not to upset the layer of settled beads. Add a volume of ice-cold binding buffer equal to the starting volume. Bring the beads up into suspension for several seconds, and then allow them to settle by gravity at 4°C. Remove the majority of the wash buffer by aspiration and repeat the wash.

5. Transfer the beads to a suitable column. Wash with 20 bed volumes of binding buffer (4°C).

The column is now ready for more extensive washing or for elution (p. 547).

## Titrating

A titration should be performed as a first step in estimating the ratio of column matrix needed to bind a given amount of antigen. This can be handled in a manner similar to immunoprecipitations (see Chapter 11), where an equal volume of the antibody–bead matrix is added to samples containing increasing concentrations of the antigen. The slurry is mixed at 4°C for 1 hr and then processed as for a standard immunoprecipitation. This will yield a rough idea of the volume of column matrix needed to collect the desired amount of antigen. If the supernatants from the binding reaction are assayed for the presence of the antigen, the extent of antigen depletion also can be determined.

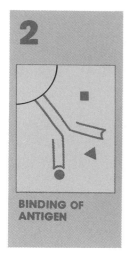

**BINDING OF ANTIGEN**

## *BINDING IN SOLUTION—BIOTIN STREPTAVIDIN–AGAROSE*

Immunoaffinity purification using a biotin streptavidin–agarose matrix relies on a different set of parameters from conventional immunoaffinity methods. Purified antibodies are biotinylated and then bound to the antigen in solution. This allows very high concentrations of antibodies to be used, thus increasing the usefulness of low-affinity antibodies. The antigen–antibody complexes are passed through a streptavidin–agarose column and allowed to bind. The biotin–streptavidin affinity is approximately $10^{14}$ mol$^{-1}$, well above that needed to collect the antibody quantitatively. This step also creates a high local concentration of antibodies on the column. Therefore, any antigen that leaches from the column under mild washing conditions has a better chance of rebinding to a nearby antibody rather than being lost in the wash buffer. The biotin–streptavidin interaction is suitably strong and will not be broken by any but the strongest elution conditions. One disadvantage of this system is that, as the biotinylated antibody cannot be removed from the column, the column will not be reusable for a similar solution-collection protocol.

1. Prior to the purification, prepare biotinylated antibodies (p. 340).

2. To a preparation of the antigen, add sufficient biotinylated antibody to bind the majority of the antigen. This may be determined using a series of titrations in small-scale reactions.

3. Incubate at 4°C for 1 hr. Agitation is not needed.

4. At 4°C, pass the antigen–antibody solution through a volume of uncoupled agarose beads equivalent to or larger than the volume of the streptavidin–agarose beads that will be used to collect the antibody–antigen complexes. This should lower the nonspecific background.

5. At 4°C, collect the biotinylated antibodies by passing the antigen–antibody solution through a streptavidin–agarose column at 10 ml/hr.

6. Wash the column with 20 bed volumes of an appropriate buffer (p. 549).

The column is now ready for more extensive washing or for elution (p. 547).

## NOTES

i. The extent of biotinylation of the antibody is important to the recovery of the antigen. If all the antibody molecules have not been biotinylated, not all of the antibody–antigen complexes will be retained on the streptavidin column. To test this, pass a sample of antibodies through a streptavidin–agarose column and check the flow through for the antibody.

ii. An effective method to remove protein aggregates is to allow the antibodies to bind to the streptavidin beads in solution and then to settle by gravity at 4°C. Remove the majority of the binding buffer by aspiration, being careful not to upset the layer of sedimented beads. Add a volume of ice-cold binding buffer equal to the starting volume. Bring the beads up into suspension for several seconds, and then allow them to settle by gravity at 4°C. Remove the majority of the wash buffer by aspiration and repeat the wash. Then collect in the column.

**BINDING OF ANTIGEN**

## BINDING IN THE COLUMN

The simplest method to bind the antigen to the antibody–beads is collect the beads in a column and pass the antigen solution down the column. The antigen solution must be free of particulate matter. Centrifugation at 100,000g for 30 min will clear most large particles from the preparation.

1. Transfer the antibody–beads to a suitable column. Wash with 20 bed volumes of PBS.

2. (**Optional**) Pass the antigen solution through a volume of uncoupled agarose beads equivalent to or larger than the volume of the antibody–beads. This should lower the nonspecific background.

3. Apply the antigen solution to the column. Either pass the solution through the column three times by gravity or once with a flow rate of approximately 2 ml/hr (this will need to be controlled by a pump).

4. Wash the column with 20 bed volumes of binding buffer.

The column is now ready for more extensive washing or for elution (p. 547).

## Eluting Antigens from Immunoaffinity Columns

The actual elution of an antigen from an immunoaffinity column is quick and easy. Most of the work is spent in designing and testing the conditions used to achieve an effective elution. Three types of elution are possible. The antigen–antibody interactions can be broken by (1) treating with harsh conditions, (2) adding a saturating amount of a small compound that mimics the binding site, and/or (3) treating with an agent that induces an allosteric change that releases the antigen.

The most commonly used elution procedure relies on breaking the bonds between the antibody and antigen. The elutions may be harsh, denaturing the antibody and the antigen, or mild, leaving both the antigen and antibody in active states. The relative harshness of the elution conditions will depend on the antibody in use. The conditions for elution can be determined only by empirical tests. If affinity-purified polyclonal antibodies are being used, the conditions for elution of the antigen will be identical to those used to purify the antibodies. In other cases, developing the elution conditions will be done by testing a series of buffers. Commonly used elution conditions are listed in Table 13.4. Strategies for these types of elutions are discussed on p. 551.

When using an antibody raised against a synthetic peptide, the antigen can sometimes be eluted by adding a large molar excess of the synthetic peptide. As the antibody–antigen interaction disassociates, the rebinding of the antigen to the column is inhibited by binding to the excess of the peptide. Although this approach is very attractive for a number of reasons, including the mild elution conditions, getting efficient elution is often difficult. Critical variables include the rate of dissociation of the antigen–antibody binding and the relative affinity of antibody for the peptide compared to the whole antigen. As the antibodies have been raised against the peptide, they usually will bind efficiently to the peptide, so the major problem will normally be the off-rate. For elution, the most effective protocols will place the peptide in contact with the antigen–antibody column until a new equilibrium is reached. With high-affinity interactions, this will take a significant length of time or an enormous excess of peptide. To test these elution conditions, add increasing amounts of peptide to small aliquots of the antigen–antibody complex in separate tubes. Mix these tubes for several hours. Centrifuge the beads and, without washing, test both the supernatants and beads for the levels of antigen. These conditions can then be scaled up for elution. In general, elution of anti-peptide columns using batch rather than column techniques is more efficient, because the time needed to reach equilibrium can be adjusted easily. Column elution will often yield a broad and slow elution profile, making the eluted protein less useful for further work. As discussed below, this is not true for standard immunoaffinity columns.

A third method for elution, which can be used only for certain proteins, relies on allosteric changes induced by treating the antigen–antibody column with appropriate ligands. Antibodies that recognize allosteric changes or bind to one member of a protein–protein complex that is dissociated by allosteric changes can be used to prepare pure preparations of proteins. As these types of elution protocols will vary from antigen to antigen, they are not discussed in any more detail here, but they can provide an effective and elegant method of elution.

**TABLE 13.4**
**Elution Reagents**

| Reagent | Suggested Buffers | Pre-elution wash | Collection buffer | Column reusable? |
|---------|-------------------|------------------|-------------------|------------------|
| High pH | 100 mM triethylamine, pH 11.5 | 10 mM phosphate, pH 8.0 | 1/20 vol 1 M phosphate, pH 6.8 | Often |
|  | 100 mM phosphate acid, pH 12.5 | 10 mM phosphate, pH 8.0 | 1/20 vol 1 M phosphate, pH 6.8 | Occasionally |
| Low pH | 100 mM glycine, pH 2.5 | 10 mM phosphate, pH 6.8 | 1/20 vol 1 M phosphate, pH 8.0 | Often |
|  | 100 mM glycine, pH 1.8 | 10 mM phosphate, pH 6.8 | 1/20 vol 1 M phosphate, pH 8.0 | Occasionally |
| High salt | 5 M LiCl, 10 mM phosphate, pH 7.2 | 10 mM phosphate, pH 7.2 | None required | Often |
|  | 3.5 MgCl$_2$, 10 mM phosphate, pH 7.2 | 10 mM phosphate, pH 7.2 | None required | Often |
| Ionic detergents | 1% SDS | Neutral wash buffer | None required | No |
|  | 1% DOC | Neutral wash buffer | None required | Seldom |
| Dissociating agents | 2 M urea | Neutral wash buffer | None required | Occasionally |
|  | 8 M urea | Neutral wash buffer | None required | No |
|  | 2 M guanidine HCl | Neutral wash buffer | None required | Seldom |
| Chaotropic agents | 3 M thiocyanate | Neutral wash buffer | None required | Seldom |
| Organic solvents | 10% dioxane | Neutral wash buffer | None required | Often |
|  | 50% ethylene glycol, pH 11.5 | Neutral wash buffer | None required | Often |
|  | 50% ethylene glycol, pH 8.0 | Neutral wash buffer | None required | Often |
| Water | | Neutral wash buffer | None required | Often |

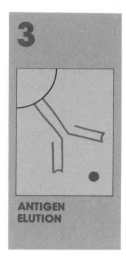

**3**

**ANTIGEN ELUTION**

## *ELUTING THE ANTIGEN*

1. After washing the column, change the buffer to the pre-elution buffer if necessary, (Table 13.4, p. 549) by passing 10 bed volumes through the column. The pre-elution buffer allows a quicker change to the elution conditions and should give a sharper elution profile.

2. Using a stepwise elution, sequentially pass samples of elution buffer (0.5 bed volumes/step is normally appropriate) through the column. Collect each fraction in separate tubes. If either high or low pH is used to elute the column, the collection tubes should contain a neutralizing buffer (see Table 13.4). When using a gradient elution, pass the gradient of increasingly harsher elution buffer through the column.

3. Check each tube for the presence of the antigen. Combine tubes with high concentrations. (Protein concentrations can be determined by the techniques discussed on pp. 670–675.)

4. Return the column to the starting buffer by passing 20 column volumes through the matrix. Add 0.01% merthiolate for long-term storage (4°C).

## NOTE

i. After the elution of the antigen, the eluent will often need to be changed by dialysis or by chromatography on a desalting column.

## Strategies for Testing Elution Conditions

Developing the best elution conditions is an empirical task; potential elution buffers are tested on small-scale columns to determine the proper conditions. There are no good shortcuts nor any guaranteed useful buffers. The best strategy is to test as many antibodies and elution conditions as possible.

An ideal elution profile will yield a sharp peak with complete release of the antigen following treatment with a single buffer. Three general points should be considered prior to choosing an appropriate starting point. First, for what techniques will the antigen be used? For many protein chemistry techniques, the antigens will not need to be native, while for most enzymatic studies the elution condition should be as gentle as possible. Second, will the antibody column be reused? If so, mild conditions should be used. Third, make sure the last wash buffer (called the pre-elution buffer in this chapter) is compatible with the elution buffer. For example, the buffering capacity of the pre-elution buffer should be low enough that a sharp change in pH will follow the front of the elution buffer.

1. If trying for the gentlest elution conditions, start with acid conditions first, then check basic elution buffers. If these conditions do not elute the antigen, try others. A general order to check the various conditions would be: acid, pH 3–1.5; base pH 10–12.5; $MgCl_2$, 3–5 M; LiCl 5–10 M; water; ethylene glycol 25–50%; dioxane 5–20%; thiocyanate 1–5 M; guanidine 2–5 M; urea 2–8 M; SDS 0.5% to 2%.

2. Use combinations of eluting agents. Antigens that are difficult to elute under the standard elution conditions normally are held by several different types of bonds. Using a combination of several elution conditions can be effective in overcoming these problems.

3. If the antigen is difficult to elute under simple conditions, changing antibodies may be easier than testing all conditions.

4. Elutions normally are more effective when columns are used than when eluting in batch. As the buffer passes down the column, any antigen–antibody interaction that breathes will be washed away from the remaining antibodies and cannot be rebound. In batch conditions, this is not true. Consequently, batch elutions normally require harsher conditions to elute antigens than column elution.

5. Try gradient elutions to generate a sharp peak of released antigen.

6. Avoid dithiothreitol and other reducing agents, as they will break the disulfide linkages between the heavy and light chains.

7. While checking different elution conditions, any buffers that fail to elute the antigen should be considered as good candidates for wash buffers.

8. Some noneluting buffers may, in fact, drive the antibody–antigen equilibrium toward complex formation. For example, high salt conditions favor hydrophobic interactions. These conditions could be used to make collecting the antigen on the antibody–matrix easier.

# 14

# IMMUNOASSAYS

Immunoassays are one of the most powerful of all immunochemical techniques. They employ a wide range of methods to detect and quantitate antigens or antibodies and to study the structure of antigens. With the appropriate assay, they can be remarkably quick and easy, yielding information that would be difficult to determine by other techniques. This chapter is divided into four sections, presenting (1) an overview of the different types of immunoassay, (2) a guide to help in choosing which assay to use, (3) the protocols themselves, and (4) the design of immunoassays.

## TYPES OF IMMUNOASSAYS

There are many variations on the ways in which immunoassays can be performed, and they can be classified on the basis of many different criteria. In this manual, immunoassays are classified on the basis of methodology. Within each group, the principle and the order of the steps are similar. However, the variable that is being tested may change. For example, by changing certain key conditions, an assay can be altered to determine either antigen or antibody level. The steps are similar, but the assays yield different results. Although this organization provides a good structure to discuss technical strategies and problems, it is not always apparent which assays should be used for which applications, particularly to the novice. The section starting on p. 557 is designed to act as a guide to determine which assays may be appropriate for different applications.

As described here, the three classes of immunoassays are (1) the antibody capture assays, (2) the antigen capture assays, and (3) the two-antibody sandwich assays. The general procedures for these three types of assays are depicted in the cartoons in Figure 14.1, 14.2, and 14.3.

**ANTIBODY CAPTURE ASSAYS**

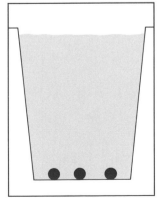
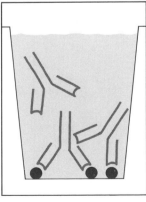
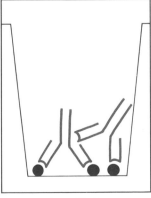

Bind Antigen
to Solid Phase

Bind Antibody
to Antigen

Quantitate
Antibody

**FIGURE 14.1**
Antibody-capture assay.

**TWO-ANTIBODY ASSAYS**

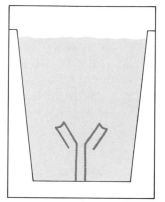

Bind First
Antibody to
Solid Phase

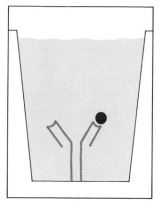

Bind Antigen
to Antibody

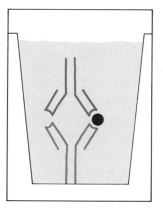

Bind Second
Antibody to
Antigen

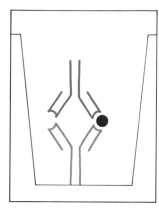

Quantitate
Second Antibody

**FIGURE 14.2**
Two-antibody assays.

In an antibody capture assay, the antigen is attached to a solid support, and labeled antibody is allowed to bind. After washing, the assay is quantitated by measuring the amount of antibody retained on the solid support. In an antigen capture assay, the antibody is attached to a solid support, and labeled antigen is allowed to bind. The unbound proteins are removed by washing, and the assay is quantitated by measuring the amount of antigen that is bound. In a two-antibody sandwich assay, one antibody is bound to a solid support, and the antigen is allowed to bind to this first antibody. The assay is quantitated by measuring the amount of a labeled second antibody that can bind to the antigen. See Silman and Katchalski (1966) for an early review of the use of immobilized antibodies and antigens.

**ANTIGEN CAPTURE ASSAYS**

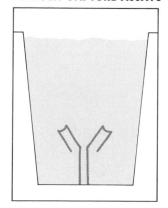

Bind Antibody
to Solid Phase

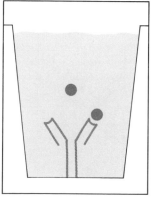

Bind Antigen
to Antibody

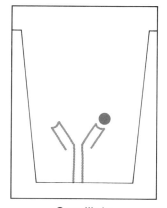

Quantitate
Antigen

**FIGURE 14.3**
Antigen-capture assay.

## ■ DECIDING WHERE TO START

Any immunoassay can be performed with four variations; the assay can be done in antibody excess, in antigen excess, as an antibody competition, or as an antigen competition. Assays done in antibody excess or as antigen competitions are used to detect and quantitate antigens, while antigen excess or antibody competition assays are used to detect and quantitate antibodies. The four variations within the three classes yields a total of 12 possible immunoassays; however, not every combination leads to a useful immunoassay, and only six of the possibilities are commonly used. Representative protocols for each of these six assays are given on pp. 561–590.

In principle, the only factors that need to be considered in choosing the correct design of an immunoassay are the types of antibodies that are available (polyclonal antibodies, affinity-purified polyclonal antibodies, a single monoclonal antibody, or two or more monoclonal antibodies) and whether pure or impure antigen is available. Table 14.1 lists the possible choices based on these two variables and groups the assays in an order showing their general usefulness.

**TABLE 14.1**
**Choosing an Assay Protocol**

| To measure | Type of antibody available | Type of antigen available | Assay choices[a] |
|---|---|---|---|
| Antigen Presence or Quantity | Polyclonal antibodies | Pure | 1. Antigen capture (Ag competition), p. 586, 588<br>2. Antibody capture (Ag competition), p. 570 |
| | | Impure | 1. Antibody capture (Ab excess), p. 574<br>Other assays possible, but<br>need secondary technique |
| | Affinity-purified polyclonal antibodies | Pure | 1. Two-antibody sandwich, p. 580<br>2. Antigen capture (Ag competition), pp. 586, 588<br>3. Antibody capture (Ag competition), p. 570 |
| | | Impure | 1. Two-antibody sandwich, p. 580<br>2. Antibody capture (Ab excess), p. 574 |
| | One monoclonal antibody | Pure | 1. Antigen capture (Ag competition), pp. 586, 588<br>2. Antibody capture (Ag competition), p. 570 |
| | | Impure | 1. Antibody capture (Ab excess), p. 574<br>Other assays possible, but may<br>need secondary technique |
| | Two or more monoclonal antibodies | Pure | 1. Two-antibody sandwich, p. 580<br>2. Antigen capture (Ag competition), pp. 586, 588<br>3. Antibody capture (Ag competition), p. 570 |
| | | Impure | 1. Two-antibody sandwich, p. 580<br>2. Antibody capture (Ab excess), p. 574 |
| Antibody Presence or Quantity[b] | Polyclonal antibodies | Pure | 1. Antibody capture (Ag excess), p. 564 |
| | | Impure | Need secondary technique |
| | Affinity-purified polyclonal antibodies | Pure | 1. Antibody capture (Ag excess), p. 564 |
| | | Impure | Need secondary technique |
| | One monoclonal antibody | Pure | 1. Antibody capture (Ag excess), p. 564 |
| | | Impure | Need secondary technique |
| | Two or more monoclonal antibodies | Pure | 1. Antibody capture (Ag excess), p. 564 |
| | | Impure | Need secondary technique |

[a]The choice within each group are listed in order of suggested preference.
[b]For most assays used to measure antibody levels, the antibody is detected using an anti-immunoglobulin antibody. Thus, the antibody becomes the antigen.

# ■ Detecting and Quantitating Antigens

To detect and quantitate antigens, the most useful method is the two-antibody sandwich assay (p. 579). These assays are quick and reliable and can be used to determine the relative levels of most protein antigens. However, they require either two monoclonal antibodies that bind to independent sites on the antigen or affinity-purified polyclonal antibodies. An example of the two-antibody sandwich assay is found on p. 580.

If two monoclonal antibodies or affinity-purified polyclonal antibodies are not available, the next most useful assays for quantitating antigen are competition assays. For a competition assay, a sample of pure or nearly pure antigen is needed. There are two possible choices for assays using antigen competition. First, for an antigen capture assay (p. 585), the antigen is labeled and a constant amount of labeled antigen is mixed with the test solution which contains an unknown amount of the antigen. The solutions are then allowed to bind to a subsaturating amount of antibody bound to a solid phase. High levels of antigen in the test solution will reduce the amount of labeled antigen that can bind. Second, for an antibody capture assay (p. 563), a sample of pure or partially pure antigen is bound to the solid support. The antigen in the test solution is mixed with a preparation of labeled antibody and both are added to the solid support. High levels of antigen in the test solution will block the binding of the labeled antibody to the solid phase.

If pure or partially pure antigen is not available, the next most useful assay for antigen detection will be an antibody capture assay. Here the test solution is bound to a solid phase and a saturating amount of labeled antibody is used to detect the antigen. Two further factors must be considered. Because of their specificity, monoclonal antibodies or affinity-purified polyclonal antibodies will be more accurate than polyclonal antibodies when the antigen is rare. Similarly, as the antigen becomes more abundant, the specificity of the antibodies becomes less important.

Last, if rare antigens must be detected without the use of a pure sample of the antigen, the assay must be combined with a secondary technique. The secondary technique is used to distinguish the antigen from the background. Commonly used secondary techniques include immunoprecipitation, immunoblotting, and cell staining.

## ▧ Detecting and Quantitating Antibodies

If pure or nearly pure antigen is available, antibody presence and level can be determined using a variation of the antibody capture assay (p. 563). Using this assay, purified or partially purified antigen is bound to a solid support. The test solution containing an unknown amount of antibody is allowed to bind to the antigen–matrix. The amount of antibody bound to the solid phase is determined by incubation with a labeled secondary reagent such as an anti-immunoglobulin antibody.

In many cases it is easiest to determine the presence or level of antibodies by considering them to be a special class of antigens and to use anti-immunoglobulin antibodies to detect them. In these cases, many of the assays discussed above can be used to detect these "antigens".

### Secondary Techniques

When using impure sources of antigens for immunoassays, antibody detection may need to be supplemented with a secondary technique such as cell staining, immunoprecipitation, or immunoblotting. All immunochemical techniques have a certain background, and as the abundance of a particular antigen decreases relative to the other proteins, the ability of the antibodies to distinquish the correct antigen from the background is reduced. This is particularly true for polyclonal antibodies. When more specific antibodies are used, the amount of background decreases, and less pure antigen sources can be used. In general, rare antigens will always need a secondary technique to differentiate the signal from the noise. Any technique that allows the antigen to be identified specifically among the background can be used as a secondary technique.

# PROTOCOLS FOR IMMUNOASSAYS

The different techniques for immunoassays have been divided into three classes: (1) antibody capture assays, (2) antigen capture assays, and (3) two-antibody sandwich assays. Antibody capture assays are the most versatile. They can be used to determine both antigen or antibody levels and to compare antibody binding sites. Two-antibody sandwich assays are one of the best techniques for determining the presence and quantity of antigens. Antigen capture assays can be used for many types of studies, but are most useful for antigen detection and quantitation.

Two types of detection systems are commonly used for immunoassays. These are iodinated reagents and enzyme-labeled reagents. Assays that use iodinated reagents are easier to quantitate than enzyme-labeled reagents, while enzyme asssys will often yield a quicker result. The various detection methods are discussed in detail on p. 591.

Either iodinated or enzyme-labeled reagents can be used for direct or indirect methods. When using direct detection methods, the antibody or antigen is purified and labeled using one of the techniques described on p. 319. For indirect detection, a labeled secondary reagent that will bind specifically to an antibody is used. The secondary reagent can be labeled using one of the methods discussed on p. 319, but in most cases it is simpler to purchase these labeled reagents from commercial sources. A third variation that uses properties of both direct and indirect detection is the biotin–streptavidin system. Here, the antigen or antibody is purified and labeled with biotin (p. 340). The biotinylated reagent is detected by binding with streptavidin that has been labeled with iodine or an enzyme. Protocols for using all of these types of detection methods are given on p. 591.

## SUMMARY

### Antibody Capture Immunoassays

Detection limit is approximately 0.1–1.0 fmole, about 0.01–0.1 ng

- Rapid and easy
- Determines antigen presence and quantity
- Quantitative
- Sensitivity dependent on specific activity of antigen and avidity of antibody
- Needs high-avidity antibody

|  | Polyclonal Antibodies | Affinity-purified Polyclonal Antibodies | Monoclonal Antibodies | Pooled Monoclonal Antibodies |
|---|---|---|---|---|
| Signal strength | Fair | Excellent | Antibody dependent (poor to excellent) | Excellent |
| Specificity | Fair, but high background | Excellent | Excellent, but some cross-reactions | Excellent |
| Good Features | Availability | Specificity Signal strength | Specificity Unlimited supply Purity | Signal strength Specificity Unlimited supply Purity |
| Bad Features | Background Spurious activities Often need secondary techniques | Availability Limited supply | Need moderate to high avidity antibody | Availability |

**ANTIBODY CAPTURE ASSAYS**

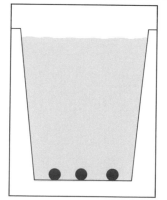
Bind Antigen to Solid Phase

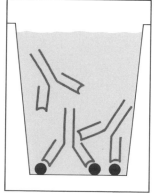
Bind Antibody to Antigen

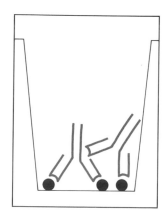
Quantitate Antibody

## ■ Antibody Capture Assays*

Antibody capture assays can be used to detect and quantitate antigens or antibodies and can be used to compare the epitopes recognized by different antibodies. The general protocol is simple; an unlabeled antigen is immobilized on a solid phase, and the antibody is allowed to bind to the immobilized antigen. The antibody can be labeled directly or can be detected by using a labeled secondary reagent that will specifically recognize the antibody. The amount of antibody that is bound determines the strength of the signal.

The three factors that will affect the sensitivity of a labeled antibody assay are (1) the amount of antigen that is bound to the solid phase, (2) the avidity of the antibody for the antigen, and (3) the type and number of labeled moieties used to label the antibody. First, the amount of antigen that is immobilized to the solid phase can be adjusted easily within the capacity of the chosen matrix by dilution or concentration of the antigen solution. Higher levels can only be achieved by changing the type of support. Polyvinylchloride (PVC) and nitrocellulose are the two most commonly used supports. PVC is normally used as a microtiter plate, while nitrocellulose is commonly used in sheets. Nitrocellulose will bind approximately 1000-fold more protein than PVC per unit of surface area. (See p. 605 for other possible matrices.) The second factor that will affect the sensitivity of an antibody capture assay is the avidity of the antibody for the immobilized antigen. Here, if other antibodies are available, they can be substituted for the original. If other antibodies are not available, the strength of the binding to the solid-phase antigen can often be increased by using a higher concentration of antigen. See Chapter 3 for a discussion of the advantages of bivalent binding and local concentration effects. Third, the type of label that is used and the specific activity of the labeled reagent will affect the sensitivity. Antibodies usually are labeled with either iodine (p. 324), enzymes (p. 342), or biotin (p. 340). The number of labeled molecules that are bound to the antibody can be varied to adjust the sensitivity of detection.

*Described originally by Miles and Hales (1968a,b).

## ANTIBODY CAPTURE ASSAYS—DETECTING AND QUANTITATING ANTIBODIES USING ANTIGEN EXCESS ASSAYS

When labeled antibody assays are performed with excess antigen on the solid phase (i.e., enough to saturate all the available antibody), the presence and level of antibodies in a test solution can be measured. To perform the assay, a preparation of purified or partially purified antigen is bound to a solid substrate. The test solution or dilution of the test solution containing an unknown amount of antibody is added. Antibodies in the test solution are allowed to bind to the immobilized antigen, and unbound antibodies are removed by washing. The presence of the bound antibodies is detected using a labeled secondary reagent. This reagent could be labeled anti-immunoglobulin antibodies, protein A, or protein G (Chapter 15). Because the assay is performed using the indirect technique, the antibodies in the test solution never need to be purified. To quantitate the levels of antibody in the test solution, a titration of the test solution is performed, thus yielding the relative titers of the antibodies.

The degree of purity of the antigen solution needed for this assay will depend on the specificity of the antibodies in the test solution. Polyclonal sera will contain extraneous antibodies that will bind and score against some antigen preparations. Often these can be distinguished by titrating the sera. Because of their specificity, solutions of monoclonal antibodies can be tested using less pure sources of antigen.

1. Prior to the assay, prepare the secondary reagent. Suitable secondary reagents include anti-immunoglobulin antibodies, protein A, or protein G. These labeled reagents can be purchased from commercial sources with many different labels or can be prepared as described on p. 319.

2. The most widely used solid phase for these assays is a PVC microtiter plate. Cut the plate to the correct size for the number of assays.

3. Bind the pure or partially pure antigen to the bottom of the wells by adding 50 $\mu$l of antigen solution to each well. PVC will bind approximately 100 ng/well (300 ng/cm$^2$). If maximal binding is required, use at least 1 $\mu$g/well (20 $\mu$g/ml). Although this is well above the capacity of the PVC, the binding will occur more rapidly, and the binding solution can be saved and used again. If the antigen is rare or expensive, lower amounts of antigen can be added, but the sensitivity of the assay will decrease. If no other buffer is dictated, dilutions of antigen should be done in PBS. Avoid extraneous proteins and detergents which will compete with the antigen for binding to the plate.

4. Incubate for 2 hr at room temperature in a humid atmosphere.

5. Wash the plate twice with PBS. A 500-ml squirt bottle is convenient. The antigen solution and washes can be removed by flicking the plate over a suitable waste container.

6. The remaining sites for protein binding on the PVC plate must be saturated by incubating with blocking buffer. Fill the wells to the top with 3% BSA/PBS with 0.02% sodium azide. Incubate for 2 hr to overnight in a humid atmosphere at room temperature.

7. Wash the plates twice with PBS.

8. Add 50 $\mu$l of the antibody test solution to each well. Incubate for 2 hr at room temperature in a humid atmosphere.

9. For quantitation, titrations of the test solution should be assayed (see p. 566). All antibody dilutions should be done in 3% BSA/PBS with 0.02% sodium azide.

10. Remove the unbound antibody by four washes with PBS.

11. Add the labeled secondary reagent. Use 50 $\mu$l/well. Incubate for 2 hr at room temperature.

12. For accurate quantitations, the amount of labeled secondary reagent must be in excess (see p. 566). All dilutions should be done in blocking buffer (3% BSA/PBS with 0.02% sodium azide). Azide should not be used with horseradish peroxidase detection systems.

13. Remove the unbound secondary reagent by washing four times with PBS.

**Determine and quantitate the amount of labeled secondary reagent (p. 591).**

## NOTES

i. If the amount of antigen bound to the plate is too low to produce a strong signal, the solid support can be changed to nitrocellulose, which can bind approximately 1000 times more protein per surface area. Nitrocellulose can be purchased either as sheets or as a filter sealed to the bottom of microtiter wells.

ii. Any of blocking solutions on p. 496, Table 12.2, can be substituted for the 3% BSA solution.

iii. Antigen-coated plates can be stored after the blocking step (step 6) either in PBS with 0.02% sodium azide at 4°C for 1 week or indefinitely at −20°C after removing the blocking solution.

iv. For some applications or with some antibodies, the signal strength can be increased by using longer incubation times. For rapid semi-quantitative assays, the incubation times can be shortened to between 30 min and 1 hr. The blocking step must still be 2 hr or more.

## MAKING THE ASSAY QUANTITATIVE

To compare the levels of antibody in different samples, prepare serial dilutions of each antibody test solution in blocking buffer. Perform the remainder of the assay as above. To determine the relative amounts of antibody, compare the midpoints of the titration curves. These assays will provide a relative measure of specific antibody concentration. To ensure that the assays are accurate, the amount of secondary labeled reagent that is used must be in excess. Nonsaturating amounts of labeled secondary reagent will yield good, but not excellent comparisons. The level of secondary antibody needed for saturation can be determined by titrating the secondary reagent, diluted in blocking buffer, on plates with saturating amounts of the primary antibody.

## OTHER APPLICATIONS

- **Class/subclass determination**   By changing the secondary reagent to class- or subclass-specific anti-immunoglobulin antibodies, the class or subclass of a particular monoclonal antibody can be determined easily.

- **Hybridoma screen**   This technique is used as one of the most common methods for screening hybridoma tissue culture supernatants after fusion. See Chapter 6 for an example.

## ANTIBODY CAPTURE ASSAYS—COMPARING ANTIBODY
## BINDING SITES USING AN ANTIBODY COMPETITION ASSAY

When antibody capture assays are performed using the antibody competition variation, the binding sites for two monoclonal antibodies can be compared. To use this assay, a sample of purified or partially purified antigen is bound to a solid support. Then two monoclonal antibodies are added, one labeled and one unlabeled. If the labeled antibody and the unlabeled antibody bind to separate and discrete sites on the antigen, the labeled antibody will bind to the same level whether or not the competing antibody is present. However, if the sites of interaction are identical or overlapping, the unlabeled antibody will compete, and the amount of labeled antibody bound to the antigen will be lowered. If the unlabeled antibody is present in excess, no labeled antibody will bind. With this assay, one monoclonal antibody will need to be purified and labeled, but none of the antibodies used as competitors need be purified.

Because only monoclonal antibodies are used in this technique, the antigen preparation need not be particularly pure. Samples containing as little as 1% pure antigen may be used, provided there are no cross-reacting proteins in the preparation.

1. Prior to the assay, purify and label each of the monoclonal antibodies that will be studied. Purification techniques are discussed in Chapter 8 and labeling techniques in Chapter 9.

2. The most widely used solid phase for this assay is a PVC microtiter plate. Cut the plate to the correct size for the number of assays.

3. Bind the standard antigen solution to the bottom of the wells by adding 50 $\mu$l of antigen solution to each well (20 $\mu$g/ml). PVC will bind approximately 100 ng/well (300 ng/cm$^2$). If maximal binding is required, use at least 1 $\mu$g/well. Although this is well above the capacity of the PVC, the binding will occur more rapidly, and the binding solution can be saved and used again. Avoid extraneous proteins, detergents, or other compounds that will lower the binding capacity of the PVC. If the choice of buffer is not dictated by experimental design, dilutions should be done in PBS.

    The amount of antigen bound to the wells should be high enough to produce an easily detectable antibody binding signal, but high concentrations also have some disadvantages. As the amount of antigen bound to the solid phase increases, the amount of competitor that is needed will increase. The amount of antigen should be titrated to the lowest needed to achieve a strong signal, thus increasing the sensitivity of the assay.

4. Incubate for 2 hr at room temperature in a humid atmosphere.

5. Wash the plate twice with PBS. A 500-ml squirt bottle is convenient. The antigen solution or washes can be removed by flicking the plate over a suitable waste container.

6. The remaining sites for protein binding on the PVC plate must be saturated by incubating with blocking buffer. Fill the wells to the top with 3% BSA/PBS with 0.02% sodium azide. Incubate for 2 hr to overnight in a humid atmosphere at room temperature.

7. Wash the plate twice with PBS.

8. To each well add a mixture of two antibodies, one labeled and one unlabeled. Incubate for 2 hr at room temperature in a humid atmosphere.

   All antibody dilutions should be done in 3% BSA/PBS with 0.02% sodium azide. Buffers with azide should not be used when the detection system depends on horseradish peroxidase.

   To optimize this assay, the amount of labeled antibody should be titrated and used at a subsaturating level. These values can be determined in preliminary tests. For accurate quantitations, the amount of cold competitor should be titrated, and the midpoints of the competition curves compared.

9. Remove the unbound antibodies by four washes with PBS.

**Determine and quantitate the amount of labeled antibody bound to the plate (p. 591).**

## NOTES

i. If the amount of antigen bound to the plate is too low to produce a strong signal, the solid support can be changed to nitrocellulose, which can bind approximately 1000 times more protein per surface area. Nitrocellulose can be purchased either as sheets or sealed to the bottom of microtiter wells. See p. 606 for special handling procedures.

ii. Any of blocking solutions on p. 496, Table 12.2, can be substituted for the 3% BSA solution.

iii. Antigen-coated plates can be stored after the blocking step (step 6) either in PBS with 0.02% sodium azide at 4°C for 1 week or indefinitely at −20°C after removing the blocking solution.

iv. For some applications or with some antibodies, the signal strength can be increased by using longer incubation times. For rapid semi-quantitative assays, the incubation times can be shortened to between 30 min and 1 hr. The blocking step must still be 2 hr or more.

## MAKING THE ASSAY QUANTITATIVE

Competition between monoclonal antibodies for binding to an antigen can be quantitated in this assay. For this application, all monoclonal antibodies must be purified. A standard curve is established by titrating one monoclonal antibody against itself, that is, the same antibody is used for both the label and the competitor. The capacity of other unlabeled monoclonal antibodies to inhibit the binding of the labeled antibody to the plate is titrated. The results are plotted, and the concentrations necessary to achieve 50% inhibition of binding are compared.

## OTHER APPLICATIONS

* **Screening hybridoma fusions for epitope-specific antibodies**  This assay can be used to screen a hybridoma fusion for epitope-specific antibodies. The assay is best used as a complement to a preliminary screening procedure. In the preliminary screen, antibodies that bind to the antigen can be identified. Then, those monoclonal antibodies that bind antigen in a primary screen are tested in this antibody competition assay.

* **Quantitating antibodies specific for a particular epitope in polyclonal sera**  If the competitor used in these assays is changed from an unlabeled monoclonal antibody to an unlabeled polyclonal antibody, the level of antibodies specific for that epitope in the polyclonal serum can be determined.

## ANTIBODY CAPTURE ASSAYS—DETECTING AND QUANTITATING ANTIGENS USING ANTIBODY EXCESS ASSAYS

Using an antibody capture assay with the antigen competition variation, the presence and level of an antigen can be determined quickly. For this assay, purified or partially purified antigen is bound to the solid phase. A sample of the test solution, containing an unknown concentration of antigen, is added together with a labeled antibody specific for the antigen. Any antigen in the test solution will compete with the immobilized antigen for binding with the labeled antibody. After the unbound proteins are removed by washing, the assay is quantitated by the amount of labeled antibody that is bound to the solid phase. When high concentrations of antigen are found in the test solution, no labeled antibody will bind to the plate. When little or no antigen is present in the test solution, high amounts of labeled antibody will bind. To quantitate the levels of antigen, dilutions of the test solution are assayed. Comparing the titration curves for each solution will yield the relative levels of antigen.

The degree of purity needed for the immobilized antigen is determined by the specificity of the antibody. Polyclonal sera will contain extraneous antibodies that will bind and score against some impure antigen preparations. Often these can be distinguished by titrating the sera. Because of their specificity, solutions of monoclonal antibodies or affinity-purified polyclonal antibodies can be used with less pure sources of antigen.

One problem that may be encountered using this technique is that the labeled antibody may bind preferentially to the antigen on the plate. Quantitating this assay relies on a comparison of two avidities, the avidity of the antibody for the immobilized antigen and the avidity of the antibody for the antigen in solution. These avidities may not be the same. Because the antigen on the plate will be present in a high local concentration, the antibody may be able to bind bivalently to these antigens (see Chapter 3). These types of bivalent interactions cannot occur in solution when the antigen is monovalent. Not all immobilized antigens will promote bivalent binding, but many will. Problems generated by differences in avidity can be detected by comparing standard curves using known concentrations of antigens. If these assays give anomalous results, several steps can be used to help. First, try using lower concentrations of the antigen on the plate, thus decreasing the local concentration and decreasing the chances of bivalent binding. Second, premix the test solution with the labeled antibody and incubate for 30 min to 1 hr before adding to the plate. Third, the problem can be eliminated by using labeled Fab fragments (p. 626).

1. Prior to the assay, purify and label the primary antibody. The primary antibody can be labeled with iodine (p. 324), biotin (p. 340), or an enzyme (p. 342).

2. The most widely used solid phase for these assays is a polyvinylchloride (PVC) microtiter plate. Cut the plate to the correct size for the number of assays.

3. Prepare a standard solution of antigen. Avoid extraneous proteins and detergents.

4. Bind a sample of the standard antigen solution to the bottom of the wells by adding 50 $\mu$l of antigen solution to each well. PVC will bind approximately 100 ng/well (300 ng/cm$^2$). If maximal binding is required, use at least 1 $\mu$g/well (20 $\mu$g/ml). Although this is well above the capacity of the PVC, the binding will occur more rapidly, and the binding solution can be saved and used again. If the antigen is rare or expensive, lower concentrations can be used. Unless another buffer is dictated by experimental conditions, dilutions can be done in PBS.

   To optimize the assay, the amount of antigen on the plate should be adjusted to a level that will just bind all of the input labeled antibody but still give a satisfactory signal strength. See p. 573.

5. Incubate for 2 hr at room temperature in a humid atmosphere.

6. Wash the plate twice with PBS. A 500-ml squirt bottle is convenient. The antigen solution or washes can be removed by flicking the plate over a suitable waste container.

7. The remaining sites for protein binding on the PVC plate must be saturated by incubating with blocking buffer. Fill the wells to the top with 3% BSA/PBS with 0.02% sodium azide. Incubate for 2 hr to overnight in a humid atmosphere at room temperature.

8. Wash the plates twice with PBS.

9. Add 50 $\mu$l of the antigen test solution together with the labeled antibody to each well. Incubate for 2 hr at room temperature in a humid atmosphere. These two solutions can be combined ahead of time and added together, or the antigen test solution can be added to the wells first, followed by the labeled antibody. However, the labeled antibody solution should not be added first.

The amount of labeled antibody that is added will ultimately determine the signal strength; however, the antibody should not be added in excess over the binding capacity of the plate (see p. 573).

All antibody and antigen dilutions should be done in 3% BSA/PBS with 0.02% sodium azide. Azide should not be included in the dilution buffer if horseradish peroxidase is used as the detection reagent.

10. Remove the unbound antibody and antigen by four washes with PBS.

**Determine and quantitate the amount of labeled antibody bound to the plate (p. 591).**

## NOTES

i. If the amount of antigen bound to the plate is too low to produce a strong signal, the solid phase can be changed to nitrocellulose, which can bind approximately 1000 times more protein per surface area. Nitrocellulose can be purchased either as sheets or sealed to the bottom of microtiter wells. See p. 606 for special handling procedures.

ii. Any of blocking solutions on p. 496 can be substituted for the 3% BSA solution.

iii. Antigen-coated plates can be stored after the blocking step (step 7) either in PBS with 0.02% sodium azide at 4°C for 1 week or indefinitely at −20°C after removing the blocking solution.

iv. This assay may also be performed using an indirect detection method. The primary antibody is not labeled, but the remainder of the assay is performed as above. After the final wash (step 10), add a labeled secondary reagent. Incubate for 2 hr at room temperature, wash, and quantitate.

## MAKING THE ASSAY QUANTITATIVE

To compare the levels of antigen in different samples, prepare serial dilutions of each antigen test solution in blocking buffer. Perform the remainder of the assay as above. To determine the relative amounts of antigen, compare the midpoints of the titration curves. To determine the absolute amount of antigen, compare these values with those obtained using known amounts of pure antigen in a standard curve.

For maximum sensitivity, first titrate the amount of standard antigen solution needed to coat the plate versus a fixed, high concentration of labeled antibody. Plot the values and select the lowest level that will yield a strong signal. Next, using plates coated with this amount of standard antigen solution, titrate the labeled antibody. Plot these values and select a level of antibody that is within the linear portion of the curve. Maximum range is obtained by choosing a point near the saturation level.

## *ANTIBODY CAPTURE ASSAYS—DETECTING AND QUANTITATING ANTIGENS USING ANTIGEN COMPETITION ASSAYS*

An antibody capture assay performed in antibody excess can be used to determine the presence and level of antigens in test solutions. The test solution is allowed to bind directly to a solid phase, and any unbound proteins are removed by washing. Then an antibody specific for the antigen is added and allowed to bind. After unbound antibodies are removed by washing, the amount of antibody adhering to the solid phase is determined by using a labeled secondary reagent. This secondary reagent could be a labeled anti-immunoglobulin antibody, protein A, or protein G. These assays cannot be used if the antigen being studied is a rare component of the test solution. In this case, the specific binding is obscured by the background. A secondary technique such as immunoblotting may be particularly valuable to help distinguish the specific signal from the background when rare antigens are being studied.

The specificity of the antibody will determine the degree of purity needed to detect the antigen. Polyclonal sera will contain extraneous antibodies that will bind and score against some antigen preparations. Often these can be distinguished by titrating the sera. Because of their specificity, monoclonal antibodies or affinity-purified polyclonal antibodies can be used to analyze less pure sources of antigen.

Making these assays quantitative is difficult for samples that contain widely varying amounts of total protein. In these cases, the total amount of protein added to the solid phase must be standardized.

1. Prior to the assay, prepare the secondary reagent. Suitable secondary reagents include anti-immunoglobulin antibodies, protein A, or protein G. These reagents can be purchased from commercial sources with many different labels or they can be prepared as described on p. 319.

2. Nitrocellulose is the most suitable support for these assays, because its high binding capacity permits the detection of less abundant antigens in the sample. If using the nitrocellulose for a sheet assay, mark the nitrocellulose into 3-mm squares using a soft lead pencil. Cut the sheet to the proper size for the number of assays to be performed. If using the nitrocellulose paper in a dot blot apparatus, cut to the dimensions of the apparatus.

3. Antigens should be prepared in neutral pH buffers without additional proteins. If no other buffers are dictated by the experimental design, use PBS. Detergents should be avoided if possible. If they must be used, test the effects of different possible detergents on antigen binding to the nitrocellulose. For sheet assays, the antigen test solutions are added to the center of individual squares (1 $\mu$l/spot). Incubate for 30 min at room temperature in a humid atmosphere. For dot blot assays, pre-wet the sheet by floating on water for 5 min. Fit the sheet into the apparatus and apply the antigen test solution to the region of nitrocellulose that is exposed. Use 30 $\mu$l/well. Incubate for 2 hr at room temperature in a humid atmosphere.

4. If using a dot blot apparatus, remove the nitrocellulose sheet. Wash the nitrocellulose with two changes of PBS.

5. Block the remaining sites for protein binding on the nitrocellulose by incubating the sheet in 3% BSA/PBS with 0.02% sodium azide for at least 2 hr at room temperature.

6. Wash the nitrocellulose sheet twice in PBS.

7. Add primary antibody at a suitable dilution to the entire sheet. Use 1 ml/cm$^2$. Incubate with agitation for 2 hr at room temperature. The amount of antibody to be added can be determined in preliminary tests. Higher amounts will increase the sensitivity and extend the range of the assay, but also will increase the possibility of detecting nonspecific antigens.

   All antibody dilutions should be done in 3% BSA/PBS with 0.02% sodium azide.

8. Remove the unbound antibody by four washes with PBS.

9. Add the labeled secondary reagent. Use 1 ml/cm$^2$. The amount of labeled secondary reagent should be determined in preliminary tests. For accurate comparisons, the secondary reagent should be used in excess. For routine comparisons, the level of secondary reagent should be adjusted to produce a easily detected signal. All dilutions should be done in blocking buffer (3% BSA/PBS with 0.02% sodium azide).

   Incubate with agitation for 2 hr at room temperature.

10. Remove the unbound antibody by washing four times with PBS for 5 min each.

Determine and quantitate the amount of labeled secondary reagent bound to the sheet (p. 591).

## NOTES

i. Any of blocking solutions on p. 496 can be substituted for the 3% BSA solution.

ii. If the antigen is impure, then a solid-phase substrate of higher capacity than nitrocellulose may be required to generate adequate signal strength. For this purpose, diazotized paper is useful as it can bind approximately 100 times more protein than nitrocellulose. With this variation, background problems caused by nonspecific binding or spurious cross-reactions may be severe.

iii. Directly labeled primary antibodies can be used in this test.

## MAKING THE ASSAY QUANTITATIVE

To compare the levels of antigen in different samples, prepare serial dilutions of each antigen test solution in PBS. Perform the remainder of the assay as above. To determine the relative amounts of antigen, compare the midpoints of the titration curves. To determine the absolute amount of antigen, compare these values with those obtained using known amounts of pure antigen in a standard curve. To ensure that the assays are accurate, the amount of primary unlabeled antibody and secondary labeled reagent must be in excess. This can be determined by titrating the reagents, diluted in blocking buffer, on sheets coated with saturating amounts of the antigen.

## OTHER APPLICATIONS

- **Column profiles**   Assays of this design are extremely useful where rapid estimates of antigen presence and quantity are required for large sample numbers. A good example of this application is the detection of antigen-positive fractions from conventional chromatography columns or gradients.

## SUMMARY

### Two-antibody Sandwich Immunoassays
Detection limit is approximately 0.1–1.0 fmole, about 0.01–0.1 ng

- Rapid and easy

- Detects antigen presence and quantity

- Quantitative

- Sensitivity dependent on specific activity of antibody and affinity of antibodies

- Needs affinity-purified polyclonal antibodies or two monoclonal antibodies with high affinity

|  | Polyclonal Antibodies | Affinity-purified Polyclonal Antibodies | Monoclonal Antibody | Multiple Monclonal Antibodies |
|---|---|---|---|---|
| Signal Strength | — | Excellent | — | Excellent |
| Specificity | — | Excellent | — | Excellent |
| Good Features | — | Multivalent | — | Unlimited supply |
| Bad Features | Cannot use polyclonal antibodies | Availability | Cannot use single antibody | Availability Need two antibodies that bind distinct epitopes |

**TWO-ANTIBODY ASSAYS**

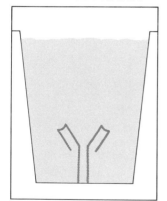

Bind First
Antibody to
Solid Phase

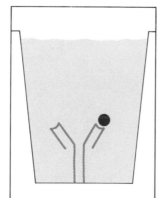

Bind Antigen
to Antibody

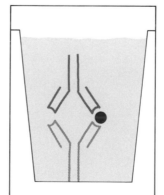

Bind Second
Antibody to
Antigen

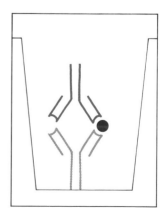

Quantitate
Second Antibody

## ▨ Two-Antibody Sandwich Assays

One of the most useful of the immunoassays is the two-antibody sandwich technique. These assays are used primarily to determine the antigen concentration in unknown samples. Two-antibody assays are quick and accurate, and if a source of pure antigen is available, the assays can be used to determine the absolute amounts of antigen in unknown samples. The assay requires two antibodies that bind to non-overlapping epitopes on the antigen. Either two monoclonal antibodies that recognize discrete sites or one batch of affinity-purified polyclonal antibodies can be used.

To use the two-antibody assay, one antibody is purified and bound to a solid phase, and the antigen in a test solution is allowed to bind. Unbound proteins are removed by washing, and the labeled second antibody is allowed to bind to the antigen. After washing, the assay is quantitated by measuring the amount of labeled second antibody that is bound to the matrix. The major advantages of this technique are that the antigen does not need to be purified prior to use and that the assays are very specific. The major disadvantage is that not all antibodies can be used.

The sensitivity of two-antibody assays is dependent on four factors: (1) the number of molecules of the first antibody that are bound to the solid phase, (2) the avidity of the first antibody for the antigen, (3) the avidity of the second antibody for the antigen, and (4) the specific activity of the labeled second antibody. First, the amount of the first antibody that is bound to the solid phase can be adjusted easily within the capacity of the chosen matrix by dilution or concentration of the antibody solution. Higher capacities can be achieved by changing the type of support. Polyvinylchloride is the most commonly used matrix, but several solid supports can be substituted to increase the linear range of the assay. The most useful alternative support is nitrocellulose, which binds approximately 1000 times more protein per surface area. (See p. 605 for other possible matrices.) The second and third factors that affect the sensitivity of the assays are the avidity of the antibodies for the antigen. These can only be altered by substitution with other antibodies. The fourth factor is the number and type of labeled moeities on the second antibody. Antibodies can be labeled conveniently with iodine, enzymes, or biotin. The choice of label is discussed on p. 591.

## DETECTING AND QUANTITATING ANTIGENS USING
## THE TWO-ANTIBODY SANDWICH ASSAY*

1. Prior to the assay, both antibody preparations need to be purified (p. 288) and one needs to be labeled (p. 319). For monoclonal antibodies, choosing which antibody to label is determined empirically. When setting up the assay, label each antibody preparation independently, and try both combinations of solid-phase and labeled antibody to determine which is best. When using affinity-purified antibodies, the same antibody preparation will be used for both sides of the assay.

2. For most applications, a PVC microtiter plate is best. Cut the plate to the correct size for the number of assays.

3. Bind the unlabeled antibody to the bottom of each well by adding approximately 50 $\mu$l of antibody solution to each well (20 $\mu$g/ml in PBS). PVC will bind approximately 100 ng/well (300 ng/cm$^2$). The amount of antibody to be used will depend on the individual assay, but if maximal binding is required, use at least 1 $\mu$g/well. Although this is well above the capacity of the well, the binding will occur more rapidly, and the binding solution can be saved and used again. In most cases, saturating amounts of the first antibody will increase the sensitivity of the assays and will not have any detrimental effects.

   Incubate for 2 hr at room temperature in a humid atmosphere.

4. Wash the wells twice with PBS. A 500-ml squirt bottle is convenient. The antibody solution or washes can be removed by flicking the plate over a suitable waste container.

5. The remaining sites for protein binding on the polyvinylchloride plate must be saturated by incubating with blocking buffer. Fill the wells to the top with 3% BSA/PBS with 0.02% sodium azide. Incubate for 2 hr to overnight in a humid atmosphere at room temperature.

6. Wash the wells twice with PBS.

7. Add 50 $\mu$l of the antigen solution to the wells. For quantitation, the antigen solution should be titrated. All dilutions should be done in the blocking buffer (3% BSA/PBS with 0.02% sodium azide). Incubate for at least 2 hr at room temperature in a humid atmosphere.

*Ling and Overby (1972); Belanger et al. (1973); Maiolini and Masseyeff (1975).

8. Wash the plate four times with PBS.

9. Add the labeled second antibody. The amount of labeled antibody to be added can be determined in preliminary experiments. For accurate quantitation, the second antibody should be used in excess. All dilutions of the labeled antibody should be done in blocking buffer (3% BSA/PBS with 0.02% sodium azide).

10. Incubate for 2 hr or more at room temperature in a humid atmosphere.

11. Wash with several changes of PBS.

**Quantitate the amount of bound labeled antibody using the proper detection method (p. 591).**

## NOTES

i. Any of blocking solutions on p. 496 can be substituted for the 3% BSA solution.

ii. Antibody-coated plates can be stored after the blocking step (step 5) either in PBS with 0.02% sodium azide at 4°C for 1 week or indefinitely at −20°C after removing the blocking solution.

iii. For some applications or with some antibodies, the signal strength can be increased by using longer incubation times. For rapid semi-quantitative assays, the incubation times can be shortened to between 30 min and 1 hr. The blocking step must still be 2 hr or more.

## MAKING THE ASSAY QUANTITATIVE

The amount of second antibody should be in excess to ensure that the assay is quantitative. The correct amount to add can be determined by titrating the amount of labeled second antibody on antigen-saturated plates. To compare the relative amounts of antigen in different test samples, prepare serial dilutions of each sample in blocking buffer. Perform the assays as described above. Plot and compare the midpoints of each titration curve. To determine absolute amounts of antigen, compare these plots to a standard curve obtained using known amounts of pure antigen.

## OTHER APPLICATIONS

- **Epitope comparisons**  This basic assay design can be adapted to perform antibody competition experiments. This is particularly useful when purified antigen is not available to establish the antibody capture/competition assay described on p. 567. The antibodies, whose binding sites are to be compared, are purified and labeled. A third antibody is used to coat the wells. Each labeled antibody is tested for its ability to bind to the antigen immobilized through this third antibody. The test can proceed if both labeled antibodies bind efficiently. For the test, varying amounts of the unlabeled, test monoclonal antibody are added to the antigen-coated wells, followed by a fixed, subsaturating amount of the labeled antibody. If the unlabeled antibody and the labeled antibody bind to overlapping sites on the antigen, then the amount of label bound to the solid phase will be reduced. If the unlabeled antibody and the labeled antibody bind to nonoverlapping sites, there will be no reduction in the binding of labeled antibody. By varying the antibody used to coat the well, many different areas of the antigen can be studied in this way, and a steric inhibition map of the antigen developed.

- **Studying protein complexes**  Two-antibody sandwich assays can be used to study a number of aspects of protein complexes. If antibodies are available to different components of a heteropolymer, a two-antibody assay can be designed to test for the presence of the complex. An antigen competition assay can be designed to measure the quantity of either component of the complex.

  Using a variation of these assays, monoclonal antibodies can be used to test whether a given antigen is multimeric. If the same monoclonal antibody is used for both the solid phase and the label, monomeric antigens cannot be detected. Such combinations, however, may detect multimeric forms of the antigen. Be cautious; in these assays negative results may be generated both by multimeric antigens held in unfavorable steric positions as well as by monomeric antigens.

## SUMMARY

### Antigen Capture Immunoassays
Detection limit is approximately 0.1–1.0 fmole, about 0.01–0.1 ng

- Rapid and easy
- Determines antigen presence and quantity
- Quantitative
- Sensitivity dependent on specific activity of antigen and avidity of antibody
- Needs high-avidity antibody

|  | Polyclonal Antibodies | Affinity-purified Polyclonal Antibodies | Monoclonal Antibodies | Pooled Monoclonal Antibodies |
|---|---|---|---|---|
| Signal Strength | Excellent | Excellent | Antibody dependent (poor to excellent) | Excellent |
| Specificity | Usually good, but some background | Excellent | Excellent, but some cross-reactions | Excellent |
| Good Features | Stable multivalent interactions | Stable, multivalent interactions | Specificity Unlimited supply | Stable, multivalent interactions Specificity Unlimited supply |
| Bad Features | Nonrenewable Background | Availability | Need high affinity | Availability |

**ANTIGEN CAPTURE ASSAYS**

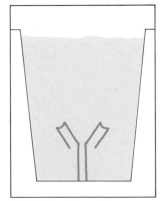

Bind Antibody
to Solid Phase

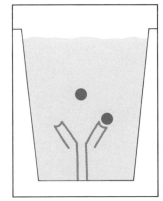

Bind Antigen
to Antibody

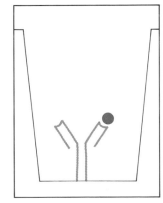

Quantitate
Antigen

## ■ Antigen Capture Assays*

Antigen capture assays are used primarily to detect and quantitate antigens. In both the variations discussed here, the amount of antigen in the test solution is determined using a competition between labeled and unlabeled antigen. Unlabeled antibody is bound to the solid phase either directly or through an intermediate protein, such as an anti-immunoglobulin antibody. The antigen is purified and labeled. A sample of the labeled antigen is mixed with the test solution containing an unknown amount of antigen, and the mixture is added to the bound antibody. The antigen in the test solution will compete with the labeled antigen for binding to the antibody–matrix. Unbound proteins are removed by washing, and the amount of labeled antigen bound to the matrix is measured. If the unknown solution contains a high concentration of antigen, it will compete effectively with the labeled antigen, and little or none of the labeled antigen will bind to the antibody. To quantitate the levels of antigen, dilutions of the test solution are performed and assayed. Comparing the titration curves for each solution will yield the relative levels of antigen.

The sensitivity of the labeled antigen assay will depend on three factors: (1) the number of antibodies that are bound to the solid phase, (2) the avidity of the antibody for the antigen, and (3) the specific activity of the labeled antigen. First, the number of antibodies that are bound to the solid phase can be adjusted within the capacity of the matrix by diluting or concentrating the antibody solution. Higher levels can be achieved by changing the type of solid-phase matrix that is used. Second, the avidity of the antibody for the antigen will affect the sensitivity of the assay, but can only be altered by changing the antibody source. Third, the choice of label for the antigen is discussed on p. 591.

The degree of purity needed for the labeled antigen will be determined by the specificity of the antibody. Polyclonal sera will contain extraneous antibodies that will bind and score against some antigen preparations. Often these can be distinguished by titrating the sera. Because of their specificity, solutions of monoclonal antibodies can be used with less pure sources of antigen.

---

*Originally developed by Yalow and colleagues. See Berson et al. (1956); Yalow and Berson (1959) for the concepts, Weiler et al. (1960) for competition assays, Wide and Porath (1966) for the use of a solid phase.

## DETECTING AND QUANTITATING ANTIGENS USING
## COMPETITION ASSAYS—VARIATION I, MICROTITER PLATES

This assay is given in two variations, here set up for microtiter plates and in the next technique for *Staphylococcus aureus* binding. The microtiter plate technique is useful when the antibody is easily purified and when there are a large number of antigen samples to test. The *S. aureus* assay is best when a small number of samples will be tested. The *S. aureus* variation also does not require that the antibody be purified.

1. Prior to the assay, label the antigen. Purified or partially purified antigen can be labeled with iodine (p. 324), enzymes (p. 342), or biotin (p. 340). Abundant antigens can also be labeled metabolically by growing cells in the presence of a radioactive precursor (p. 358).

2. Prior to the assay, purify the antibody as described on p. 288.

3. Bind the unlabeled antibody to the bottom of each well by adding approximately 50 $\mu$l of antibody solution to each well (20 $\mu$g/ml in PBS). Polyvinylchloride (PVC) will bind approximately 100 ng/well (300 ng/cm$^2$). The amount of antibody to be used will depend on the individual assay, but if maximal binding is required, use at least 1 $\mu$g/well. Although this is well above the capacity of the well, the binding will occur more rapidly, and the binding solution can be saved and used again.

   Maximal binding may not always be desirable in these assays. Higher amounts of antibody bound to the solid phase increase the level of antigen needed for competition to be detected. See p. 587 for a discussion of the parameters that need to be checked to make this assay quantitative.

   Incubate for 2 hr at room temperature in a humid atmosphere.

4. Wash the wells twice with PBS. A 500-ml squirt bottle is convenient. The antibody solution or washes can be removed by flicking the plate over a suitable waste container.

5. The remaining sites for protein binding on the polyvinylchloride plate must be saturated by incubating with blocking buffer. Fill the wells to the top with 3% BSA/PBS with 0.02% sodium azide. Incubate for 2 hr to overnight in a humid atmosphere at room temperature.

6. Wash the wells twice with PBS.

7. Add 50 $\mu$l of the antigen test solution along with the standard labeled antigen solution to the wells. Either add the test solution first, followed immediately by the labeled solution or mix in a separate plate and transfer together.

The amount of labeled antigen to be added should be sufficient to generate a strong signal, but should not greatly exceed the capacity of the antibody matrix (see p. 605). For accurate quantitation the test solution should be assayed over a number of dilutions.

All dilutions should be done in the blocking buffer (3% BSA/PBS with 0.02% sodium azide).

Incubate for 2 hr at room temperature in a humid atmosphere.

8. Wash the plate four times with PBS.

**Determine the amount of labeled antigen bound (p. 591).**

## NOTES

i. If the amount of antibody bound to the plate is too low to produce a strong signal, the solid support can be changed to nitrocellulose, which can bind approximately 1000 times more protein per surface area. Nitrocellulose can be purchased either as sheets or sealed to the bottom of microtiter wells. See p. 606 for special handling procedures.

ii. Any of blocking solutions on p. 496 can be substituted for the 3% BSA solution.

iii. Antibody-coated plates can be stored after the blocking step (step 5) either in PBS with 0.02% sodium azide at 4°C for 1 week or indefinitely at −20°C after removing the blocking solution.

## MAKING THE ASSAY QUANTITATIVE

To compare the levels of unlabeled antigen in different samples, prepare serial dilutions of each antigen test solution in blocking buffer. Perform the remainder of the assay as above. To determine the relative amounts of antigen, compare the mid points of the titration curves. To determine the absolute amount of antigen, compare these values with those obtained using known amounts of pure unlabeled antigen in a standard curve.

For maximum sensitivity, first titrate the amount of purified antibody bound to the plate versus a saturating amount of labeled antigen solution without the addition of competing unlabeled antigen. Plot the values and select the lowest level that will yield a strong signal. Next, using plates coated with this amount of antibody, titrate the labeled antigen. Plot the percentage of input antigen bound versus the total antigen input. Select a level of antigen that is within the linear portion of the curve. Maximum range is obtained by choosing a point near the saturation level.

## DETECTING AND QUANTITATING ANTIGENS USING
## COMPETITION ASSAYS—VARIATION II, S. AUREUS

In this version of the antigen capture assay, the solid-phase matrix is provided by adding *Staphylococcus aureus*. These killed bacteria have a high concentration of protein A on their surface that binds the antibody in the solution. The *S. aureus* assay is best when a small number of samples will be tested and also does not require that the antibody be purified.

1. Prior to the assay, label the antigen with iodine (p. 324). Other labels are not as useful with this assay design, because the results are quantitated from 1.5-ml conical tubes and iodine is easy to detect by counting the entire tube in a gamma-counter.

2. Mix a 50-$\mu$l sample of labeled antigen with a 50-$\mu$l sample of the unlabeled test solution in a 1.5-ml conical tube.

   The amount of labeled antigen to be added should be sufficient to generate a strong signal, but should not greatly exceed the capacity of the antibody that will be added (see p. 589). For accurate quantitation, a series of dilutions of test samples should be assayed. All dilutions should be done in 3% BSA/PBS with 0.02% sodium azide.

3. Add 100 $\mu$l of the antibody solution to the 1.5-ml conical tube. The concentration of antibody should be subsaturating to detect the competition. The preliminary tests needed to adjust the amount of antibody to add are discussed on p. 589. All dilutions should be done in 3% BSA/PBS with 0.02% sodium azide.

   Incubate at room temperature for 2 hr.

4. If the specific antibody binds protein A tightly (see p. 616), add 50 $\mu$l of a 10% suspension of fixed *S. aureus* cells (prewashed twice in PBS, see p. 620).

   If the particular antibody does not bind protein A, add 50 $\mu$l of a dilution of an anti-immunoglobulin antibody raised in a species whose IgG bind protein A (e.g., rabbits). Incubate for 1 hr, and add 50 $\mu$l of a 10% suspension of fixed *S. aureus* cells (prewashed twice in PBS, see p. 620). For complete binding of the first antibody to the solid phase, the anti-immunoglobulin antibody must be in excess. This should be titrated to determine the correct value; however, 5 $\mu$l of a high-titer rabbit sera will bind approximately 1 $\mu$g of mouse antibody. All dilutions should be done in 3% BSA/PBS with 0.02% sodium azide.

5. Incubate for 15 min at room temperature.

6. Add 1.0 ml of PBS containing 1% NP-40 to each tube. Centrifuge for 3 min.

7. Aspirate the supernatant to remove unbound antigen.

8. Determine the amount of bound antigen in the pellet by counting the entire tube in a gamma counter.

## MAKING THE ASSAY QUANTITATIVE

To compare the levels of unlabeled antigen in different samples, prepare serial dilutions of each antigen test solution in blocking buffer. Perform the remainder of the assay as above. To determine the relative amounts of antigen, compare the mid points of the titration curves. To determine the absolute amount of antigen, compare these values with those obtained using known amounts of pure unlabeled antigen in a standard curve.

For maximum sensitivity, first titrate the amount of antibody versus a saturating amount of labeled antigen solution without the addition of unlabeled antigen. Plot the values and select the lowest level that will yield a strong signal (approximately 10,000–50,000 cpm). Next, using this amount of antibody, titrate the labeled antigen. Plot the percentage of input antigen bound versus the total antigen input. Select a level of antigen that is within the linear portion of the curve. Maximum range is obtained by choosing a point near the saturation level.

## Epitope Mapping

For some purposes it is necessary to determine whether individual monoclonal antibodies raised against the same antigen bind to identical or overlapping epitopes. Three methods commonly are used to test whether antibodies recognize similar sites on protein antigens.

- **Steric competition**  To set up a steric competition immunoassay, one of the monoclonal antibodies is purified and labeled. The capacity of other monoclonal antibody preparations to block the binding of the labeled antibody to the antigen is tested. Two methods for these types of assay are described on pp. 567 and 583.

- **Linear maps**  If a DNA clone is available for a protein antigen, deletion mutations can be assayed for the production of fragments that can be recognized by an antibody. In practice, this method can only localize the binding site to a small region if the monoclonal antibody recognizes an epitope that is formed by a contiguous stretch of the primary sequence of the protein. If the antibody binds to an epitope formed by a secondary or tertiary structure, this technique will only identify a relatively large region containing the epitope. When locating the boundaries of an epitope using this method, only results that produce a positive binding should be used to localize the binding site. Lack of binding can be caused by reasons other than the loss of the epitope, such as steric interference, long-range conformation changes, or rapid degradation.

- **Random cloning of epitopes in bacteria**  If the monoclonal antibodies can recognize the antigen synthesized in bacteria, a simple and rapid method to determine the location of the epitope is randomly to clone small fragments of the coding region into an expression vector and screen for binding of the antibody to the bacterially expressed protein. Random fragments can be generated using sonication (Deininger 1983) or partial DNase I digestion (Anderson 1981) and cloned by blunt-end ligation into a suitable site of a fusion protein vector. Expression of the epitope can then be determined by screening lifts for antibody binding (see Helfman et al. 1983 or Young and Davis 1983a,b for appropriate antibody screening techniques or pp. 691 and 692). If expression is combined with DNA sequencing, an accurate location of the epitope can be determined.

## ■ Detection

All immunoassays rely on labeled antigens, antibodies, or secondary reagents for detection. These proteins can be labeled with radioactive compounds, enzymes, biotin, or fluorochromes. Of these, radioactive labeling can be used for almost all types of assays and with most variations. It is easy to quantitate and simple to detect. Enzyme-conjugated labels are particularly useful when radioactivity must be avoided or when quick results are helpful. They yield semi-quantitative results with no adjustments, but are more troublesome to use for exact quantitation. Although it is widely argued, neither is remarkably more sensitive than the other. In the future, however, it is clear that the use of new and more sensitive enzyme substrates will make these assays more valuable.

Biotin-coupled reagents usually are detected with labeled streptavidin. Streptavidin binds tightly and quickly to biotin and can be labeled with iodine or enzymes.

Of the methods used to label proteins, fluorochromes have the fewest applications for immunoassays. Although they can be very sensitive, they require rather expensive equipment to use and, unlike radiometric or enzymatic detection, no alternative methods can be used to locate and quantitate positives. In the foreseeable future, their major use will continue to be in clinical and large-volume laboratories where numerous samples need to be processed.

### Iodine-Labeled Antigens, Antibodies, or Secondary Reagents

Iodine-labeled reagents can be detected either by autoradiography or by direct counting. For autoradiographic detection the sample should be placed in direct contact with the film and placed at $-70°C$ with an intensifying screen. Results can be quantitated at a crude level by simple visual examination of the exposed film and more quantitatively by densitometric tracing. For accurate quantitation, individual samples are counted in a gamma-counter.

Antibodies can be labeled with $^{125}I$ using the techniques described on p. 324. To be useful in most of these assays, they should be labeled to specific activities of 10–100 $\mu Ci/\mu g$. Protein antigens can be labeled to similar activities using the same techniques. As a starting point, add 50,000 cpm/well for microtiter plate assays and 200,000 cpm/cm$^2$ for sheet assays.

### Biotin-Labeled Antibodies, Antigens, or Secondary Reagents

Biotin-labeled reagents normally are coupled to proteins through free amino groups; so as long as the amino groups do not form an essential structural region of the protein, biotin will not interfere with the immunoassay. The major advantage of the biotinylated reagents is that they are stable for years with little or no loss of specific activity. Methods for biotinylating protein antigens are discussed on p. 340.

Biotin-labeled reagents are detected and quantitated by using avidin or streptavidin. These proteins bind tightly to biotin, forming an essentially irreversible complex. Both avidin and streptavidin can be labeled with iodine or enzymes and are available commercially. Usually, streptavidin is the preferred ligand because of its more favorable pI.

Follow the instructions for iodine- or enzyme-labeled reagents on pp. 591 and 592 for detection protocols.

### Enzyme-Labeled Antigens, Antibodies, or Secondary Reagents*

Enzyme-labeled reagents are detected using chromogenic substrates. For assays carried out in microtiter wells, a soluble substrate that is converted to a soluble colored product is used. Enzyme levels are determined by monitoring color development in a spectrophotometer. Many commercial spectrophotometers are now available that will directly read samples in 96-well microtiter plates. For these assays, most workers substitute polystyrene plates for polyvinylchloride, as polystyrene can be read accurately in 96-well spectrophotometers. The polystyrene plates are handled identically to the polyvinylchloride (PVC) plates, except they cannot be cut. Antibodies can be purchased commercially or they can be labeled using the techniques on p. 342. Antigens can be labeled using similar techniques.

For assays carried out on nitrocellulose or diazotized paper, a soluble substrate that yields an insoluble colored product at the site of reaction is used. These results can be evaluated by visual inspection and quantitated by reflection densitometry. Suitable substrates of both types are listed below for the commonly used enzyme labels. For a starting point, use 5–50 ng of antibody-conjugate/well for microtiter plates assays (1/1000–1/10,000 dilutions of most commercial stocks) or 20–200 ng/cm$^2$ for sheet assays (1/100–1/1,000 dilutions of commercial stocks).

> **Caution** Some chromogenic substrates may be carcinogenic. Consult your local authorities for handling and disposal procedures.

*Engvall and Perlmann (1972); Van Weemen and Schuurs (1971).

## HORSERADISH PEROXIDASE—SOLUBLE PRODUCT

The most sensitive chromogenic substrate for detection of horseradish peroxidase (HRP)-labeled reagents is tetramethylbenzidine (TMB).

1. Prior to the detection, dissolve 0.1 mg of 3',3',5',5'-tetramethylbenzidine (TMB) in 0.1 ml of dimethylsulfoxide. Add 9.9 ml of 0.1 M sodium acetate (pH 6.0). Filter through Whatman No. 1 or equivalent. Add hydrogen peroxide to a final concentration of 0.01%. This is enough substrate for two 96-well microtiter plates.

   Hydrogen peroxide is normally supplied as a 30% solution. After opening, it should be stored at 4°C, where it is stable for 1 month.

2. After the final wash in PBS, add 50 $\mu$l of the substrate solution to each microtiter well.

3. Incubate for 10–30 min at room temperature. Positives appear pale blue.

4. Add 50 $\mu$l of 1 M $H_2SO_4$, to every well. Positives now appear bright yellow.

5. Read the results at 450 nm.

## *HORSERADISH PEROXIDASE—INSOLUBLE PRODUCT*

Three substrates are suitable for developing dot blots stained with horseradish peroxidase-coupled antibodies. These are chloronaphthol, aminoethylcarbazole, and diaminobenzidine.

### Chloronaphthol

4-Chloro-1-naphthol (chloronaphthol) is relatively insensitive but gives an intense blue-black reaction product in an easily controlled reaction. Assays developed with this reagent are photographed readily, but the color fades noticeably on storage. Background binding of the product to the filter is low.

1. Prepare the stock solution of 4-chloro-1-naphthol. Dissolve 0.3 gram of chloronaphthol in 10 ml of absolute ethanol. The chloronaphthol stock is stable at −20°C for at least 1 year.

2. Just prior to developing the assay, add 0.1 ml of chloronaphthol stock to 10 ml of 50 mM Tris (pH 7.6). A white precipitate will form.

3. Remove the precipitate by filter through Whatman No. 1 filter paper (or equivalent).

4. Add 10 $\mu$l of 30% $H_2O_2$. $H_2O_2$ is generally supplied as a 30% solution and should be stored at 4°C, where it will last about 1 month.

5. Place the washed nitrocellulose sheet in a suitable container. Add 10 ml of substrate solution per $15 \times 15$-cm$^2$ membrane. Develop at room temperature with agitation until the spots are suitably dark. A typical incubation would be approximately 30 min.

6. To stop the reaction, remove the $H_2O_2$ by rinsing with PBS.

### Aminoethylcarbazole

3-Amino-9-ethylcarbazole (AEC) yields an aesthetically pleasing red reaction product. It is slightly more sensitive than chloronaphthol but harder to photograph.

1. Prepare the stock solution of AEC. Dissolve 0.4 gram of 3-amino-9-ethylcarbozole in 100 ml of dimethylformamide. The AEC stock is stable at room temperature for at least 1 year.

2. Just prior to developing the assay, add 0.67 ml of AEC stock to 10 ml of 0.1 M sodium acetate (pH 5.2) with stirring. A small precipitate may form.

3. Filter through Whatman No. 1 filter paper (or equivalent).

4. Add 10 $\mu$l of 30% $H_2O_2$. $H_2O_2$ is generally supplied as a 30% solution and should be stored at 4°C, where it will last about 1 month.

5. Place the washed nitrocellulose sheet in a suitable container. Add 10 ml of substrate solution per $15 \times 15$-cm$^2$ membrane. Develop at room temperature with agitation until the spots are suitably dark. A typical incubation would be approximately 30 min.

6. To stop the reaction, remove the $H_2O_2$ by rinsing with PBS.

**Diaminobenzidine**

Diaminobenzidine (DAB, 3,3′,4,4′-tetraaminobiphenyl) is an exceptionally sensitive substrate, yielding a brown reaction product. One drawback to DAB is that it reacts so fast that it is very easy to overdevelop giving a high background.

1. Just prior to developing the assay, dissolve 6 mg of 3,3′-diaminobenzidine (use DAB tetrahydrochloride) in 10 ml of 50 mM Tris (pH 7.6). A small precipitate may form.

2. Filter through Whatman No. 1 filter paper (or equivalent).

3. Add 10 $\mu$l of 30% $H_2O_2$. $H_2O_2$ is generally supplied as a 30% solution and should be stored at 4°C, where it will last about 1 month.

4. Place the washed sheet in a suitable container. Add 10 ml of substrate solution per $15 \times 15$-cm$^2$ membrane. Develop at room temperature with agitation until the spots are suitably dark. A typical incubation would be approximately 1–5 min.

5. To stop the reaction, remove the $H_2O_2$ by rinsing with PBS.

### Diaminobenzidine with Metal Ion Enhancement (DAB/Metal)

The sensitivity DAB can be enhanced by adding cobalt or nickel ions to the substrate solution. In the presence of these ions, the product is slate black in color and more easily photographed than the unenhanced product. One drawback to DAB is that it reacts so fast that it is very easy to overdevelop, giving a high background.

1. Just prior to developing the assay, dissolve 6 mg of 3,3'-diaminobenzidine (use DAB tetrahydrochloride) in 9 ml of 50 mM Tris (pH 7.6). A small precipitate may form. Add 1 ml of 0.3% (wt/vol) $NiCl_2$ or $CoCl_2$.

2. Filter through Whatman No. 1 filter paper (or equivalent).

3. Add 10 $\mu$l of 30% $H_2O_2$. $H_2O_2$ is generally supplied as a 30% solution and should be stored at 4°C, where it will last about 1 month.

4. Place the washed sheet in a suitable container. Add 10 ml of substrate solution per $15 \times 15$-$cm^2$ membrane. Develop at room temperature with agitation until the spots are suitably dark. A typical incubation would be approximately 1–5 min.

5. To stop the reaction, remove the $H_2O_2$ by rinsing with PBS.

## ALKALINE PHOSPHATASE—SOLUBLE PRODUCT

Probably the best enzyme–substrate combination for sensitive quantitation of 96-well microtiter immunoassays is alkaline phosphatase–nitrophenyl phosphate (PNPP).

1. Prior to developing the enzyme label, dissolve 10 mg of *p*-nitrophenyl phosphate (PNPP) in 10 ml of 10 mM diethanolamine (pH 9.5) containing 0.5 mM $MgCl_2$. This is enough substrate for two 96-well microtiter plates.

2. After the final wash with PBS, wash the plate twice with 10 mM diethanolamine (pH 9.5) containing 0.5 mM $MgCl_2$.

3. Add 50 μl of the substrate solution to each microtiter well.

4. Incubate for 10–30 min at room temperature.

5. Add 50 μl of stop solution, 0.1 M EDTA, to each well. Positives appear bright yellow.

Read the plate at 405 nm.

## ALKALINE PHOSPHATASE—INSOLUBLE PRODUCT

The bromochloroindolyl phosphate–nitro blue tetrazolium (BCIP/NBT) substrate generates an intense black-purple precipitate at the site of enzyme binding. The substrate solution is stable in the absence of enzyme. The reaction proceeds at a steady rate, thus allowing accurate control of the development of the reaction. This allows the relative sensitivity to be controlled by the length of incubation. The BCIP/NBT substrate characteristically produces sharp spots with very little background coloring of the membrane.

Coupled reagents should be prepared with eukaryotic alkaline phosphatase, as this enzyme is readily inactivated with EDTA. The bacterial enzyme is difficult to stop, causing overdevelopment and leading to high background.

1. Prior to developing the assay, prepare the three stock solutions. NBT: Dissolve 0.5 gram of NBT in 10 ml of 70% dimethylformamide. BCIP: Dissolve 0.5 gram of BCIP (disodium salt) in 10 ml of 100% dimethylformamide. Alkaline phosphatase buffer: 100 mM NaCl, 5 mM $MgCl_2$, 100 mM diethanolamine (pH 9.5).

   All stocks are stable at 4°C for at least 1 yr.

2. Just prior to developing the assay, prepare fresh substrate solution. Add 66 $\mu$l of NBT stock to 10 ml of alkaline phosphatase buffer. Mix well and add 33 $\mu$l of BCIP stock. Use within 1 hr.

3. After the final wash, rinse the sheet twice with 100 mM NaCl, 5 mM $MgCl_2$, 100 mM diethanolamine (pH 9.5). Add 10 ml of substrate solution per 15 × 15-$cm^2$ membrane. Develop at room temperature with agitation until the spots are suitably dark. A typical incubation would be approximately 30 min.

4. To stop the reaction, rinse with PBS containing 20 mM EDTA, chelating the $Mg^{2+}$ ions.

## DESIGNING IMMUNOASSAYS

In addition to the specific assay protocols described above, many different types of immunoassays are possible. The unique specificity of the antibody–antigen interaction and the ease with which each can be labeled permit a wide spectrum of assay designs. The techniques described above are commonly used because they have succeeded in a wide variety of applications. In special circumstances, alternative designs may be more appropriate. The design of these assays will be dictated by the specific circumstances of the individual study. Common techniques that may be useful include different binding supports, different detection methods, and different assay geometries. Suggestions for these are discussed below.

## ▪ Assay Geometry

A series of summary diagrams illustrates the range of available designs (Fig. 14.4).

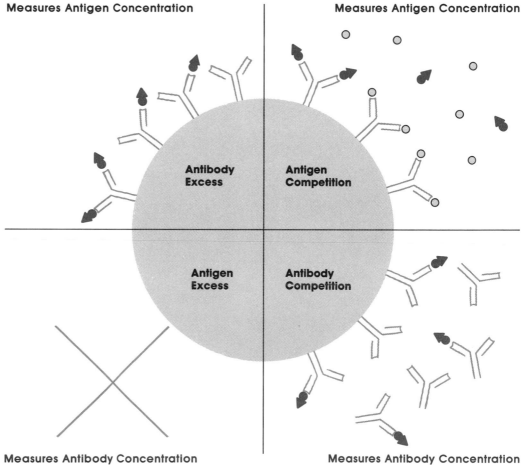

**FIGURE 14.4a**

Antigen-capture assays, direct detection.

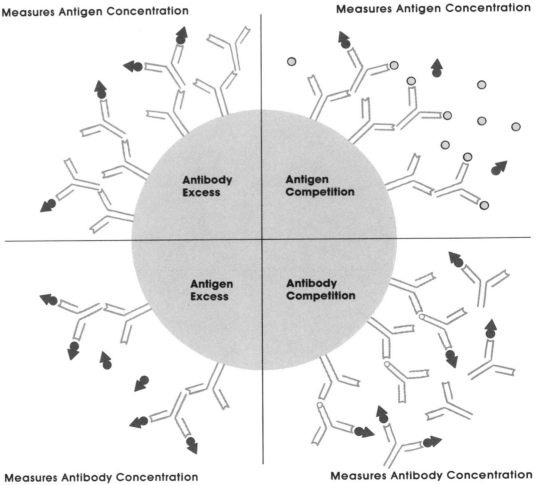

**FIGURE 14.4b**
Antigen-capture assays, indirect detection.

Measures Antigen Concentration                    Measures Antigen Concentration

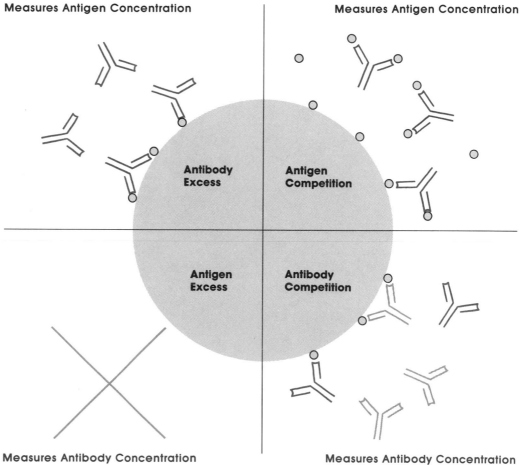

Measures Antibody Concentration              Measures Antibody Concentration

**FIGURE 14.4c**
Antibody-capture assays, direct detection.

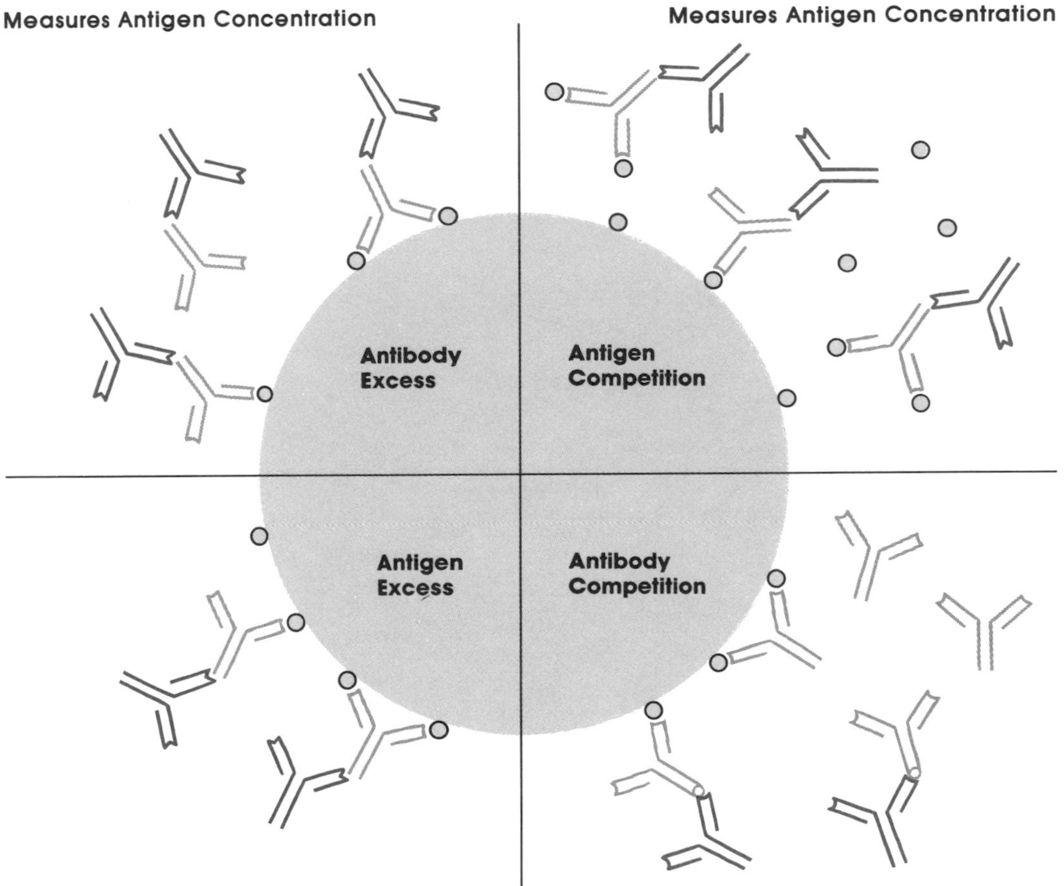

**Measures Antigen Concentration**

**Measures Antigen Concentration**

Antibody
Excess

Antigen
Competition

Antigen
Excess

Antibody
Competition

**Measures Antibody Concentration**

**Measures Antibody Concentration**

**FIGURE 14.4d**

Antibody-capture assays, indirect detection.

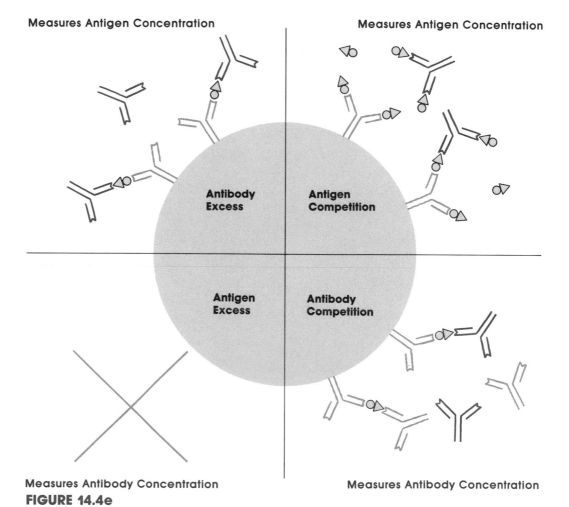

Measures Antigen Concentration     Measures Antigen Concentration

Antibody
Excess

Antigen
Competition

Antigen
Excess

Antibody
Competition

Measures Antibody Concentration     Measures Antibody Concentration

**FIGURE 14.4e**
Two-antibody assays.

## ■ Solid-Phase Matrices for Immunoassays

To allow the easy manipulation of the antibody–antigen complexes during the steps of an immunoassay, one component of the reaction is often bound to a solid substrate. The commonly used solid phases are listed in Table 14.2. Depending on the type of support and on the desired conditions, the solid support can be used in a number of different forms. Commonly, these will be microtiter plates where each well serves as a separate incubation chamber, flat sheets or membranes where the crosstalk between different reactions is limited by diffusion rates, or beads where the individual reactions are performed in separate tubes. Many of the individual matrix materials can be purchased in several of these forms, and the materials for each solid phase are discussed separately below.

**TABLE 14.2**
**Solid Supports for Immunoassays**

| Support | Forms available | Method of binding | Protein capacity |
|---|---|---|---|
| Nitrocellulose | Membranes Microtiter wells | Noncovalent | $100 \ \mu g/cm^2$ |
| Polyvinylchloride | Microtiter plates, Sheets | Noncovalent | $300 \ ng/cm^2$ |
| Polystyrene | Beads, Microtiter plates | Noncovalent | $300 \ ng/cm^2$ |
| Diazotized paper | Sheets | Covalent, through free amino groups | $>10 \ mg/cm^2$ |
| Activated beads, numerous attachments possible | Beads | Covalent, through free amino groups | $10 \ mg/ml$ |
| *S. aureus,* Protein A beads | Beads | Noncovalent, although cross-linking possible | $20 \ mg/ml$, but antibody linkage only |

## NITROCELLULOSE*

Both antigens and antibodies can be attached to nitrocellulose by incubating the proteins with the membrane for short periods of time at room temperature. Nitrocellulose binds proteins by noncovalent bonds, but the types of interactions are not clearly understood. However, for many purposes and applications, the bonds are stable, thus making nitrocellulose one of the most useful of the solid phases for immunoassays. The antibodies and antigens are not oriented with respect to their binding sites, and therefore some binding sites will be obscured. Nitrocellulose is available either as sheets or sealed to the bottom of wells in microtiter plates.

1. Cut the nitrocellulose to the appropriate size. Multiple assays can be performed on the same sheet by spacing each assay center approximately 3 to 4 mm apart. This is far enough to avoid cross-contamination between different samples due to diffusion (with 1 $\mu$l of sample). Nitrocellulose paper can be used in a dot blot apparatus by cutting it to the dimensions of the apparatus.

2. The concentration of the antibody or antigen solution will be determined by the individual assay conditions. Nitrocellulose will bind approximately 100 $\mu$g/cm$^2$. The protein can be bound in many different buffers and under many different conditions. Binding buffers should not contain extraneous proteins or detergents such as Tween 20 (low concentrations of SDS, NP-40, or Triton X-100 are fine). If no other factors need to be considered, do the binding in 10 mM sodium phosphate (pH 7) or PBS.

   If the antigen or antibody is plentiful, it can be bound to the entire sheet, regardless of how the paper will be used. Use 0.1 ml/cm$^2$. Incubate for 2 hr or overnight at room temperature in a humid atmosphere.

   If the antigen or antibody is scarce or expensive, bind it only to the areas where the test solution will be applied. This can be done in two ways, by direct application or by using a dot blot apparatus. When using a dot blot apparatus, place the sheet in the apparatus, and apply the antigen or antibody to just the region of the nitrocellulose that is exposed. Use 30 $\mu$l/well. If not using a dot blot apparatus, mark the sheet with a soft lead pencil to show the area where the assay solutions will be added. These spots should be at least 3 mm apart. Add the antigen or antibody at 1 $\mu$l/spot. Incubate for 30 min to 1 hr in a humid atmosphere.

*Sharon et al. (1979).

3. Block the remaining sites for protein binding on the nitrocellulose by incubating the sheet in blocking buffer for 2 hr at room temperature or overnight at 4°C.

There are a number of possible blocking buffers. The two most useful are 3% BSA/PBS with 0.02% sodium azide or Blotto/Tween (5% nonfat dry milk, 0.2% Tween 20, and 0.02% sodium azide in PBS). The advantages and disadvantages of these and other blocking buffers are listed on p. 496.

## NOTES

i. Nitrocellulose is the matrix of choice for dot blot assays. If high protein binding or noncovalent linkage is needed for dot blots, use diazotized paper (p. 609).

ii. Nitrocellulose can be prepared up to 1 week before the assay and stored at 4°C in PBS with 0.02% sodium azide. Sheets may be frozen at this stage and stored at −70°C. Alternatively, many antigens survive indefinitely on dry sheets of nitrocellulose.

## *POLYVINYLCHLORIDE OR POLYSTYRENE**

Polyvinylchloride (PVC) or polystyrene will bind antibodies or antigens when the proteins are simply incubated with the plastic. The bonds that hold the proteins are noncovalent, but the exact types of interactions are not known. Because these plastics can be molded into many shapes, they are a convenient solid support for many immunoassays. They are most commonly used in the 96-well microtiter plate form, where the wells provide convenient reaction chambers for immunoassays. Because PVC can be cut easily, it is an excellent matrix for radioimmunoassays in which quantitation is done by counting individual wells. Because it is not translucent, enzyme assays that will be quantitated by a plate reader should be performed in polystyrene and not PVC plates. Polystyrene is also available in many different bead sizes, where it can be used conveniently to bind antibodies or antigens for bead-based assays.

1. Bind the antibody or antigen to the bottom of each well by adding approximately 50 $\mu$l of protein solution in each well. Incubate for 2 hr at room temperature in a humid atmosphere. PVC and polystyrene will bind approximately 100 ng/well (300 ng/cm$^2$).

   The binding can be done at pHs between 6 and 9 at temperatures between 4°C and 37°C. For convenience, the binding normally is done at room temperature at neutral pH. Avoid detergents and extraneous proteins.

2. After the antigen or antibody is bound, the remaining sites on the plate must be saturated by incubating the sheet with any of the blocking buffers described on p. 496. Fill the wells to the top with blocking buffer and incubate at room temperature for at least 2 hr. If no experimental reasons suggest otherwise, use Blotto/Tween (5% nonfat dry milk, 0.2% Tween 20, 0.02% sodium azide in PBS) or 3% BSA/PBS with 0.02% sodium azide.

### NOTE

i. Plates may be prepared up to 1 week before the assay. Store in PBS with sodium azide at 4°C. Alternatively, most antibodies and some antigens survive indefinite storage at −20°C after shaking the plate dry.

*Adapted from Catt and Tregear (1967).

## DIAZOTIZED PAPER*

Diazotized paper can be used for dot blot assays in which the amount of protein bound to the sheet must be particularly high. The linkage to the paper is covalent through free amino groups on the antigen or antibody. There are a number of commercially available sources for diazotized paper. The most stable of the derivatized papers is amino-phenylthioether cellulose (APT paper). Diazotized paper is prepared by treating the APT paper with an acidic nitrite solution converting it to diazophenylthioether (DPT) paper. The coupling is then done by adding the protein solution to the paper.

1. Cut the APT paper to the appropriate size. If using the paper for multiple assays on the same sheet (dot blots), each assay should be approximately 3–4 mm apart. This is far enough to avoid cross-contamination between the different assays from diffusion of 1 $\mu$l of sample. If using the diazotized paper in a dot blot apparatus, cut to the dimensions of the apparatus.

2. Activate the paper by following the manufacturer's instructions or as below. All reactions should be done on ice.

   Wash the APT paper twice with ice-cold distilled water. Drain well and add ice-cold 1.2 M HCl (0.3 ml/cm$^2$). Add 0.3 ml of 10 mg/ml sodium nitrite (prepared fresh) per 10 ml of HCl. Incubate on ice with occasional rocking for 30 min. Wash twice with ice-cold distilled water. The paper is now ready for binding.

3. Immediately, bind the antigen or antibody to the activated paper by incubating the protein solution with the paper for 1 hr at room temperature. Diazotized paper will bind more than 10 mg/cm$^2$. The protein must be bound in a buffer that has no free amino groups. Binding buffers should not contain extraneous proteins. If no other factors need to be considered, do the binding in 100 mM sodium phosphate (pH 7).

4. Block the remaining sites for protein binding on the paper by incubating the sheet in 0.2 M ethanolamine (pH 7) for 2 hr at room temperature or overnight at 4°C.

## NOTE

i. Diazotized antibody or antigen paper can be prepared up to 1 year before the assay and stored at 4°C in PBS with 0.02% sodium azide.

*Renart et al. (1979) and Reiser and Wardale (1981) based on Alwine et al. (1977).

## ACTIVATED BEADS

A number of different types of beads can be activated by chemical treatment to produce an appropriate binding site for proteins with free amino groups. The various types of these activated beads are discussed in detail in Chapter 13 (p. 528). The major advantage when using activated beads is that the antigen or antibody is attached covalently, thus ensuring a stable reactive phase. The disadvantages are caused by the use of the beads which makes washing of numerous samples a time consuming and tedious process. Consult Chapter 13 for the proper binding and blocking conditions.

## PROTEIN A BEADS

In some assays protein A beads may be used to bind antibodies to a solid substrate. The protein A has a high affinity for the Fc region of some antibodies and thus can be used to bind antibodies without interfering with the antigen combining site. Protein A and its use are discussed in Chapter 15 (p. 616).

## ■ Alternative Detection Methods

### Detection with Microscopic Particles

Purified antigens or antibodies can be attached readily to a range of particles such as red blood cells or colored latex beads. Antibodies labeled in this way can be used to detect antigens on dot blots, in microtiter plates, or on immunoblots. While this form of labeling is not as sensitive or quantitative as those described above, it does have a number of advantages. No special apparatus is required, and the results can be scored by visual inspection. Because the label does not need to be developed with a substrate or counted, the protocols are simplified and the results obtained rapidly from very large numbers of samples.

### Agglutination

When particles as described above are coated with specific antibody, adding a solution containing multivalent antigens to a suspension of the particles will lead to cross-linking and aggregation (agglutination). These aggregates are detected by simple visual inspection. Similarly, suspensions of antigen-coated particles are agglutinated by solutions containing specific antibody. Agglutination assays can be established using essentially all the assay geometries described above.

# 15 REAGENTS

A number of specialized reagents for studying and handling antibodies have been developed. This chapter summarizes the preparation and use of four of these, including: (1) bacterial cell wall proteins that bind to antibodies, (2) anti-immunoglobulin antibodies, (3) proteolytic fragments of antibodies, and (4) acetone powders.

## ■ BACTERIAL CELL WALL PROTEINS THAT BIND ANTIBODIES

During early studies of the structure of the bacterial cell wall of *Staphylococcus aureus*, workers noticed that one fraction isolated during purification of the individual components of the wall contained a protein that would bind to rabbit and human antibodies. Surprisingly, this protein, named protein A, bound to antibodies isolated not only from animals that had been immunized with protein A but also from animals who had never seen this antigen. Early suggestions of a "natural" immunity against *S. aureus* were incorrect, because Forsgren and Sjöquist (1966) showed that protein A actually bound to the Fc portion of the antibody molecule. Similar types of proteins are found in other bacteria, in particular streptococci from the A, C, or G strains. The best studied of these is a polypeptide known as protein G. Although protein A and protein G have been studied extensively, no biological explanation for their strong affinity for antibodies has been found. Nonetheless, their high affinity for antibodies has made them important and useful reagents (reviewed by Goding 1978; Langone 1982).

## ▪ Protein A

Protein A is a 42,000-dalton polypeptide that is a normal constituent of the cell wall of *S. aureus*. Approximately 98% of all *S. aureus* isolates contain this protein. The interaction between protein A and antibodies has been studied in great detail, and the binding is very well understood, both at a practical and at a structural level. Protein A has four potential binding sites for antibodies; however, only two of them can be used at one time. Nevertheless, protein A is clearly bifunctional, allowing multimeric complexes to be formed. The binding site on the antibody molecule is found in the second and third constant regions of the heavy-chain polypeptides. This means that there are at least two binding sites on any antibody molecule for protein A, and this also provides an excellent arrangement for multimeric complex formation (Fig. 15.1).

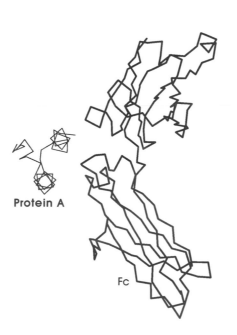

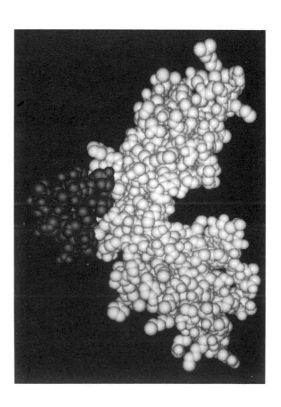

**FIGURE 15.1**
Protein A binding to Fc domain.

Cocrystals of the Fc domain and protein A have been resolved to 2.9 Å (Deisenhofer 1981). There are extensive hydrophobic interactions with both the second and third constant regions of the Fc domain. Because the Fc domains of different classes and subclasses of antibodies are different, it is not surprising that the affinity of protein A for antibodies varies with the class, subclass, and species of the immunoglobulin. When using polyclonal sera, normally it is sufficient to consider the avidity of protein A for all antibodies from that species together. Table 15.1 lists affinities from several commonly used species. In practice, sera from humans, donkeys, rabbits, dogs, pigs, and guinea pigs can be used without worry for all tests that rely on protein A. Most immunochemical assays will not be affected by using polyclonal antibodies from mice, cows, or horses. However, depending on the type of assay, antibodies from sheep, goats, rats, or chickens will often need a second antibody layer to ensure quantitative binding (see p. 622).

**TABLE 15.1**
**Protein A/G Affinities for Antibodies from Various Species**

| Species | Affinity for protein A[a] | Affinity for protein G[b] |
|---|---|---|
| **Human** | + + + + | + + + + |
| **Horse** | + + | + + + + |
| **Cow** | + + | + + + + |
| **Pig** | + + + | + + + |
| **Sheep** | + / − | + + |
| **Goat** | − | + + |
| **Rabbit** | + + + + | + + + |
| **Chicken** | − | + |
| **Hamster** | + | + + |
| **Guinea pig** | + + + + | + + |
| **Rat** | + / − | + + |
| **Mouse** | + + | + + |

[a]Kronvall et al. (1970); Richman et al. (1982).
[b]Kronvall (1973); Åkerström et al. (1985); Åkerström and Björck (1986).

When using monoclonal antibodies, more care needs to be taken if protein A binding will play a role in the assay. Table 15.2 summarizes the affinity of protein A and antibodies from the commonly used subclasses. Subclasses that are more difficult to use include human $IgG_3$, mouse $IgG_1$, and all rat subclasses except $IgG_{2c}$. In all these cases, protein A may need to be supplemented with a second bridging antibody layer (p. 622) or be substituted with another reagent.

On the practical side, three characteristics of protein A have made its use in immunochemical studies particularly valuable. First, because the binding site on the antibody is found in the Fc region, the interaction with protein A does not change the ability of the antibody to combine with the antigen. Second, even a highly denatured protein A molecule is easily renatured. Protein A treated with 4 M urea, 4 M thiocyanate, 6 M guanidine HCl, or pH 2.5 can be renatured to full binding capacity. The third important characteristic of protein A is that, although the affinity for antibody is high, the antibody–antigen bond can be broken effectively by lowering the pH. These characteristics have made protein A extremely useful in the study of antibodies and antibody–antigen interactions.

**TABLE 15.2**
**Protein A/G Affinities for Various Monoclonal Antibodies**

| Antibody | Affinity for protein A[a] | Affinity for protein G[b] |
|---|---|---|
| **Human IgG$_1$** | + + + + | + + + + |
| **Human IgG$_2$** | + + + + | + + + + |
| **Human IgG$_3$** | − | + + + + |
| **Human IgG$_4$** | + + + + | + + + + |
| **Rat IgG$_1$** | − | + |
| **Rat IgG$_{2a}$** | − | + + + + |
| **Rat IgG$_{2b}$** | − | + + |
| **Rat IgG$_{2c}$** | + | + + |
| **Mouse IgG$_1$** | + | + + + + |
| **Mouse IgG$_{2a}$** | + + + + | + + + + |
| **Mouse IgG$_{2b}$** | + + + | + + + |
| **Mouse IgG$_3$** | + + | + + + |

[a]Langone (1978); Ey et al. (1978).
[b]Björck and Kronvall (1984); Åkerström and Björck (1986).

Protein A has been used in a wide variety of applications in immuno-chemical assays. When coupled to a radioactive, enzymatic, or fluorescent tag, it is an excellent reagent to locate antibodies with high affinity for protein A. All of these reagents are available commercially, although the coupling can be done using any of the techniques for labeling antibodies describing on p. 319. The pH dependence and specificity of the antibody–protein A interaction have made protein A columns one of the favored methods for antibody purification (p. 309). In addition, one valuable approach to collecting immune complexes has been the use of protein A bound to solid substrates. Originally, this was described for the *S. aureus* bacterium itself (Kessler 1975, 1976). In this case, bacteria carrying high concentrations of protein A on the cell surface are used to collect antibody–antigen complexes. The bacteria provide a solid phase that is manipulated easily. In recent years, this approach has been applied to other solid phases by coupling purified protein A to many types of secondary targets.

Although the cost of protein A or its derivatives has been high, the gene for protein A has been cloned in several laboratories, and purified protein A is now available from more economical bacterial over-expression systems. In practice, this should mean that the cost will continue to drop in the next few years, making other applications feasible.

Fixed *S. aureus* for collecting immune complexes is now available commercially, but can be prepared conveniently using the procedure below.

## PREPARING S. AUREUS FOR COLLECTING IMMUNE COMPLEXES*

In this procedure, the protein A molecules found on the surface of *S. aureus* Cowan I (SAC; ATCC 12598, NCTC 8530) are fixed by treating the cells with formaldehyde. The bacteria are then killed by heat treatment, and the resulting protein A solid phase can be used to bind to antibodies or antibody–antigen complexes.

SAC express approximately 80,000 functional antibody binding sites on each bacterium. However, the capacity of SAC will vary depending on the species, class, and subclass of the antibody being used. For high-affinity antibodies such as those from rabbits, humans, or $IgG_{2a}$ subclass of mice, 10 $\mu$l of a 10% SAC suspension (1 $\mu$l of SAC pellet) will bind approximately 10–15 $\mu$g of antibody at full capacity. In practice, the binding of antibodies to protein A is normally driven by an excess of SAC, so these exact volumes may not be used in any particular assay; however, these figures do point out the high capacity of this reagent.

> **Caution** *S. aureus* is a pathogenic organism and should be handled using good microbiological techniques.

1. Streak a blood agar plate for single colonies of the *S. aureus* Cowan I bacteria. Use a single colony to inoculate a 50-ml culture of Penassay media (0.15% beef extract, 0.15% yeast extract, 0.5% peptone, 0.1% glucose, 0.35% NaCl, 0.37% $K_2HPO_4$, and 0.13% $KH_2PO_4$). Grow overnight to saturation with aeration.

2. Use the 50-ml culture to inoculate one 500-ml culture of Penassay media in a 2-liter flask. Grow to saturation with aeration.

3. Harvest the bacteria by centrifugation in a preweighed centrifuge bottle at 7000*g*. Remove the spent media and dispose of it, using proper techniques to kill any bacteria left in the supernatant. Resuspend the cell pellet using a wide-bore pipet in 100 ml of PBS with 0.02% sodium azide.

4. Spin and wash the cells as before. After the second wash, weigh the bottle with the bacterial cell pellet and calculate the yield. Good growth should yield approximately 6 grams wet weight/500 ml culture. Resuspend the bacteria to 10% wt/vol in PBS with 0.02% sodium azide.

*Kessler 1981.

5. Combine the bacteria from each flask and place them in a flask or beaker in which they can be stirred rapidly at room temperature. To the stirring suspension, slowly add formaldehyde to a final concentration of 1.5%. Stir for 90 min at room temperature. (Formaldehyde is commonly sold as a 37% solution.)

6. Spin the cells at 7000g and resuspend the pellet at 10% wt/vol in PBS with 0.02% sodium azide. The bacteria are now ready for heat treatment. Prepare an 80°C water bath. The SAC suspension is placed in a glass beaker or flask to allow good heat transfer. The depth of the SAC suspension in the container is kept at approximately 2 cm. Place a thermometer in the SAC. Incubate the SAC at 80°C with constant agitation. Monitor the temperature and continue the incubation for 5 min after the temperature passes 75°C. Transfer to an ice bath to cool. The *S. aureus* are now killed and are nonpathogenic.

7. Spin and wash the SAC as before. Repeat.

8. Prepare a 10% vol/vol suspension of the SAC. Dispense in convenient volumes and freeze at −70°C. SAC is stable at −70°C for over 1 year. SAC stored at −20°C will aggregate and become more difficult to use.

## NOTE

i. Resuspension of large SAC pellets is difficult. A small blender used at low speeds will help. Appropriate precautions with the SAC aerosols should be taken, however.

## ▨ Protein G

Protein G is a 30,000- to 35,000-dalton protein isolated from the cell wall of β-hemolytic streptococci of the C or G strains (Kronvall 1973). Less is known about protein G than about protein A, primarily because protein G was not purified until 1984 (Björck and Kronvall 1984). The interest in protein G and in other less well-characterized bacterial cell wall proteins is that these proteins have different affinities for antibodies compared to protein A. Many antibodies from species or subclasses that do not bind well to protein A bind very well to protein G. Tables 15.1 and 15.2 summarize the affinity of protein G for a spectrum of antibody sources.

Like protein A, the gene for protein G has been cloned recently, and protein G is now available in purified form, isolated from bacterial overexpression systems. This should allow more convenient applications and should speed the collection of information on its use in immunochemical tests. One disadvantage concerning protein G is that it contains a second binding site that will interact with albumin (Björck et al. 1987). Recently, the antibody binding domain has been separated from the albumin domain by genetic manipulations, and this should increase the usefulness of protein G.

## ▨ ANTI-IMMUNOGLOBULIN ANTIBODIES

When purified antibodies from one species are injected into a distant species, the antibodies are often recognized as foreign antigens and elicit a strong humoral response. This characteristic can be used to generate specific anti-immunoglobulin reagents. Using anti-immunoglobulin antibodies solves many of the difficulties in locating and quantitating antibodies when these molecules are found in complex mixtures of serum or tissue culture proteins. The specificity of the anti-immunoglobulin antibodies can be used in many of the same ways and with many of the same advantages as anti-antigen antibodies. One of the major advantages of using these anti-immunoglobulin reagents is that with careful preparation they can be made species, class, subclass, or domain specific. Therefore, anti-immunoglobulin antibodies are prepared for two purposes, either as general reagents used to bind all antibodies or as selective reagents used to identify one type of antibody molecule.

Using antibodies as immunogens has proven to be one of the best experimental systems for studying how a region of a protein is recognized as foreign and how it elicits a humoral response. In practical terms, if a region of the injected antibody molecule is identical to or is sufficiently similar to the corresponding region of the antibodies from the host animal, this region will not be seen as foreign and will not elicit a response. Therefore, the regions that can be recognized as foreign in the generation of anti-immunoglobulin antibodies are determined by the genetic differences between the animal from which the immunogen is prepared and the immunized animal. In addition, as the antibodies become more dissimilar, the number of regions that will be recognized as foreign increases.

In general, animals from different genuses will respond to a number of epitopes on both the heavy and light chains, and these immunogens will induce a strong humoral response. The epitopes recognized in these types of immunizations are described as *xenotypic*. Immunoglobulins isolated from one strain of a species and injected into another strain will induce a response to *allotypic* epitopes. When immunoglobulins are isolated from one inbred strain and injected into the same strain, the epitopes that are recognized are known as *idiotypic* epitopes. Idiotypic epitopes are associated with the hypervariable regions that form the antigen combining site of a particular antibody (see Chapters 2 and 3). Idiotypic epitopes are a subset of the allotypic epitopes which in turn are a subset of the xenotypic epitopes.

One important aspect in the generation of an anti-immunoglobulin response that is not obvious from the above discussion is how similar antibodies are from species to species. This becomes apparent when analyzing the cross-reactions of anti-immunoglobulin antibodies. Although a region may be seen as foreign between one species and another (i.e., the region is different), this does not mean that the same region is not conserved in other species. For example, in a high-titered rabbit anti-mouse immunoglobulin serum, as many as 10% of the antibodies will cross-react with human antibodies. In many instances, these cross-reactions will not lead to any technical problems, but in some cases, such as staining tissue sections, these cross-reactions will ruin an experiment. These types of cross-reactions can be found at all levels between different subclasses, classes, light chains, and species. Eliminating these cross-reactions is one of the problems in preparing antibodies that are specific for different species, different regions of the antibody molecule, or different types of heavy or light chain. To isolate antibodies with these specificities, three steps may be taken. First, the immunization should be targeted to the desired region or polypeptide chain by only injecting this portion of the antibody. Second, the resulting antibodies must then be depleted of all the cross-reacting antibodies by adsorption on samples of the cross-reacting reagents. The last step is to purify the antibodies by binding to the region of interest and then eluting the purified specific antibodies (p. 313).

## *PREPARING ANTI-IMMUNOGLOBULIN ANTIBODIES*

The major consideration when preparing anti-immunoglobulin antibodies is the host in which the antibodies will be raised. This decision should be made on the ease of using the resulting antibodies. In general, although antibodies can be raised in closely related species, most immunizations are done in distant species. Unless there are extenuating circumstances, larger animals yield more sera and should be used if possible. Most of these immunizations are done in rabbits, goats, or sheep. Of these, if protein A will be used in any of the assays, only rabbits produce antibodies that bind protein A with high affinity and are, consequently, the animal of choice. Rabbits are commonly used to prepare anti-mouse, -rat, and -human immunoglobulin antibodies, while goat, sheep, or pigs are used to produce anti-rabbit immunoglobulin antibodies.

Anti-immunoglobulin antibodies are available commercially, and unless large quantities are needed or special antibody couplings are planned, it is seldom worthwhile to prepare these reagents yourself. However, since they are relatively easy to prepare, a suitable protocol is outlined below.

1. Prepare a large batch of antibodies. See p. 288 for appropriate purification protocols. Dissolve the immunoglobulins at approximately 2 mg/ml in PBS.

2. At the time of each injection, make an emulsion of 500 $\mu$l of the immunoglobulin solution with 500 $\mu$l of Freund's adjuvant. The first injection should be done with complete Freund's; all subsequent injections should be done with incomplete.

   The emulsion can be prepared by taking the 500 $\mu$l of each suspension into separate 1-ml syringe barrels. These two solutions are immiscible but with mixing will form an emulsion. This is most easily done by pushing the mixture between the two 1-ml syringes via a two- or three-way valve as shown in Figure 6.5, p. 158). Remove all air from both syringes before connecting to the valve. Make sure the connections are tightened securely. Depress the plunger on the syringe containing the aqueous antigen solution first. Continue to push the mixture between the two syringes until it becomes difficult (approximately 10–20 times). Push the emulsion into one syringe. Remove the valve and the other syringe, and place a 23-gauge needle on the syringe. Displace any air that may be held in the barrel or in the needle by carefully depressing the plunger until the emulsion reaches the needle.

3. Inject the emulsion into multiple subcutaneous sites in a female New Zealand white rabbit. For most purposes, several rabbits should be injected at the same time to lessen the impact of animal-to-animal differences.

4. Take sample test bleeds from the rabbits 14 days after each injection. Compare the titers using one of the standard quantitative tests or in the tests in which they will be used.

5. Let the rabbits rest for 6 weeks between injections.

6. When the titers have increased to usable levels, each rabbit should be bled to obtain approximately 50 ml of blood. Continue to repeat the injections while collecting the serum 14 days after each injection.

7. Rabbits can continue to produce good anti-immunoglobulin antibodies for over a year. When the last boost has been made, sacrifice the rabbit and collect all the blood. The final bleed will yield approximately 100 ml of serum.

## NOTES

i. For protocols in each of the steps above, consult Chapter 5.

ii. This protocol can be modified to produce antibodies that are specific for the antibodies from a particular species, class, or subclass, the Fc or Fab region, or the particular light chain. Begin the immunizations with the desired region or polypeptide chain. After a good immune response has developed, the resulting antibodies are first adsorbed against other regions or polypeptides to remove cross-reacting activities. Then the antibodies are affinity-purified on the starting immunogen (p. 313).

# PROTEOLYTIC FRAGMENTS OF ANTIBODIES

The use of an intact antibody molecule in some immunochemical techniques introduces certain problems. For example, using multivalent antibodies to identify cell-surface antigens on mammalian cells often will lead to capping and internalization of the antigen. Also, many cells have receptors for binding to the Fc portion of antibodies. These problems and others like them can be overcome by using fragments of the antibody. With carefully chosen conditions, antibodies can be digested with proteases to release the various protein domains, thus allowing the use of small fragments of the antibody molecule for various immunochemical tests. Antibodies can be conveniently thought of as having three protein domains, two identical antigen binding sites called Fab fragments and an effector domain known as the Fc fragment. More information on the structure of the antibody molecule is found in Chapter 2.

Antibodies usually are fragmented by partial digestion with either papain or pepsin (Fig. 15.2). Papain treatment will produce two Fab fragments and one Fc fragment. Pepsin treatment can be used to release the two antigen binding domains still bound together, a fragment known as a F(ab)$_2$. Fab and F(ab)$_2$ fragments both can be used in immunochemical tests where the presence of the Fc fragment will produce anomalous results. Fab fragments are used when the multivalent binding of antibodies must be avoided.

Rabbit and human IgG molecules exhibit only a relatively few cleavage sites for papain or pepsin digestion and therefore are relatively easy to handle in these techniques. Mouse monoclonal antibodies from the different subclasses show a wide degree of variation and therefore preliminary tests will need to be run to assess each digestion protocol. For an excellent review of the digestion of mouse monoclonal antibodies see Parham (1986).

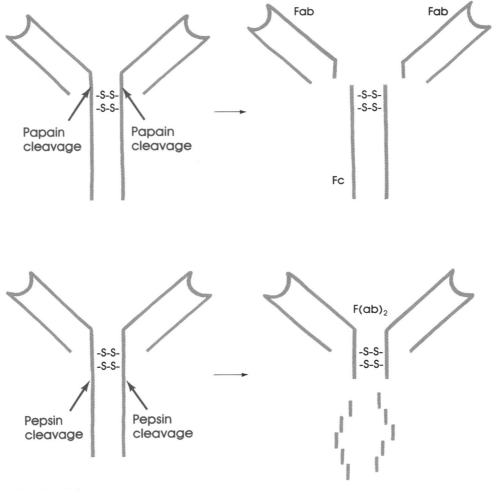

**FIGURE 15.2**
Papain and pepsin cleavage of antibodies.

## *DIGESTING ANTIBODIES WITH PAPAIN TO ISOLATE Fab FRAGMENTS*

The principal sites of papain cleavage are found on the amino-terminal side of the disulfide bonds that hold the two heavy chains together (see Fig. 15.2). Therefore, digestion with papain releases two antigen binding domains and one Fc fragment. Human, rabbit, and mouse $IgG_1$ antibodies are relatively resistant to cleavage at secondary sites, and therefore determining the correct conditions for papain treatment is relatively easy. The mouse immunoglobulin subclasses fall into this order on their resistance to secondary digestion, $IgG_1 > IgG_{2a} > IgG_3 > IgG_{2b}$.

1. Prior to large-scale digestion, run small preliminary tests to determine and verify the amount of papain to use and the length of digestion. All digestions can be assayed by comparing the resultant antibody fragments on SDS-polyacrylamide gels (p. 636). Run one sample with dithiothreitol and one without. Nonreduced IgGs migrate at approximately 150,000 daltons, nonreduced Fab fragments migrate at approximately 50,000, and nonreduced Fc fragments at approximately 25,000. Under reducing conditions, IgG yields two bands, one at about 50,000 and one at 25,000; Fab fragments yield a doublet of bands at about 25,000; and Fc fragments yield one band that runs at about 25,000, but normally slower than the Fab doublet.

2. Prepare a solution of IgG at approximately 5 mg/ml in 100 mM sodium acetate (pH 5.5). Add 1/20 volume of cysteine from a 1 M stock. Add 1/20 volume of EDTA from a 20 mM stock. Final concentrations are 50 mM cysteine and 1 mM EDTA.

3. Add 10 μg of papain for each milligram of antibody. Seal the tube, mix well, and place in a 37°C water bath.

4. Incubate for the appropriate length of time. See step 1. For rabbit, human, and mouse $IgG_1$ antibodies, 8–12 hr is a suggested starting point. For others, the incubation time may be 1–6 hr.

5. Add iodoacetamide to a final concentration of 75 mM. Incubate at room temperature for 30 min.

6. The Fab fragments must be purified from any remaining intact antibodies. If the antibodies bind to protein A, the intact antibodies and Fc fragments can be removed by chromatography on protein A (p. 309). For antibodies that do not bind to protein, a two-column purification is suggested. First, separate on the basis of size using a gel filtration column (G100 or equivalent is appropriate) and then by ion exchange on DEAE (p. 301).

## DIGESTING ANTIBODIES WITH PEPSIN TO ISOLATE F(ab)₂ FRAGMENTS

Pepsin cleaves antibodies on the carboxy-terminal side of the disulfide bonds that hold the two heavy chains together (see Figure 15.2) and therefore digestion produces bivalent fragments with both antigen binding sites intact known as the $F(ab)_2$ fragments. The remaining antibody domain, which is related to the Fc fragment, may or may not contain secondary sites for pepsin cleavage. Often this domain is digested quickly to yield small proteolytic products. Human, rabbit, and mouse $IgG_1$ antibodies are relatively resistant to cleavage at secondary sites within the $F(ab)_2$, and, therefore, determining the correct conditions for pepsin treatment is relatively easy. The mouse subclasses fall into this order on their resistance to secondary digestion, $IgG_1 > IgG_{2a} > IgG_3 > IgG_{2b}$.

1. Prior to large-scale digestion, run smaller tests to determine and verify the amount of pepsin to use and the length of digestion. All digestions can be assayed by comparing the resultant antibody fragments on SDS-polyacrylamide gels (p. 636). Run one sample with dithiothreitol and one without. Nonreduced IgGs migrate at approximately 150,000 daltons, nonreduced $F(ab)_2$ fragments migrate at approximately 110,000, and nonreduced Fc fragments at approximately 25,000. Under reducing conditions, IgG yields two bands, one at about 50,000 and one at 25,000, $F(ab)_2$ fragments yield a doublet of bands at about 25,000, and Fc fragments yield one band that runs at about 25,000, but normally behind the Fab doublet.

2. Prepare a solution of IgG at approximately 5 mg/ml in 100 mM sodium citrate (pH 3.5).

3. Add 5 μg of papain for each milligram of antibody. Seal the tube, mix well, and place in a 37°C water bath.

4. Incubate for the appropriate length of time. See step 1. For rabbit, human, and mouse $IgG_1$ antibodies 12–24 hr is a suggested starting point. For others, the incubation time may be considerably shorter.

5. Stop the reaction by adding 1/10 volume of 3.0 M Tris (pH 8.8). Centrifuge at 10,000*g* for 30 min.

6. The F(ab)$_2$ fragments must be purified from any remaining intact antibodies. If the antibodies bind to protein A, the intact antibodies and Fc fragments can be removed by chromatography on protein A (p. 309). For antibodies that do not bind to protein A, first, separate on an ion-exchange column of DEAE (p. 301). Then, if necessary, separate on the basis of size using a gel filtration column (G100 or equivalent is appropriate).

## NOTE

i. For mouse IgG$_{2b}$, use elastase in place of pepsin. Exchange the citrate buffer for Tris buffer at pH 8.8. Stop by the addition of phenylmethyl sulfonyl fluoride (see p. 677).

## ▪ ADSORPTION TO REMOVE NONSPECIFIC BINDING

Many antibody solutions contain a small proportion of antibodies that will bind either nonspecifically to certain antigens (e.g., charged residues) or specifically to contaminating antigens in the preparation (e.g., low levels of anti-bacterial antibodies present in many sera). Although extensive purification could be used to separate these activities, a simple way to remove or to disguise these activities is to block their binding by including a saturating amount of competitor protein that does not contain the antigen of interest. Traditionally, this has been done by either adding a source of nonspecific protein (i.e., BSA, etc.) or including a more complex suspension of proteins. This second approach can be achieved by adding a suspension of proteins prepared by acetone precipitation. Useful acetone powders to have on hand include mammalian liver powder, bacterial whole-cell powder, or yeast whole-cell powder.

## PREPARING ACETONE POWDERS

1. Prepare a fine suspension of tissue or cells in saline (0.9% NaCl). If a tissue is used, either discard any fibrous material by teasing it away from the remainder of the tissue or homogenize the whole tissue by placing it in a blender. (One gram of tissue should be resuspended in approximately 1.0 ml of saline.)

2. Transfer the tissue/saline suspension to ice for 5 min.

3. Add 8 ml of acetone ($-20°C$) per 2 ml of cell suspension. Mix vigorously. Incubate at $0°C$ for 30 min with occasional vigorous mixing.

4. Collect the precipitate by centrifugation at 10,000g for 10 min. Remove and discard the supernatant.

5. Resuspend the pellet with fresh acetone ($-20°C$) and mix vigorously. Allow to sit at $0°C$ for 10 min.

6. Spin at 10,000g for 10 min. Transfer the pellet to a clean piece of filter paper. Spread the precipitate and allow to air-dry at room temperature. As it dries, continue to spread and disperse the pellet.

7. After the powder is dry, transfer it to an airtight container. Remove any large pieces that will not break into a fine powder.

   Yield is approximately 10–20% of the original wet weight.

   To use acetone powders, add to a final concentration of 1%. Incubate for 30 min at $4°C$. Spin at 10,000g for 10 min. Use the supernatant as a source of antibodies for your assay.

# Appendix I

# ELECTROPHORESIS

## SDS-POLYACRYLAMIDE GEL ELECTROPHORESIS*

1. If using a commercial apparatus, assemble the gel plates following the manufacturer's instructions. If not, arrange gel plates as below. The plates should be clean and free of detergents. Silicon rubber tubing can be placed between the glass plates to make a good, water-tight seal.

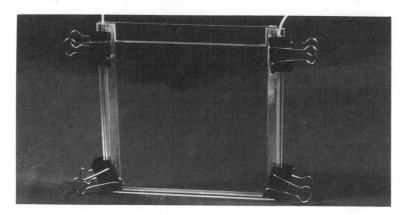

2. Prepare the acrylamide solution for the separating gel using the values given on p. 639.

3. Pour the acrylamide solution between the plates. The meniscus of the acrylamide solution should be far enough below the top of the notched plate to allow for the length of the teeth on the comb plus 1 cm. Carefully overlay the acrylamide solution with isobutanol or water. For gels made with acrylamide concentrations lower than 8% use water; for gels of 10% or greater use isobutanol (or water-saturated isobutanol). The overlaying solution creates a barrier to oxygen, which inhibits the polymerization of the acrylamide. Place the gel in a vertical position at room temperature.

4. After the gel has set (about 20–30 min), pour off the overlay and wash the top of the separating gel several times with distilled water. Drain well or dry with the edge of a paper towel.

5. Prepare the acrylamide solution for the stacking gel from the values on p. 639. Pour the solution for the stack directly onto the polymerized separating gel. Place the appropriate comb into the gel solution, being careful not to trap any bubbles. To clean the comb, wash with $H_2O$ and ethanol. Place the gel in a vertical position at room temperature. The stacking gel will set in approximately 10 min.

6. Heat samples in sample buffer at 85°C for 10 min.

*Laemmli (1970), based on Smithies (1955), Raymond and Weintraub (1959), Davis (1964), Ornstein (1964), Summers et al. (1965).

7. After the stacking gel has set, carefully remove the comb. Wash the wells immediately with distilled water to remove unpolymerized acrylamide. Straighten the teeth of the stacking gel, if necessary, and place in an appropriate gel box with running buffer in the bottom reservoir. Add running buffer to top reservoir. Any bubbles caught between the plates at the bottom of the gel can be removed by squirting running buffer through a syringe fitted with a bent needle.

8. Load samples into the bottom of the wells. The samples can be conveniently loaded using either: (1) a microliter syringe (washed by pipetting in the bottom reservoir buffer between each sample), (2) a micropipetter fitted with a long narrow tip, or (3) Teflon tubing fixed to the end of a microliter syringe (24-gauge, thin-walled spaghetti tubing fits snugly on 22-gauge needles; discard each piece after a single use).

9. Start electrophoresis at 100–125 V. After the dye front has moved into the separating gel, increase to 200 V. The plates will become warm to the touch, but should not be hot. If the plates are too warm to touch easily, lower the voltage.

10. When the dye front reaches the bottom of the gel, turn off the power pack. Remove the gel plates and gently pry the plates apart. Use a spatula or similar tool to separate the plates (not at an ear). Cut a corner from the bottom of the gel that is closest to the number 1 well.

**The gels are now ready for fixing (p. 655), staining (p. 649), transfer (p. 486), fluorography (p. 647), or autoradiography (p. 657).**

## NOTES

i. Spacers can be glued directly onto the glass plates. Two good choices of glues are 3M Scotch epoxy 2216 B/A or quick-drying CA9. Be sure both the gel plates and the spacers are clean before glueing. Clamp the spacers tightly to the plates before the glue hardens. This will prevent the hardened glue from increasing the thickness of the gel.

ii. If protein samples are resuspended in sample buffer lacking any reducing agents and the gels are run as normal, proteins that are linked by disulfides will run as discrete multimers. Comparing samples, plus and minus reduction, can then be used to study protein complexes held by disulfide bridges. **Caution** $\beta$-Mercaptoethanol or dithiothreitol will diffuse between lanes, and if samples with and without reducing agents are to be compared, they should be separated by several lanes or run on different gels.

iii. Effective separation of some proteins can be enhanced by varying the ratio of the bis-acrylamide cross-linker ($N,N'$-methylene-bis-acrylamide) to the acrylamide monomer. The gel formulations described on p. 639

yield 30% total acrylamide composed of 29.2% monomer to 0.8% cross-linker, or a ratio of 36.5:1. Higher ratios will allow most proteins to migrate faster and also will promote easier transfer for immunoblots. Lower ratios yield firmer gels and slower rates of migration. Not all proteins will be affected equally, and changing the ratio of monomer to cross-linker can be used to gain better separation between some proteins.

iv. When proteins are difficult to elute from gels, the cross-linker can be changed to ones that can be cleaved by simple chemical modifications. Common cleavable cross-linkers are BAC (*N,N'*-bis-acrylylcystamine) and DATD (*N,N'*-diallyltartardiamide). See Hansen (1976) and Hansen et al. (1980) for more information on BAC or Anker (1970) for more information on DATD.

v. When studying two or more proteins with widely differing molecular weights, gels can be prepared with higher acrylamide concentrations at the bottom of the gel compared with the top. These gradient gels can be prepared from any combination of acrylamide concentrations and are poured by using a gradient maker using standard recipes. Be certain to clean the gradient maker before using and to wash the gradient maker before the remaining acrylamide polymerizes in the apparatus.

vi. In addition to the Tris/glycine buffers suggested here, many different combinations of buffers can be used. See p. 640 for a partial list or Jovin et al. (1970) for a more detailed list.

vii. For greater reproducibility from day to day, the acrylamide, bis-acrylamide, buffer, and water can be prepared in large batches and either stored at 4°C for 1 month or frozen in aliqouts and used indefinitely. Remove the required amount, warm to room temperature, and add ammonium persulfate and TEMED immediately before use.

## Solutions for Tris/Glycine SDS-Polyacrylamide Gel Electrophoresis

| 6% | 5 ml | 10 ml | 15 ml | 20 ml | 25 ml | 30 ml | 40 ml | 50 ml |
|---|---|---|---|---|---|---|---|---|
| H₂O | 2.7 | 5.3 | 8.0 | 10.6 | 13.3 | 15.9 | 21.1 | 26.5 |
| 30% Acrylamide mix[a] | 1.0 | 2.0 | 3.0 | 4.0 | 5.0 | 6.0 | 8.0 | 10.0 |
| 1.5 *M* Tris (pH 8.8) | 1.3 | 2.5 | 3.8 | 5.0 | 6.3 | 7.5 | 10.0 | 12.5 |
| 10% SDS[b] | 0.05 | 0.1 | 0.15 | 0.2 | 0.25 | 0.3 | 0.4 | 0.5 |
| 10% APS[c] | 0.05 | 0.1 | 0.15 | 0.2 | 0.25 | 0.3 | 0.4 | 0.5 |
| TEMED[d] | 0.004 | 0.008 | 0.012 | 0.016 | 0.02 | 0.024 | 0.032 | 0.04 |

| 8% | 5 ml | 10 ml | 15 ml | 20 ml | 25 ml | 30 ml | 40 ml | 50 ml |
|---|---|---|---|---|---|---|---|---|
| H₂O | 2.3 | 4.6 | 7.0 | 9.3 | 11.6 | 13.9 | 18.6 | 23.2 |
| 30% Acrylamide mix | 1.3 | 2.7 | 4.0 | 5.3 | 6.7 | 8.0 | 10.7 | 13.4 |
| 1.5 *M* Tris (pH 8.8) | 1.3 | 2.5 | 3.8 | 5.0 | 6.3 | 7.5 | 10.0 | 12.5 |
| 10% SDS | 0.05 | 0.1 | 0.15 | 0.2 | 0.25 | 0.3 | 0.4 | 0.5 |
| 10% APS | 0.05 | 0.1 | 0.15 | 0.2 | 0.25 | 0.3 | 0.4 | 0.5 |
| TEMED | 0.003 | 0.006 | 0.009 | 0.012 | 0.015 | 0.018 | 0.024 | 0.03 |

| 10% | 5 ml | 10 ml | 15 ml | 20 ml | 25 ml | 30 ml | 40 ml | 50 ml |
|---|---|---|---|---|---|---|---|---|
| H₂O | 2.0 | 4.0 | 5.9 | 7.9 | 9.9 | 11.9 | 15.8 | 20 |
| 30% Acrylamide mix | 1.7 | 3.3 | 5 | 6.7 | 8.3 | 10.0 | 13.3 | 16.6 |
| 1.5 *M* Tris (pH 8.8) | 1.3 | 2.5 | 3.8 | 5.0 | 6.3 | 7.5 | 10.0 | 12.5 |
| 10% SDS | 0.05 | 0.1 | 0.15 | 0.2 | 0.25 | 0.3 | 0.4 | 0.5 |
| 10% APS | 0.05 | 0.1 | 0.15 | 0.2 | 0.25 | 0.3 | 0.4 | 0.5 |
| TEMED | 0.002 | 0.004 | 0.006 | 0.008 | 0.01 | 0.012 | 0.016 | 0.02 |

| 12% | 5 ml | 10 ml | 15 ml | 20 ml | 25 ml | 30 ml | 40 ml | 50 ml |
|---|---|---|---|---|---|---|---|---|
| H₂O | 1.7 | 3.3 | 5.0 | 6.6 | 8.3 | 9.9 | 13.2 | 16.4 |
| 30% Acrylamide mix | 2.0 | 4.0 | 6.0 | 8.0 | 10.0 | 12.0 | 14.0 | 20.0 |
| 1.5 *M* Tris (pH 8.8) | 1.3 | 2.5 | 3.8 | 5.0 | 6.3 | 7.5 | 10.0 | 12.5 |
| 10% SDS | 0.05 | 0.1 | 0.15 | 0.2 | 0.25 | 0.3 | 0.4 | 0.5 |
| 10% APS | 0.05 | 0.1 | 0.15 | 0.2 | 0.25 | 0.3 | 0.4 | 0.5 |
| TEMED | 0.002 | 0.004 | 0.006 | 0.008 | 0.01 | 0.012 | 0.016 | 0.02 |

| 15% | 5 ml | 10 ml | 15 ml | 20 ml | 25 ml | 30 ml | 40 ml | 50 ml |
|---|---|---|---|---|---|---|---|---|
| H₂O | 1.2 | 2.3 | 3.5 | 4.6 | 5.7 | 6.9 | 9.2 | 11.4 |
| 30% Acrylamide mix | 2.5 | 5.0 | 7.5 | 10.0 | 12.5 | 25.0 | 20.0 | 25.0 |
| 1.5 *M* Tris (pH 8.8) | 1.3 | 2.5 | 3.8 | 5.0 | 6.3 | 7.5 | 10.0 | 12.5 |
| 10% SDS | 0.05 | 0.1 | 0.15 | 0.2 | 0.25 | 0.3 | 0.4 | 0.5 |
| 10% APS | 0.05 | 0.1 | 0.15 | 0.2 | 0.25 | 0.3 | 0.4 | 0.5 |
| TEMED | 0.002 | 0.004 | 0.006 | 0.008 | 0.01 | 0.012 | 0.016 | 0.02 |

| STACK | 1 ml | 2 ml | 3 ml | 4 ml | 5 ml | 6 ml | 8 ml | 10 ml |
|---|---|---|---|---|---|---|---|---|
| H₂O | 0.68 | 1.4 | 2.1 | 2.7 | 3.4 | 4.1 | 5.5 | 6.8 |
| 30% Acrylamide mix | 0.17 | 0.33 | 0.5 | 0.67 | 0.83 | 1.0 | 1.3 | 1.7 |
| 1.0 *M* Tris (pH 6.8) | 0.13 | 0.25 | 0.38 | 0.5 | 0.63 | 0.75 | 1.0 | 1.25 |
| 10% SDS | 0.01 | 0.02 | 0.03 | 0.04 | 0.05 | 0.06 | 0.08 | 0.1 |
| 10% APS | 0.01 | 0.02 | 0.03 | 0.04 | 0.05 | 0.06 | 0.08 | 0.1 |
| TEMED | 0.001 | 0.002 | 0.003 | 0.004 | 0.005 | 0.006 | 0.008 | 0.01 |

[a]Commonly 29.2% acrylamide and 0.8% *N,N'*-methylene-bis-acrylamide; [b]Sodium dodecyl sulfate; [c]Ammonium persulfate; [d]*N,N,N',N'*-Tetramethylethylenediamine.

## Alternative Buffer Systems for Discontinuous Polyacrylamide Gel Electrophoresis

| System | Running buffer | Separating gel buffer | Stacking gel buffer | Sample buffer[a] |
|---|---|---|---|---|
| SDS/phosphate[b] | 100 mM sodium phosphate (pH 7.2) 0.2% SDS | 100 mM sodium phosphate (pH 7.2) 0.2% SDS | No stack | 100 mM sodium phosphate (pH 7.2) 1% SDS 10% glycerol 0.001% bromphenol blue |
| Urea/glycerol[c] | 50 mM Tris, 100 mM glycine (pH 8.7) | 0.375 M Tris, 40% glycerol (pH 9.1) | 0.12 M Tris, 8.5 M urea (pH 6.8) | 0.12 M Tris (pH 6.8) 8.5 M urea 0.001% bromphenol blue |
| Tris/borate[d] | 65 mM Tris, 0.1% SDS, (pH to 8.3 with boric acid) | 0.375 M Tris, (pH to 8.3 with sulfuric acid) | 75 mM Tris, (pH to 8.3 with sulfuric acid) | 75 mM Tris, 2% SDS 10% glycerol 0.001% bromphenol blue (pH to 8.3 with sulfuric acid) |

[a]All samples buffers can be supplemented with 100 mM dithiothreitol, if desired.
[b]Weber and Osborn (1969).
[c]Sobieszek and Jertschin (1986).
[d]From J. Brugge (pers. comm.).

## PARTIAL PROTEOLYTIC PEPTIDE MAPS—V8 PROTEASE*

Treating identical samples of a protein with the same protease, but at different concentrations, will generate a series of products ranging in size from fully digested to undigested. When the products of these individual reactions are run side-by-side on lanes of a one-dimensional SDS-polyacrylamide gel, a ladder of digestion products is displayed that gives a diagnostic fingerprint of the polypeptide backbone. When two proteins are identical or closely related, these partial proteolytic maps can be used to confirm their relationship. Although partial proteolytic maps can confirm that two proteins are related, only minor changes in the polypeptide sequence are needed to produce a vastly different pattern. Therefore, partial maps should only be used to test for identity or close relationships.

To prepare samples for enzyme treatment, run multiple lanes of the antigen on a preliminary gel. Each band will then become the substrate for a different digest, normally being exposed to one of a range of enzyme concentrations.

1. Run the preliminary SDS-polyacrylamide gel as normal. **If the protein bands of interest are radiolabeled:** rinse the gel in water and dry without fixing. Place labels near three corners of the dry gel. Mark the labels with radioactive ink to serve as guides when aligning the film after developing. Expose the gel to X-ray film as normal. Develop the film and use the film as a template to locate and excise the bands of interest. **If the proteins of interest are not labeled:** locate the bands by staining the gel with one of the Coomassie or silver staining methods on pp. 649–653. Dry as normal, and excise the bands of interest.

2. Remove the backing paper from each slice with a scapel.

3. Prepare a polyacrylamide gel as described on p. 636 but with these changes: (1) include 1 mM EDTA in all the gel mixes, and (2) make the stacking gel about twice the normal depth (about 3 cm from the bottom of the well to the top of the separating gel). Usually a 12–15% separating gel is appropriate. After the stacking gel has polymerized, transfer to the gel tank. Add running buffer to the bottom chamber and remove any bubbles between the plates at the bottom of the gel.

4. Fill the wells of the stacking gel with 0.1% SDS, 1 mM EDTA, 2.5 mM dithiothreitol, and 0.125 M Tris (pH 6.8).

5. Push the dried gel slices into the appropriate wells. Align the gel slice along the bottom of the well. Incubate for 10 min.

6. Add running buffer to the top chamber.

*Cleveland (1977); Fischer (1983).

7. Overlay the slice with 20 $\mu$l of 20% glycerol, 0.1% SDS, 1 mM EDTA, 2.5 mM dithiothreitol, and 0.125 M Tris (pH 6.8).

8. Overlay the 20% glycerol mix with 10 $\mu$l of 10% glycerol, 0.1% SDS, 1 mM EDTA, 2.5 mM dithiothreitol, 0.001% bromphenol blue, and 0.125 M Tris (pH 6.8) containing a dilution of V8 protease. In preliminary tests, 10-fold dilutions between 50 ng and 5 $\mu$g of protease per lane are used to establish the range of protease concentrations that will be used in the final experiment.

9. Run the gel at 125 V until the bromphenol blue forms a sharp line and has migrated about two-thirds of the distance into the stacking gel.

10. Turn off the power pack and allow the proteins to digest for 30 min.

11. Turn on the power pack and run the gel until the bromphenol blue reaches the bottom of the separating gel.

12. Turn off the power pack. Disassemble the gel plates and process as needed for the various samples (radioactive samples for autoradiography or fluorography and nonradioactive samples for staining).

## NOTE

i. Depending on the protein, V8 protease can be substituted with other proteases or chemical cleavage agents using a similar technique.

## TWO-DIMENSIONAL ISOELECTRIC FOCUSING/ SDS-POLYACRYLAMIDE GEL ELECTROPHORESIS*

Prior to electrophoresis, prepare the following solutions:

*Urea/NP-40 solution*

| | |
|---|---|
| Urea | 120 grams |
| NP-40 | 4.2 ml |
| Distilled $H_2O$ | 90 ml |

*Acrylamide*

| | |
|---|---|
| Acrylamide | 9.0 grams |
| Bisacrylamide | 0.54 gram |
| Distilled $H_2O$ | 30 ml |

*Ampholyte mixture*

| | |
|---|---|
| Ampholyte, pH 6.0–8.0 (or other narrow range) | 80% |
| Ampholyte, pH 3.5–10.0 (or other wide range) | 20% |

*Ammonium persulfate*    10%

*Phosphoric acid*    34%

*Sodium hydroxide*    2.4%

*Sample buffer*

| | |
|---|---|
| Urea | 9 M |
| NP-40 | 4% |
| $\beta$-Mercaptoethanol | 2% |

For high reproducibility between different gels, all these solutions should be dispensed in appropriate volumes, sealed, and frozen at −20°C.

### Sample Preparation

1. For sample preparation, immune complexes collected on SAC or protein A beads (p. 466) can be resuspended in 25 $\mu$l of sample buffer. Samples of other proteins can be diluted into sample buffer. Dried proteins can be resuspended in sample buffer.

2. Heat at 50°C for 30 min. Spin to remove the solid-phase matrix or any debris. Carefully transfer to clean tubes. Ready for loading. Samples may be stored at −20°C prior to electrophoresis.

*Knowles (1987).

**First Dimension**

1. To prepare the first dimension (isoelectric focusing gel), add 3.0 ml of the acrylamide solution (at room temperature) to 18.8 ml of the urea/NP-40 solution (the urea/NP-40 solution can be heated to 37°C to increase the solubility of the urea). Mix gently. Add 1.25 ml of the Ampholyte solution (room temperature), and mix gently. Add 75 $\mu$l of 10% ammonium persulfate and 20 $\mu$l of TEMED. Yields a final volume of 23.15 ml.

2. Pour the first-dimension tube gels. There are many different two-dimensional gel apparatuses. Follow the manufacturer's instructions for pouring the first dimension. If many different samples will be compared, an apparatus such as the MEGA System ISO 20 from Health Products Inc. (Pierce) is suitable.

3. After the first dimension has polymerized (about 15 min for the initial stages), allow the gel to sit at room temperature for a further 30 min.

4. Remove the tube gels from the casting apparatus. Wash the bottom ends of the tube gels with distilled water. Insert the tube gels into the electrophoresis apparatus. The bottom running buffer is prepared by diluting 1 volume of 34% phosphoric acid to 400 volumes of distilled $H_2O$. Check to ensure that no bubbles are found at the interface of the bottom of the tube gel and the running buffer. Completely assemble the apparatus and add the top running buffer. Top running buffer can be prepared by adding 1 volume of 2.4% NaOH to 30 volumes of distilled $H_2O$. Using a syringe fitted with a 1.5-in. needle, wash the top of each tube gel with top running buffer. This will remove any unpolymerized acrylamide and any air bubbles.

5. Carefully layer the samples onto the top surface of the isoelectric focusing gel. Approximately 20 $\mu$l/gel is appropriate. Samples can be applied using either: (1) a microliter syringe (washed by pipetting in distilled $H_2O$ between each sample), (2) a micropipetter fitted with a long narrow tip, or (3) Teflon tubing fixed to the end of a microliter syringe (24-gauge, thin-walled spaghetti tubing fits snugly on 22-gauge needles; discard each piece after a single use).

6. Attach the electrodes to the isoelectric focusing apparatus. The positive (red) electrode to the bottom and the negative (black) electrode to the top. Consult the manufacturer's instruction for the appropriate running conditions.

7. After the run is complete, turn off the power pack and disassemble the apparatus. Remove the tube gel.

8. Remove the gels from the glass tubes by applying even gentle pressure from the 1-ml syringe filled with distilled water applied to one end of the tube. As the gel emerges, be careful to align it in such a way that the positive and negative ends are known.

The isoelectric focusing gel can be immediately loaded on the second dimension or stored at $-70°C$.

### Second Dimension

1. Cast an SDS-polyacrylamide slab gel using the protocol described on p. 636. The choice of acrylamide percentage will depend on the protein under study. The separating gel is poured and allowed to polymerize as usual. **For isoelectric focusing gels of 1 to 2 mm in diameter:** the separating gel should be poured to within approximately 1 mm of the top. Overlay with isobutanol. After polymerization, go directly to step 3. The tube gel will be placed directly onto the separating gel, with no stacking gel. **For thicker isoelectric focusing gels:** a stacking gel is needed. The meniscus of the separating gel should be approximately 1 cm below top. For these gels go on to step 2.

2. Wash the top of the separating gel with $H_2O$. Drain well or dry with the edge of paper towel. Prepare the stacking gel solution as described on p. 639. Pour on top of the separating gel. The stacking gel solution should come to approximately 1 mm below the top of the glass plates. Overlay with isobutanol-saturated $H_2O$. Allow the gel to polymerize (approximately 15–30 min).

3. Wash the top of the gel with water. Drain well or dry with the edge of a paper towel. Prepare a 0.5% agarose solution in $1×$ Laemmli gel running buffer (p. 684). Add bromphenol blue to 0.001%. Melt in a boiling water bath or in a microwave oven. Cool to 50°C. Layer over the separating gel, completely filling the remaining space. Carefully lay the isoelectric focusing gel into the agarose, being careful to avoid any bubbles. Mark the positive end. Allow the agarose to harden at room temperature (approximately 10 min).

4. Place the gel plates in an electrophoresis tank and run as per normal.

The gels can be processed for staining, autoradiography, fluorography, or blotting as per normal.

### NOTE

i. Many protocols will need good internal markers both for molecular weights and pI. Commercially available prestained markers are the best that are available at present.

## ALKYLATION OF PROTEINS PRIOR TO RUNNING ON A GEL*

1. If using proteins from immunoprecipitations, wash immune complexes as usual. Remove the final wash as completely as possible. If using other sources of proteins, transfer them to 20 mM phosphate (pH 7.0), or resuspend the proteins directly in 20 $\mu$l of water. Protein concentrations should be under 1 mg/ml.

2. Add 20 $\mu$l of 1.5% SDS with 20 mM dithiothreitol.

3. Heat to 85°C for 10 min. If using *Staphylococcus aureus* Cowan I from an immunoprecipitation, spin for 2 min at 4°C. Carefully transfer the supernatant to a fresh tube.

4. Add 10 $\mu$l of freshly prepared 0.2 M *N*-ethylmaleimide (12.5 mg/ml of water; this should be a saturated solution). Make up the solution just prior to using. The 10-min, 85°C incubation is a good time to try to get the NEM into solution. An ultrasonic water bath helps, but repeated pipetting is sufficient. As the solution is saturated, complete dissolution is impossible. Alternatively, dissolve the NEM in 20 mM sodium acetate (pH 6.0).

5. Incubate for 60 min on ice.

6. Add 10 $\mu$l of 50% glycerol, 0.5 M $\beta$-mercaptoethanol with 0.001% bromphenol blue.

The sample is now ready for electrophoresis.

*L. Crawford (pers. comm.); Riordan and Vallee (1972).

## *PPO FLUOROGRAPHY**

> **Caution**   PPO may be carcinogenic. Wear heavy rubber gloves, and dispose of the waste using methods approved by your local authorities.

1. After electrophoresis or staining, place the gel in destain (7% glacial acetic acid, 25% methanol) for 30 min with shaking. Carefully pour off the destain, rinse in dimethylsulfoxide (DMSO) waste solution (see step 3 for the origin of DMSO waste) and drain. Add five gel volumes of DMSO waste solution.

2. Incubate for 45 min with shaking at room temperature. Discard the DMSO waste and add five gel volumes of fresh DMSO.

3. Incubate for 45 min with shaking at room temperature. Decant the DMSO into the DMSO waste solution container. Add five gel volumes of 22% (wt/vol) diphenyloxazole (PPO) in DMSO.

4. Incubate for 45 min with shaking at room temperature. Decant the PPO/DMSO back into its container. For routine work this solution can be used again. (A 500-ml stock can be used for 10–15 gels when [$^{35}$S]methionine is the radioisotope).

5. Place the PPO-impregnated gel in a gentle flow of water. It should immediately turn white as the PPO begins to come out of solution.

6. Wash under a gentle stream of fresh water for 30 min to 1 hr. Remove the gel and dry as normal.

   Fluorographed gels should be exposed at −70°C (p. 657).

## NOTES

i. For gels using $^3$H radioisotopes, film should be preflashed to maintain a linear relationship between amount of radioactive decay and film darkening. See Laskey and Mills (1975) for details. An excellent description of fluorography can be found in a pamphlet available from Amersham plc written by Ron Laskey (Radioisotope detection by fluorography and intensifying screens; Review 23).

ii. Several companies supply commercial alternatives to PPO. They are simpler to use, but are considerably more costly than PPO.

* Bonner and Laskey (1974).

## COUNTING $^{35}$S, $^{14}$C, OR $^3$H FROM POLYACRYLAMIDE GEL SLICES

1. Using an autoradiogram as a template, excise the region of the gel that contains the radioactive protein of interest. This can be most easily done from a dried, but unfixed, gel.

2. Place the gel slice in a counting vial.

3. Swell the gel slice in a minimal amount of water (20 $\mu$l per 1 × 5-mm slice) for 30 min at room temperature.

4. **Either:** Dissolve the gel slice in 1.0 ml of 19 parts 30% $H_2O_2$, 1 part 14.8 M ammonium hydroxide by incubation at 37°C overnight.* Add aqueous-based scintillant.

   **Or:** Add 10 ml of freshly prepared ammonia/solubilizer/scintillant [prepared by adding 0.5 ml of 14.8 M ammonium hydroxide to 10 ml of NCS solubilizer (Amersham) or Protosol (NEN), mix, and add to 100 ml of toluene based scintillant]. Incubate overnight at room temperature with shaking.**

5. Count.

## COUNTING $^{32}$P OR $^{125}$I FROM POLYACRYLAMIDE GEL SLICES

1. Using an autoradiogram as a template, excise the region of the gel that contains the radioactive protein of interest.

2. Place the gel slice in a counting vial. $^{32}$P samples can be counted using Cerenkov counting, and $^{125}$I samples in a gamma-counter.

### NOTE

i. The efficiency of Cerenkov counting will depend on the volume of the sample, and comparisons should be made from similar sized gel slices.

*Bonner and Laskey (1974).
**Ward et al. (1970).

## COOMASSIE BLUE STAINING—STANDARD METHOD

1. After electrophoresis, transfer the gel to a clean glass or plastic container. Add 5 gel volumes of 0.25% Coomassie brilliant blue, R-250, 50% methanol, and 10% acetic acid.

2. Incubate 4 hr to overnight at room temperature with shaking.

3. Remove the stain and save. The staining solution can be used many times (20–40) before replacing.

4. Destain the gel by successive incubations in 5% methanol, 7.5% acetic acid at room temperature with shaking.

   Sensitivity is 0.1–0.5 $\mu$g per track.

### NOTES

i. Staining and destaining time can be substantially reduced by using hot stain or destaining solutions. Heat the stain or destain in microwave oven or a water bath (about 50–60°C). Staining times can be shortened to approximately 20 min, while destaining will take approximately 1–2 hr.

ii. Adding foam rubber, wool, or other components that will bind the dye will speed destaining times, particularly when destaining at room temperature.

## COOMASSIE BLUE STAINING—QUICK METHOD

1. Place the gel in a clean glass or plastic dish. Add approximately 5 gel volumes of 0.25% Coomassie brilliant blue R-250 in 50% trichloroacetic acid.

2. Incubate at room temperature for 20 min with shaking.

3. Drain the stain and save (it is good for numerous stainings). Wash the gel with water to remove the stain that has not penetrated the gel.

4. Add several gel volumes of destain (25% methanol, 7% acetic acid). Change as needed. Warming the gel will help speed destaining, as will including a sponge, a paper towel, or any material that will bind the stain in the destain solution.

   Sensitivity is 1.0 $\mu$g per track.

## *COOMASSIE BLUE STAINING—MAXIMAL SENSITIVITY*

1. Prior to staining, the Coomassie brilliant blue G-250 (a dimethylated derivative of Coomassie brilliant blue R-250) must be purified. Commercial sources are contaminated with compounds that increase the background. Dissolve 4 grams of dye in 250 ml of 7.5% acetic acid. Heat to 70°C. Add 44 grams of ammonium sulfate. After the ammonium sulfate has dissolved, cool the mixture to room temperature. Recover the dye by filtration or centrifugation at 3000g for 10 min. Remove the supernatant and allow the dye to dry.

2. To prepare the staining solution, dissolve 30 grams of ammonium sulfate in 500 ml of 2% phosphoric acid. After the ammonium sulfate has dissolved *completely*, add 0.5 gram of the purified Coomassie brilliant blue G-250 dissolved in 10 ml of water. Store at room temperature.

3. After electrophoresis, transfer the gel to a clean glass or plastic container. Add 5 gel volumes of 12.5% trichloroacetic acid.

4. Incubate for 1 hr or longer at room temperature with shaking.

5. Drain well. Shake the Coomassie brilliant blue G-250 stain to disperse large colloids. Add to the gel.

6. Incubate overnight at room temperature with shaking.

7. Pour the stain back into the stock bottle. The stain solution can be used for multiple stainings. Wash the gel with water or standard destain solution and observe.

8. To fix the colloidal stain to the proteins, add 5 gel volumes of 20% ammonium sulfate and incubate overnight at room temperature with shaking. After fixing the gel, it can be stained again for added sensitivity. Repeat the staining from step 5.

Sensitivity is 0.05–0.1 $\mu$g per track.

## *SILVER STAINING OF GELS—AMMONIACAL SILVER STAINING*

1. Wear gloves and use only clean glassware. Fingerprints will stain, and dirty glassware will affect the sensitivity of these reactions.

2. After running a standard gel, place in 5 gel volumes of 50% ethanol, 10% acetic acid for 30 min to overnight with shaking.

3. Remove the 50% ethanol, 10% acetic acid and rinse the gel in deionized water. Add 5 gel volumes of 20% ethanol and incubate for 30 min with shaking.

4. Remove the ethanol and add 5 gel volumes of 20% ethanol. Incubate for 30 min with shaking.

5. Remove the ethanol and move to a fume hood. Add 5 gel volumes of 5% glutaraldehyde prepared in deionized water. Incubate at room temperature for 30 min with shaking in a fume hood.

> **Caution** Glutaraldehyde is harmful. Wear gloves and work only in the fume hood.

6. Remove the glutaraldehyde and rinse the gel with deionized water. Add 5 gel volumes of 20% ethanol. Incubate for 20 min at room temperature with shaking. This step and the following steps do not need to be performed in the fume hood.

7. Remove the 20% ethanol and repeat the ethanol wash twice (step 6).

8. After the last ethanol wash, rinse the gel with deionized water and add 5 gel volumes of deionized water. Incubate for 10 min at room temperature with shaking.

9.  Remove the water wash and add four gel volumes of a freshly prepared ammonia/silver solution. For 100 ml: Add 1.4 ml of 14.8 M $NH_4OH$ to 100 ml of water. Add 190 $\mu$l of 10 N NaOH. Place this on a vortex and slowly add 1 ml of a fresh prepared solution of $AgNO_3$ (0.8 gram of $AgNO_3$ in 1 ml of water). Add the $AgNO_3$ dropwise. A precipitate will appear, but will dissolve quickly.

    Incubate for 30 min at room temperature with shaking.

10. Remove the ammonia/silver solution, and wash the gel with several changes of deionized water over 20 min. Precipitate the silver in the used ammonia/silver solution by adding HCl.

11. Remove the water and add five gel volumes of freshly prepared 0.005% citric acid, 0.019% formaldehyde (diluted from a commercial 37% solution). Mix gently. The bands will start to develop within a few minutes. As the background begins to change, pour off the developer and wash the gel with water. A little practice will tell you when to stop this step. Stop the reaction by incubating the gel in 10% acetic acid, 20% ethanol.

    Sensitivity is about 1–10 ng per band.

## NOTE

i. Several companies supply good silver staining kits.

## *SILVER STAINING OF GELS—NEUTRAL SILVER STAINING*

1. Wear gloves and use only clean glassware. Fingerprints will stain, and dirty glassware will affect the sensitivity of these reactions.

2. After running a standard gel, place in 5 gel volumes of 30% ethanol, 10% acetic acid for 3 hr to overnight with shaking.

3. Remove the ethanol/acetic acid solution, and add 5 gel volumes of 30% ethanol. Incubate for 30 min at room temperature with shaking. Repeat the 30% ethanol wash.

4. Remove the ethanol solution and add 10 gel volumes of deionized water. Incubate for 10 min at room temperature with shaking. Repeat the water wash twice.

5. Remove the water and add 5 gel volumes of 0.1% $AgNO_3$ solution (diluted from a 20% stock). Incubate for 30 min at room temperature with shaking.

6. Remove the $AgNO_3$ solution and wash the gel for 20 sec under a stream of deionized water.

7. Add 5 gel volumes of 2.5% sodium carbonate, 0.02% formaldehyde. Incubate at room temperature with shaking. Bands should begin to appear after several min. Incubate until the background begins to darken.

8. Stop the reaction by washing in 1% acetic acid. Wash with several changes of deionized water of 10 min each. **Optional:**  A thin film of gray may develop on the gel surface. This can be removed by a short wash in a photographic reducer such as Farmer's (0.5% for 30 sec).

   Sensitivity is 1–10 ng per band.

## NOTES

i. To intensify the silver staining, repeat steps 5–8.

ii. Several companies supply good silver staining kits.

## COPPER STAINING OF GELS*

A useful alternative to the more traditional staining methods using Coomassie brilliant blue or silver is copper staining. Incubating gels in a solution of $CuCl_2$ allows the formation of a white opaque precipitate apparently involving both Tris and SDS. Protein bands remain clear, leaving a negative image of the polypeptide separation pattern. The proteins are not fixed in the gel and can be eluted by simple removal of the Cu ions by chelation with EDTA. This method is particularly useful for the rapid localization of protein bands either for immunization or further protein chemistry studies. The stained patterns are as easy to photograph as Coomassie or silver stained gels.

1. After electrophoresis, the gels should be washed briefly with distilled water. Several changes over 30 sec. Longer washes will begin to allow the Tris or SDS to elute and should be avoided.

2. Place the gel in a glass or plastic tray. Add at least 5 gel volumes of 0.3 M $CuCl_2$.

3. Incubate at room temperature with agitation for 5 min. Longer incubation times may be suitable with thicker gels. As the $CuCl_2$ enters the gel, a white precipitate will form in the regions of the gel that do not contain proteins.

4. Wash the gel for several minutes with distilled water. Observe against a dark background.

   Sensitivity is approximately 10–100 ng per band for 0.5-mm gels and ~1 μg per band for 1-mm gels.

## NOTE

i. The staining can be reversed by incubating in 0.25 M EDTA, 0.25 M Tris (pH 9.0).

*Lee et al. (1987).

## FIXING GELS

For many purposes, proteins separated by electrophoresis need to be precipitated or bound within the gel. Collectively, these techniques are described as fixation. Gels of all percentages and widths can be fixed by washing with acids or alcohols. This is commonly done by soaking a gel at room temperature in 12.5% trichloroacetic acid or destain (7% acetic acid, 25% methanol). If bromphenol blue is used as a tracking dye, it can also act as a pH indicator to determine whether the acid has diffused into the gel. The bromphenol blue will turn yellow below pH 5.0. Incubate for 10 minutes after the blue turns to yellow. If no tracking dye has been used or if it has been run off the bottom of the gel, fix for 30 min for gels up to 1 mm thick, 45 min for gels 1–1.5 mm, and 1 hr for gels over 1.5 mm.

## DRYING GELS

1. Wash the gel briefly in water.

2. Place the gel onto a piece of plastic wrap with the cut corner marking the number 1 lane in the lower right hand side.

3. Carefully cut the plastic wrap to a size just larger than the gel. A sharp scalpel is useful, if the countertop resists cutting.

4. Place a piece of absorbent paper (Whatman 3MM or equivalent, cut larger than the gel, but smaller than the size of the gel dryer) onto the gel. If the paper is dry, the gel will stick to the paper, making handling easier.

5. Place the gel with the plastic wrap and paper on a second piece of absorbent paper in a gel dryer. The order should be plastic wrap, gel, paper, paper, and then the vacuum source. Apply a good vacuum and gentle heat.

   For most percentages of gels the vacuum produced from a water aspirator or a vacuum pump is sufficient for drying; however, house vacuums often will be too slow for most applications. Heat can be applied from either the top or bottom, but higher-percentage gels will crack less often when the heat is applied to the top (plastic wrap side).

6. Dry for 30 min to 2 hr (30 min for minigels, 2 hr for 2 mm or wider gels). Release the vacuum to the gel and remove.

   The gel is now ready for exposure to film or storage.

## *REHYDRATING DRIED POLYACRYLAMIDE GELS*

1. Peel the paper from the dried gel. Starting at one corner, peel the paper from the back of the gel, working toward the center. Some paper will remain on the gel and eventually the paper will tear. Switch to another corner and repeat. Continue until all of the surrounding paper and much of the backing paper is removed.

2. Submerge the gel completely in a large container of water. As the gel begins to rehydrate, it will pull away from the remaining backing paper.

3. As the gel rehydrates it will shrink severely. After several minutes the remaining backing paper can be removed from the gel.

4. After the paper has been removed, incubate at room temperature with gentle shaking for 30 min. The gel will begin to swell within a few minutes.

5. After the gel has reached its normal size, it can be transferred into any appropriate buffer. Continued incubation in water will cause the gel to swell to a size larger than the original starting size. Maintaining the gel in 25% methanol will hold the gel at approximately its normal size.

## AUTORADIOGRAPHY AND FLUOROGRAPHY

### Autoradiography

1. Place the dried gel or blot in direct contact with the emulsion of an X-ray film. Kodak XAR or Fuji RX films are suitable. Blots with $^{14}$C or $^{35}$S markers or samples must be oriented in the proper direction with the correct side facing the film to ensure proper exposure.

2. Expose in a sealed, light-proof container for the appropriate length of time at room temperature.

3. Develop using the manufacturer's instructions.

### Fluorography (PPO-Impregnated Gels)

1. Place the gel or blot in direct contact with the emulsion of the X-ray film.

2. Expose in a sealed, light-proof container for the appropriate length of time at −70°C.

3. Develop using the manufacturer's instructions.

### Exposure with Intensifying Screens ($^{32}$P or $^{125}$I)*

1. Place a piece of X-ray film on a calcium tungstate intensifying screen. Place the gel or blot in direct contact with the emulsion of the film.

2. Expose in a sealed, light-proof container for the appropriate length of time at −70°C.

3. Develop using the manufacturer's instructions.

*Swanstrom and Shank (1978).

### Sensitivities of Autoradiography Versus Fluorography

| Isotope | Autoradiography cpm needed for overnight band | Fluorography method | Fluorography cpm needed for overnight band | Enhancement |
|---|---|---|---|---|
| $^3$H | $>10^7$ | PPO-impregnated gel −70°C | 3000 | 1000 |
| $^{14}$C | 2000 | PPO-impregnated gels −70°C | 200 | 10 |
| $^{32}$P | 50 | Screen, −70°C | 10 | 5 |
| $^{35}$S | 1000 | PPO-impregnated gels −70°C | 100–200 | 5–10 |
| $^{125}$I | 100 | Screen, −70°C | 10 | 10 |

# Appendix II
# PROTEIN TECHNIQUES

## Ammonium Sulfate Saturation Tables

| Starting concentration | \multicolumn{14}{c}{Final concentration} | | | | | | | | | | | | | |
|---|---|---|---|---|---|---|---|---|---|---|---|---|---|---|
| | 10% | 20% | 25% | 30% | 35% | 40% | 45% | 50% | 55% | 60% | 65% | 70% | 75% | 80% |
| 0% | 56 | 114 | 144 | 176 | 209 | 243 | 277 | 313 | 351 | 390 | 430 | 472 | 516 | 561 |
| 10% | — | 57 | 86 | 118 | 150 | 183 | 216 | 251 | 288 | 326 | 365 | 406 | 449 | 494 |
| 20% | | — | 29 | 59 | 91 | 123 | 155 | 189 | 225 | 262 | 300 | 340 | 382 | 424 |
| 25% | | | — | 30 | 61 | 93 | 125 | 158 | 193 | 230 | 267 | 307 | 348 | 390 |
| 30% | | | | — | 30 | 62 | 94 | 127 | 162 | 198 | 235 | 273 | 314 | 356 |
| 35% | | | | | — | 31 | 63 | 94 | 129 | 164 | 200 | 238 | 278 | 319 |
| 40% | | | | | | — | 31 | 63 | 97 | 132 | 168 | 205 | 245 | 285 |
| 45% | | | | | | | — | 32 | 65 | 99 | 134 | 171 | 210 | 250 |
| 50% | | | | | | | | — | 33 | 66 | 101 | 137 | 176 | 214 |
| 55% | | | | | | | | | — | 33 | 67 | 103 | 141 | 179 |
| 60% | | | | | | | | | | — | 34 | 69 | 105 | 143 |

Values given are the number of grams to be added to 1 liter of solution to change the ammonium sulfate concentration from the starting concentration to final concentration. All values are adjusted for changes in volume at room temperature. The saturation of ammonium sulfate does not vary significantly between 4°C and 25°C, so the values given here can normally be used at both temperatures. Saturated ammonium sulfate is 4.1 M at 25°C (add 761 grams to 1 liter of distilled H$_2$O).

## Proteins Used as Molecular Weight Standards

| Protein | Relative molecular weight | Calculated molecular weight | Native molecular weight/S value |
|---|---|---|---|
| Thyroglobulin | 340,000 | | 660,000/19S |
| Myosin | 205,000 | | 470,000 |
| | 21,000 | | |
| | 19,000 | | |
| | 17,000 | | |
| $\alpha_2$-Macroglobulin | 170,000 | 160,798 | 820,000/19S |
| $\beta$-Galactosidase | 116,000 | 116,365 | 540,000 |
| Phosphorylase-b | 94,000 | | |
| Phosphorylase-a | 92,500 | | 370,000 |
| Transferrin, Human | 80,000 | 75,190 | 80,000 |
| Serum albumin, Bovine | 68,000 | 66,322 | 68,000 |
| Catalase, Bovine Liver | 60,000 | 57,592 | 230,000 |
| Pyruvate Kinase, Muscle | 57,000 | | 220,000 |
| IgG | | | |
|   Heavy Chain | 55,000 | | 150,000 |
|   Light Chain | 22,000 | | |
| Glutamate Dehydrogenase | 53,000 | 55,567 | |
| Leucine Aminopeptidase | 53,000 | 51,698 | 300,000 |
| Fumerase | 49,000 | | 200,000 |
| Ovalbumin | 45,000 | 42,755 | |
| Alcohol Dehydrogenase, Liver | 41,000 | 39,572 | 80,000 |
| Aldolase, Rabbit Muscle | 40,000 | 39,216 | 160,000 |
| Creatine Kinase, Rabbit Muscle | 40,000 | 43,117 | 80,000 |
| Alcohol Dehydrogenase, Yeast | 36,000 | 36,735 | 140,000 |
| Glyceraldehyde Phosphate Dehydrogenase | 36,000 | 35,880 | |
| Lactate Dehydrogenase, Pig | 36,000 | 36,481 | 140,000 |
| Tropomyosin, Rabbit | 36,000 | 32,840 | 65,000 |
| Pepsinogen, Pig | 35,000 | 39,538 | |
| Carbonic Anhydrase | 29,000 | 29,343 | 180,000 |
| Chymotrypsinogen a, Bovine | 24,000 | 25,669 | |
| Trypsinogen, Bovine | 24,000 | 23,996 | |
| Trypsin Inhibitor, Soybean | 20,100 | 20,097 | |
| Apoferritin | 18,500 | | 440,000 |
| $\beta$-Lactoglobulin, Bovine | 18,000 | 18,283 | 35,000 |
| Myoglobin | 17,000 | 17,080 | |
| Hemoglobin | 16,000 | | 64,000 |
| Lysozyme, Chicken | 14,300 | 14,296 | |
| $\alpha$-Lactalbumin, Bovine | 14,200 | 14,188 | |
| Cytochrome $c$ | 12,400 | | |
| Aprotinin | 6,500 | | |

## Amino Acids

| Amino acid | Single-letter code | Triple-letter code | Molecular weight (pH 7) | Side chain pK | $\alpha$-NH$_2$ pK | $\alpha$-COOH pK |
|---|---|---|---|---|---|---|
| **Alanine** | A | Ala | 89 | | 9.87 | 2.35 |
| **Arginine** | R | Arg | 174 | 13.2 | 9.09 | 2.18 |
| **Asparagine** | N | Asn | 132 | | 8.8 | 2.02 |
| **Aspartic Acid** | D | Asp | 133 | 3.65 | 9.6 | 1.88 |
| **Cysteine** | C | Cys | 121 | 8.33 | 10.78 | 1.71 |
| **Glutamic Acid** | E | Glu | 147 | 4.25 | 9.67 | 2.19 |
| **Glutamine** | Q | Gln | 146 | | 9.13 | 2.17 |
| **Glycine** | G | Gly | 75 | | 9.6 | 2.34 |
| **Histidine** | H | His | 155 | 6.0 | 8.97 | 1.78 |
| **Isoleucine** | I | Ile | 131 | | 9.76 | 2.32 |
| **Leucine** | L | Leu | 131 | | 9.6 | 2.36 |
| **Lysine** | K | Lys | 146 | 10.28 | 8.9 | 2.2 |
| **Methionine** | M | Met | 149 | | 9.21 | 2.28 |
| **Phenylalanine** | F | Phe | 165 | | 9.24 | 2.58 |
| **Proline** | P | Pro | 115 | | 10.6 | 1.99 |
| **Serine** | S | Ser | 105 | | 9.15 | 2.21 |
| **Threonine** | T | Thr | 119 | | 9.12 | 2.15 |
| **Tryptophan** | W | Trp | 204 | | 9.39 | 2.38 |
| **Tyrosine** | Y | Tyr | 181 | 10.1 | 9.11 | 2.2 |
| **Valine** | V | Val | 117 | | 9.72 | 2.29 |

| Hydropathy values | | Solubility | Amino acid |
|---|---|---|---|
| Hopp and Woods[a] | Kyte and Doolittle[b] | (g/100 ml of $H_2O$) | |
| −0.5 | −1.8 | 15.8 | Alanine |
| 3.0 | 4.5 | 71.8 | Arginine |
| 0.2 | 3.5 | 2.4 | Asparagine |
| 3.0 | 3.5 | 0.42 | Aspartic acid |
| −1.0 | −2.5 | freely | Cysteine |
| 3.0 | 3.5 | 0.72 | Glutamic acid |
| 0.2 | 3.5 | 2.6 | Glutamine |
| 0.0 | 0.4 | 22.5 | Glycine |
| −0.5 | 3.2 | 4.19 | Histidine |
| −1.8 | −4.5 | 3.36 | Isoleucine |
| −1.8 | −3.8 | 2.37 | Leucine |
| 3.0 | 3.9 | 66.6 | Lysine |
| −1.3 | −1.9 | 5.14 | Methionine |
| −2.5 | −2.8 | 2.7 | Phenylalanine |
| 0.0 | 1.6 | 154 | Proline |
| 0.3 | 0.8 | 36.2 | Serine |
| −0.4 | 0.7 | freely | Threonine |
| −3.4 | 0.9 | 1.06 | Tryptophan |
| −2.3 | −1.3 | 0.038 | Tyrosine |
| −1.5 | −4.2 | 5.6 | Valine |

[a]Hopp and Woods (1981).
[b]Kyte and Doolittle (1982).

**Genetic Code**

| | | Second nucleotide | | | | | | | | |
|---|---|---|---|---|---|---|---|---|---|---|
| | | U | | C | | A | | G | | |
| **U** | | UUU | Phe | UCU | Ser | UAU | Tyr | UGU | Cys | U |
| | | UUC | Phe | UCC | Ser | UAC | Tyr | UGC | Cys | C |
| | | UUA | Leu | UCA | Ser | UAA | End[a] | UGA | End[c] | A |
| | | UUG | Leu | UCG | Ser | UAG | End[b] | UGG | Trp | G |
| **C** | | CUU | Leu | CCU | Pro | CAU | His | CGU | Arg | U |
| | | CUC | Leu | CCC | Pro | CAC | His | CGC | Arg | C |
| | | CUA | Leu | CCA | Pro | CAA | Gln | CGA | Arg | A |
| | | CUG | Leu | CCG | Pro | CAG | Gln | CGG | Arg | G |
| **A** | | AUU | Ile | ACU | Thr | AAU | Asn | AGU | Ser | U |
| | | AUC | Ile | ACC | Thr | AAC | Asn | AGC | Ser | C |
| | | AUA | Ile | ACA | Thr | AAA | Lys | AGA | Arg | A |
| | | AUG | Met | ACG | Thr | AAG | Lys | AGG | Arg | G |
| **G** | | GUU | Val | GCU | Ala | GAU | Asp | GGU | Gly | U |
| | | GUC | Val | GCC | Ala | GAC | Asp | GGC | Gly | C |
| | | GUA | Val | GCA | Ala | GAA | Glu | GGA | Gly | A |
| | | GUG | Val | GCG | Ala | GAG | Glu | GGG | Gly | G |

*First nucleotide* (left margin)    *Third nucleotide* (right margin)

[a] Ochre codon.
[b] Amber codon.
[c] Opal codon.

**Mitochondrial Genetic Code**

| | | Second nucleotide | | | | | | | | |
|---|---|---|---|---|---|---|---|---|---|---|
| | | **U** | | **C** | | **A** | | **G** | | |
| First nucleotide | **U** | UUU | Phe | UCU | Ser | UAU | Tyr | UGU | Cys | U |
| | | UUC | Phe | UCC | Ser | UAC | Tyr | UGC | Cys | C |
| | | UUA | Leu | UCA | Ser | UAA | End | UGA | Trp | A |
| | | UUG | Leu | UCG | Ser | UAG | End | UGG | Trp | G |
| | **C** | CUU | Leu[a] | CCU | Pro | CAU | His | CGU | Arg | U |
| | | CUC | Leu[a] | CCC | Pro | CAC | His | CGC | Arg | C |
| | | CUA | Leu[a] | CCA | Pro | CAA | Gln | CGA | Arg | A |
| | | CUG | Leu[a] | CCG | Pro | CAG | Gln | CGG | Arg | G |
| | **A** | AAU | Ile | ACU | Thr | AAU | Asn | AGU | Ser | U |
| | | AUC | Ile | ACC | Thr | AAC | Asn | AGC | Ser | C |
| | | AUA | Met[b] | ACA | Thr | AAA | Lys | AGA | End[c] | A |
| | | AUG | Met | ACG | Thr | AAG | Lys | AGG | End[c] | G |
| | **G** | GUU | Val | GCU | Ala | GAU | Asp | GGU | Gly | U |
| | | GUC | Val | GCC | Ala | GAC | Asp | GGC | Gly | C |
| | | GUA | Val | GCA | Ala | GAA | Glu | GGA | Gly | A |
| | | GUG | Val | GCG | Ala | GAG | Glu | GGG | Gly | G |

[a] Thr in yeast.
[b] Ile in yeast.
[c] Arg in yeast.

## Amino Acid and Codon Usage

| Amino Acid | Amino acid usage[a] | | Codon | Codon usage[b] | |
|---|---|---|---|---|---|
| | *E. coli*[c] | human[d] | | *E.coli*[c] | human[d] |
| **Alanine** | 11.1 | 7.0 | GCU | 0.33 | 0.31 |
| | | | GCC | 0.18 | 0.40 |
| | | | GCA | 0.28 | 0.17 |
| | | | GCG | 0.21 | 0.12 |
| **Leucine** | 7.9 | 10.4 | UUA | 0.11 | 0.05 |
| | | | UUG | 0.11 | 0.09 |
| | | | CUU | 0.12 | 0.11 |
| | | | CUC | 0.12 | 0.22 |
| | | | CUA | 0.03 | 0.07 |
| | | | CUG | 0.72 | 0.46 |
| **Valine** | 7.5 | 6.2 | GUU | 0.37 | 0.13 |
| | | | GUC | 0.12 | 0.27 |
| | | | GUA | 0.28 | 0.09 |
| | | | GUG | 0.23 | 0.50 |
| **Glycine** | 7.2 | 5.7 | GGU | 0.46 | 0.15 |
| | | | GGC | 0.40 | 0.44 |
| | | | GGA | 0.06 | 0.17 |
| | | | GGG | 0.08 | 0.24 |
| **Arginine** | 6.5 | 5.0 | CGU | 0.46 | 0.09 |
| | | | CGC | 0.32 | 0.19 |
| | | | CGA | 0.05 | 0.10 |
| | | | CGG | 0.06 | 0.15 |
| | | | AGA | 0.08 | 0.24 |
| | | | AGG | 0.03 | 0.23 |
| **Lysine** | 6.4 | 7.0 | AAA | 0.72 | 0.45 |
| | | | AAG | 0.28 | 0.55 |
| **Serine** | 6.3 | 8.1 | UCU | 0.27 | 0.17 |
| | | | UCC | 0.21 | 0.26 |
| | | | UCA | 0.13 | 0.11 |
| | | | UCG | 0.14 | 0.07 |
| | | | AGU | 0.11 | 0.11 |
| | | | AGC | 0.14 | 0.29 |
| **Isoleucine** | 6.1 | 2.9 | AUU | 0.39 | 0.23 |
| | | | AUC | 0.52 | 0.64 |
| | | | AUA | 0.08 | 0.13 |

[a]Amino acid usage given in percentage of the total amino acid composition.
[b]Fractional use of each codon for a particular amino acid.

| Amino Acid | Amino acid usage[a] | | Codon | Codon usage[b] | |
|---|---|---|---|---|---|
| | E. coli[c] | human[d] | | E.coli[c] | human[d] |
| **Threonine** | 5.8 | 5.6 | ACU | 0.36 | 0.20 |
| | | | ACC | 0.38 | 0.47 |
| | | | ACA | 0.09 | 0.21 |
| | | | ACG | 0.17 | 0.12 |
| **Glutamic Acid** | 5.5 | 7.3 | GAA | 0.67 | 0.40 |
| | | | GAG | 0.33 | 0.60 |
| **Aspartic Acid** | 5.2 | 4.9 | GAU | 0.48 | 0.38 |
| | | | GAC | 0.52 | 0.62 |
| **Glutamine** | 4.2 | 4.5 | CAA | 0.31 | 0.26 |
| | | | CAG | 0.69 | 0.74 |
| **Asparagine** | 3.5 | 3.5 | AAU | 0.29 | 0.34 |
| | | | AAC | 0.71 | 0.66 |
| **Proline** | 3.5 | 4.9 | CCU | 0.14 | 0.24 |
| | | | CCC | 0.11 | 0.41 |
| | | | CCA | 0.20 | 0.24 |
| | | | CCG | 0.54 | 0.11 |
| **Methionine** | 2.2 | 1.8 | AUG | 1.00 | 1.00 |
| **Phenylalanine** | 3.6 | 4.5 | UUU | 0.50 | 0.35 |
| | | | UUC | 0.50 | 0.65 |
| **Tyrosine** | 2.6 | 3.6 | UAU | 0.54 | 0.47 |
| | | | UAC | 0.46 | 0.53 |
| **Histidine** | 2.5 | 2.5 | CAU | 0.64 | 0.42 |
| | | | CAC | 0.36 | 0.58 |
| **Tryptophan** | 1.2 | 1.3 | UGG | 1.00 | 1.00 |
| **Cysteine** | 1.1 | 3.4 | UGU | 0.45 | 0.30 |
| | | | UGC | 0.55 | 0.70 |

[c]Grantham et al. (1981).
[d]Lathe (1985).

## Log Odds Matrix For Relationships Between Protein Sequences (MDM$_{78}$)[a]

| | A | R | N | D | C | Q | E | G | H | I | L | K | M | F | P | S | T | W | Y | V |
|---|---|---|---|---|---|---|---|---|---|---|---|---|---|---|---|---|---|---|---|---|
| A | 18 | | | | | | | | | | | | | | | | | | | |
| R | −15 | 61 | | | | | | | | | | | | | | | | | | |
| N | 2 | 0 | 20 | | | | | | | | | | | | | | | | | |
| D | −4 | −13 | 21 | 39 | | | | | | | | | | | | | | | | |
| C | −2 | −36 | −36 | −51 | 119 | | | | | | | | | | | | | | | |
| Q | −4 | 13 | 8 | 16 | −54 | 40 | | | | | | | | | | | | | | |
| E | 3 | −11 | 14 | 34 | −53 | 25 | 38 | | | | | | | | | | | | | |
| G | 13 | −26 | 3 | 6 | −34 | −12 | 2 | 48 | | | | | | | | | | | | |
| H | −14 | 16 | 16 | 7 | −34 | 29 | 7 | −21 | 65 | | | | | | | | | | | |
| I | −5 | −20 | −18 | −24 | −23 | −20 | −20 | −26 | −24 | 45 | | | | | | | | | | |
| L | −19 | −30 | −29 | −40 | −60 | −10 | −34 | −41 | −21 | 24 | 59 | | | | | | | | | |
| K | −12 | 34 | 10 | 1 | −54 | 7 | −1 | −17 | 0 | −19 | −29 | 47 | | | | | | | | |
| M | −11 | −4 | −17 | −26 | −52 | −10 | −21 | −28 | 22 | 37 | 4 | 22 | 43 | | | | | | | |
| F | −35 | −45 | −35 | −56 | −43 | −47 | −54 | −48 | −18 | 10 | 18 | −53 | 2 | 91 | | | | | | |
| P | 11 | −2 | −5 | −10 | −28 | 2 | −6 | −5 | −2 | −20 | −25 | −11 | −21 | −46 | 59 | | | | | |
| S | 11 | −3 | 7 | 3 | 0 | −5 | 0 | 11 | −8 | −14 | −28 | −2 | −16 | −32 | 9 | 16 | | | | |
| T | 12 | −9 | 4 | −1 | −22 | −8 | −4 | 0 | −13 | 1 | −17 | 0 | −6 | −31 | 3 | 13 | 26 | | | |
| W | −58 | 22 | −42 | −68 | −78 | −48 | −70 | −70 | −28 | −51 | −18 | −35 | −42 | 4 | −56 | −25 | −52 | 173 | | |
| Y | −35 | −42 | −21 | −43 | 3 | −40 | −43 | −52 | −1 | −9 | −9 | −44 | −24 | 70 | −49 | −28 | −27 | −2 | 101 | |
| V | 2 | −25 | −17 | −21 | −19 | −19 | −18 | −14 | −22 | 37 | 19 | −24 | 18 | −12 | −12 | −10 | 3 | −62 | −25 | 43 |

[a]Dayhoff (1978). Evolutionary relationship of amino acids. See Dayhoff for description of the calculations.

## Accepted Amino Acid Substitutions[a]

| Second Amino Acid \ First amino acid | A | R | N | D | C | Q | E | G | H | I | L | K | M | F | P | S | T | W | Y | V |
|---|---|---|---|---|---|---|---|---|---|---|---|---|---|---|---|---|---|---|---|---|
| A | — | 2.7 | 4.8 | 7.4 | 12 | 6.2 | 13 | 32 | 2.3 | 4.5 | 6.7 | 3 | 5 | 2.9 | 29 | 22 | 25 | 0 | 3.9 | 18 |
| R | 0.82 | — | 0.75 | 0 | 3.6 | 8.1 | 0 | 0.55 | 11 | 2.0 | 1.2 | 25 | 2.9 | 1 | 5.6 | 3.9 | 0.84 | 34 | 0.58 | 1 |
| N | 3.0 | 1.5 | — | 26 | 0 | 3.4 | 4.4 | 8.6 | 24 | 2.4 | 2.6 | 17 | 0 | 1 | 2.3 | 12 | 7.1 | 3.8 | 7 | 0.65 |
| D | 4.2 | 0 | 23 | — | 0 | 5.1 | 39 | 8.9 | 4.6 | 0.88 | 0 | 4.5 | 0 | 0 | 0.84 | 2.8 | 2.4 | 0 | 0 | 0.85 |
| C | 0.91 | 0.9 | 0 | 0 | — | 0 | 0 | 0.55 | 1.1 | 1.1 | 0 | 0 | 0 | 0 | 0.84 | 3.3 | 4.2 | 0 | 5.8 | 2.5 |
| Q | 2.6 | 11 | 2.2 | 3.7 | 0 | — | 20 | 1.7 | 26 | 0.54 | 5.2 | 7.8 | 3.4 | 0 | 7.8 | 1.3 | 1.6 | 0 | 0 | 1.3 |
| E | 7.3 | 0 | 4.1 | 40 | 0 | 28 | — | 6.2 | 2.5 | 2.4 | 1.1 | 5.5 | 1.2 | 0 | 3.4 | 2.5 | 1.3 | 0 | 1.9 | 1.8 |
| G | 16 | 0.9 | 6.9 | 7.8 | 3.6 | 2.0 | 5.3 | — | 1.1 | 0 | 1.2 | 3.2 | 1.2 | 2.5 | 4.1 | 13 | 2.1 | 0 | 0 | 4.8 |
| H | 0.58 | 9.3 | 10 | 2.1 | 3.6 | 1.6 | 1.1 | 0.55 | — | 0.2 | 2.8 | 1.2 | 0 | 2.9 | 4.2 | 0.7 | 0.59 | 3.8 | 7.8 | 1.5 |
| I | 1.8 | 2.7 | 1.6 | 0.63 | 6.1 | 0.54 | 1.7 | 0 | 3.2 | — | 18 | 2.3 | 9.8 | 13 | 0.59 | 0.57 | 5.4 | 0 | 2.5 | 33 |
| L | 2.6 | 1.5 | 1.6 | 0 | 0 | 5.0 | 0.7 | 0.94 | 4.3 | 17 | — | 2.1 | 36 | 24 | 3.6 | 0.9 | 2.2 | 16 | 4.5 | 15 |
| K | 1.6 | 43 | 14 | 4.1 | 0 | 9.9 | 4.9 | 3.3 | 2.4 | 20 | 2.7 | — | 16 | 0 | 3.6 | 4.8 | 8.4 | 0 | 1.9 | 0.85 |
| M | 0.80 | 1.5 | 0 | 0 | 0 | 1.3 | 0.33 | 3.9 | 0 | 3.8 | 14 | 4.8 | — | 2.5 | 0.34 | 0.57 | 1.2 | 0 | 0 | 3.8 |
| F | 0.55 | 0.63 | 0 | 0 | 0 | 0 | 0 | 0.94 | 2.2 | 6.1 | 12 | 0 | 2.9 | — | 0.59 | 1.1 | 0.42 | 13 | 51 | 0.5 |
| P | 9.5 | 6.0 | 1.2 | 0.48 | 3.6 | 6.2 | 1.9 | 2.7 | 5.4 | 0.49 | 3 | 2.3 | 0.69 | 1 | — | 7.7 | 3.1 | 0 | 0 | 2.5 |
| S | 21 | 12.3 | 19 | 4.7 | 42 | 3.1 | 4.1 | 25 | 2.8 | 1.4 | 2.2 | 8.9 | 3.4 | 5.9 | 23 | — | 29 | 22 | 4.3 | 2.1 |
| T | 16 | 1.8 | 7.5 | 2.7 | 3.6 | 2.5 | 1.7 | 2.8 | 1.5 | 8.7 | 3.6 | 11 | 8.3 | 1.5 | 6.1 | 20 | — | 0 | 4.5 | 9.3 |
| W | 0 | 2.4 | 0.13 | 0 | 0 | 0 | 0 | 0 | 0.32 | 0 | 0.91 | 0 | 0 | 1.5 | 0 | 0.49 | 0 | — | 1.2 | 0 |
| Y | 0.55 | 0.26 | 1.6 | 0 | 11 | 0 | 0 | 0 | 4.3 | 0.88 | 1.6 | 0.53 | 0 | 38 | 0 | 0.63 | 0.97 | 7.6 | — | 0.85 |
| V | 10 | 1.8 | 0.57 | 0.82 | 12 | 1.8 | 1.8 | 5.3 | 3.2 | 45 | 21 | 0.9 | 13 | 1.5 | 4.2 | 1.2 | 7.8 | 0 | 3.3 | — |
| [b] Total | 364 | 111 | 227 | 208 | 28 | 149 | 211 | 182 | 93 | 148 | 142 | 188 | 58 | 68 | 119 | 349 | 238 | 8 | 51 | 200 |

[a]Calculated from Dayhoff (1978). Related proteins derived from 34 super families and grouped into 71 evolutionary trees were compared for accepted mutations at a single site. A total of 1572 changes were used to assemble the data base. The frequency for an amino acid to change to another amino acid is given as a percentage of the total changes detected for that amino acid. Also listed are the total changes used to determine the percentage.

[b]Total number of amino acids tested

## Common Protein Sequence Motifs

| | | |
|---|---|---|
| **Nucleotide Binding** | | |
| ATP binding, kinases | L/I/V–G–X–G–X–F/Y–G–X–L/I/V<br>often followed 8 to 20<br>amino acids later by<br>–V–A–V/I–K–X–L/I–K/R | Barker and Dayhoff (1982)<br>Kamps et al. (1984)<br>Bairoch and Claverie (1988) |
| ATP binding, type A | G–X–X–X–X–G–K–T/S–X–X–X–X–X–X–I/V | Walker et al. (1982) |
| GTP binding?, G proteins | G–X–X–X–X–G–K–S/T | Robishaw et al. (1986) |
| ATP binding, type B | R/K–X–X–X–G–X–X–X–L–(HYDROPHOBIC)$_4$–D | Walker et al. (1982) |
| **Phosphotransferases** | | |
| In addition to ATP<br>binding site above,<br>these enzymes may<br>contain this sequence | H–X–D–F/L/I/M/V/W/Y–X–X–X–(F/L/I/M/V/W/Y)$_3$ | Brenner (1987) |
| | L/I/V–H/Y–X–D–F/I/L/M/V/Y–X–X–X–N–X–(F/I/L/M/V)$_2$ | Bairoch and Claverie (1988) |
| **Zinc Finger** | Y/F–X–C–(X)$_{2-4}$–C–(X)$_3$–F/Y–(X)$_5$–L–(X)$_n$–H–(X)$_{3-4}$–H | Miller et al. (1986) |
| **Modification Sites** | | |
| N-glycosylation | N–X–T(S) | Marshall (1972) |
| Phosphorylation by | | |
| cAMP kinase | R–R–X–S/T–X | Feramisco et al. (1980) |
| Casein kinase II | S/T–D/E–D/E–D/E | |
| Tyrosine kinases | D/E–X–Y–X–D/E | |
| **Nuclear Localization** | | |
| SV40 T antigen | P–K–K–K–R–K–V | Kalderon et al. (1984) |

X can equal any amino acid.

**David's Life Chart II**

| Molecular weight (daltons) | 1 μg | 1 nmole |
|---|---|---|
| 100 | 10 nmoles or 6 × 10^15 molecules | 0.1 μg |
| 1,000 | 1 nmole or 6 × 10^14 molecules | 1 μg |
| 10,000 | 100 pmoles or 6 × 10^13 molecules | 10 μg |
| 20,000 | 50 pmoles or 3 × 10^13 molecules | 20 μg |
| 30,000 | 33 pmoles or 2 × 10^13 molecules | 30 μg |
| 40,000 | 25 pmoles or 1.5 × 10^13 molecules | 40 μg |
| 50,000 | 20 pmoles or 1.2 × 10^13 molecules | 50 μg |
| 60,000 | 17 pmoles or 10^13 molecules | 60 μg |
| 70,000 | 14 pmoles or 8.4 × 10^12 molecules | 70 μg |
| 80,000 | 12 pmoles or 7.2 × 10^12 molecules | 80 μg |
| 90,000 | 11 pmoles or 6.6 × 10^12 molecules | 90 μg |
| 100,000 | 10 pmoles or 6 × 10^12 molecules | 100 μg |
| 120,000 | 8.3 pmoles or 5 × 10^12 molecules | 120 μg |
| 140,000 | 7.1 pmoles or 4.3 × 10^12 molecules | 140 μg |
| 160,000 | 6.3 pmoles or 3.7 × 10^12 molecules | 160 μg |
| 180,000 | 5.3 pmoles or 3.2 × 10^12 molecules | 180 μg |
| 200,000 | 5 pmoles or 3 × 10^12 molecules | 200 μg |

## *PROTEIN QUANTITATION—BRADFORD\**

This assay is relatively accurate for most proteins, except for small basic polypeptides such as ribonuclease or lysozyme. It is also hampered by detergent concentrations over about 0.2% (e.g., Triton X-100, SDS, NP-40).

1. Prior to the assay, prepare the Bradford dye concentrate. Dissolve 100 mg of Coomassie brilliant blue G-250 in 50 ml of 95% ethanol. Add 100 ml of concentrated phosphoric acid. Add distilled water to a final volume of 200 ml. The dye is stable at 4°C for at least 6 months. This dye concentrate is also available commercially from BioRad.

2. Prepare a series of protein samples for a standard curve. Use a protein as similar in its properties to your sample as possible (i.e., if doing antibody concentrations, use purified antibody). If your sample is unknown, use antibody. The standard curve will be linear between about 20 and 150 $\mu$g in 100 $\mu$l.

3. Prepare your test samples in 100 $\mu$l of the same buffer used for the standard curve (PBS is fine).

4. Dilute the concentrated dye binding solution 1 in 5 with distilled water. Filter if any precipitate develops.

5. Add 5 ml of diluted dye binding solution to each sample. Allow the color to develop for at least 5 min but not longer than 30 min. The red dye will turn blue as it binds protein. Read the absorbance at 595 nm.

## NOTE

i. BSA gives a value about twofold higher than its weight for Bradford dye binding assays.

\*Bradford (1976).

## PROTEIN QUANTITATION—BRADFORD SPOT TEST*

This assay is particularly useful for testing column fractions to locate the protein eluate (e.g., testing affinity column eluates).

1. Prior to the assay, prepare the Bradford dye concentrate. Dissolve 100 mg of Coomassie brilliant blue G-250 in 50 ml of 95% ethanol. Add 100 ml of concentrated phosphoric acid. Add distilled water to a final volume of 200 ml. The dye is stable at 4°C for at least 6 months. The dye concentrate is available commercially from BioRad.

2. Remove 8 $\mu$l of each protein sample to be tested and transfer to a strip of parafilm or Saran wrap. Be sure to include a sample of the elution buffer to give you a background reading.

3. Add 2 $\mu$l of concentrated Bradford dye solution to each sample. Mix by pipetting.

4. The samples that contain protein will turn blue within approximately 2 min. The sensitivity of this assay is about 10 $\mu$g/ml. The reaction is very sensitive to the presence of detergents at concentration over about 0.2%, but is not affected by KCl, NaCl, $MgCl_2$, EDTA, and only slightly by Tris.

*Adapted from Bradford (1976).

## *PROTEIN QUANTITATION — COOMASSIE SPOT TEST\**

This assay is particularly useful for testing column fractions to locate the protein eluate (e.g., testing affinity column eluates).

1. Rule Whatman 3MM paper or equivalent into 4-mm squares.

2. Spot 5 $\mu$l of sample on a square.

3. Dry the samples under a constant air flow. A hair dryer will dry the spots in approximately 1 min or less.

4. Dip the paper in 0.25% Coomassie brilliant blue in 10% methanol, 7% acetic acid for 30 sec.

5. Wash the paper 3–5 min in 10% methanol, 7% acetic acid. Dry.

\*C. Anderson (pers. comm.).

## PROTEIN QUANTITATION—UV DETECTION*

Absorbance of UV irradiation by proteins is the quickest of all methods for quantitating protein solutions. Readings are often performed at either 280 nm or 205 nm. The absorbance maximum at 280 nm is due primarily to the presence of tyrosine and tryptophan. Absorbance at 205 nm is due primarily to the peptide bond although other amino acids also contribute. In addition to the speed with which absorbance readings can be made, one major advantage of UV quantitation is that the none of the sample is destroyed in determining the concentration.

### Pure Protein Solutions

1. Read the absorbance versus a suitable control at 280 nm.

2. For antibodies and BSA, use the table to calculate the concentration. A very rough approximation for other proteins is 1 absorbance unit is equal to 1 mg/ml.

| Protein | $A_{280}$ (for 1 mg/ml) |
|---------|--------------------------|
| IgG     | 1.35                     |
| IgM     | 1.2                      |
| BSA     | 0.7                      |

### Protein Solutions Contaminated with Nucleic Acids

1. Read the absorbance versus a suitable control at 280 nm and 260 nm or 280 nm and 205 nm.

2. Calculate the approximate concentration using one of the equations below:

Protein concentration (in mg/ml) = $(1.55 \times A_{280}) - (0.76 \times A_{260})$

Protein concentration (in mg/ml) = $A205/(27 + A_{280}/A_{205})$

*Layne (1957) and Peterson (1983).

## *PROTEIN QUANTITATION — BICINCHONINIC ACID\**

1. The two reagents needed for the bicinchoninic acid assays are available from Pierce. Mix 1 volume of reagent A to 50 volumes of reagent B.

2. Adjust the protein samples to 100 $\mu$l. Prepare standard of BSA at 100, 50, 25, and 12.5 $\mu$g/100 $\mu$l.

3. Add 2 ml of the combined reagent to each sample. Incubate at room temperature for 2 hr or 37°C for 30 min.

4. Read the samples versus an appropriate blank at 562 nm.

*Smith et al. (1985).

## PROTEIN QUANTITATION — LOWRY*

1. Prior to the assay dissolve 10 grams of $NaCO_3$ in 500 ml of distilled water. Dissolve 0.5 gram of $CuSO_4 \cdot 5H_2O$ and 1 gram of Na tartrate in 500 ml of distilled water. Place the copper/tartrate solution on a magnetic stirrer and slowly add the carbonate. Store with refrigeration. This solution is stable for over 1 year.

2. The Folin–Ciocalteu phenol reagent can be prepared in the laboratory, but it can be purchased economically from numerous suppliers.

3. Combine 1 volume of the copper/tartrate/carbonate solution to 2 volumes of 5% SDS and 1 volume of 0.8 M NaOH. Stable at room temperature for 2 weeks. Label as Reagent A.

4. Combine 1 volume of the 2 N Folin–Ciocalteu phenol reagent with 5 volumes distilled water. Store in an amber bottle at room temperature. Stable for months. Label as Reagent B.

5. Samples (of 5–100 $\mu$g) should be adjusted to 1 ml by adding water. Prepare standards of BSA containing 100, 50, 25, and 12.5 $\mu$g/ml.

6. Add 1.0 ml of Reagent A to each protein sample. Mix and incubate for 10 min at room temperature.

7. Add 0.5 ml of Reagent B and mix immediately. Incubate at room temperature for 30 min.

8. Read the absorbance at 750 nm. Prepare a standard curve and compute the protein concentration.

*Lowry et al. (1951); Peterson (1983).

## Proteases

| Protease | Class | Cleavage site | Known inhibitors |
|---|---|---|---|
| Bromelain | Thiolprotease | No specificity | $\alpha_2$-Macroglobulin, TPCK, TLCK, alkylation |
| Chymotrypsin | Serine protease | After F, T, or Y | Aprotinin, PMSF, TPCK, $\alpha_2$-macroglobulin |
| Collagenase | Metalloprotease | After X in P–X–G–P | EDTA, reducing agents |
| Dispase | Metalloprotease | No specificity | EDTA |
| Endoproteinase Arg–C | Serine protease | After R | $\alpha_2$-Macroglobulin, TLCK |
| Endoproteinase Glu–C (S. aureus V8) | Serine protease | After E, or after D or E | $\alpha_2$-Macroglobulin |
| Endoproteinase Lys–C | Serine protease | After K | TLCK, aprotinin, leupeptin |
| Factor Xa | Serine protease | After R | PMSF |
| Ficin | Thiolprotease | No specificity | TPCK, TLCK, $\alpha_2$-macroglobulin |
| Kallikrein | Serine protease | After some R | Leupeptin, aprotinin |
| Papain | Thiolprotease | On long incubation, broad specificity | TPCK, TLCK, leupeptin, $\alpha_2$-macroglobulin, alkylating agents |
| Pepsin | Acid protease | Broad specificity | Pepstatin |
| Plasmin | Serine protease | After K or R | PMSF, TLCK, aprotinin, $\alpha_2$-macroglobulin |
| Pronase | Mixture | No specificity | None known |
| Proteinase K | Serine protease | Broad specificity | PMSF |
| Subtilisin | Serine protease | Broad specificity | PMSF, $\alpha_2$-macroglobulin |
| Thrombin | Serine protease | After R | TLCK, PMSF, leupeptin, aprotinin, $\alpha_2$-macroglobulin |
| Trypsin | Serine protease | After K or R | TLCK, PMSF, aprotinin, $\alpha_2$-macroglobulin |

**Protease Inhibitors***

| Inhibitor | Protease target | Effective concentrations | Stock solution | Comments |
|---|---|---|---|---|
| **Aprotinin** | Serine proteases | 0.1–2 µg/ml | 10 mg/ml in PBS | Avoid repeated freezing |
| **EDTA** | Metalloproteases | 0.5–2 mM | 500 mM in $H_2O$, pH 8.0 | |
| **Leupeptin** | Serine and thiolproteases | 0.5–2 µg/ml | 10 mg/ml in $H_2O$ | |
| **α-Macroglobulin** | Broad spectrum | 1 unit/ml | 100 units/ml in PBS | Avoid reducing agents |
| **Pepstatin** | Acid proteases | 1 µg/ml | 1 mg/ml in methanol | |
| **PMSF** | Serine proteases | 20–100 µg/ml | 10 mg/ml in isopropanol | Add fresh at each step |
| **TLCK** | Trypsin | 50 µg/ml | 1 mg/ml in 50 mM acetate, pH 5.0 | Chymotrypsin unaffected |
| **TPCK** | Chymotrypsin | 100 µg/ml | 3 mg/ml in ethanol | Trypsin unaffected |

*Derived from Boehringer Mannheim Biochemicals (1987).

## PREPARING DIALYSIS TUBING

1. Cut the dialysis tubing to appropriate lengths. For most work, lengths of 10, 20, and 30 cm are useful.

2. Place the tubing in a large volume of 5 mM EDTA, 200 mM sodium bicarbonate.

3. Boil for 5 min. Pour off the EDTA/bicarbonate wash. Rinse briefly with deionized water. Add a large volume of 5 mM EDTA, 200 mM sodium bicarbonate and boil for 5 min.

4. Discard the second wash. Rinse the tubing thoroughly with deionized water. Add a large volume of deionized water and cover (aluminum foil is fine).

5. Autoclave for 10 min on liquid cycle. Store at 4°C. For most purposes, the addition of 0.02% sodium azide will not interfere with the dialysis, and this will eliminate the chances of microbial growth.

To use, wear gloves and remove an appropriate length of tubing. Wash the tubing extensively, both inside and out, in a stream of deionized or distilled water.

## TCA PRECIPITATION—FILTRATION

1. To measure the amount of incorporation of a radiolabeled precursor into protein, transfer a standard volume of each sample (normally 5 $\mu$l) to 100 $\mu$l of 1 mg/ml of BSA in a 1.5-ml conical tube.

2. Add 1 ml of ice-cold 10% trichloroacetic acid (TCA) to each tube. Mix.

3. Incubate on ice for 30 min.

4. Place glass fiber filters into a suitable filtration device. Apply a gentle vacuum and prewet by passing a few milliliters of 10% TCA through the filter.

5. Add the sample dropwise to the filter using the center of the filter as a target for adding the drops. The precipitated proteins will be trapped by the filter, but the unincorporated precursors will pass through.

6. Wash the filter twice with 10% TCA using a few milliliters for each wash.

7. Wash the filter twice with 95% ethanol using a few milliliters for each wash.

8. Remove the filters and allow to dry. Placing the filters on a sheet of aluminum foil under an infrared light will speed drying.

   Add the *dry* filters to counting vials, add scintillant, and count.

## NOTE

i. Adding a nonradioactive form of the labeled molecule to the TCA solution will lower the background. For example, add cold methionine to TCA to lower background of [$^{35}$S]methionine.

## *TCA PRECIPITATION — SPOTTING*

1. To measure the amount of incorporation of a radiolabeled precursor into protein, mix a standard volume of each sample (normally 5 μl) with 5 μl of 10 mg/ml of BSA.

2. Apply the mixed sample to the center of a dry glass fiber filter. Place the filters in a shallow tray or petri dish. Filters should be marked prior to use.

3. Carefully flood the dish with ice-cold 10% trichloroacetic acid (TCA).

4. Incubate on ice or at 4°C for 30 min.

5. Drain the TCA from the dish and add fresh 10% TCA. Incubate for 5 min with shaking at room temperature. Drain the TCA and repeat twice.

6. Drain the TCA and add 95% ethanol. Incubate for 5 min with shaking at room temperature.

7. Remove the filters and allow to dry. Placing the filters on a sheet of aluminum foil under an infrared light will speed drying.

   Add the *dry* filters to counting vials, add scintillant, and count.

## NOTE

i. Adding a nonradioactive form of the labeled molecule to the TCA solution will lower the background. For example, add cold methionine to TCA to lower background of [$^{35}$S]methionine.

**Chromogenic Substrates Yielding Water-Soluble Products**

| Enzyme | Substrate | Abbreviation | Starting color | Final color | Absorbance peak (in nm) |
|---|---|---|---|---|---|
| **Horseradish Peroxidase** | 2,2′-Azinodi[ethylbenzthiazoline] sulfonate[a] | ABTS | Clear | Green | 410 |
| | | | | | 650 |
| | o-Phenylene diamine[b] | OPD | Clear | Brown | 492 |
| | 3,3′,5,5′-Tetramethylbenzidine[c] | TMB | Clear | Yellow | 450 |
| **Alkaline Phosphatase** | p-Nitrophenol phosphate[d] | PNPP | Clear | Yellow | 405 |
| **β-Galactosidase** | o-Nitrophenyl-β-d-galactophyranoside[e] | ONPG | Clear | Yellow | 410 |

[a]Porstmann (1981).
[b]Voller et al. (1979).
[c]Holland et al. (1974); Hardy and Heimer (1977).
[d]Snyder et al. (1972).
[e]Craven (1965).

## Chromogenic Substrates Yielding Water-Insoluble Products

| Enzyme | Substrate | Abbreviation | Starting color | Final color | Soluble in alcohol? |
|---|---|---|---|---|---|
| **Horseradish Peroxidase** | Diaminobenzidene[a] | DAB | Clear | Brown | No |
| | Diaminobenzidene with nickel enhancement[b] | DAB/nickel | Clear | Grey/black | No |
| | 3-Amino-9-ethylcarbazole[c] | AEC | Clear | Red | Yes |
| | 4-Chloro-1-naphthol[d] | — | Clear | Blue | Yes |
| **Alkaline Phosphatase** | Naphthol-AS-BI-phosphate/fast red TR[e] | NABP/FR | Clear | Red | Yes |
| | Naphthol-AS-MX-phosphate/fast red TR[f] | NAMP/FR | Clear | Red | No |
| | Naphthol-AS-BI-phosphate/new fuchsin[g] | NABP/NF | Clear | Red | No |
| | Bromochloroindolyl phosphate/ nitroblue tetrazolium[h] | BCIP/NBT | Clear | Purple | — |
| | 5-Bromo-4-chloro-3-indolyl-β-d-galactopyranoside[i] | BCIG | Clear | Blue | No |
| **β-Galactosidase** | Naphthol AS-BI-β-d-galactopyranoside[j] | NABG | Clear | Red | No |

[a]Graham and Karnovsky (1966).
[b]Hsu and Soban (1982).
[c]Graham et al. (1964).
[d]Nakane (1968).
[e]Pearse (1968); Ponder and Wilkinson (1981).
[f]Amersham handbook.
[g]Stutte (1967); Malik and Dayman (1982).
[h]McGadey (1970); Leary et al. (1983).
[i]Bondi et al. (1982).
[j]Gossrau (1973).

# Appendix III

## GENERAL INFORMATION

## Commonly Used Buffers

| Buffer | Synonyms | pK | Molecular weight |
|---|---|---|---|
| Phosphate ($pK_1$) | | 2.12 | 98.0, free acid |
| Glycine-HCl | | 2.34 | 111.53 |
| Citrate ($pK_1$) | | 3.14 | 192.1, free acid |
| Formate | | 3.75 | 68.0, Na salt |
| Carbonate ($pK_1$) | | 3.76 | 106.0 |
| Succinate ($pK_1$) | | 4.19 | 118.1, free acid |
| Acetate | | 4.75 | 82.0, Na salt |
| Citrate ($pK_2$) | | 4.76 | 214.1, Na salt |
| Succinate ($pK_2$) | | 5.57 | 162.1, diNa salt |
| MES | | 6.15 | 195.2, hydrate |
| Carbonate ($pK_2$) | Bicarbonate | 6.36 | 84.0, bicarbonate |
| Citrate ($pK_3$) | | 6.39 | 294.1, triNa salt, dihydrate |
| PIPES | | 6.8 | 302.4 |
| ACES | | 6.9 | 182.2 |
| MOPS | | 7.2 | 209.7 |
| HEPES | | 7.55 | 238.3 |
| Phosphate ($PK_2$) | | 7.21 | 120.0, Na salt, monobasic |
| TES | | 7.7 | 229.2 |
| Barbital | Barbitone Veronal | 7.78 | 128.1, barbituric acid |
| Triethanolamine | | 7.8 | 149.2 |
| TRICINE | | 8.15 | 179.2 |
| TRIS | | 8.3 | 121.1 |
| BICINE | | 8.35 | 163.2 |
| Glycylglycine | | 8.4 | 132.1 |
| Borate | | 9.24 | 201.2, Na tetraborate |
| CHES | | 9.5 | 207.3 |
| Ethanolamine | | 9.5 | 61.1 |
| Glycine-NaOH | | 9.6 | 97.1, Na salt hydrate |
| CAPS | | 10.4 | 221.3 |
| Triethylamine | | 10.7 | 101.2 |
| Phosphate ($pK_3$) | | 12.3 | 141.9, Na salt, dibasic |

## Preparation of Stock Solutions

| Solution | Method of preparation | Comments |
|---|---|---|
| **Phosphate-buffered Saline (PBS)** | Dissolve 8.0 grams of NaCl, 0.2 gram of KCl, 1.44 grams of Na$_2$HPO$_4$, and 0.24 gram of KH$_2$PO$_4$ in 800 ml of distilled water. Adjust the pH to 7.2. Adjust the volume to 1 liter. Dispense in convenient volumes and sterilize by autoclaving. Store at room temperature. | |
| **Tris-buffered Saline (TBS) (25 mM Tris)** | Dissolve 8.0 grams of NaCl, 0.2 gram of KCl, and 3 grams of Tris base in 800 ml of distilled water. Adjust the pH to 8.0 with 1 M HCl. Adjust the volume to 1 liter. Dispense in convenient volumes and sterilize by autoclaving. Store at room temperature. | |
| **10% Sodium Azide** | Dissolve 10 grams of sodium azide in 100 ml of distilled water. Store at room temperature. | **Caution** Sodium azide is poisonous. Avoid inhalation of the powder and use caution when handling the liquid. |
| **3% Bovine Serum Albumin in Phosphate-buffered Saline (3% BSA/PBS)** | Add 3 grams of bovine serum albumin (Fraction V) to 100 ml of PBS. Allow to dissolve. Add 0.2 ml of 10% sodium azide. Store at 4°C. | Do not use sodium azide for experiments using live organisms. |
| **1 M Tris** | Dissolve 121 grams of Tris base to 800 ml of distilled water. Adjust to the desired pH by adding concentrated HCl. Adjust the volume to 1000 ml with distilled water. Dispense in convenient volumes and sterilize by autoclaving. Store at room temperature. | |
| **500 mM EDTA** | Add 186 grams of disodium ethylene diamine tetraacetate·2H$_2$O to 500 ml of distilled water. Add NaOH to adjust the pH to 8.0 and to allow the EDTA to dissolve. Dispense in convenient volumes and sterilize by autoclaving. Store at room temperature. | |
| **100% (wt/vol) Trichloro-acetic Acid (TCA)** | Add 227 ml of distilled water to a 500-gram bottle of TCA. | |

**30% Acrylamide Mix**

Dissolve 29.2 grams of acrylamide (electrophoresis grade) and 0.8 grams of $N,N'$-methylene-bis-acrylamide (electrophoresis grade) in 80 ml of distilled water. Adjust the volume to 100 ml. Store at room temperature.

**Caution** Acrylamide is a potent neurotoxin in the unpolymerized state. Wear gloves and a mask during preparation.

**1.5 M Tris (pH 8.8)**

Dissolve 181.5 grams of Tris base in 800 ml of distilled water. Adjust the pH to 8.8 with concentrated HCl. Adjust the volume to 1 liter. Dispense in convenient volumes and sterilize by autoclaving. Store at room temperature.

**1.0 M Tris (pH 6.8)**

Dissolve 12.1 grams of Tris base in 80 ml of distilled water. Adjust the pH to 6.8 with concentrated HCl. Adjust the volume to 100 ml. Sterilize by autoclaving. Store at room temperature.

**10% Sodium Dodecyl Sulfate (SDS)**

Dissolve 10 grams of SDS in 100 ml of distilled water. Store at room temperature.

**10% Ammonium Persulfate (APS)**

Dissolve 0.5 gram of ammonium persulfate (electrophoresis grade) in 5 ml of distilled water. Store at 4°C.

Make fresh solution weekly.

**2× Laemmli Sample Buffer**

Add 4 ml of 10% SDS, 2 ml of glycerol, and 1.2 ml of 1 M Tris (pH 6.8) to 2.8 ml of distilled water. Add 0.01% bromphenol blue as a tracking dye. Store at room temperature.

To prepare 1× sample buffer, mix 5 parts 2×, 4 parts water, and 1 part 1 M DTT.

**1 M Dithiothreitol**

Dissolve 5 grams of DTT in 32 ml of distilled water. Dispense in 1-ml aliquots. Store at −20°C.

**10× Laemmli Running Buffer**

To a 10-liter carboy, add 8 liters of distilled water, 303 grams of Tris base, 1442 grams glycine, and 100 grams of SDS. After all the chemicals have dissolved, adjust the pH to 8.3. Adjust the volume to 10 liters with water. Store at room temperature.

To prepare 1× runing buffer, dilute the 10× stock 1 in 10 with distilled water.

**Destain**

To a 10-liter carboy add 2.5 liters of methanol and 700 ml of glacial acetic acid. Adjust the volume to 10 liters with water. Store at room temperature.

## Concentrations of Commercial Liquids

| Compound | Molecular weight | Molarity | pH of dilute solutions | | |
|---|---|---|---|---|---|
| | | | 1 M | 0.1 M | 0.01 M |
| Acetic Acid, Glacial | 60.05 | 17.4 | 2.4 | 2.9 | 3.4 |
| Formic Acid | 46.02 | 23.4 | | | |
| Hydrochloric Acid, 38% | 36.47 | 11.6 | 0.1 | 1.1 | 2.02 |
| Nitric Acid, 70% | 63.02 | 16 | | | |
| Phosphoric Acid | 98.0 | 18.1 | | 1.5 | |
| Sulfuric Acid | 98.08 | 18 | | | |
| Ammonium Hydroxide | 35.0 | 14.8 | | | |
| Ethanolamine, 99% | 61.08 | 16.5 | | | |
| Triethylamine | 101.19 | 7.16 | | | |
| Formaldehyde, 37% | 30.3 | 12.2 | | | |
| Hydrogen Peroxide, 30% | 34.02 | 8.8 | | | |
| $\beta$-Mercaptoethanol | 78.13 | 14.4 | | | |

## Strains of Laboratory Mice

| Strain | Abbreviation | Coat color | Class II haplotype* | Comments |
|---|---|---|---|---|
| A/J | A | White | k | |
| AKR | AK | White | k | |
| BALB/c | C | White | d | Isogenic for most myeloma parents |
| CBA | CB | Agouti | k | |
| C3H | C3 | Agouti | k | |
| C57BL/6 | B6 | Black | b | Common parent for transgenic $F_2$ |
| C57BL/10 | B10 | Black | b | Many H2 congenics on this background |
| DBA/2 | D2 | Dilute brown | d | Common parent for transgenic $F_2$ |
| NZB | | Black | d | Good for anti-"self" antibodies |
| SJL | | White | s | common parent for transgenic $F_2$ |
| 129 | — | White or light chinchilla | b | Parent of many teratocarcinomas |

The standard method to describe an $F_1$ cross is to list the two mice used for the mating followed by the designation "$F_1$." The female is always given first, followed by the male. Therefore, a B6D2F$_1$ is a C57BL/6 female crossed with a DBA/2 male.

*Klein et al. (1983).

## ■ DETERGENTS

Detergents* form one class of polar lipids, characterized by their solubility in water. They have a bipartite structure with one hydrophobic portion and one hydrophilic portion. The presence of these two groups makes detergents useful for lysis of lipid membranes, solubilization of antigens, and washing of immune complexes.

A wide variety of chemicals can be classified as detergents. Of these, the most useful for biological studies fall into two groups based on the structure of their hydrophobic regions. Most biologically important detergents have hydrophobic regions with an 8- to 16-member alkyl chain (either with or without a phenolic moeity) or a structure that resembles the aromatic hydrophobic group of bile salts. The hydrophobic region leads to the solubilization of lipid bilayers, and only hydrophobic regions that are effective in this dissolution are useful for most biological applications.

Detergents are classified further by the type of hydrophilic group. These groups can be anionic, cationic, amphoteric, or nonionic. Examples of useful members of each of the four classes are shown on p. 689. In general, nonionic and amphoteric detergents are less denaturing to protein antigens than ionic detergents. Of the ionic detergents, sodium cholate and sodium deoxycholate are less denaturing than other ionic detergents.

Many of the properties of detergents can be described by three values: the critical micelle concentration, the micelle molecular weight, and the hydrophile–lipophile balance.

- **Critical micelle concentration (CMC)**   The CMC is the concentration at which monomeric detergent molecules join to form micelles. Below this concentration, detergent molecules are found predominately as monomers; above the CMC, the detergent molecules form micelles.
- **Micelle molecular weight**   Micelles formed by a particular detergent will have a characteristic molecular weight. Detergents with large micelle molecular weights, such as the nonionic detergents, will be difficult to dialyze.
- **Hydrophile–lipophile balance (HLB)**   The HLB gives a numerical value to the overall hydrophilic properties of the detergent. Values above 7 indicate that the detergent is more soluble in water than oils. For most biological applications, values of 12.5 and higher are needed. Although there are many exceptions, the HLB values can be used to give a general view of how denaturing a detergent will be. Values between 12 and 16 are relatively nondenaturing, while values above 20 indicate increasing denaturing possibilities. However, this correlation often does not apply outside of the detergents considered here, and will differ from one protein to another.

*Reviewed in Helenius and Simons (1975), Helenius et al. (1979), Furth (1980).

Both the CMC and the micelle molecular weight will vary in different buffers. All of the values given in the table on p. 689 will vary with the addition of salts or changes in temperature or pH. In general, adding salt will lower the CMC and raise the micelle size. The addition of as little as 100 mM NaCl may induce dramatic changes in these values. The extent of change in CMC or micelle molecular weight caused by varying the temperature or pH will depend on the various detergents, but two detergents that are drastically affected by temperature or pH are SDS (temperatures below 20°C will often lead to crystallization) and DOC (insoluble below about pH 7.5).

## DETERGENT REMOVAL

The ease with which detergents can be removed from protein solutions will depend on a number of variables, including: (1) the properties of the detergent itself, (2) the hydrophobic/hydrophilic characteristics of the protein, and (3) the other components of the buffer.

- Ionic detergents with relatively low micelle size and high CMC. Dilute as much as possible and dialyze. Add mixed bed resin to dialysate to increase exchange rate.

- Ionic detergents. Add urea to 8 M, then bind detergent to an ion-exchange column. Protein flows through in 8 M urea. Dialyze to remove urea.

- Ionic detergents. Gel filtration on G25 column. For some proteins, equilibrate column in another detergent below its CMC.

- Amphoteric detergents. Dilute if possible, dialyze.

- Nonionic detergents. Gel filtration on G200 column. For some proteins, equilibrate column in another detergent below the CMC.

- Nonionic detergents. Dilute if possible, dialyze extensively against DOC, then slowly remove DOC by dialysis.

- Nonionic detergents. Velocity sedimentation into sucrose without detergent.

- Nonionic detergent. Bind protein to affinity matrix or ion-exchange column, wash extensively to remove detergent, then elute protein. For some proteins, equilibrate column in another detergent below the CMC.

## Properties of Many Commonly Used Detergents

| Detergent | Abbrev | Commercial name | Molecular weight | Relative purity | CMC (grams/100 ml) | Micelle molecular weight | HLB number | Comments |
|---|---|---|---|---|---|---|---|---|
| **Anionic** | | | | | | | | |
| Sodium dodecylsulfate | SDS | | 288 | Homogeneous | 0.24 | 18,000 | 40 | Excellent solubilization, highly denaturing |
| Sodium dodecyl-N-sarcosinate | | | 293 | Homogeneous | | | 24.7 | Excellent solubilization, strong denaturant |
| Sodium cholate | | | 431 | Homogeneous | 0.57 | 1,800 | 18 | Keep pH above 8.0, moderately denaturing |
| Sodium deoxycholate | DOC | | 433 | Homogeneous | 0.2 | 4,200 | 16 | Keep pH above 8.0, avoid divalent cations moderately denaturing |
| **Cationic** | | | | | | | | |
| Cetyltrimethylammonium bromide | CTAB | | 364 | Homogeneous | 0.033 | 62,000 | | Highly denaturing |
| **Amphoteric** | | | | | | | | |
| 3-[(Cholamidopropyl)-dimethyl ammonio]-1-propanesulfonate | CHAPS | | 651 | Homogeneous | 0.5 | | | Fair solubilization, weakly denaturing, easily dialyzed |
| **Nonionic** | | | | | | | | |
| Polyoxyethylene (10) cetyl alcohol | | Brij 56 | 683 | Heterogeneous | 0.00014 | 130,000 | 12.9 | Good solubilization |
| Polyoxyethylene (20) cetyl alcohol | | Brij 58 | 1120 | Heterogeneous | 0.008 | 82,000 | 15.7 | |
| Polyoxyethylene (23) lauryl alcohol | | Brij 35 | 1200 | Heterogeneous | 0.58 | 49,000 | | |
| Polyoxyethylene (4-5) p-t-octyl phenol | | Triton X-45 | 405 | Heterogeneous | 0.0044 | | 10.4 | Selective extraction |
| Polyoxyethylene (7-8) p-t-octyl phenol | | Triton X-114 | 537 | Heterogeneous | 0.011 | | 12.4 | Selective extraction |
| Polyoxyethylene (9) p-t-octyl phenol | | Nonidet P-40 | 603 | Heterogeneous | 0.017 | 90,000 | 13.1 | Good solubilization, weakly denaturing |
| Polyoxyethylene (9-10) p-t-octyl phenol | | Triton X-100 | 625 | Heterogeneous | 0.016 | 90,000 | 13.5 | Good solubilization, weakly denaturing |
| Polyoxyethylene (9-10) nonylphenol | | Triton N-101 | 642 | Heterogeneous | 0.005 | 66,000 | 13.4 | Good solubilization, weakly denaturing |
| Polyoxyethylene (20) sorbitol monolaurate | | Tween 20 | 1230 | Heterogeneous | 0.006 | | 16.7 | Good solubilization, mildly denaturing |
| Polyoxyethylene (20) sorbitol monopalmitate | | Tween 40 | 1280 | Heterogeneous | 0.003 | | 15.6 | good washing |
| Polyoxyethylene (20) sorbitol monooleate | | Tween 80 | 1310 | Heterogeneous | 0.0013 | | 15.0 | |
| Octyl-β-glucoside | | | 292 | Homogeneous | 0.7 | | | Fair solubilization, easily dialyzed |

# Appendix IV

# BACTERIAL EXPRESSION

Most bacteria expression systems use inducible promoters to help control the level of expression of the foreign protein. Many foreign proteins are toxic to bacteria, and many of the recent improvements in bacterial expression systems have involved the development of better systems for inducible control. If toxicity is a problem with the protein under study, using systems that have lower levels of baseline expression will help overcome these difficulties. One particularly valuable system that has come into use recently is the T7 expression system of Studier and colleagues (Rosenberg et al. 1987; Studier and Moffatt 1986). Expression systems such as these are becoming increasingly important for the production of protein antigens in bacteria.

Methodologies for the production of the bacterial expression vectors are discussed in detail in a number of techniques manuals published recently. We recommend Sambrook et al. (1989).

## SCREENING FOR EXPRESSION IN BACTERIA—SDS LYSIS

1. Prepare lifts on nitrocellulose filters from bacterial colonies or from lambda plaques using standard techniques (see Sambrook et al. 1989, for examples of appropriate techniques). After the bacteria have grown to small colonies they are ready to lyse. Lifts from lambda plaques are ready to use immediately.

2. Prepare a 70°C water bath or a 70°C oven. Place two sheets of absorbent paper (Whatman 3MM or equivalent) in the bottom of a glass tray. Add enough 1% SDS to soak the paper thoroughly and leave a shallow puddle on top of the paper.

3. Transfer the tray to the 70°C bath or to the oven. Allow 15 min for the temperature to rise to 70°C.

4. Remove the tray. Quickly place the filters on the absorbent paper with the colony or plaque side up.

5. Return the tray to 70°C. Incubate for 30 min.

6. Remove the tray and allow to cool. Carefully remove the filters and float on a 3% BSA/PBS solution containing 0.02% sodium azide for several minutes. After the nitrocellulose filters have been wet thoroughly, submerge them in the BSA solution and incubate at room temperature with shaking for 2 hr or longer. The bacterial colonies will have long strings of DNA extending from the filters. These can either be ignored or removed by gentle rubbing with a gloved hand after the 2-hr incubation.

7. Wash the filters two times for 5 min each with PBS.

8. To locate colonies or phage that are expressing a particular protein, add the antibody in 3% BSA/PBS with 0.02% sodium azide (3 ml/100 cm$^2$). Use approximately 1 ml of tissue culture supernatant, 5 $\mu$l of ascites solution, or 10 $\mu$l of polyclonal sera per 100 cm$^2$. Incubate for 2 hr with shaking at room temperature.

9. Wash with four changes of PBS for 5 min each. Locate the antibodies using any of the detection techniques discussed in Chapter 12.

If using this assay to test different antibodies for binding, grow the bacteria as a lawn for harvesting or for high density infection with phage. Follow the remainder of the protocol as described, except cut the filter in appropriate sized pieces after blocking. Spot 1 $\mu$l of tissue culture supernatants or dilutions of ascites or polyclonal antibodies onto the filter. Incubate for 30 min at room temperature in a humid atmosphere. Wash and detect as above.

## SCREENING FOR EXPRESSION IN BACTERIA—CHLOROFORM LYSIS*

1. Prepare lifts on nitrocellulose filters from bacterial colonies or from lambda plaques using standard techniques (see Sambrook et al. 1989, for examples of appropriate techniques). After the bacteria have grown to small colonies they are ready to lyse. Lifts from lambda plaques are ready to incubate with block directly (step 5).

2. Prepare a chamber filled with chloroform vapor. Any glass container with a wide mouth and lid is suitable; a beaker with a top of aluminum foil is fine. Add 1–2 cm of chloroform to the bottom. Seal the top and leave at room temperature for 30 min.

3. Suspend the filters from small bulldog clips or paper clips on a narrow bar that is longer than the width of the mouth of the chloroform container. Remove the top and suspend the filters in the chloroform vapor. Replace the top.

4. Incubate for 30 min at room temperature.

5. Remove the filters and float on a 3% BSA/PBS solution containing 0.02% sodium azide for several minutes. After the nitrocellulose filters have been wet thoroughly, submerge them in the BSA solution and incubate at room temperature with shaking for 2 hr or longer. The bacterial colonies leave deposits of cellular debris on the filter. These can be removed by gentle rubbing with a gloved hand after the 2-hr incubation.

6. Wash the filters two times for 5 min each with PBS.

7. To locate colonies or phage that are expressing a particular protein, add the antibody in 3% BSA/PBS with 0.02% sodium azide (3 ml/100 cm$^2$). Use approximately 1 ml of tissue culture supernatant, 5 $\mu$l of ascites solution, or 10 $\mu$l of polyclonal sera per 100 cm$^2$. Incubate for 2 hr with shaking at room temperature.

8. Wash with four changes of PBS for 5 min each. Locate the antibodies using any of the detection techniques discussed in Chapter 12.

If using this assay to test different antibodies for binding, grow the bacteria as a lawn for harvesting or for high-density infection with phage. Follow the remainder of the protocol as described, except cut the filter in appropriately sized pieces after blocking. Spot 1 $\mu$l of tissue culture supernatants or dilutions of ascites or polyclonal antibodies onto the filter. Incubate for 30 min at room temperature in a humid atmosphere. Wash and detect as above.

*Helfman et al. (1983).

λgt11

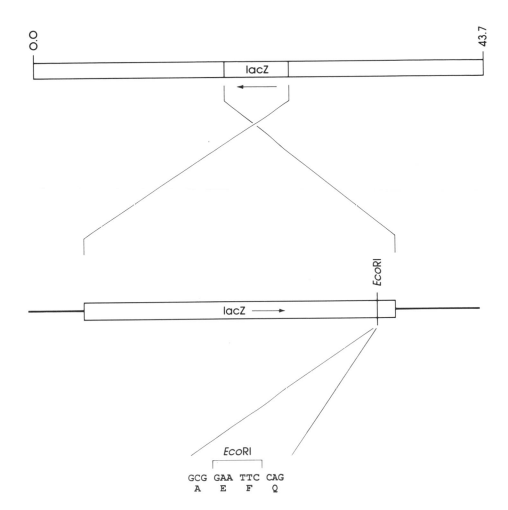

From Young and Davis (1983).

## T7 vectors

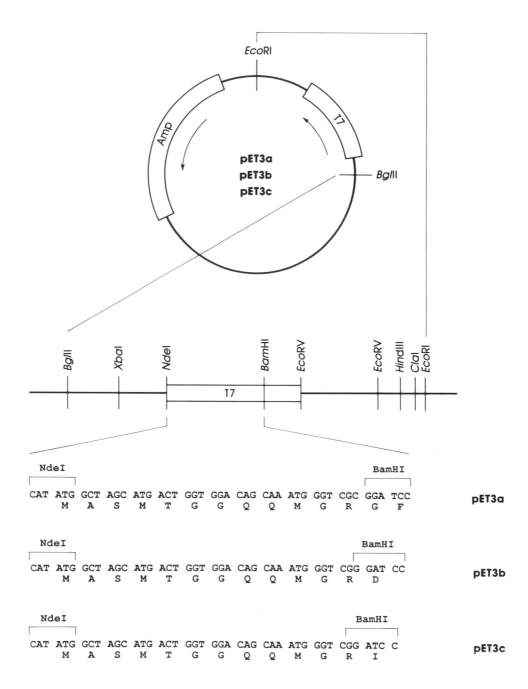

From Rosenberg et al. (1987).

# Trp fusion vectors

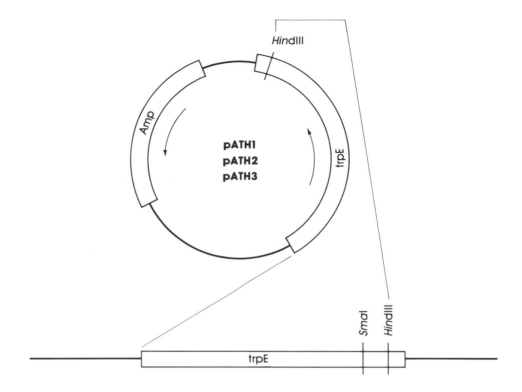

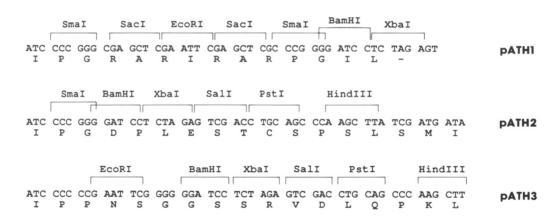

From Dieckmann and Tzagoloff (1985) and A. Tzagoloff (pers. comm.).

β-Gal fusion vectors

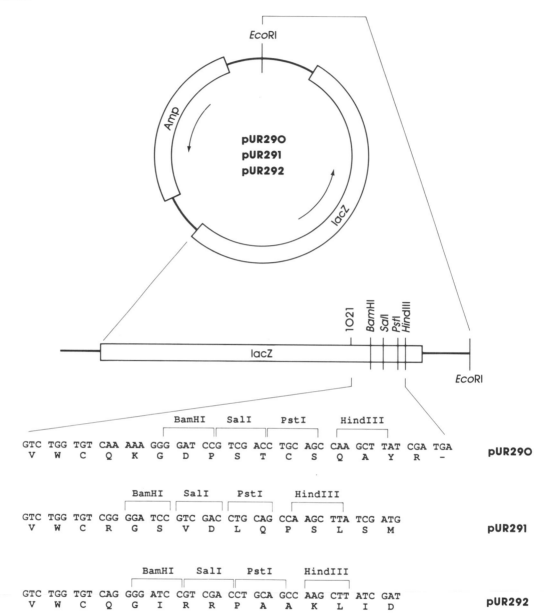

| | | | BamHI | | SalI | | PstI | | HindIII | | | | |
|---|---|---|---|---|---|---|---|---|---|---|---|---|---|
| GTC | TGG | TGT | CAA | AAA | GGG | GAT | CCG | TCG | ACC | TGC | AGC | CAA GCT TAT CGA TGA | |
| V | W | C | Q | K | G | D | P | S | T | C | S | Q A Y R – | **pUR290** |

| | | | BamHI | | SalI | | PstI | | HindIII | | | | |
|---|---|---|---|---|---|---|---|---|---|---|---|---|---|
| GTC | TGG | TGT | CGG | GGA | TCC | GTC | GAC | CTG | CAG | CCA | AGC | TTA TCG ATG | |
| V | W | C | R | G | S | V | D | L | Q | P | S | L S M | **pUR291** |

| | | | BamHI | | SalI | | PstI | | HindIII | | | | |
|---|---|---|---|---|---|---|---|---|---|---|---|---|---|
| GTC | TGG | TGT | CAG | GGG | ATC | CGT | CGA | CCT | GCA | GCC | AAG | CTT ATC GAT | |
| V | W | C | Q | G | I | R | R | P | A | A | K | L I D | **pUR292** |

# REFERENCES

Adler, F.L. and L.T. Adler. 1980. Passive hemagglutination and hemolysis for estimation of antigens and antibodies. *Methods Enzymol.* **70:** 455–466.

Åkerström, B. and L. Björck. 1986. A physicochemical study of protein G, a molecule with unique immunoglobulin G-binding properties. *J. Biol. Chem.* **261:** 10240–10247.

Åkerström, B., T. Brodin, K. Reis, and L. Björck. 1985. Protein G: A powerful tool for binding and detection of monoclonal and polyclonal antibodies. *J. Immunol.* **135:** 2589–2592.

Allen, L. 1967. Lymphatics and lymphoid tissues. *Annu. Rev. Physiol.* **29:** 197–224.

Allen, P.M., D.J. McKean, B.N. Beck, J. Sheffield, and L.H. Glimcher. 1985. Direct evidence that a class II molecule and a simple globular protein generate multiple determinants. *J. Exp. Med.* **162:** 1264–1274.

Alwine, J.C., D.J. Kemp, and G.R. Stark. 1977. Method for detection of specific RNAs in agarose gels by transfer to diazobenzyloxymethyl-paper and hybridization with DNA probes. *Proc. Natl. Acad. Sci.* **74:** 5350–5354.

Amersham International. 1984. *Radioisotope detection by fluorography and intensifying screens* (ed. R.A. Laskey). Amersham International, England.

Amkraut, A.A., J.S. Garvey, and D.H. Campbell. 1966. Competition on haptens. *J. Exp. Med.* **124:** 293–306.

Anderer, F.A. 1963. Preparation and properties of an artificial antigen immunologically related to tobacco mosaic virus. *Biochim. Biophys. Acta* **71:** 246–248.

Anderer, F.A. and H.D. Schlumberger. 1965. Properties of different artificial antigens immunologically related to tobacco mosaic virus. *Biochim. Biophys. Acta* **97:** 503–509.

Anderson, S. 1981. Shotgun DNA sequencing using cloned DNase I-generated fragments. *Nucleic Acids Res.* **9:** 3015–3027.

Anker, H.S. 1970. A solubilizable acrylamide gel for electrophoresis. *FEBS Lett.* **7:** 293.

Arnon, R., E. Maron, M. Sela, and C.B. Anfinsen. 1971. Antibodies reactive with native lysozyme elicited by a completely synthetic antigen. *Proc. Natl. Acad. Sci.* **68:** 1450–1455.

Atassi, M.Z. and A.F.S.A. Habeeb. 1972. Reaction of proteins with citraconic anhydride. *Methods Enzymol.* **25:** 546–553.

Atassi, M.Z. and R.G. Webster. 1983. Localization, synthesis, and activity of an antigenic site on influenza virus hemagglutinin. *Proc. Natl. Acad. Sci.* **80:** 840–844.

Avrameas, S. 1969. Coupling of enzymes to proteins with glutaraldehyde. Use of the conjugates for the detection of antigens and antibodies. *Immunochemistry* **6:** 43–52.

Avrameas, S. 1972. Enzyme markers: Their linkage with proteins and use in immuno-histochemistry. *Histochem. J.* **4:** 321–330.

Avrameas, S. and J. Uriel. 1966. Methode de marquage d'antigen et danticorps avec des enzymes et son application en immunodiffusion. *Cr. Acad. Sci. D* **262:** 2543–2545.

Avrameas, S. and T. Ternynck. 1969. The cross-linking of proteins with glutaraldehyde and its use for the preparation of immunoadsorbents. *Immunochemistry* **6:** 53–66.

Avrameas, S. and T. Ternynck. 1971. Peroxidase labelled antibody and Fab conjugates with enhanced intracellular penetration. *Immunochemistry* **8:** 1175–1179.

Axén, R., J. Porath, and S. Ernback. 1967. Chemical coupling of peptides and proteins to polysaccharides by means of cyanogen halides. *Nature* **214:** 1302–1304.

Bairoch, A. and J.-M. Claverie. 1988. Sequence patterns in protein kinases. *Nature* **331:** 22.

Barker, W.C. and M.O. Dayhoff. 1982. Viral *src* gene products are related to the catalytic chain of mammalian cAMP-dependent protein kinase. *Proc. Natl. Acad. Sci.* **79:** 2836–2839.

Bassiri, R.M., J. Dvorak, and R.D. Utiger. 1979. Thyrotropin-releasing hormone. In *Methods of hormone radioimmunoassay*, 2nd edition. (B.M. Jaffee and H.R. Behrman), pp. 46–47. Academic Press, New York.

Batteiger, B., W.J. Newhall, and R.B. Jones. 1982. The use of Tween 20 as a blocking agent in the immunological detection of proteins transferred to nitrocellulose membranes. *J. Immunol. Methods* **55:** 297–307.

Bauminger, S. and M. Wilchek. 1980. The use of

carbodiimides in the preparation of immunizing conjugates. *Methods Enzymol.* **70**:151–159.

Bayer, E.A. and M. Wilchek. 1980. The use of avidin-biotin complex as a tool in molecular biology. *Methods Biochem. Anal.* **26**: 1–45.

Bayer, E.A., M. Wilchek, and E. Skutelsky. 1976. Affinity cytochemistry: The localization of lectin and antibody receptors on erythrocytes via the avidin-biotin complex. *FEBS Lett.* **68**: 240–244.

Bazin, H. 1982. In *Protides of the biological fluids* (ed. H. Peters and L. Pergamon). Pergamon Press, New York.

Beachy, E.H., J.M. Seyer, J.B. Dale, W.A. Simpson, and A.H. Kang. 1981. Type-specific protective immunity evoked by synthetic peptide of *Streptococcus pyogenes* M protein. *Nature* **292**: 457–459.

Becker, J.M. and M. Wilchek. 1972. Inactivation by avidin of biotin-modified bacteriophage. *Biochim. Biophys. Acta* **264**: 165–170.

Beisiegel, U. 1986. Review: Protein blotting. *Electrophoresis* **7**: 1–18.

Belanger, L., C. Sylvestre, and D. Dufour. 1973. Enzyme-linked immunoassay for alpha-fetoprotein by competitive and sandwich procedures. *Clin. Chim. Acta* **48**: 15–18.

Berenbaum, M.C. 1959. The autoradiographic localization of intracellular antibody. *Immunology* **2**: 71–83.

Berkower, I., G.K. Buckenmeyer, and J.A. Berzofsky. 1986. Molecular mapping of a histocompatibility-restricted immunodominant T cell epitope with synthetic and natural peptides: Implications for T cell antigenic structure. *J. Immunol.* **136**: 2498–2503.

Berman, J.W. and R.S. Basch. 1980. Amplification of the biotin-avidin immunofluorescence technique. *J. Immunol. Methods* **36**: 335–338.

Berson, S.A., R.S. Yalow, A. Bauman, M.A. Rothschild, and K. Newerly. 1956. Insulin-I[131] metabolism in human subjects: demonstration of insulin binding globulin in the circulation of insulin treated subjects. *J. Clinical Invest.* **35**: 170–190.

Bethell, G.S., J.S. Ayers, W.S. Hancock, and M.T.W. Hearn. 1979. A novel method of activation of cross-linked agaroses with 1,1′-carbonyldiimidazole which gives a matrix for affinity chromatography devoid of additional charged groups. *J. Biol. Chem.* **254**: 2572–2574.

Bittner, M., P. Kupferer, and C.F. Morris. 1980. Electrophoretic transfer of proteins and nucleic acids from slab gels to diazobenzyloxymethyl cellulose or nitrocellulose sheets. *Anal. Biochem.* **102**: 459–471.

Björck, L. and G. Kronvall. 1984. Purification and some properties of streptococcal protein G, a novel IgG-binding reagent. *J. Immunol.* **133**: 969–974.

Björck, L., W. Kastern, G. Lindahl, and K. Wideback. 1987. Streptococcal protein G, expressed by streptococci or by *Escherichia coli*, has separate binding sites for human albumin and IgG. *Mol. Immunol.* **24**: 1113–1122.

Blakeslee, D. 1977. Immunofluorescence using dichlorotriazinylamino-fluorescein (DTAF). II. Preparation, purity and stability of the compound. *J. Immunol. Methods* **17**: 361–364.

Blakeslee, D. and M.G. Baines. 1976. Immunofluorescence using dichlorotriazinylaminofluorescein (DTAF). I. Preparation and fractionation of labelled IgG. *J. Immunol. Methods* **13**: 305–320.

Boehringer Mannheim Biochemicals. 1987. *Protease inhibitors.* Boehringer Mannheim, Indianapolis, Indiana.

Bolton, A.E. and W.M. Hunter. 1973. The labelling of proteins to high specific radioactivities by conjugation to a [125]I-containing acylating agent. *Biochem. J.* **133**: 529–539.

Bondi, A., G. Chieregatti, V. Eusebi, E. Fulcheri, and G. Bussolati. 1982. The use of β-galactosidase as a tracer in immunocytochemistry. *Histochemistry* **76**: 153–158.

Bonner, W.M. and R.A. Laskey. 1974. A film detection method for tritium-labelled proteins and nucleic acids in polyacrylamide gels. *Eur. J. Biochem.* **46**: 83–88.

Borras-Cuesta, F., A. Petit-Camurdan, and Y. Fedon. 1987. Engineering of immunogenic peptides by co-linear synthesis of determinants recognized by B and T cells. *Eur. J. Immunol.* **17**:1213–1215.

Bowen, B., J. Steinberg, U.K. Laemmli, and H. Weintraub. 1980. The detection of DNA-binding proteins by protein blotting. *Nucleic Acids Res.* **8**: 1–20.

Boyd, G.W. and W.S. Peart. 1968. Production of high-titre antibody against free angiotensin 2 *Lancet* **2**: 129.

Brada, D. and J. Roth. 1984. "Golden blot"—Detection of polyclonal and monoclonal antibodies bound to antigens on nitrocellulose by protein A-gold complexes. *Anal. Biochem.* **142**: 79–83.

Bradford, M.M. 1976. A rapid and sensitive method for the quantitation of microgram quantities of protein utilizing the principle of protein-dye binding. *Anal. Biochem.* **72**: 248–254.

Brenner, S. 1987. Phosphotransferase sequence homology. *Nature* **329**: 21.

Bukovsky, J. and R.H. Kennett. 1987. Simple and rapid purification of monoclonal antibodies from cell culture supernatants and ascites fluids by hydroxylapatite chromatography on analytical and preparative scales. *Hybridoma* **6**: 219–228.

Bullock, G.R. and P. Petrusz. 1982. *Techniques in Immunocytochemistry.* vol. 1. Academic Press, Orlando, Florida.

Burnette, W.N. 1981. "Western blotting": Electrophoretic transfer of proteins from sodium dodecyl sulfate-polyacrylamide gels to unmodified nitrocellulose and radiographic detection with antibody and radioiodinated protein A. *Anal. Biochem.* **112:** 195–203.

Burridge, K. 1976. Changes in cellular glycoproteins after transformations: Identification of specific glycoproteins and antigens in sodium dodecyl sulfate gels. *Proc. Natl. Acad. Sci.* **73:** 4457–4461.

Campbell, D.H., E. Luescher, and L.S. Lerman. 1951. Immunologic adsorbents. I. Isolation of antibody by means of a cellulose-protein antigen. *Proc. Natl. Acad. Sci.* **37:** 575–578.

Carroll, S.B. and A. Laughon. 1987. Production and purification of polyclonal antibodies to the foreign segment of β-galactosidase fusion proteins. In *DNA cloning: A practical approach* (ed. D.M. Glover), vol. 3, pp. 89–111. IRL Press, Oxford, Washington, D.C.

Catt, K. and G.W. Tregear. 1967. Solid-phase radioimmunoassay in antibody-coated tubes. *Science* **153:** 1570–1572.

Cease, K.B., I. Berkower, J. York-Jolley, and J.A. Berzofsky. 1986. Cell clones specific for an amphipathic helical region of sperm whale myoglobin show differing fine specificities for synthetic peptides. *J. Exp. Med.* **164:** 1779–1784.

Chase, M.W. 1967. Production of antiserum. *Methods Immunol. Immunochem.* **1:** 197–209.

CIBA Foundation. 1986. *Synthetic peptides as antigens* (ed. R. Porter and J. Whelan). John Wiley and Sons, Avon, England.

Civin, C.I. and M.L. Banquerigo. 1983. Rapid, efficient cloning of murine hybridoma cells in low gelaton temperature agarose. *J. Immunol. Methods* **61:** 1–8.

Cleveland, D.W., S.G. Fischer, M.W. Kirschner, and U.K. Laemmli. 1977. Peptide mapping by limited proteolysis in sodium dodecyl-sulfate and analysis by gel electrofocusing. *J. Biol. Chem.* **252:** 1102–1106.

Contreras, M.A., W.F. Bale, and I.L. Spar. 1983. Iodine monochloride (IC1) iodination techniques. *Methods Enzymol.* **92:** 277–292.

Coons, A.H. 1956. Histochemistry with labeled antibody. *Int. Rev. Cytol.* **5:** 1–23.

Coons, A.H. and M.H. Kaplan. 1950. Localization of antigen in tissue cells. II. Improvements in a method for the detection of antigen by means of fluorescent antibody. *J. Exp. Med.* **91:** 1–13.

Coons, A.H., H.J. Creech, and R.N. Jones. 1941. Immunological properties of an antibody containing a fluorescent group. *Proc. Soc. Exp. Biol. Med.* **47:** 200–205.

Craven, G.R., E. Steers, Jr., and C.B. Anfinsen. 1965. Purification, composition, and molecular weight of the β-galactosidase of *Escherichia coli* K12. *J. Biol. Chem.* **240:** 2468–2477.

Cross, F.R., E.A. Garber, D. Pellman, and H. Hana-

fusa. 1984. A short sequence in the p60$^{src}$ N terminus is required for p60$^{src}$ myristylation and membrane association and for cell transformation. *Mol. Cell. Biol.* **4:** 1834–1842.

Cuatrecasas, P., M. Wilchek, and C.B. Anfinsen. 1968. Selective enzyme purification by affinity chromatography. *Biochemistry* **61:** 636–643.

Cuello, A.C., J.V. Priestley, and C. Milstein. 1982. Immunocytochemistry with internally labeled monoclonal antibodies. *Proc. Natl. Acad. Sci.* **79:** 665–669.

Danscher, G. 1981. Localization of gold in biological tissue. A photochemical method for light and electronmicroscopy. *Histochemistry* **71:** 81–88.

Danscher, G. and J.O.R. Nörgaard. 1983. Light microscopic visualization of colloidal gold on resin-embedded tissue. *J. Histochem. Cytochem.* **31:** 1394–1398.

Davis, B.J. 1964. Disc electrophoresis. II. Method and application to human serum proteins. *Ann. N.Y. Acad. Sci.* **121:** 404–427.

Dayhoff, M.O. 1978. *Atlas of protein sequence and structure*, vol. 5 (suppl. 3). National Biomedical Research Foundation.

de St. Groth, F. and D. Scheidegger. 1980. Production of monoclonal antibodies: Strategy and tactics. *J. Immunol. Methods* **35:** 1–21.

Deininger, P.L. 1983. Random subcloning of sonicated DNA: Application to shotgun DNA sequence analysis. *Anal. Biochem.* **129:** 216–223.

Deisenhofer, J. 1981. Crystallographic refinement and atomic models of a human Fc fragment and its complex with fragment B of protein A form *Staphylococcus aureus* at 2.9- and 2.8-Å resolution. *Biochemistry* **20:** 2361–2370.

Diano, M., A. Le Bivic, and M. Hirn. 1987. A method for the production of highly specific polyclonal antibodies. *Anal. Biochem.* **166:** 224–229.

Dieckmann, C.L. and A. Tzagoloff. 1985. Assembly of the mitochondrial membrane system. *CBP6*, a yeast nuclear gene necessary for synthesis of cytochrome *b*. *J. Biol. Chem.* **260:** 1513–1520.

Dienes, L. 1936. The specific immunity response and the healing of infectious diseases. *Arch. Pathol.* **21:** 357–386.

Dienes, L. and E.W. Schoenheit. 1930. Certain characteristics of the infections processes in connection with the influence exerted in the immunity response. *J. Immunol.* **19:** 41–61.

Doolittle, R.F., ed. 1976. *Of URFs and ORFs: A primer on how to analyze derived amino acid sequences.* University Science, Mill Valley, California.

Dreesman, G.R., Y. Sanchez, I. Ionescu-Matiu, J.T. Sparrow, H.R. Six, D.L. Peterson, F.B. Hollinger, and J.L. Melnick. 1982. Antibody to hepatitis B surface antigen after a single inoculation of uncoupled synthetic HBsAG peptides. *Nature* **295:** 158–160.

Druguet, M. and M.B. Pepys. 1977. Enumeration of lymphocyte populations in whole peripherial blood with alkaline phosphatase-labelled reagents. A method for routine clinical use. *Clin. Exp. Immunol.* **29**: 162–167.

Ellouz, F., A. Adam, R. Ciorbaru, and E. Lederer. 1974. Minimal structural requirements for adjuvant activity of bacterial peptidoglycan derivatives. *Biochem. Biophys. Res. Commun.* **59**: 1317–1325.

Engvall, E. and P. Perlmann. 1971. Enzyme-linked immunosorbent assay (ELISA) quantitative assay of immunoglobulin G. *Immunochemistry* **8**: 871–879.

Engvall, E. and P. Perlmann. 1972. Enzyme-linked immunosorbent assay, ELISA. III. Quantitation of specific antibodies by enzyme-labeled anti-immunoglobulin in antigen-coated tubes. *J. Immunol.* **109**: 129–135.

Ey, P.L., S.J. Prowse, and C.R. Jenkin. 1978. Isolation of pure $IgG_1$, $IgG_{2a}$, and $IgG_{2b}$ immunoglobulins from mouse serum using protein A-Sepharose. *Biochemistry* **15**: 429–436.

Farr, A.G. and P.K. Nakane. 1981. Immunohistochemistry with enzyme labeled antibodies: A brief review. *J. Immunol. Methods* **47**: 129–144.

Faulk, W.P. and G.M. Taylor. 1971. An immunocolloid method for the electron microscope. *Immunochem.* **8**: 1081–1083.

Feramisco, J.R., D.B. Glass, and E.G. Krebs. 1980. Optimal spatial requirements for the location of basic residues in peptide substrates for the cyclic AMP-dependent protein kinase. *J. Biol. Chem.* **255**: 4240–4245.

Fischer, S.G. 1983. Peptide mapping in gels. *Methods Enzymol.* **100**: 424–430.

Forsgren, A. and J. Sjöquist. 1966. "Protein A" from *S. aureus*. I. Pseudo-immune reaction with human γ-globulin. *J. Immunol.* **97**: 822–827.

Foung, S.K.H., D.T. Sasaki, F.C. Grumet, and E.C. Engleman. 1982. Production of functional human T-T hybridomas in selection medium lacking aminopterin and thymidine. *Proc. Natl. Acad. Sci.* **79**: 7484–7488.

Fraker, P.J. and J.C. Speck, Jr. 1978. Protein and cell membrane iodinations with a sparingly soluble chloroamide, 1,3,4,6-tetrachloro-3a,6a-diphenylglycoluril. *Biochem. Biophys. Res. Commun.* **80**: 849–857.

Francis, M.J., G.Z. Hastings, A.D. Syred, B. McGinn, F. Brown, and D.J. Rowlands. 1987. Non-responsiveness to a foot-and-mouth disease virus peptide overcome by addition of foreign helper T-cell determinants. *Nature* **300**: 168–170.

Freund, J. 1956. The mode of action of immunologic adjuvants. *Adv. Tuberc. Res.* **7**: 130–148.

Freund, J. and K. McDermott. 1942. Sensitization to horse serum by means of adjuvants. *Proc. Soc. Exp. Biol. Med.* **49**: 548–553.

Furth, A.J. 1980. Review: Removing unbound detergent from hydrophobic proteins. *Anal. Biochem.* **109**: 207–215.

Galfre, G. and C. Milstein. 1981. Preparation of monoclonal antibodies: Strategies and procedures. *Methods Enzymol.* **73**: 3–46.

Galfre, G., C. Milstein, and B. Wright. 1979. Rat × rat hybrid myelomas and a monoclonal anti-Fd portion of mouse IgG. *Nature* **277**: 131–133.

Galfre, G., S.C. Howe, C. Milstein, G.W. Butcher, and J.C. Howard. 1977. Antibodies to major histocompatibility antigens produced by hybrid cell lines. *Nature* **266**: 550–552.

Gefter, M.L., D.H. Margulies, and M.D. Scharff. 1977. A simple method for polyethylene glycol-promoted hybridization of mouse myeloma cells. *Somatic Cell Genet.* **3**: 231–236.

Gershoni, J.M. and G.E. Palade. 1982. Electrophoretic transfer of sodium dodecyl sulfate-polyacrylamide gels to a positively charged membrane filter. *Anal. Biochem.* **124**: 396–405.

Gershoni, J.M. and G.E. Palade. 1983. Review: Protein blotting: Principles and applications. *Anal. Biochem.* **131**: 1–15.

Gersten, D.M. and J.J. Marchalonis. 1978. A rapid, novel method for the solid-phase derivatization of IgG antibodies for immune-affinity chromatography. *J. Immunol. Methods* **24**: 305–309.

Gillette, R.W. 1987. Alternatives to pristane priming for ascitic fluid and monoclonal antibody production. *J. Immunol. Methods* **88**: 21–23.

Giloh, H. and J.W. Sedat. 1982. Fluorescence microscopy: Reduced photobleaching of rhodamine and fluorescein protein conjugates by *n*-propyl gallate. *Science (Abstr.)* **217**: 1252–1255.

Glenney, J. 1986. Antibody probing of western blots which have been stained with india ink. *Anal. Biochem.* **156**: 315–319.

Glenney, J.R., Jr., P. Glenney, and K. Weber. 1982. Erythroid spectrin, brain fodrin, and intestinal brush border proteins (TW-260/240) are related molecules containing a common calmodulin-binding subunit bound to a variant cell type-specific subunit. *Proc. Natl. Acad. Sci.* **79**: 4002–4005.

Glenny, A.T., C.G. Pope, H. Waddington, and U. Wallace. 1926. Immunological notes. *J. Pathol.* **29**: 31–40.

Goding, J.W. 1976. Conjugation of antibodies with fluorochromes: Modifications to the standard methods. *J. Immunol. Methods* **13**: 215–226.

Goding, J.W. 1978. Use of staphylococcal protein A as an immunological reagent. *J. Immunol. Methods* **20**: 241–253.

Goding, J.W. 1987. *Monoclonal antibodies: Principles and practice.* 2nd ed. Academic Press, London.

Goebel, W.F. 1938. Chemo-immunological studies on conjugated carbohydrate-proteins XLL: The immunological properties of an artificial antigen containing cellobiuronic acid. *J. Exp. Med.* **68**: 469.

Goldstein, L., N.R. Klinman, and L.A. Manson. 1973. A microtest radioimmunoassay for noncytotoxic tumor-specific antibody to cell-surface antigens. *J. Natl. Cancer Inst.* **51**: 1713–1715.

Good, M.F., W.L. Maloy, M.N. Lunde, H. Margalit, J.L. Cornette, G.L. Smith, B. Moss, L.H. Miller, and J.A. Berzofsky. 1987. Construction of synthetic immunogen: Use of new T-helper epitope on malaria circumsporozoite protein. *Science* **235**: 1059–1062.

Goodfriend, T.L., L. Levine, and G.D. Fasman. 1964. Antibodies to Bradykinin and angiotensin: A use of carbodiimides in immunology. *Science* **144**: 1344–1346 (Abstr.).

Gordon, J., B. Rose, and A.H. Sehon. 1958. Detection of "non-precipitating" antibodies in sera of individuals allergic to ragweed pollen by an in vitro method. *J. Exp. Med.* **108**: 37–51.

Gossrau, R. 1973. Splitting of naphthol AS-BI $\beta$-galactopyranoside by acid $\beta$-galactosidase. *Histochemie* **37**: 89–91.

Graham, R.C., Jr. and M.J. Karnovsky. 1966. The early stages of absorption of injected horseradish peroxidase in the proximal tubules of mouse kidney ultrastructural cytochemistry by a new technology. *J. Histochem. Cytochem.* **14**: 291–302.

Graham, R.C., Jr., U. Lundholm, and M.J. Karnovsky. 1964. Cytochemical demonstration of peroxidase activity with 3-amino-9-ethylcarbazole. *J. Histo. Chem.* **13**: 150–152.

Grantham, R., C. Gautier, M. Gouy, M. Jacobzone, and R. Mercier. 1981. Codon catalog usage is a genome strategy modulated for gene expressivity. *Nucleic Acids Res.* **9**: r43–r74.

Green, N., H. Alexander, A. Olson, S. Alexander, T.M. Shinnick, J.G. Sutcliffe, and R.A. Lerner. 1982. Immunogenic structure of the influenza virus hemagglutinin. *Cell* **28**: 477–487.

Green, N.M. 1963. Avidin. I. The use of [$^{14}$C]biotin for kinetic studies and for assay. *Biochem. J.* **89**: 585–591.

Greenwood, F.C., W.M. Hunter, and J.S. Glover. 1963. The preparation of $^{131}$I-labelled human growth hormone of high specific radioactivity. *Biochem. J.* **89**: 114–123.

Guesdon, J.-L. and S. Avrameas. 1976. Polyacrylamide-agarose beads for the preparation of effective immunoabsorbents. *J. Immunol. Methods* **11**: 129–133.

Guesdon, J.-L., T. Ternynck, and S. Avrameas. 1979. The use of avidin-biotin interaction of immunoenzymatic techniques. *J. Histochem. Cytochem.* **27**: 1131–1139.

Guillet, J.G., M.-Z. Lai, T.J. Briner, J.A. Smith, and M.L. Gefter. 1986. Interaction of peptide antigens and class II major histocompatibility complex antigens. *Nature* **324**: 260–262.

Gurvich, A.E. and G.I. Drizlikh. 1964. Use of antibodies on an insoluble support for specific detection of radioactive antigens. *Nature* **5**: 648–649.

Hackett, C.J., B. Dietzschold, W. Gerhard, B. Ghrist, R. Knorr, D. Gillessen, and F. Melchers. 1983. Influenza virus site recognized by a murine helper T cell specific for H1 strains. *J. Exp. Med.* **158**: 294–302.

Hancock, K. and V.C.W. Tsang. 1983. India ink staining of proteins on nitrocellulose paper. *Anal. Biochem.* **133**: 157–162.

Hansen, J.N. 1976. Electrophoresis of ribonucleic acid on a polyacrylamide gel which contains disulfide cross-linkages. *Anal. Biochem.* **76**: 37–44.

Hansen, J.N., B.H. Pheiffer, and J.A. Boehnert. 1980. Chemical and electrophoretic properties of solubilizable disulfide gels. *Anal. Biochem.* **105**: 192–301.

Hansen, P.A. 1967. Spectral data of fluorescent tracers. *Acta Histochem.* (suppl. 1) **7**: 167–180.

Hardy, H. and L. Heimer. 1977. A safer and more sensitive substitute for diamino-benzidine in the light microscopic demonstration of retrograde and anterograde axonal transport of HRP. *Neurosci. Lett.* **5**: 235–240.

Hawkes, R. 1986. The dot immunobinding assay. *Methods Enzymol.* **121**: 484–491.

Hawkes, R., E. Niday, and J. Gordon. 1982. A dot-immunobinding assay for monoclonal and other antibodies. *Anal. Biochem.* **119**: 142–147.

Hearn, M.T.W. 1987. 1,1'-Carbonyldiimidazole-mediated immobilation of enzymes and affinity ligands. *Methods Enzymol.* **135**: 102–117.

Hearn, M.T.W., E.L. Harris, G.S. Bethell, W.S. Hancock, and J.A. Ayers. 1981. Application of 1,1'-carbonyldiimidazole-activated matrices for the purification of proteins. III. The use of 1,1'-carbonyldiimidazole-activated agaroses in the biospecific affinity chromatographic isolation of serum antibodies. *J. Chromatogr.* **218**: 509–518.

Heggeness, M.H. and J.F. Ash. 1977. Use of the avidin-biotin complex for the localization of actin and myosin with fluorescence microscopy. *J. Cell Biol.* **73**: 783–788.

Heimer, G.V. and C.E.D. Taylor. 1974. Improved mountant for immunofluorescence preparations. *J. Clin. Pathol.* **27**: 254–256.

Heitzmann, H. and F.M. Richards. 1974. Use of the avidin-biotin complex for specific staining of biological membranes in electron microscopy. *Proc. Natl. Acad. Sci.* **71**: 3537–3541.

Helenius, A. and K. Simons. 1975. Solubilization of membranes by detergents. *Biochim. Biophys. Acta* **415**: 29–79.

Helenius, A., D.R. McCaslin, E. Fries, and C. Tanford. 1979. Properties of detergents. *Methods Enzymol.* **56**: 734–749.

Helfman, D.M., J.R. Feramisco, J.C. Fiddes, G.P.

Thomas, and S.H. Hughes. 1983. Identification of clones that encode chicken tropomysin by direct immunological screening of a cDNA expression library. *Proc. Natl. Acad. Sci.* **80:** 31–35.

Herbert, W.J. and F. Kristensen. 1986. Laboratory animal techniques for immunology. In *Handbook of experimental immunology* (ed. D.M. Weir et al.), chap. 133, pp. 133.1–133.36. Blackwell Scientific, Oxford.

Herbrink, P., F.J. Van Bussel, and S.O. Warnaar. 1982. The antigen spot test (AST): A highly sensitive assay for the detection of antibodies. *J. Immunol. Methods* **48:** 293–298.

Higgins, R.C. and M.E. Dahmus. 1979. Rapid visualization of protein bands in preparative SDS-polyacrylamide gels. *Anal. Biochem.* **93:** 257–260.

Hoare, D.G. and D.E. Koshland, Jr. 1967. A method for the quantitative modification and estimation of carboxylic acid groups in proteins. *J. Biol. Chem.* **242:** 2447–2453.

Hofmann, K., F.M. Finn, and Y. Kiso. 1978. Avidin-biotin affinity columns. General methods for attaching biotin to peptides and proteins. *J. Amer. Chem. Soc.* **100:** 3585-3590.

Holgate, C.S., P. Jackson, P.N. Cownen, and C.C. Bird. 1983. Immunogold–silver staining: New method of immunostaining with enhanced sensitivity. *J. Histochem. Cytochem.* **31:** 938–944.

Holland, V.R., B.C. Saunders, F.L. Rose, and A.L. Walpole. 1974. Safer substitute for benzidine in detection of blood. *Tetrahedron* **30(18):** 3299.

Hoogenraad, N.J. and C.J. Wraight. 1986. The effect of pristane on ascites tumor formation and monoclonal antibody production. *Methods Enzymol.* **121:** 375–381.

Hopp, T.P. and K.R. Woods. 1981. Prediction of protein antigenic determinants from amino acid sequences. *Proc. Natl. Acad. Sci.* **78:** 3824–3828.

Hopp, T.P. and K.R. Woods. 1983. A computer program for predicting protein antigenic determinants. *Mol. Immunol.* **20:** 483–489.

Horibata, K. and A.W. Harris. 1970. Mouse myelomas and lymphomas in culture. *Exp. Cell Res.* **60:** 61–77.

Horisberger, M. 1979. Evaluation of colloidal gold as a cytochemical marker for transmission and scanning electron microscopy. *Biol. Cell.* **36:** 253–258.

Horisberger, M. and J. Rosset. 1977. Collodial gold, a useful marker for transmission and scanning electron microscopy. *J. Histo. Cytochem.* **25:** 295–305.

Howard, A.N. and F. Wild. 1957. The reactions of diazonium compounds with amino acids and proteins. *Biochem. J.* **65:** 651–659.

Hsu, S.-M., L. Raine, and H. Fanger. 1981. Use of avidin-biotin-peroxidase complex (ABC) in immunoperoxidase techniques: A comparison between ABC and unlabeled antibody (PAP) procedures. *J. Histochem. Cytochem.* **29:** 577–580.

Hsu, U.-M. and E. Soban. 1982. Color modification of diaminobenzidine (DAB) precipitation by metallic ions and its application for double immunohistochemistry. *J. Histochem. Cytochem.* **30:** 1079–1082.

Hsu, Y.-H. 1984. Immunogold for detection of antigen on nitrocellulose paper. *Anal. Biochem.* **142:** 221–225

Hubbard, A.L. and Z.A. Cohn. 1972. The enzymatic iodination of the red cell membrane. *J. Cell Biol.* **55:** 290–405.

Huet, J., A. Sentenac, and P. Fromageot. 1982. Spot-immunodetection of conserved determinants in eukaryotic RNA polymerases. *J. Biol. Chem.* **257:** 2613–2618.

Hunter, R., F. Strickland, and F. Kézdy. 1981. The adjuvant activity of nonionic block polymer surfactants. I. The role of hydrophile-kipophile balance. *J. Immunol.* **127:** 1244–1250.

Hunter, W.M. and F.C. Greenwood. 1962. Preparation of iodine-131 labelled human growth hormone of high specific activity. *Nature* **194:** 495–496.

Hurwitz, J.L., E. Heber-Katz, C.J. Hackett, and W. Gerhard. 1984. Characterization of the murine $T^H$ response to influenza virus hemagglutinin: Evidence for three major specificities. *J. Immunol.* **133:** 3371–3377.

Jackson, D.C., J.M. Murray, D.O. White, C.N. Fagan, and G.W. Tregear. 1982. Antigenic activity of a synthetic pepetide comprising the "loop" region of influenza virus hemagglutinin. *Virology* **120:** 273–276.

Jensenius, J.C. and A.F. Williams. 1974. The binding of anti-immunoglobulin antibodies to rat thymocytes and thoracic duct lymphocytes. *Eur. J. Immunol.* **4:** 91–97.

Johnson, A.G., S. Gaines, and M. Landy. 1956. Studies on the O antigen of *Salmonella typhosa.* V. Enhancement of antibody response to protein antigens by the purified lipopolysaccharide. *J. Exp. Med.* **103:** 225–246.

Johnson, D.A., J.W. Gautsch, J.R. Sportsman, and J.H. Elder. 1984. Improved technique utilizing nonfat dry milk for analysis of proteins and nucleic acids transferred to nitrocellulose. *Gene Anal. Tech.* **1:** 3–8.

Johnson, G.D. and G.M. de C. Nogueira Araujo. 1981. A simple method of reducing the fading of immunofluorescence during microscopy. *J. Immunol. Methods* **43:** 349–350.

Johnson, G.D., R.S. Davidson, K.C. McNamee, G. Russell, D. Goodwin, and E.J. Holborow. 1982. Fading of immunofluorescence during microscopy: A study of the phenomenon and its remedy. *J. Immunol. Methods* **55:** 213–242.

Jovin, T.M., M.L. Dante, and A. Chrambach. 1970. Multiphasic buffer systems output, PB 196085-19609, *National Tech. Info. Service.*

Kalderon, D., B.L. Roberts, W.D. Richardson, and A.E. Smith. 1984. A short amino acid sequence able to specify nuclear location. *Cell* **39**: 499-509.

Kamps, M.P., S.S. Taylor, and B.M. Sefton. 1984. Direct evidence that oncogenic tyrosine kinases and cyclic AMP-dependent protein kinase have homologous ATP-binding sites. *Nature* **310**: 589–592.

Katz, M.E., R.M. Maizels, L. Wicker, A. Miller, and E.E. Sercarz. 1982. Immunological focusing by the mouse major histocompatibility complex: Mouse strains confronted with distantly related lysozymes confine their attention to very few epitopes. *Eur. J. Immunol.* **12**: 535–540.

Kawamoto, T., J.D. Sato, D.B. McClure, and G.II. Sato. 1986. Serum-free medium for the growth of NS-1 mouse myeloma cells and the isolation of NS-1 hybridomas. *Methods Enzymol.* **121**: 266–277.

Kearney, J.F., A. Radbruch, B. Liesegang, and K. Rajewsky. 1979. A new mouse myeloma cell line that has lost immunoglobulin expression but permits the construction of antibody-secreted hybrid cell lines. *J. Immunol.* **123**: 1548–1550.

Kendall, C., I. Ionescu-Matiu, and G.R. Dressman. 1983. Utilization of the biotin/avidin system to amplify the sensitivity of enzyme-linked immunosorbent assay (ELISA). *J. Immunol. Methods* **56**: 329–339.

Kennett, R.H. 1978. Cell fusion. *Methods Enzymol.* **58**: 345–359.

Kessler, S.W. 1975. Rapid isolation of antigens from cells with a staphylococcal protein A-antibody adsorbent: Parameters of the interaction of antibody-antigen complexes with protein A. *J. Immunol.* **115**: 1617–1623.

Kessler, S.W. 1976. Cell membrane antigen isolation with the staphylococcal protein A-antibody adsorbent. *J. Immunol.* **117**: 1482–1490.

Kessler, S.W. 1981. Use of protein A-bearing staphylococci for the immunoprecipitation and isolation of antigens from cells. *Methods Enzymol.* **73**: 442–458.

Kilmartin, J.V. and A.E.M. Adams. 1984. Structural rearrangements of tubulin and actin during the cell cycle of the yeast *Saccharomyces*. *J. Cell Biol.* **98**: 922–933.

Kilmartin, J.V., B. Wright, and C. Milstein. 1982. Rat monoclonal antitubulin antibodies derived by using a new nonsecreting rat cell line. *J. Cell Biol.* **93**: 576–582.

Kitagawa, T. and T. Aikawa. 1976. Enzyme coupled immunoassay of insulin using a novel coupling reagent. *J. Biochem.* **79**: 233–236.

Kleid, D.G., D. Yansura, B. Small, D. Dowbenko, D.M. Moore, M.J. Grubman, P.D. McKercher, D.O. Morgan, B.H. Robertson, and H.L. Bachrach. 1981. Cloned viral protein vaccine for foot-and-mouth disease: Responses in cattle and swine. *Science* **214**: 1125–1129.

Klein, J., F. Figueroa, and C.S. David. 1983. *H-2* haplotypes, genes and antigens: Second Listing. II. The *H-2* complex. *Immunogenetics* **17**: 553–596.

Knowles, R.W. 1987. Two-dimensional gel analysis of transmembrane proteins. In *Histocompatibility testing* (ed. B. Dupont), pp. 1–45. Springer Verlag, New York.

Knudsen, K.A. 1985. Proteins transferred to nitrocellulose for use as immunogens. *Anal. Biochem.* **147**: 285–288.

Köhler, G. and C. Milstein. 1975. Continuous cultures of fused cells secreting antibody of predefined specificity. *Nature* **256**: 495–497.

Köhler, G. and C. Milstein. 1976. Derivation of specific antibody-producing tissue culture and tumor lines by cell fusion. *Eur. J. Immunol.* **6**: 511–519.

Köhler, G., S.C. Howe, and C. Milstein. 1976. Fusion between immunoglobulin-secreting and nonsecreting myeloma cell lines. *Eur. J. Immunol.* **6**: 292–295.

Kohn, J. and M. Wilchek. 1984. The use of cyanogen bromide and other novel cyanylating agents for the activation of polysaccharide resins. *Appl. Biochem. Biotechnol.* **9**: 285–304.

Korn, A.H., S.H. Feairheller, and E.M. Filachione. 1971. Glutaraldehyde: Nature of the reagent. *J. Mol. Biol.* **65**: 525–529.

Kovář, J. and F. Franěk. 1986. Serum-free medium for hybridoma and parental myeloma cell cultivation. *Methods Enzymol.* **121**: 277–292.

Kronvall, G. 1973. A surface component in group A, C, and G streptococci with non-immune reactivity for immunoglobulin G. *J. Immunol.* **111**: 1401–1406.

Kronvall, G., U.S. Seal, J. Finstad, and R.C. Williams, Jr. 1970. Phylogenetic insight into evolution of mammalian Fc fragment of γG globulin using staphylococcal protein A. *J. Immunol.* **104**: 140–147.

Kuno, H. and H.K. Kihara. 1967. Simple microassay of protein with membrane filter. *Nature* **215**: 974.

Kyhse-Andersen, J. 1984. Electroblotting of multiple gels: A simple apparatus without buffer tank for rapid transfer of proteins from polyacrylamide to nitrocellulose. *J. Biochem. Biophys. Methods* **10**: 203–209.

Kyte, J. and R.F. Doolittle. 1982. A simple method for displaying the hydropathic character of a protein. *J. Mol. Biol.* **157**: 105–132.

Lacy, M.J. and E.W. Voss, Jr. 1986. A method to induce immune polyclonal ascites fluid in BALB/c mice using Sp2/O-Ag14 cells. *J. Immunol. Methods* **87**: 169–177.

Laemmli, E.K. 1970. Cleavage of structural proteins during the assembly of the head of bacteriophage T4. *Nature* **227**: 680–685.

Lane, D.P. and E.B. Lane. 1981. A rapid antibody assay system for screening hybridoma cultures. *J. Immunol. Methods* **47**: 303–307.

Langone, J.J. 1978. [$^{125}$I]protein A: A tracer for general use in immunoassay. *J. Immunol. Methods* **24:** 269–285.

Langone, J.J. 1980. Radioiodination by use of the Bolton-Hunter and related reagents. *Methods Enzymol.* **70:** 221–247.

Langone, J.J. 1982. Protein A of *Staphylococcus aureus* and related immunoglobulin receptors produced by streptococci and pneumonococci. *Adv. Immunol.* **32:** 157–251.

Lasch, J. and R. Koelsch. 1978. Enzyme leakage and multipoint attachment of agarose-bound enzyme preparations. *Eur. J. Biochem.* **82:** 181–186.

Laskey, R.A. and A.D. Mills. 1975. Quantitative detection of $^3$H and $^{14}$C in polyacrylamide gels by fluorography. *Eur. J. Biochem.* **56:** 335–341.

Lathe, R. 1985. Synthetic oligonucleotide probes deduced from amino acid sequence data. Theoretical and practical considerations. *J. Mol. Biol.* **183:** 1–12.

Layne, E. 1957. Spectrophotometric and turbidimetric methods for measuring proteins. *Methods Enzymol.* **3:** 447–454.

Leary, J.J., D.J. Brigati, and D.C. Ward. 1983. Rapid and sensitive colorimetric method for visualizing biotin-labeled DNA probes hybridized to DNA or RNA immobilized on nitrocellulose: Bio-blots. *Proc. Natl. Acad. Sci.* **80:** 4045–4049.

Leclerc, C., G. Przewlocki, M.-P. Schutze, and L. Chedid. 1987. A synthetic vaccine constructed by copolymerization of B and T cell determinants. *Eur. J. Immunol.* **17:** 269–273.

Lee, C., A. Levin, and D. Branton. 1987. Copper staining: A five-minute protein stain for sodium dodecyl sulfate-polyacrylamide gels. *Anal. Biochem.* **166:** 308–312.

Leppard, K., N. Totty, M. Waterfield, E. Harlow, J. Jenkins, and L. Crawford. 1983. Purification and partial amino acid sequence analysis of the cellular tumour antigen, p53, from mouse SV40-transformed cells. *EMBO J.* **2:** 1993–1999.

Lerner, R.A. 1982. Tapping the immunological repertoire to produce antibodies of predetermined specificity. *Nature* **299:** 592–596.

Lerner, R.A. 1984. Antibodies of predetermined specificity in biology and medicine. *Adv. Immunol.* **36:** 1–44.

Lerner, R.A., N. Green, H. Alexander, F.-T. Liu, J.G. Sutcliffe, and T.M. Shinnick. 1981. Chemically synthesized peptides predicted from the nucleotide sequence of the hepatitis B virus genome elicit antibodies reactive with the native envelope protein of Dane particles. *Proc. Natl. Acad. Sci.* **78:** 3404–3407.

Lieberman, R., J.O.A. Douglas, and N. Mantel. 1960. Production in mice of ascitic fluid containing antibodies induced by *Staphylococcus*- or *Salmonella*-adjuvant mixtures. *J. Immunol.* **84:** 514–429.

Ling, C.M. and L.R. Overby. 1972. Prevalence of hepatitis B virus antigen as revealed by direct radioimmune assay with $^{125}$I-antibody. *J. Immunol.* **109:** 834–841.

Littlefield, J.W. 1964. Selection of hybrids from matings of fibroblasts in vitro and their presumed recombinants. *Science* **145:** 709.

Liu, F.-T., M. Zinnecker, T. Hamaoka, and D.H. Katz. 1979. New procedures for preparation and isolation of conjugates of proteins and a synthetic copolymer of D-amino acids and immunochemical characterization of such conjugates. *Biochemistry* **18:** 690–697.

Livingstone, A.M. and C.G. Fathman. 1987. The structure of T-cell epitopes. *Annu. Rev. Immunol.* **5:** 477–501.

Lowry, O.H., N.J. Rosebrough, A.L. Farr, and R.J. Randall. 1951. Protein measurement with the Folin phenol reagent. *J. Biol. Chem.* **193:** 265–275.

Maiolini, R. and R. Masseyeff. 1975. A sandwich method of enzymoimmunoassay. I. Application to rat and human alpha-fetoprotein. *J. Immunol. Methods* **8:** 223–234.

Malik, N.J. and M.E. Daymon. 1982. Improved double immunoenzyme labelling using alkaline phosphatase and horseradish peroxidase. *J. Clin. Pathol.* **35:** 1092–1094.

March, S.C., I. Parkh, and P. Cuatrecasas. 1974. A simplified method for cyanogen bromide activation of agarose for affinity chromatography. *Anal. Biochem.* **60:** 149–152.

Marchalonis, J.J. 1969. An enzymic method for the trace iodination of immunoglobulins and other proteins. *Biochem. J.* **113:** 299–305.

Marshall, R.D. 1972. Glycoproteins. *Annu. Rev. Biochem.* **41:** 673–675.

Marston, F.A.O. 1987. The purification of eukaryotic polypeptides expressed in *Escherichia coli*. In *DNA cloning: A practical approach* (ed. D.M. Glover), vol. 3, pp. 59–88. IRL Press, Oxford, Washington, D.C.

Mason, T.E., R.F. Phifer, S.S. Spicer, R.A. Swallow, and R.B. Dreskin. 1969. An immunoglobulin-enzyme bridge method for localizing tissue antigens. *J. Histochem. Cytochem.* **17:** 563–569.

Mazia, D., G. Schatten, and W. Sale. 1975. Adhesion of cells to surfaces coated with polylysine. *J. Cell Biol.* **66:** 198–200.

McConahey, P.J. and F.J. Dixon. 1980. Radioiodination of proteins by the use of the chloramine-T method. *Methods Enzymol.* **70:** 210–213.

McFarlane, A.S. 1958. Efficient trace-labelling of proteins with iodine. *Nature* **182:** 53.

McGadey, J. 1970. A tetrazolium method for nonspecific alkaline phosphatase. *Histochemie* **23:** 180–184.

Meek, G.A. 1976. *Practical electron microscopy for biologists* 2nd edition. Wiley-Interscience, England.

Merrifield, R.B. 1963. Solid phase peptide synthesis. I. The synthesis of a tetrapeptide. *Science* **85:** 2149–2154.

Miles, L.E.M. and C.N. Hales. 1968a. The preparation and properties of purified $^{125}$I-labelled antibodies to insulin. *Biochem. J.* **108:** 611–618.

Miles, L.E.M. and C.N. Hales. 1968b. Labelled antibodies and immunological assay systems. *Nature* **219:** 186–189.

Milich, D.R., D.L. Peterson, G.G. Leroux-Roels, R.A. Lerner, and F.V. Chisari. 1985. Genetic regulation of the immune response to hepatitis B surface antigen (HBsAg). VI. T cell fine specificity. *J. Immunol.* **134:** 4203–4211.

Miller, J., A.D. McLachlan, and A. Klug. 1985. Repetitive zinc-binding domains in the protein transcription factor IIIA from *Xenopus* oocytes. *EMBO J.* **4:** 1609–1614.

Moore, G.R. and R.J.P. Williams. 1980. Comparison of the structures of various eukaryotic ferricytochromes c and ferrocytochromes and their antigenic differences. *Eur. J. Biochem.* **103:** 543–550.

Morris, R.J. and A.F. Williams. 1975. Antigens on mouse and rat lymphocytes recognized by rabbit antiserum against rat brain: The quantitative analysis of a xenogeneic antiserum. *Eur. J. Immunol.* **5:** 274–282.

Morrison, M. 1980. Lactoperoxidase-catalyzed iodination as a tool for investigation of proteins. *Methods Enzymol.* **70:** 214–220.

Morrison, M. and G.R. Schonbaum. 1976. Peroxidase-catalyzed halogenation. *Annu. Rev. Biochem.* **45:** 861–887.

Mueller, U.W., C.S. Hawes, and W.R. Jones. 1986. Monoclonal antibody production by hybridoma growth in Freund's adjuvant primed mice. *J. Immunol. Methods* **87:** 193–196.

Nakane, P.K. 1968. Simultaneous localization of multiple tissue antigens using the peroxidase-labeled antibody method: A study of pituitary glands of the rat. *J. Histochem. Cytochem.* **16:** 557–560.

Nakane, P.K. and A. Kawaoi. 1974. Peroxidase-labeled antibody: A new method of conjugation. *J. Histochem. Cytochem.* **22:** 1084–1091.

Nakane, P.K. and G.B. Pierce, Jr. 1967a. Enzyme-labeled antibodies for the light and electron microscopic localization of tissue antigens. *J. Cell Biol.* **33:** 307–318.

Nakane, P.K. and G.B. Pierce, Jr. 1967b. Enzyme-labeled antibodies: Preparations and applications for the localization of antigens. *J. Histochem. Cytochem.* **14:** 929–930.

Nanba, K., E.S. Jaffe, R.C. Braylan, E.J. Soban, and C.W. Bernard. 1977. Alkaline phosphatase-positive malignant lymphoma. A subtype of B-cell lymphomas. *Amer. J. Clin. Pathol.* **68:** 535–542.

Nilsson, B.O., P.C. Svalander, and A. Larsson. 1987. Immunization of mice and rabbits by intrasplenic deposition of nanogram quantities of protein attached to Sepharose beads or nitrocellulose paper strips. *J. Immunol. Methods* **99:** 67–75.

Nilsson, K. and K. Mosbach. 1980. *p*-Toluenesulfonyl chloride as an activating agent of agarose for the preparation of immobilized affinity ligands and proteins. *Eur. J. Biochem.* **112:** 397–402.

Nilsson, K. and K. Mosbach. 1981. Immobilization of enzymes and affinity ligands to various hydroxyl group carrying supports using highly reactive sulfonyl chlorides. *Biochem. Biophys. Res. Commun.* **102:** 449–457.

Nilsson, K. and K. Mosbach. 1984. Immobilization of ligands with organic sulfonyl chlorides. *Methods Enzymol.* **104:** 56–69.

Nilsson, K. and K. Mosbach. 1987. Tresyl chloride-activated supports for enzyme immobilization. *Methods Enzymol.* **135:** 65–78.

Nilsson, K., O. Norrlöw, and K. Mosbach. 1981. *p*-Toluenesulfonyl chloride as an activating agent of agarose for the preparation of immobilized affinity ligands and proteins. Optimization of conditions for activation and coupling. *Acta Chem. Scand. Ser. B* **35:** 19–27.

O'Sullivan, M.J., E. Gnemmi, D. Morris, G. Chieregatti, A.D. Simmonds, M. Simmons, J.W. Bridges, and V. Marks. 1979. Comparison of two methods of preparing enzyme-antibody conjugates: Application of these conjugates for enzyme immunoassay. *Anal. Biochem.* **100:** 100–108.

Oi, V.T., A.N. Glazer, and L. Stryer. 1982. Fluorescent phycobiliprotein conjugates for analyses of cells and molecules. *J. Cell Biol.* **93:** 981–986.

Olmsted, J.B. 1981. Affinity purification of antibodies from diazotized paper blots of heterogeneous protein samples. *J. Biol. Chem.* **256:** 11955–11957.

Ornstein, L. 1963. Disc electrophoresis. I. Background and theory. *Ann. N.Y. Acad. Sci.* **121:** 321–349.

Osborn, M. and K. Weber. 1982. Immunofluorescence and immunocytochemical procedures with affinity purified antibodies: Tubulin-containing structures. *Methods Cell Biol.* **24:** 97–132.

Ouchterlony, Ö. 1949. Antigen-antibody reactions in gels. *Ark. Kemi. Mineral Geol.* **26:** 1.

Parham, P. 1986. Preparation and purification of active fragments from mouse monoclonal antibodies. In *Cellular immunology*. 4th ed. (ed. D.M. Weir), vol. 1, chap. 14. Blackwell Scientific Publications, California.

Pearse, A.G.E. 1968. Alkaline phosphatases. In *Histochemistry: Theoretical and applied*, 3rd edition, vol. 1, pp. 517–521. Churchill Livingston, Edinburgh.

Pearson, B., P.L. Wolf, J. Vazquez. 1963. A comparative study of a series of new indolyl compounds to localize β-Galactosidase in tissues. *Lab. Invest.* **12:** 1249–1259.

Peferoen, M., R. Huybrechts, and A. De Loof. 1982. Vacuum-blotting: A new simple and efficient transfer of proteins from sodium dodecyl sulfate–polyacrylamide gels to nitrocellulose. *FEBS Lett.* **145**: 369–372.

Peterson, G.L. 1983. Determination of total protein. *Methods Enzymol.* **91**: 95–119.

Polak, J.M. and S. Van Noorden. 1983. *Immunocytochemistry: Practical applications in pathology and biology* John Wright & Sons, Bristol, England.

Ponder, B.A. and M.M. Wilkinson. 1981. Inhibition of endogenous tissue alkaline phosphatase with the use of alkaline phosphatase conjugates in immunohistochemistry. *J. Histochem. Cytochem.* **29**: 981–984.

Pontecorvo, G. 1975. Production of mammalian somatic cell hybrids by means of polyethylene glycol treatment. *Somatic Cell Genet.* **1**: 397–400.

Poole, T.B. 1987. *The UFAW handbook on the care and management of laboratory animals* 6th edition. Longman Scientific and Technical, London, England.

Porath, J. and R. Axén. 1976. Immobilizaton of enzymes to Agar, Agarose, and sephadex supports. *Methods Enzymol.* **44**: 19–45.

Porstmann, B., T. Porstmann, and E. Nugel. 1981. Comparison of chromogens for the determination of horseradish peroxidase as a marker for enzyme immunoassay. *J. Clin. Chem. Clin. Biochem.* **19**: 435–439.

Potter, M. 1972. Immunoglobulin-producing tumors and myeloma proteins of mice. *Physiol. Rev.* **52**: 631–719.

Raymond, S. and L. Weintraub. 1959. Acrylamide gel as a supporting medium for zone electrophoresis. *Science* **130**: 711.

Reichlin, M. 1980. Use of glutaraldehyde as a coupling agent for proteins and peptides. *Methods Enzymol.* **70**: 159–165.

Reiser, J. and J. Wardale. 1981. Immunological detection of specific proteins in total cell extracts by fractionation in gels and transfer to diazophenylthioether paper. *Eur. J. Biochem.* **114**: 569–575.

Renart, J., J. Reiser, and G.R. Stark. 1979. Transfer of proteins from gels to diazobenzyloxymethylpaper and detection with antisera: A method for studying antibody specificity and antigen structure. *Proc. Natl. Acad. Sci.* **76**: 3116–3120.

Richman, D.D., P.H. Cleveland, M.N. Oxman, and K.M. Johnson. 1982. The binding of staphylococcal protein A by the sera of different animal species. *J. Immunol.* **128**: 2300–2305.

Riggs, J.L., R.J. Seiwald, J.H. Burckhalter, C.M. Downs, and T.G. Metcalf. 1958. Isothiocyanate compounds as fluorescent labeling agents for immune serum. **34**: 1081–1096.

Riordan, J.F. and B.L. Vallee. 1972. Reactions with *N*-ethylmaleimide and *p*-mercuribenzoate. *Methods Enzymol.* **25**: 449–456.

Robinson, M.N., A.P. Boswell, Z.-X. Huang, C.G.S. Eley, and G.R. Moore. 1983. The conformation of eukaryotic cytochrome *c* around residues 39, 57, 59 and 74. *Biochem. J.* **213**: 687–700.

Robishaw, J.D., D.W. Russell, B.A. Harris, M.D. Smigel, and A.G. Gilman. 1986. Deduced primary structure of the $\alpha$ subunit of the GTP-binding stimulatory protein of adenylate cyclase. *Proc. Natl. Acad. Sci.* **83**: 1251–1255.

Rosenberg, A.H., B.N. Lade, D.-S. Chui, S.-W. Lin, J.J. Dunn, and F.W. Studier. 1987. Vectors for selective expression of cloned DNAs by T7 RNA polymerase. *Gene* **56**: 125–135.

Roth, J., M. Bendayan, and L. Orci. 1978. Ultrastructural localization of intracellular antigens by the use of protein A-gold complex. *J. Histochem. Cytochem.* **26**: 1074–1081.

Russ, C., I. Callegaro, B. Lanza, and S. Ferrone. 1983. Purification of IgG monoclonal antibody by caprylic acid precipitation. *J. Immunol. Methods* **65**: 269–271.

Rüther, U. and B. Müller-Hill. 1983. Easy identification of cDNA clones. *EMBO J.* **2**: 1791–1794.

Salmon, S.E., G. Mackey, and H.H. Fudenberg. 1969. "Sandwich" solid phase radioimmunoassay for the quantitative determination of human immunoglobulins. *J. Immunol.* **103**: 129–137.

Sambrook, J., E.F. Fritsch, and T. Maniatis. 1989. *Molecular Cloning: A laboratory manual.* 2nd ed. Cold Spring Harbor Laboratory, Cold Spring Harbor, New York.

Sanderson, C.J. and D.V. Wilson. 1971. A simple method for coupling proteins to insoluble polysaccharides. *Immunology* **20**: 1061–1065.

Schlager, S.I. 1980. Radioiodination of cell surface lipids and proteins for use in immunological studies. *Methods Enzymol.* **70**: 252–265.

Schneider, C., R.A. Newman, D.R. Sutherland, U. Asser, and M.F. Greaves. 1982. A one-step purification of membrane proteins using a high efficiency immunomatrix. *J. Biol. Chem.* **257**: 10766–10769.

Schwartz, R.H., B.S. Fox, E. Fraga, C. Chen, and B. Singh. 1985. The T lymphocyte response to cytochrome c. V. Determination of the minimal peptide size required for stimulation of T cell clones and assessment of the contribution of each residue beyond this size to antigenic potency. *J. Immunol.* **135**: 2598–2608.

Scouten, W.H. 1987. A survey of enzyme coupling techniques. *Methods Enzymol.* **135**: 30–65.

Sefton, B.M., I.S. Trowbridge, and J.A. Cooper. 1982. The transforming proteins of Rous sarcoma virus, Harvey sarcoma virus and Abelson virus contain tightly bound lipid. *Cell* **31**: 465–474.

Sela, M. 1966. Immunological studies with synthetic polypeptides. *Adv. Immunol.* **5**: 29–129.

Sela, M. 1969. Antigenicity: Some molecular aspects. *Science* **166**: 1365–1374.

Sharon, J., S.L. Morrison, and E.A. Kabat. 1979. Detection of specific hybridoma clones by replica immunoadsorption of their secreted antibodies. *Proc. Natl. Acad. Sci.* **76**: 1420–1424.

Shastri, N., A. Oki, A. Miller, and E.E. Sercarz. 1985. Distinct recognition phenotypes exist for T cell clones specific for small peptide regions of proteins. *J. Exp. Med.* **162**: 332–345.

Sherman, F., G.R. Fink, and J.B. Hicks. 1986. *Laboratory course manual for methods in yeast genetics*. Cold Spring Harbor Laboratory, Cold Spring Harbor, New York.

Shimonkevitz, R., S. Colon, J.W. Kappler, P. Marrack, and H.M. Grey. 1984. Antigen recognition by H-2-restricted T cells. II. A tryptic ovalbumin peptide that substitutes for processed antigen. *J. Immunol.* **133**: 2067–2074.

Showe, M.K., E. Isobe, and L. Onorato. 1976. Bacteriophage T4 prehead proteinase. II. Its cleavage from the product of gene *21* and regulation in phage-infected cells. *J. Mol. Biol.* **107**: 55–69.

Shulman, M., C.D. Wilde, and G. Köhler. 1978. A better cell line for making hybridomas secreting specific antibodies. *Nature* **276**: 269–270.

Silman, I.H. and E. Katchalski. 1966. Water-soluble derivatives of enzymes, antigens, and antibodies. *Annu. Rev. Biochem.* **35**: 873–908.

Sigel, M.B., Y.N. Sinha, and W.P. VanderLaan. 1983. Production of antibodies by inoculation into lymph nodes. *Methods Enzymol.* **93**: 3–12.

Silverton, E.W., M.A. Navia, and D.R. Davies. 1977. Three-dimensional structure of an intact human immunoglobulin. *Proc. Natl. Acad. Sci.* **74**: 5140–5144.

Simanis, V. and D.P. Lane. 1985. An immunoaffinity purification procedure for SV40 large T antigen. *Virology* **144**: 88–100.

Smith, D.E. and P.A. Fisher. 1984. Identification, developmental regulation, and response to heat shock of two antigenically related forms of a major nuclear envelope protein in *Drosophila* embryos: Application of an improved method for affinity purification of antibodies using polypeptides immobilized on nitrocellulose blots. *J. Cell Biol.* **99**: 20–28.

Smith, P.K., R.I. Krohn, G.T. Hermanson, A.K. Mallia, F.H Gartner, M.D. Provenzano, E.K. Fujimoto, N.M. Goeke, B.J. Olson, and D.C. Klenk. 1985. Measurement of protein using bicinchoninic acid. *Anal. Biochem.* **150**: 76–85.

Smithies, O. 1955. Zone electrophoresis in starch gels: Group variations in the serum proteins of normal human adults. *Biochemistry* **61**: 629–641.

Snyder, S.L., I. Wilson, and W. Bauer. 1972. The subunit composition of *Escherichia coli* alkaline phosphatase in 1 M tris. *Biochim. Biophys. Acta* **258**: 178–187.

Sobieszek, A. and P. Jertschin. 1986. Urea-glycerol-acrylamide gel electrophoresis of acidic low molecular weight muscle proteins: Rapid determination of myosin light chain phosphorylation in myosin, actomyosin and whole muscle samples. *Electrophoresis* **7**: 417–425.

Spira, G., A. Bargellesi, R.R. Pollack, H.L. Aguila, and M.T. Scharff. 1985. The generation of better monoclonal antibodies through somatic mutation. In *Hybridoma technology in the biosciences and medicine* (ed. T.A. Springer), pp. 77–88. Plenum Press, New York.

Spouge, J.L., H.R. Guy, J.L. Cornette, H. Margalit, K. Cease, J.A. Berzofsky, and D. DeLisi. 1987. Strong conformational propensities enhance T cell antigenicity. *J. Immunol.* **138**: 204–212.

Springer, T.A. 1985. *Hybridoma technology in the biosciences and medicine*. Plenum Press, New York.

Steinbuch, M. and R. Audran. 1969. The isolation of IgG from mammalian sera with the aid of caprylic acid. *Arch. Biochem. Biophys.* **134**: 279–284.

Sternberger, L.A. 1979. *Immunochemistry* 2nd edition. John Wiley and Sons, New York.

Sternberger, L.A., P.H. Hardy, Jr., J.J. Cuculis, and H.G. Meyer. 1970. The unlabeled antibody enzyme method of immunohistochemistry: Preparation and properties of soluble antigen-antibody complex (horseradish peroxidase-antihorseradish peroxidase) and its use in identification of spirochetes. *J. Histochem. Cytochem.* **18**: 315–333.

Studier, F.W. and B.A. Moffatt. 1986. Use of bacteriophage T7 RNA polymerase to direct selective high-level expression of cloned genes. *J. Mol. Biol.* **189**: 113–130.

Stulting, R.D. and G. Berke. 1973. Nature of lymphocyte-tumor interaction. A general method for cellular immunoabsorption. *J. Exp. Med.* **137**: 932–942.

Stutte, H.J. 1967. Hexazotiertes triamino-tritolyl-methanchlorid (Neufuchsin) als Kupplungssalz in der fermenthistochemie. *Histochemie* **8**: 327–331.

Summers, D.F., J.V. Maizel, Jr., and J.E. Darnell, Jr. 1965. Evidence for virus-specific noncapsid proteins in poliovirus-infected HeLa cells. *Proc. Natl. Acad. Sci.* **54**: 505–513.

Sutcliffe, J.G., T.M. Shinnick, N. Green, F.-T. Liu, H.L. Niman, and R.A. Lerner. 1980. Chemical synthesis of a polypeptide predicted from nucleotide sequence allows detection of a new retroviral gene product. *Nature* **287**: 801–805.

Swanstrom, R. and P.R. Shank. 1978. X-ray intensifying screens greatly enhance the detection by autoradiography of the radioactive isotopes $^{32}$P and $^{125}$I. *Anal. Biochem.* **86**: 184–192.

Tabachnick, M. and H. Sobotka. 1959. Azoproteins. I. Spectrophotometric studies of amino acid azo derivatives. *J. Biol. Chem.* **234**: 1726–1730.

Tabachnick, M. and H. Sobotka. 1960. Azoproteins. II. A spectrophotometric study of the coupling of diazotized arsanilic acid with proteins. *J. Biol. Chem.* **235**: 1051–1054.

Taggart, R.T. and I.M. Samloff. 1982. Stable antibody-producing murine hybridomas. *Science* **219**: 1228–1230.

Tainer, J.A., E.D. Getzoff, H. Alexander, R.A. Houghten, A.J. Olson, and R.A. Lerner. 1984. The reactivity of anti-peptide antibodies is a function of the atomic mobility of sites in a protein. *Nature* **312**: 127–134.

Ternynck, T. and S. Avrameas. 1972. Polyacrylamide-protein immunoadsorbents prepared with glutaraldehyde. *FEBS Lett.* **23**: 24–28.

The, T.H. and T.E.W. Feltkamp. 1970a. Conjugation of fluorescein isothiocyanate to antibodies. I. Experiments on the conditions of conjugation. *Immunology* **18**: 865–873.

The, T.H. and T.E.W. Feltkamp. 1970b. Conjugation of fluorescein isothiocyanate to antibodies. II. A reproducible method. *Immunology* **18**: 875–881.

Tijssen, P. and E. Kurstak. 1984. Highly efficient and simple methods for the preparation of peroxidase and active peroxidase-antibody conjugates for enzyme immunoassays. *Anal. Biochem.* **136**: 451–457.

Towbin, H. and J. Gordon. 1984. Immunoblotting and dot immunobinding—Current status and outlook. *J. Immunol. Methods* **72**: 313–340.

Towbin, H., T. Staehelin, and J. Gordon. 1979. Electrophoretic transfer of proteins from polyacrylamide gels to nitrocellulose sheets: Procedure and some applications. *Proc. Natl. Acad. Sci.* **76**: 4350–4354.

Tung, A.S., S.-T. Ju, S. Sato, and A. Nisonoff. 1976. Production of large amounts of antibodies in individual mice. *J. Immunol.* **116**: 676–681.

Updyke, T.V. and G.L. Nicolson. 1984. Immunoaffinity isolation of membrane antigens with biotinylated monoclonal antibodies and immobilized streptavidin matrices. *J. Immunol. Methods* **73**: 83–95.

Updyke, T.V. and G.L. Nicolson. 1986. Immunoaffinity isolation of membrane antigens with biotinylated monoclonal antibodies and streptavidin-agarose. *Methods Enzymol.* **121**: 717–725.

Vaitukaitis, J.L. 1981. Production of antisera with small doses of immunogen: Multiple intradermal injections. *Methods Enzymol.* **73**: 46–52.

Vaitukaitis, J., J.B. Robbins, E. Nieschlag, and G.T. Ross. 1971. A method for producing specific antisera with small doses of immunogen. *J. Clin. Endocrinol. Metab.* **33**: 988–991.

Van Weemen, B.K. and A.H.W.M. Schuurs. 1971. Immunoassay using antigen—enzyme conjugates. *FEBS Lett.* **15**: 232–236.

Voller, A., D. Bidwell, and A. Bartlett. 1976. Microplate enzyme immunoassays for the immuno-diagnosis of virus infections. In *Manual of clinical immunology*, chap. 69, pp. 506–512.

Voller, A., D.E. Bidwell, and A. Bartlett. 1979. *The enzyme linked immunosorbent assay (ELISA): A guide with abstracts of microplate applications.* Dynatech Laboratories, Alexandria, Virginia.

Walker, J.E., M. Saraste, M.J. Runswick, and N.J. Gay. 1982. Distantly related sequences in the $\alpha$- and $\beta$-subunits of ATP synthase, myosin, kinases and other ATP-requiring enzymes and a common nucleotide binding fold. *EMBO J.* **1**: 945–951.

Walter, G. 1986. Production and use of antibodies against synthetic peptides. *J. Immunol. Methods* **88**: 149–161.

Walter, G., K.-H. Scheidtmann, A. Carbone, A.P. Laudano, and R.F. Doolittle. 1980. Antibodies specific for the carboxy- and amino-terminal regions of simian virus 40 large tumor antigen. *Proc. Natl. Acad. Sci.* **77**: 5197–5200.

Ward, S., D.L. Wilson, and J.J. Gilliam. 1970. Methods for fractionation and scintillation counting of radioisotope-labeled polyacrylamide gels. *Anal. Biochem.* **38**: 90–97.

Weber, K. and M. Osborn. 1969. The reliability of molecular weight determinations by dodecyl sulfate-polyacrylamide gel electrophoresis. *J. Biol. Chem.* **244**: 4406–4412.

Weiler, R.J., D. Hofstra, A. Szentivanyi, R. Blaisdell, and D.W. Talmage. 1960. The inhibition of labeled antigen precipitation as a measure of serum $\gamma$-globulin. *J. Immunol.* **85**: 130–137.

Weir, D.M., L.A. Herzenberg, C. Blackwell, and L.A. Herzenberg, eds. 1986. *Handbook of experimental immunology.* 4th ed. Mosby, St. Louis, Missouri.

Welsh, K.I., G. Dorval, and H. Wigzell. 1975. Rapid quantitation of membrane antigens. *Nature* **254**: 67–69.

Westhof, E., D. Altschuh, D. Moras, A.C. Bloomer, A. Mondragon, A. Klug, and M.H.V. Van Regenmortel. 1984. Correlation between segmental mobility and the location of antigenic determinants in proteins. *Nature* **311**: 123–126.

Wide, L. and J. Porath. 1966. Radioimmunoassay of proteins with the use of Sephadex-coupled antibodies. *Biochim. Biophys. Acta* **130**: 257–260.

Wilchek, M., T. Miron, and J. Kohn. 1984. Affinity chromatography. *Methods Enzymol.* **104**: 3–55.

Williams, A.F. 1977. Differentiation antigens of the lymphocyte cell surface. *Contemp. Top. Mol. Immunol.* **6**: 83–116.

Wrigley, N.G., E.B. Brown, and J.J. Skehel. 1983. Electron microscopic evidence for axial rotation and inter-domain flexibility of the Fab regions of immunoglobulin G. *J. Mol. Biol.* **169**: 771–774.

Yalow, R.S. and S.A. Berson. 1959. Assay of plasma insulin in human subjects of immunological methods. *Nature* **184**: 1648–1649.

Yen, T.S.B. and R.E. Webster. 1982. Translational

control of bacteriophage f1 gene II and gene X proteins by gene V protein. *Cell* **29:** 337–345.

Young, C.R., H.E. Schmitz, and M.Z. Atassi. 1983. Antibodies with specificities to preselected protein regions evoked by free synthetic peptides representing protein antigenic sites or other surface locations: Demonstration wth myoglobin. *Mol. Immunol.* **20:** 567–570.

Young, R.A. and R.W. Davis. 1983a. Yeast RNA polymerase II genes: Isolation with antibody probes. *Science* **222:** 778–782.

Young, R.A. and R.W. Davis. 1983b. Efficient isolation of genes by using antibody probes. *Proc. Natl. Acad. Sci.* **80:** 1194–1198.

# INDEX